Aesthetic and Regenerative Gynecology

Preeti Jindal • Narendra Malhotra
Shashi Joshi
Editors

Aesthetic and Regenerative Gynecology

Springer

Editors
Preeti Jindal
The Touch, Advanced Obstetrics
IVF & Cosmetic Gynae Centre
Mohali, India

Narendra Malhotra
Global Rainbow Healthcare
Agra, Uttar Pradesh, India

Shashi Joshi
City Clinic - International
Mahboula, Kuwait

ISBN 978-981-16-1742-3 ISBN 978-981-16-1743-0 (eBook)
https://doi.org/10.1007/978-981-16-1743-0

This Springer imprint is published by the registered company Springer Nature Singapore Pte Ltd.
The registered company address is: 152 Beach Road, #21-01/04 Gateway East, Singapore 189721, Singapore

Dedicated to all my teachers, patients and well-wishers

Acknowledgement

Any endeavour big or small is not complete without the help of numerous behind the scene well-wishers. I will like to convey special thanks to my solid support system—my husband Dr Ravul Jindal who not only ushered me into a new and intriguing field of aesthetic and regenerative Gynecology—a new path I was afraid of to tread; but also supports me in all my endeavours.

Special thanks to my parents for imbibing me with virtue and knowledge. No words are enough to thank my loving children—Adit and Kush for bearing with my long hours of work and absentee from their special moments.

Thanks to all authors for sparing valuable time and sharing their knowledge, experience and penning down the chapters which helped in creating this beautiful book. Thanks to Dr Isha without whose contribution this book could not have been completed.

Hope it inspires and guides all those who want to follow this wonderful field.

Enjoy gaining knowledge and shine

Fond love

Preeti Jindal

Contents

Editors and Contributors

Chief Editor

Preeti Jindal, MD, MRCOG, DNB, FICOG The Touch, Advanced Obstetrics, IVF & Cosmetic Gynae Centre, Mohali, India

Co-Editors

Narendra Malhotra, MD, FICOG, FICMCH, FRCOG, FICS Global Rainbow Healthcare, Agra, India

Shashi Joshi City Clinic, Mahboula, Kuwait

Contributors

Biplav Agarwal, MBBS Gurgaon, India

Madhuri Agarwal, MD, DDV Mumbai, India

Ragini Agrawal, MS, FICOG, FICMCH Pratiksha Hospital, Dr Ragini's Urogynae, Laser & Cosmetic Gynecology Centre, Gurgaon, India

Alex Bader, MD, FAAOCG, FAAC Bader Medical Institute of London, London, UK

Rashmi Bagga Department of Obstetrics and Gynecology, Post Graduate Institute of Medical Education and Research, Chandigarh, India

Shemi Bansal, MBBS, DIPGO Global Rainbow Healthcare, Agra, India

Asawari Bapat, DPB (DCP), PGDHHM Nashville, TN, USA

Neerja Bharti Department of Anaesthesia, PGIMER, Chandigarh, India

Hitesh Bhatt, MD, LLB Mumbai, India

Bruno Boccioli, PhD Perugia, Italy

Piyush Chaudhary Mohali, India

Carolyn DeLucia, MD Hillsborough, NJ, USA

Shabjot Dhillon, MBBS Mohali, India

Ronen Gold, MD Department of Uroygnecology and Pelvic Floor Rehabilitation, Aya Medical Advanced Gynecology, Tel Aviv-Yafo, Israel

Service for Obstetric Anal Sphincter Injury at Birth, Lis Women's Hospital, Tel Aviv Medical Center, Tel Aviv Medical School, Tel Aviv-Yafo, Israel

Surajit Gorai, MBBS, MD, DNB, MNAMS Department of Dermatology, Apollo Hospitals, Kolkata, India

Sonam Goyal, MBBS The Touch, Advanced Obstetrics, IVF & Cosmetic Gynae Centre, Mohali, India

Madhu Gupta, MD, PhD Department of Community Medicine and School of Public Health, Postgraduate Institute of Medical Education and Research, Chandigarh, India

Krishna Hari Indore, India

A. Jayanthi Mala, MS Department of Plastic Surgery, CMC, Ludhiana, India

Ravul Jindal, MS (PGI), DNB, FRCS (UK), FAMS Fortis Hospital, Venous Association of India (VAI), Mohali, India

Rakesh Kain, BSc, MBBS, MA (Psychology), MS AIIMS, Rishikesh, India

Ridhi Kalra The Touch, Advanced Obstetrics, IVF & Cosmetic Gynae Centre, Mohali, India

Kiranjit Kaur, MPH Department of Community Medicine and School of Public Health, Postgraduate Institute of Medical Education and Research, Chandigarh, India

Taranvir Kaur Mohali, India

Amit Kumar Department of Anaesthesia and Critical Care, Fortis Hospital, Mohali, India

Isha Kundal, MBBS The Touch, Advanced Obstetrics, IVF & Cosmetic Gynae Centre, Mohali, India

Koushik Lahiri, MBBS, DVD (CAL), FIAD, FFAADV Department of Dermatology, Apollo Hospitals, Kolkata, India

Ting Song Lim, MD (Dermatology) Department of Aesthetic Medicine, Clique Clinic, Kuala Lumpur, Malaysia

Luv Luthra Mohali, India

Jaideep Malhotra, MD, FICMCH, FICOG, FICS, FMAS Global Rainbow Healthcare, Agra, India

Neharika Malhotra, MD, FICMCH, FMAS Global Rainbow Healthcare, Agra, India

Vivek Malhotra, MD Beau Visage Skin Clinic, Chandigarh, India

Ksenija Selih Martinec Kallista Medical Center, Domžale, Slovenia

Francesco Merelli, MD Arezzo, Italy

Poonam Mishra, MBBS, DGO, DNB, FAM, FCG Lucknow, India

Preethi B. Nayak Cutis Academy of Cutaneous Sciences, Bengaluru, Karnataka, India

Anusha H. Pai, MD Derma-Care, Mangalore, India

Ganesh S. Pai, MD, DVD, FAAD Derma-Care, Mangalore, India

Vidya Pancholia, MD Indore, India

Purva Pande, MBBS, MD Dr. Purva's Skin Clinic, Panchkula, Haryana, India

A. B. Prabhu, MS (General Surgery), MCh Plastic, Cosmetic & Reconstructive Surgery, Fortis Hospital, Mohali, India

Avitus John Raakesh Prasad, DDVL S. P. Derma Center, Madurai, India

Sumitu Sahdev, MBBS The Touch, Advanced Obstetrics, IVF & Cosmetic Gynae Centre, Mohali, India

Rashmi Sarkar, MD Lady Hardinge Medical College, New Delhi, India

Sejal K. Shah, MD, DVD, DDV Mumbai, India

Geetendra Sharma, MD, LLB Ahmedabad, India

Manpreet Sharma, MBBS, MS (OBS & GYN) Global Rainbow Healthcare, Agra, India

Neena Singla, DGO, DHM Department of Community Medicine and School of Public Health, Postgraduate Institute of Medical Education and Research, Chandigarh, India

Rimpi Singla Department of Obstetrics and Gynecology, Post Graduate Institute of Medical Education and Research, Chandigarh, India

Karin Staab, MD Universidad de Chile, Santiago, Chile

Adarsh Chandra Swami Department of Anaesthesia and Critical Care, Fortis Hospital, Mohali, India

James D'silva The Cosmetic Surgery Institute, Mumbai, India

Mohan Thomas, MD, FACS The Cosmetic Surgery Institute, Mumbai, India

Ohad Toledano, MSc Tel Aviv, Israel

Introduction to the Rising field of Aesthetic and Regenerative Gynecology

1

Preeti Jindal and Isha Kundal

1.1 Introduction

Cosmetic Gynecology is becoming one of the fastest-growing branches in women's healthcare. It is very interesting, intriguing and futuristic branch that spans over fields of gynecology, dermatology, urogynecology, urology, vascular and plastic surgery. In fact, dentist fraternity is also showing lots of interest in this field. It was called cosmetic gynecology as it originally included **cosmetic procedures** to enhance the aesthetic appearance of the female genital region. Presently along with cosmesis; it includes **functional vulvo-vaginal repairs** to restore anatomy and physiology following recurrent trauma of childbirth, menopause and ageing. It covers not only women genitalia but complete female aesthetics from head to toe. In fact, it is time for all concerned specialities to join hands and take this field to newer heights to obtain optimal results.

1.2 Why is There a Sudden Interest in this Field?

Studies have shown that with rising awareness and increasing accessibility to Internet; women are becoming more aware of the beauty of their intimate parts. According to a 1997 survey, it is evident that 30% of all visitors to porn sites are women and more women watch porn on mobile phones than men as shown in Fig. 1.1 [1]. Hence it is but natural for women to desire for perfect body and mimic the appearances of their favourite models.

Another reason for increased interest in this field is due to increase in longevity of life. Everyone wants to reverse ageing and maintain youth. The world is moving rapidly towards rejuvenation. Nowadays, an average woman is expected to spend one-third of her life in postmenopausal phase. The usefulness for vaginal rejuvenation, non-invasive management of urinary incontinence, treatment of genitourinary syndrome of menopause and aesthetic upliftment cannot be understated. All these needs have lead to the sudden growth in field of cosmetic and rejuvenative gynecology and all efforts should be made by young aspiring clinicians to learn about it so that they can do justice to their patients.

1.3 History

Cosmetic is a Greek adjective "kosmetikos" which means—to adorn [2]. According to Merriam Webster thesaurus's first known use of this word was in1638. It may be used as a noun to refer to cosmetic preparation for external use.

P. Jindal (✉) · I. Kundal
The Touch, Advanced Obstetrics, IVF & Cosmetic Gynae Centre, Mohali, India

P. Jindal et al. (eds.), *Aesthetic and Regenerative Gynecology*,
https://doi.org/10.1007/978-981-16-1743-0_1

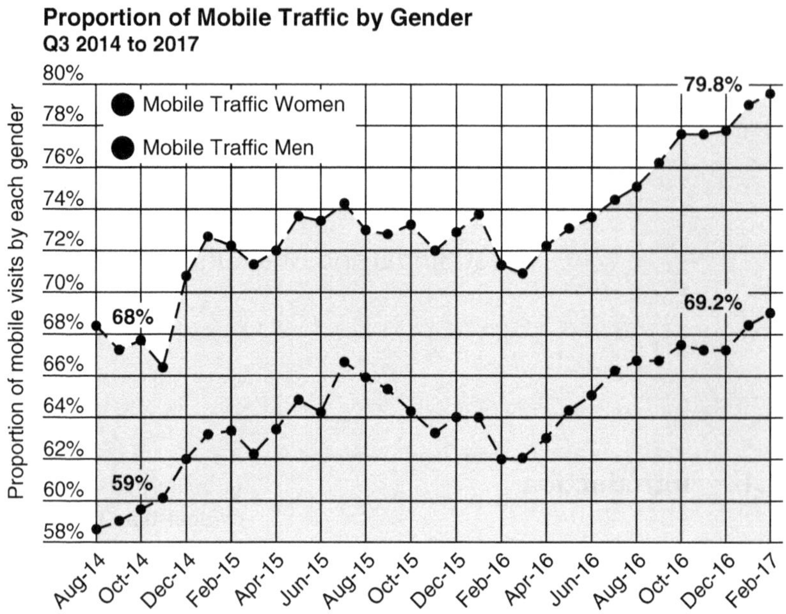

Fig. 1.1 Graph showing more women watching porn than men

Since time immemorial people have tried to improve beauty or aesthetics. **Sushruta** (Fig. 1.2), an Indian physician is considered as the world's first plastic surgeon who described the surgical treatment of a wide variety of ailments, including reconstructive procedures as early as 600 BC [3]. He was born 150 years before Hippocrates. His book "Sushruta Samhita" mentions labiaplasty.

In olden times smoke was used in newly wed women to enhance vaginal sexuality and tightness and also to kill abnormal flora leading to infection. Even today an old Javanese tradition of preparing the vagina by smoke and steam to tighten and rejuvenate is practised (RatusV tradition [4]. Indian mythology also describes similar practises. Interestingly, warming or heating actually is the basis of all **energy-based devices (EBD)** used nowadays for rejuvenation where energy is used to raise temperature of the tissue to an optimal level, which leads to remodelling of elastin and collagen causing rejuvenation of that part.

Vaginal rejuvenation is over 1000-year old. The work of female physician, **Trotula de Ruggiero** (1050 AD), a teacher whose main interest was to alleviate the suffering of women is believed to have first described Vaginoplasty [5]. Greek literature in the early first and second century AD by physician **Soranus** of Ephesus mentions female cosmetic procedures. In the sixth century AD, **Musico** did Latin translation of his 4 volume treatise entitled "gynecology" and describes clitoral surgeries to enhance or diminish sexual pleasures. In third century AD, **Philumenos** of Alexandria described the excision of hypertrophied clitoris because it was considered not appealing, in fact ugly and disgraceful. His work was abstracted by the physician **Aetios** of Amida in his fourth century work *Sixteen Books on Medicine*. Labiaplasty was performed by Greek physician **Paulos** of Aegina as early as in seventh century AD.

These procedures always were not done for aesthetics. The practice of excising the *nonhypertrophied* clitoris originated in Egypt in an effort to prevent any desire for coitus in premarital girls and not out of any cosmetic motivation [6]. Unfortunately such varied practises of female genital mutilation are still currently practised in some parts of the world and its correction also comes under the domain of cosmetic gynecology and plastic surgery.

Rati Rahasya (translated in English as Secrets of Love) an ancient Indian book written by **Kokkoka**, an Indian poet in eleventh–twelfth century on female sexuality mentions vaginal rejuve-

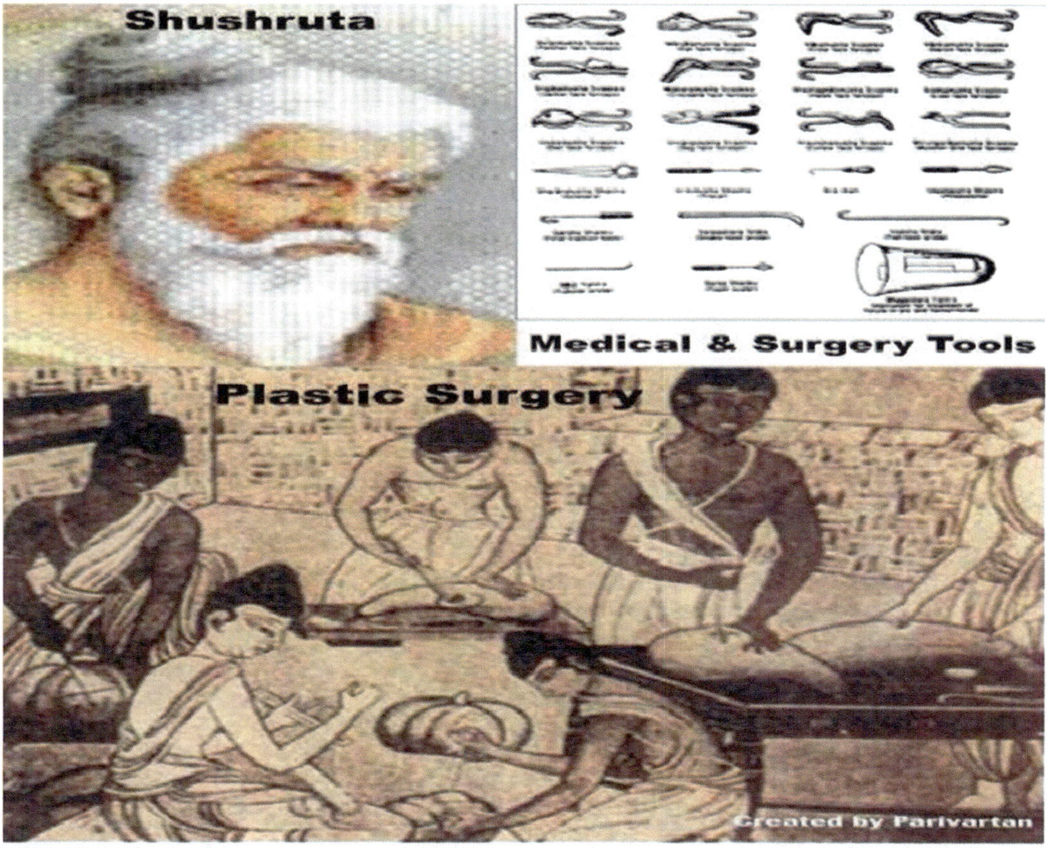

Fig. 1.2 Sushruta; Father of surgery

nation and Yoni shastra to contract or enlarge the vagina (nari kunjara Chapter 11 verse 3) [7].

In Renaissance period—**Pierre dionis** (1643–1718), a parisian described nymphaeplasty, i.e. labiaplasty in his Coursd'operations de chirurgie (1707) based on his late 1600 work (iscgmedia.com; history of cosmetic gyne). In 1905, **Barton Hirst** described vulval reconstructive surgeries and **Jeffcoate** published these in detail in 1957. Many textbooks of gynecology in the last century have given in detail vaginoplasty, labial reconstruction, and vulvodynia under different headings.

Recently in 2008 **Jamie McCartney**, a professional artist living in Brighton, England made a ten-panelled wall sculpture of plaster cast taken from 400 volunteers' genitalia—"The Great Wall of Vagina", to demonstrate diversity in appearances of the vulva; reviving new interest in female genital cosmesis (Fig. 1.3).

1.4 Present Scenario

Energy-based devices like lasers, radiofrequency, highly focused USG (HIFU), carboxytherapy, and many more advancements have made it possible to treat conditions previously considered untreatable or difficult to treat. Patients benefit from improved results, painless therapies, walk-in procedures and less cost as compared to conventional surgery. In the last few years, the main focus of research and development of medical lasers has been on laser hair removal, the treatment of vascular lesions including leg veins, and vision correction. But now there has been a revolution in the last two decades and the focus has shifted to cosmetic gynecology.

In fact, non-invasive management of non-neurological causes of urinary incontinence as highlighted in this chapter by the editor with help

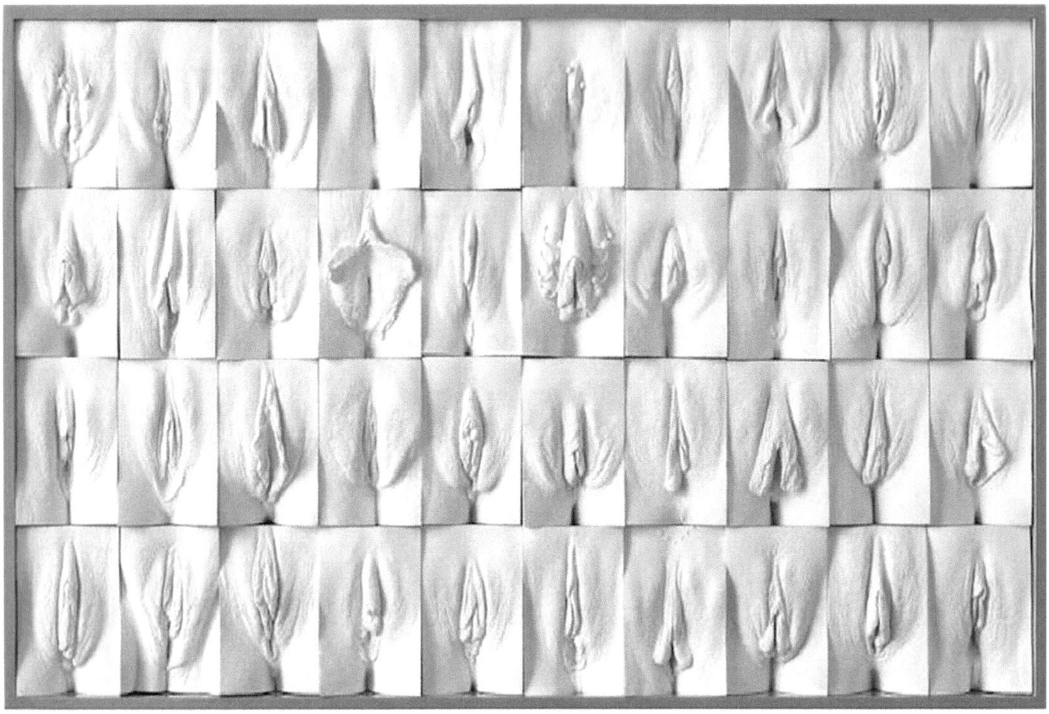

Fig. 1.3 The Great wall of vagina; McCartney (2008)

of laser, HIFEM (Emsella chair of BTL), carboxy, radiofrequency, PRP therapy, fillers etc. has revolutionised the management of this distressing problem [8].

North America holds the major share of this growing market due to factors such as high prevalence of the aged population, high disposable income, increased awareness of aesthetics, and presence of sophisticated infrastructure [9]. The American Society for Aesthetic Plastic Surgery reported an increase of 446% in cosmetic procedures since 1997 and an overall increase of 8% in 2007, with a 17% increase in men undertaking cosmetic surgery [10].

Asia-Pacific is expected to be the second fastest-growing market because of prominent growth factors such as rising disposable income, increasing awareness among the women population base, improving healthcare infrastructure and larger opportunities for physicians (Fig. 1.4). Europe is the third-largest market of cosmetic gynecology, propelling growth owing to increasing number of vaginal rejuvenation procedures providing centres and trained specialists.

The desire to look beautiful is something which all women have right from time immemorial. Although it is not possible to define the ideal aesthetic genitalia, patient-specific techniques chosen based on the patient's anatomy and applied with a realistic approach can increase patient satisfaction and reduce complication rates. Not only aesthetics but due to aging vaginal dryness, urinary leakage and genitourinary syndrome of menopause make it necessary that we offer these new technologies to women. Laser though not FDA (July 30, 2018, FDA statement) approved for vaginal rejuvenation (till date of writing of this chapter) has been found useful in these conditions by many users [11]. It is questionable that if FDA approves laser to be used on face and in vagina for surgical purposes (destruction of abnormal or pre-cancerous cervical, vaginal tissues and condylomas, warts; then why it is not approved for vaginal rejuvenation). The thickness of vaginal mucosa is around 4 mm and penetration of CO_2 laser is only 50–125 micrometres. Other lasers also penetrate much lesser. Hence, it is considered to be very safe modality

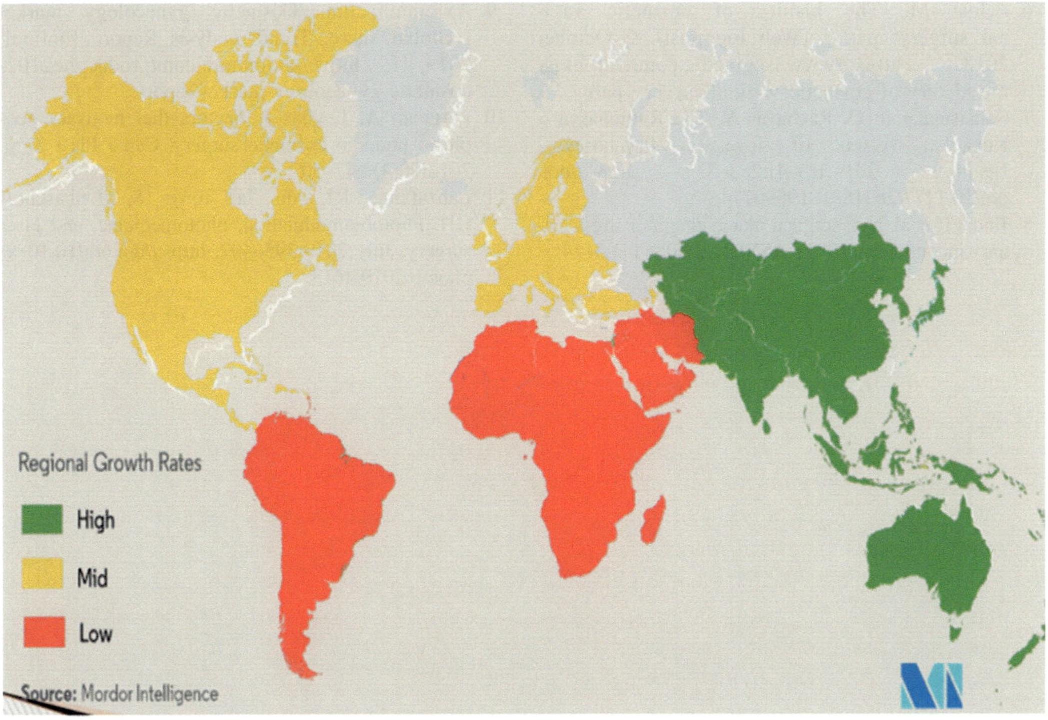

Fig. 1.4 Medical laser market—Growth rate by region (2018)

provided used with adequate knowledge and training. In fact with more and more studies citing the benefits of vaginal lasing with very few side effects it is speculated that it is a matter of time that EBDs will get the required approvals. HIFEM technology is FDA approved for male and female non-invasive urinary incontinence treatment. Other energy-based devices are also being regularly used for the treatment of these conditions with good patient satisfaction rates (as you will read further in the following chapters by various experts) and minimal observed complications.

As this field is relatively new and we are at that point in history that major advancements are occurring in this branch; it therefore becomes the responsibility of teachers as well as students alike to learn more about this field and offer evidence-based treatment to women to improve their quality of life. At the same time as it is a new field, we also have to be very careful and adhere to strict guidelines so that no harm is done to patients. Medicolegally also clinicians have to safeguard themselves with appropriate informed consents.

Bibliography

1. Dailymail.com, Erica Tempesta for. Women reveal the biggest mistakes that men make in the bedroom. Daily Mail Online, Associated Newspapers. 13 November 2020. www.dailymail.co.uk/femail/article-8947809/Women-reveal-biggest-mistakes-men-make-bedroom.html.
2. Lerner C. History of feminine beautification. Arch Dermatol Syphilol. 1932;26(6):1022–31. https://doi.org/10.1001/archderm.1932.01450031024008.
3. Bhagyashri S, Lalitkumar V. Concept of cosmetology in ancient India with special reference to Sushruta Samhita. Int A Med J. 2018;6(4):893–4.
4. Benefits of V-Spa Ratus. n.d. http://ratusv.blogspot.com/2013/04/benefits-of-v-spa-ratus.html.
5. Cavallo P, Proto MC, Patruno C, Del Sorbo A, Bifulco M. The first cosmetic treatise of history. A female point of view. Int J Cosmet Sci 2008 Apr;30(2):79–86. https://doi.org/10.1111/j.1468-2494.2007.00414.x. PMID: 18377616.

6. Pelosi M. The history of cosmetic vaginal surgery: part I [Web log post]. 2 October 2013. https://www.iscgmedia.com/iscg-blog/the-history-of-cosmetic-vaginal-surgery-part-i.
7. Somasundaram O, Raghavan V. The Ratirahasya—Kukkoka: secrets of love. J Psychosexual Health. 2019;1(3–4):283–5. https://doi.org/10.1177/2631831819890774.
8. Jindal P, et al. Non surgical management of stress urinary incontinence (SUI). ASWH. 2020;2(11):51–4.
9. Axiommrc.com. Cosmetic gynecology market | Global share, Trend analysis Report. [online]. 2019. https://axiommrc.com/product/hc2102-cosmetic-gynecology-market-report.
10. Furnham A, Levitas J. Factors that motivate people to undergo cosmetic surgery. Can J Plast Surg. 2012;20(4):e47–50.
11. Lanzafame RJ, de la Torre S, Leibaschoff GH. Photobiomodulation, photomedicine, and laser surgery. July 2019:395–407. https://doi.org/10.1089/photob.2019.4618.

2 Epidemiological Perspective in Aesthetic and Regenerative Gynecology

Madhu Gupta, Neena Singla, and Kiranjit Kaur

2.1 Introduction

Aesthetic and regenerative gynecology is a relatively new and fast-emerging sub-speciality of gynecology. The aim is to enhance the appearance and restore the function of vulvo-vaginal region which may have got damaged during pregnancy, childbirth or tissue changes brought about by ageing [1]. These procedures are no longer the sole domain of dermatologists and plastic surgeons. With increasing empowerment and financial independence of women; many women are demanding these procedures today leading to an increasing number of Gynecologists and Urogynecologists stepping into this field. The use of minimally invasive energy-based treatments along with a range of other medical and surgical options are used to achieve the twin goals of satisfactory function and appearance. It is important to have insight into the prevalence and incidence of the problems for which aesthetic and regenerative gynecology has a big role in the management.

M. Gupta (✉) · N. Singla · K. Kaur
Department of Community Medicine and School of Public Health, Postgraduate Institute of Medical Education and Research, Chandigarh, India

2.2 Complaints for Which Aesthetic and Regenerative Treatment Is Sought

Mostly women seeking aesthetic and regenerative treatments may present with decreased sexual satisfaction on the part of self or partner, feeling of looseness in vagina or vaginal laxity, pain during coitus, decreased lubrication, dribbling of urine or urinary incontinence on coughing, laughing, sneezing, vaginal discharge which may be foul smelling or blood stained, dryness in vagina with irritation and itching. Some women may complain of hypertrophied or dark labia which may be congenital or just a normal variation leading to low self-confidence and dissatisfaction with appearance.

Events like pregnancy, childbirth, ageing, menopause or therapy for malignancy result in significant functional and physical changes in the woman's body. A loose vagina post childbirth is the most subjective and commonly self-reported complaint for which treatment may be sought. But substantial data is lacking [2]. Still in a large majority, women may seek help only because the spouse complains of decreased sexual satisfaction due to her lax vagina. In a UK study, women who attended gynecology and urogynecology clinics were evaluated through a questionnaire. More than one-third had sex-related symptoms which were told only when specifically asked, but very few complained of

P. Jindal et al. (eds.), *Aesthetic and Regenerative Gynecology*,
https://doi.org/10.1007/978-981-16-1743-0_2

vaginal laxity as a symptom [3]. Another survey by Millheiser L et al. (2010) of parous women in 25–55 years of age revealed that almost half of them felt decreased interest in sexual activity and were concerned about vaginal laxity following childbirth [4]. Endocrine changes at menopause lead to vaginal dryness, dyspareunia, low sexual desire and consequently sexual dysfunction [5]. Urinary incontinence as a result of tissue damage during childbirth or at menopause along with other symptoms of vaginal atrophy and the resultant sexual dysfunction are a cause of considerable disability and despair in many cases, leading to low self-esteem, strained relations with spouse and a poor quality of life.

In an OPD-based retrospective study done by Jindal P et al. in 2019–2020, prevalence of urinary incontinence among one thousand and eighteen women studied in North India was found to be 25.8% [6].

Satisfactory sexual function is associated with physical and mental well-being at all ages, even in the later years [7]. Decreased sexual function adversely affects relationship status with a partner and is associated with negative emotional and psychological state [8]. Similarly in a study by Kingsberg SA et al. (2013), it was observed that although the prevalence of symptoms of vulvo-vaginal atrophy is quite high in middle-aged and post-menopausal women, approximately, 50% of the women did not consult a healthcare professional, or complain about symptoms affecting their quality of life. Even healthcare providers did not bring up or discuss the subject in more than 90% of cases [9].

2.3 Problem Statement

The incidence and prevalence of female sexual dysfunction and other related disorders including vaginal laxity, dyspareunia and vaginal dryness, and their risk factors are summarised in Table 2.1.

2.3.1 Globally

A study among 29 countries across the globe was conducted to assess sexual behaviours and sexual dysfunction after 40 years among men and women. It was observed that 21% of women had lack of sexual interest, 16% had inability to reach orgasm and 16% had lubrication difficulties. Lack of sexual interest was found in more women in South East Asia and the Middle East. Dyspareunia was reported by 10% of women, with higher prevalence in South East Asia (22%) and lower in Northern Europe (5%) [10].

2.3.2 High-Income Countries

Incidence of female sexual dysfunction in one of the older studies in Finland was found to vary from 20% in females 25 years or younger to 70–80% among women in 55–74 years age group, while low sexual desire in women greater than 65 years was found to be prevalent in a range of 40–50%, in a review of literature by Kontula O, et al. (2015) [11]. In Sweden, the prevalence of sexual disabilities and problems were reported to be low sexual desire (65%), low achievement of orgasm (48%) and dyspareunia (30%) [12]. Self-reported sexual problems like lack of sexual desire, arousal and orgasm, which were accompanied by personal distress, varied from 8.9% among women greater than 64 years to 14.8% in women from 45–64 years in studies done in the USA by Shifren JL et al. (2002) [13]. Another study in the USA by Luber KM et al. (2004) found that ageing; obesity and smoking are major risk factors for stress urinary incontinence and its prevalence ranges from 4 to 35% [14]. Impact of vulvo-vaginal atrophy in post-menopausal women is seen in 45% of women aged 45 years and above [8], while in another study 63% had symptoms interfering with the enjoyment of sex [15]. Vaginal laxity is seen among younger women (35.9%) in Saudi Arabia. Though not associated with higher parity, Caesarean section was found to be protective [16].

Table 2.1 Incidence and prevalence of female sexual dysfunction (FSD) and related disorders globally

Country	Year	Author reference	Study design	Age group (year)	Prevalence[a]/Incidence[b] (%)	Risk factors
Global study	2004	Nicolosi A, et al. [9]	Cross-sectional study	40–80 (men and women)	Dyspareunia: 10 Decreased lubrication: 16 Orgasmic dysfunction: 16 At least 1 FSD: 39	–
Finland	2015	Kontula O, et al. [10]	Cross-sectional National Sex Survey (2007)	18–74	Low sexual desire: 41 Orgasm difficulty: 9 Lubrication issues 40	–
Sweden	2002	K Sjogren Fugl-Meyer et al. [11]	Cross-sectional National Survey	18–74	Lubrication: 12 Low sexual interest: 65 Low orgasmicity: 48 Vaginismus: 5 Dyspareunia: 30	–
United States of America	2014	Wysocki et al. [14]	Cross-sectional survey	45 and above	Vulvovaginal atrophy: 63	–
	2013	Kinsberg SA, et al. [8]	Cross-sectional survey	45–74	Vulvo vaginal atrophy: 38 Relationship with a partner was affected in 47	–
	2009	Santoro N, et al. [29]	Cross-sectional study	45 and above	Vulvo-vaginal atrophy: 45	–
	2008	Shifren JL, et al. [12]	Cross-sectional study	18 and above	FSD <45 years.: 10.8 45–64 years.: 14.8 >64 years.: 8.9 Low sexual desire: 38.7 Orgasmic difficulty: 20.5% Age adjusted: 43.1 for any sexual problem	Poor self-assessed health Low education level Depression Anxiety Thyroid conditions Urinary incontinence
	2004	Luber KM [13]	Review of literature	18–60	Stress urinary incontinence: Young women: 4–14 Elder women:12–35	Ageing Obesity and smoking Data regarding pregnancy and childbirth is inconsistent
Saudi Arabia	2019	Taleb S, et al. [15]	Retrospective hospital-based study	23–99	Vaginal laxity: 35.9 Stress urine incontinency: 64.4	Parity Menopause Diabetes was not associated with vaginal laxity More in vaginal delivery
Iran	2013	Jaafarpour M, et al. [16]	Cross-sectional study	18–50	FSD prevalence: 46.2 <20 years: 22 40–50 years: 75.7 Problems related to sexual desire: 45.3 Arousal: 37.5 Lubrication: 41.2 Orgasm: 42.0 Satisfaction: 44.5 Dyspareunia: 42.5	Age > 40 years Parity > = 3 Married for > = 10 years husband age > =40 years Unemployed Less educated

(continued)

Table 2.1 (continued)

Country	Year	Author reference	Study design	Age group (year)	Prevalence[a]/Incidence[b] (%)	Risk factors
India	2019	Jindal P, et al. [6]	Retrospective OPD-based study	18 and above	Stress urinary incontinence: 25.8% 76–85 years: 46.7% 66–75 years: 33.6% 15–55 years: 14.6%	Age 18 years and above Urban and rural areas Multiparous Menopause
	2016	Mishra V, et al. [18]	Cross-sectional study	20–47	FSD: 55.5 Sexual desire: 85.8 Arousal dysfunction: 91.7 Lubrication: 83.5 Orgasm: 82.3 Satisfaction: 71.7 Dyspareunia: 85.8	Age 26–30 years Middle level education Upper middle-class psychological stress Married for >16 years
	2016	Santpure A, et al. [5]	Cross-sectional study	46–65 and above	Dyspareunia & vaginal dryness 10.7 Decreased libido 55.3 Sexual activity decreased with increasing sexual dysfunction and age: 54.4 to 5.6 Willing for treatment: 2.1	Ageing Duration of menopause
	2015	Rao TS, et al. [17]	Cross sectional study: Door to door survey	18–50	More than 1 sexual problem: 44.5 FSD: 14 Dyspareunia: 2.34 Arousal dysfunction: 6.65	Age 31–50 yrs. Literacy Daily wage earners Home-makers
	2013	Singh U, et al. [20]	Cross-sectional study: Hospital-based survey	18 and above	Stress urinary incontinence 73.8 In Indian population 16.13	Age > 40 years Multi-parity Obesity Asthma Tea intake Menopause Vaginal delivery Post-hysterectomy
	2009	Singh JC, et al. [19]	Cross-sectional study: Hospital-based survey	>18	FSD: 73.2 <40 year: 60 >40 year: 90 Sexual desire: 77.2 Arousal dysfunction: 91.3 Lubrication: 96.6 Orgasm: 86.6 Dyspareunia: 64.4	Ageing Low literacy
Ghana	2010	Amidu, et al. [21]	Cross-sectional study: Prospective survey	18–58	Overall FSD prevalence: 72.8 Anorgasmia: 72.4 Dissatisfaction: 77.7	Alcohol

[a]Prevalence characterizes the proportion of a given population that at a given time has a particular condition; FSD: Female sexual dysfunction

[b]Incidence is defined as the number of new cases of a certain condition during a specific period in relation to the size of the population studied

2.3.3 Low- and Middle-Income Countries

Prevalence of FSD was reported to be the highest among Iranian women of age 40–50 years (75.7%) [17]. Problems related to sexual desire were 45.3%, arousal 37.5%, lubrication 41.2% and orgasm 42.0% among these women.

In a rural population in India, Rao et al. (2015), reported the prevalence of FSDs to be 14%, which was greater in age group 31 to 50 years. 44.5% women out of these had more than one sexual problem. Socio-economic status affects the prevalence of FSD which was found to be 17.7% in high school educated women; 14.5% in daily wage earners and 14.8% in homemakers. Arousal dysfunction was found in 6.65% which was less as compared to Western studies [18]. Among Indian women aged 20–47 years, the prevalence of FSD is found to be 55.5%, which was more in 26–30 years age group and increased with associated socio-cultural risk factors like education to middle level; upper middle class; psychological stress, and in those married for >16 years [19]. In another study by Singh JC et al. considering increasing age as a risk factor, FSD was found to be prevalent in more than 95% of women greater than 40 years [20]. Singh et al. (2013) reported 16.3% prevalence of urinary incontinence among Indian adults (18 years and above). This prevalence was observed to be higher among older population of age > 40 years, women with increased parity, history of vaginal delivery, hysterectomy; and individuals with obesity, history of tea intake and smoking [21]. Among Indian women aged 46–65 years, in spite of a decrease in sexual activity from 54.4% to 5.6% due to increasing sexual dysfunction with age, only 2.1% were willing for treatment [5]. In a study done in Ghana by Amidu et al. (2010), it was found that overall sexual dysfunction was prevalent among 72.8% of women, out of which most prevalent was dissatisfaction (77.4%) and anorgasmia (72.4%), with alcohol intake being reported as the main risk factor [22].

In most of these studies, women presented to the clinic primarily for some other problems and came out with these complaints only on being specifically asked. From these studies, it is seen that the prevalence of sexual dysfunction globally varies in the range 10%–70%, even going up to 90% in an Indian study in older women.

It is dependent on various factors like age, parity, years since marriage, educational and social status and cultural perceptions. Age was found to be the most important factor and sexual dysfunction increased with age in most studies. It was associated with distress in 10–15% of cases. Association of menopause with decreased sexuality was very high and this affected relationships with partner as well. Many women suffered from multiple disorders. Studies show that age greater than 40 years, parity of 3 or more children is a risk factor for female sexual dysfunction among women aged 18–50 years [17]. Female sexual dysfunction is one of the most under-recognised and undertreated conditions. It is not surprising that most women continue to suffer silently with their unspoken problems and hesitate to express their concerns even when specifically asked. The percentage that seeks advice is very low which may be because of embarrassment, shyness or socio-cultural taboos.

2.4 Aesthetic and Regenerative Procedures: History and Current Status

History is full of examples of Aesthetic Medicine practised from the time of ancient Egyptians, who used milk, honey, alabaster and animal oils to improve the skin texture. Indian surgeons about 2000 years ago had invented the forehead flap for reconstruction of the nose. The humans' need to improve looks and enhance beauty has not changed over the ages. With the advent of new surgical and non-surgical minimally invasive techniques, like energy based devices using fractional carbon dioxide (CO_2) lasers, fractional erbium lasers and radio-frequency devices, the field of aesthetic and regenerative female genital surgery has evolved into a highly specialised one with increasing popularity among patients and physicians. Certification and preceptorship programmes in the USA and the UK offer training to surgeons in this field. In the past few years, a number of certification training programmes for Aesthetic procedures have come up worldwide, especially in Turkey, the Middle East, Spain and South America [1]. The aesthetic and regenerative society of India (InSARG) offers such courses at very economical rates with hands-on experience.

These days a large number of younger women are seeking cosmetic procedures. Most operations are performed upon the patient's request due to a feeling of enlargement and looseness in the vagina, a desire to improve sexual function, discomfort when wearing clothes or doing fitness activities, or with an aim to increase sexual satisfaction for both herself and her partner. This field is gaining popularity steadily in the developing countries also, but is still in its infancy in these countries. In an Indian study, the number of younger females (21 to 40 years) who approached for aesthetic surgery was much more as they are sexually most active and also more aware of their appearance [23]. In another cross-sectional study done in the USA, it was found that the likelihood of undergoing aesthetic procedures in women in the age group of 45 years or more is greater (15%) as compared to those in the younger age groups (8%) due to social pressures to look younger [24]. Vaginal laxity, dryness, atrophy, pain during coitus all affect sexual function, and when repaired, sexual function improves including relationship status and mental well-being. Women with stress incontinence are low on confidence, avoid social interaction, are generally miserable with a poor quality of life. Relief in stress incontinence gives them a new lease on life. Many a time women suffering from these issues present in the psychiatry clinic for the first time with symptoms of anxiety, depression due to poor self-image and relationship problems.

Various procedures that can be performed under Aesthetic and Regenerative Gynecology may be surgical or minimally non-invasive, non-surgical techniques called energy-based devices (EBD). These apply thermal or non-thermal energy to the tissues to enhance collagen regeneration and neo-vascularisation, increased epithelial proliferation and tissue regeneration to help restore physical appearance and function.

The minimally invasive techniques are:

1. Energy-Based Devices (EBD) using CO_2 lasers, Erbium YAG laser, low-level laser therapy, radio frequency, high intensity focused ultrasound, high-intensity focused electromagnetic waves (HIFEM).
2. Chemical treatment.
3. Labial fillers like hyaluronic acid.
4. PRP or platelet-rich plasma.
5. Carboxy therapy and LED therapy.
6. Stem cell therapy.

These may be performed for cosmetic or functional indications or both as described below:

2.5 Cosmetic Indications

- ***Labiaplasty*** of Labia minora/majora which may be reduction to eliminate unwanted tissue or augmentation to create fuller and symmetrical looking labia depending upon personal or cultural preferences. This may be achieved surgically or by use of EBD. Filling is more popular in Europe as compared to the USA [1]. Labiaplasty procedures showed an increase of 23% from 2015 to 2016 as per data of The American Society for Aesthetic Plastic Surgery and is one of the most commonly performed procedures along with clitoral hood reduction [25]. Not only the USA other countries like Australia, the UK and the Middle East have all shown an increase in these procedures over the last decade. Genital Cosmetic surgery is generally not recommended for females below 18 years as full genital maturity is not normally achieved before age of 18 years [26].
- Hymenoplasty, to recreate the virginal state. Labiaplasty and hymenoplasty have ethical

issues and are considered non-medically indicated surgical procedures by many experts [26].

- ***Reduction of*** Lipodystrophy in the Mons region.
- Vaginoplasty is both cosmetic and functional and may be done surgically or non-surgically by use of fractional CO_2 lasers, fractional erbium lasers and radio frequency for tightening or rejuvenation of vagina to improve sexual function. Vaginal tightening, or vaginoplasty, refers to surgery of the vaginal entrance, deeper canal, and epithelium. This procedure is not the same as pelvic floor repair.
- ***Lightening of Vulva***: Chemical agents or CO_2 fractional laser techniques are used to achieve whitening of a hyperpigmented vulva. It is quite popular in the Middle-East and Europe and is catching up in the UK and USA. However, this technique is also not without risk. Therefore, risk and benefits to be weighed before recommendation.
- ***To remove scars*** (cosmetic/functional) in cases of Lichen sclerosus.

2.6 Functional Indications

Functionally cases with symptoms of orgasmic dysfunction, stress incontinence and vulvovaginal atrophy are relieved by laser treatment and laser radio frequency.

- Vaginal rejuvenation using radio frequency by improving vaginal blood flow which causes stimulation of collagen regeneration, connective tissue restoration and tissue tightening. Statistically significant relief in vaginal laxity [4], symptoms of atrophy, stress incontinence and improvement in sexual pleasure were reported [27].
- ***G-spot amplification*** consists of injecting hyaluronic acid or collagen filler in a special spot in the female vagina, the "G-spot", to augment and heighten sexual satisfaction. It is a matter of debate if it is really effective, as sexual pleasure depends on many other factors.

2.7 Caution

American College of Obstetrics and Gynecology (ACOG) recommends that patients should be made aware that procedures to change sexual function or appearance (except done for clinical indications such as female sexual dysfunction, pain with intercourse, vaginal prolapse and incontinence) are not medically indicated, pose a substantial risk and their safety and effectiveness have not been established [28]. Further, the US Food and Drug Administration (FDA) warns against the use of energy-based devices to perform vaginal rejuvenation of vaginal cosmetic procedures as the safety and effectiveness of these devices have not been established [29].

2.8 Conclusion

There is a substantial burden of female sexual dysfunction and related disorders among middle-aged women globally including in India. To address these aesthetic, functional and sexual concerns of women, aesthetic and regenerative cosmetology is emerging as an upcoming field. These procedures are potentially beneficial among women suffering from chronic debilitating conditions like lichen sclerosus, stress-incontinence, sexual dysfunction, scars, vulvodynia and side effects of chemotherapy. But as with any new technology, caution in use is to be advocated. ACOG recommends proper counselling of patients including risks and limitations of procedures, and informed consent should be taken before undertaking any cosmetic procedure. In experienced hands, these procedures are quite safe with a high degree of patient satisfaction and life-changing benefits. However, should preferably be advised for management of clinical conditions rather than purely cosmetic reasons.

References

1. Güneş A, Alinsod RM. A mini-review of aesthetic gynecology and leading gynecology associations' approaches to this issue. Turk Jinekoloji ve Obstet Dern Derg. 2018;15(2):105–11.
2. Vaginal Laxity. The Frequent Problem That's Infrequently Addressed. MedResults Network [Internet]. [cited 2020 Aug 24]. Available from https://www.medresultsnetwork.com/vaginal-laxity-the-frequent-problem/
3. Roos AM, Sultan AH, Thakar R. Sexual problems in the gynecology clinic: are we making a mountain out of a molehill? Int Urogynecol J. 2012;23(2):145–52.
4. Millheiser L, Kingsberg S, Pauls R. A cross-sectional survey to assess the prevalence and symptoms associated with laxity of the vaginal Introitus. Neurourol Urodyn; 2010.
5. Santpure A, Nagapurkar S, Giri P, Bhanap P. Female sexual dysfunction amongst rural postmenopausal woman. Int J Reprod Contraception, Obstet Gynecol. 2016;5(12):4385–9.
6. Jindal P, Gupta M, Isha K, Mittal M, Gulati R, Cherry M, et al. Female urinary incontinence - A revisit on prevalence of this problem in today's society, which is increasing due to increase in average life span. Acta Scientific Women's Health 3.9. 2021;3(9):27–33.
7. Lindau ST, Gavrilova N. Sex, health, and years of sexually active life gained due to good health: evidence from two US population based cross sectional surveys of ageing. BMJ. 2010;340(7746):580.
8. Leiblum SR, Koochaki PE, Rodenberg CA, Barton IP, Rosen RC. Hypoactive sexual desire disorder in postmenopausal women: US results from the Women's international study of health and sexuality (WISHeS). Menopause. 2006;13(1):46–56.
9. Kingsberg SA, Wysocki S, Magnus L, Krychman ML. Vulvar and vaginal atrophy in postmenopausal women: findings from the REVIVE (REal women's views of treatment options for menopausal vaginal changes) survey. J Sex Med. 2013;10(7):1790–9.
10. Nicolosi A, Laumann EO, Glasser DB, Moreira ED, Paik A, Gingell C. Sexual behavior and sexual dysfunctions after age 40: the global study of sexual attitudes and behaviors. Urology. 2004;64(5):991–7.
11. Kontula O. Sex life challenges: the Finnish case. Int Encycl Soc Behav Sci Second Ed. 2015 Mar;26:665–71.
12. Sjögren Fugl-Meyer K, Fugl-Meyer AR. Sexual disabilities are not singularities. Int J Impot Res. 2002;14(6):487–93.
13. Shifren JL, Monz BU, Russo PA, Segreti A, Johannes CB. Sexual problems and distress in United States women: prevalence and correlates. Obstet Gynecol. 2008;112(5):970–8.
14. Luber KM. Risk factors for stress urinary incontinence. Rev Urol. 2004;6(3):3–9.
15. Wysocki S, Kingsberg S, Krychman M. Clinical medicine insights: reproductive health Management of Vaginal Atrophy. Implications from the REVIVE Survey. 2014:23–30.
16. Talab S, Al-Badr A, AlKusayer GM, Dawood A, Bazi T. Correlates of vaginal laxity symptoms in women attending a urogynecology clinic in Saudi Arabia. Int J Gynecol Obstet [Internet]. 2019 Jun 8 [cited 2020 Aug 1];145(3):278–82. Available from https://onlinelibrary.wiley.com/doi/abs/10.1002/ijgo.12810
17. Jaafarpour M, Khani A, Khajavikhan J, Suhrabi Z. Female sexual dysfunction: prevalence and risk factors. J Clin Diagnostic Res. 2013;7(12):2877–80.
18. Rao TS, Darshan MS, Tandon A. An epidemiological study of sexual disorders in south Indian rural population. Indian J Psychiatry. 2015;57(2):150–7.
19. Mishra V, Nanda S, Vyas B, Aggarwal R, Choudhary S, Saini S. Prevalence of female sexual dysfunction among Indian fertile females. J Midlife Health. 2016;7(4):154–8.
20. Singh JC, Tharyan P, Kekre NS, Singh G, Gopalakrishnan G. Prevalence and risk factors for female sexual dysfunction in women attending a medical clinic in South India. J Postgrad Med. 2009;55(2):113–20.
21. Singh U, Agarwal P, Verma ML, Dalela D, Singh N, Shankhwar P. Prevalence and risk factors of urinary incontinence in Indian women: a hospital-based survey. Indian J Urol. 2013;29(1):31–6.
22. Amidu N, Owiredu WKBA, Woode E, Addai-Mensah O, Quaye L, Alhassan A, et al. Incidence of sexual dysfunction: a prospective survey in Ghanaian females. Reprod Biol Endocrinol. 2010;8:2–7.
23. Desai SA, Dixit VV. Audit of female genital aesthetic surgery: changing trends in India. J Obstet Gynecol India. 2018;68(3):214–20.
24. Yurteri-Kaplan LA, Antosh DD, Sokol AI, Park AJ, Gutman RE, Kingsberg SA, et al. Interest in cosmetic vulvar surgery and perception of vulvar appearance. Am J Obstet Gynecol [Internet]. 2012;207(5):428.e1–7. Available from https://doi.org/10.1016/j.ajog.2012.06.056
25. Data M. Cosmetic surgery National Data Bank Statistics. Aesthet Surg J. 2018;38:1–24.
26. Committee RE. Ethical opinion paper [Internet]. 2013. Available from https://www.rcog.org.uk/globalassets/documents/guidelines/ethics-issues-and-resources/rcog-fgcs-ethical-opinion-paper.pdf
27. Alinsod RM. Transcutaneous temperature controlled radiofrequency for orgasmic dysfunction. Lasers Surg Med. 2016;48(7):641–5.
28. Committee on Gynecologic Practice. ACOG COMMITTEE OPINION no 795:Elective Female Genital Cosmetic Surgery. Vol. 107, ACOG. 2006. p. 213.
29. FDA warns against use of energy based devices to perform 'rejuvenation' or vaginal cosmetic procedures: FDA Safety Communication. 2018. Available at https://www.fda.gov/medical-devices/safety-communications/fda-warns-against-use-energy-based-devices-perform-vaginal-rejuvenation-or-vaginal-cosmetic. Accessed on September 18, 2020.

Anatomy and Physiology in Relation to Invasive and Non-invasive Procedures in Aesthetic and Regenerative Gynecology

3

Jaideep Malhotra, Narendra Malhotra, Neharika Malhotra, Manpreet Sharma, and Shemi Bansal

3.1 Introduction

Aesthetic or cosmetic gynecology is an upcoming field with increasing demands to improve women's reproductive health and well being and enhance or restore sexual function.

The common procedures done are labioplasty, clitoral hood reduction, hymenoplasty, vaginoplasty, perineoplasty and G-Spot augmentation [1–5]. The various energy sources being used for non-invasive vaginal tightening include RF and different types of lasers [6] and other sources as described in the book (Table 3.1).

The procedures above are used both for functional and anatomical restoration of pelvic support defects and also as cosmetic (on-demand procedures) [7]. The cosmetic gynecological procedures have been shown to enhance and improve self-esteem and sexual function [8].

Table 3.1 Common procedures

Hymenoplasty	Creating an intact hymen
Vaginoplasty	Tightening of vagina/vaginal rejuvenation
Labiaplasty	Improves appearance of labia
Hoodectomy	Removes tissue covering clitoris
Monsplasty	Shaping the pubis

All the above procedures are now being classified as FCGS (Female Cosmetic Gynecological Surgery). A comprehensive review by the brain explores all aspects of FGCS [9].

Concerns have been raised against FGCS and these are viewed as female genital mutilation. Various societies of ObGyn including ACOG, RCOG, Australian and New Zealand College and Malaysian Society are now putting forwards recommendations [1–4] and policy for its members.

The InSARG (Indian Society of Aesthetic & Regenerative Gynecology) is also in process of forming policy and guidelines for Indian Obstetricians & Gynecologists and Indian patients.

3.2 Anatomy And Physiology

The external genital organs include mons pubis, labia majora, labia minora, bartholin and clitoris. The internal organs include vagina, cervix, hymen, uterus, tubes, skene glands and G spot.

Female genitalia shows a very diverse spectrum of normal anatomic variation (Table 3.1 and Fig. 3.1).

Vulva: Both urinary tract and reproductive structures form the female external genitalia, collectively called as VULVA.

It acts as sensory tissue during sexual intercourse, assists in micturition by directing the flow of urine and protects the internal female reproductive tract from infection.

J. Malhotra · N. Malhotra (✉) · N. Malhotra
M. Sharma · S. Bansal
Global Rainbow Healthcare, Agra, India

P. Jindal et al. (eds.), *Aesthetic and Regenerative Gynecology*,
https://doi.org/10.1007/978-981-16-1743-0_3

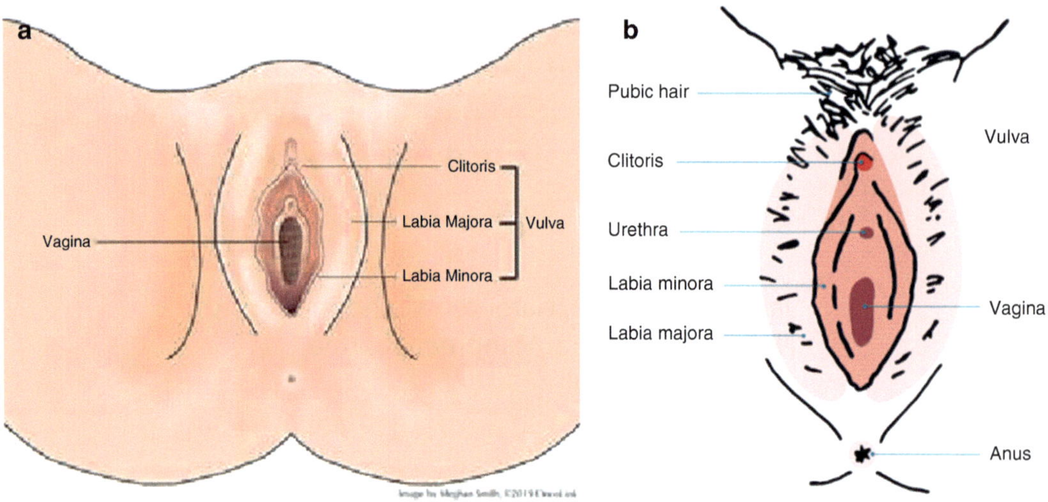

Fig. 3.1 (**a** and **b**) Anatomy of female genitalia

The vulva is made up of:

a. Mons pubis
b. Labia majora
c. Labia minora
d. Clitoris
e. Urethra
f. Vestibule
g. Bartholin's gland
h. Skene's glands
i. Vaginal opening
j. Hymen

- **Mons pubis**: The mons pubis has cushioning effect during sexual intercourse. The mons pubis also contains sebaceous glands that secrete pheromones to induce sexual attraction. The smell of secretions is highly individualized as these variations are the result of menstrual cycle, hygiene habits and body secretions. Excess skin can be addressed by removing and contouring the region to a natural-looking form with mons reduction surgery (monsplasty).
- **Labia majora**: Their function is to cover and protect the inner, more delicate and sensitive structures of the vulva, such as the labia minora, clitoris, urinary orifice and vaginal orifice. In women with enlarged labia majora its reduction known as majoraplasty can be done. With age fat of labia majora atrophies, here labial augmentation with nano fat, PRP, fillers may be offered.
- **Labia minora**: The middle region of the labia minora covers and protects the urethral orifice and vaginal orifice from the exterior environment. Appearance and shape of the labia minora has many anatomic variations, asymmetries being the commonest. Labia minora reduction surgery, or labiaplasty, is a popular surgery in the cosmetic field. Excessive hypertrophied labia minora may cause physical obstruction and here reduction surgeries can be offered.
- **Clitoris**: It is the principal female erogenous organ. Superiorly is located under a clitoral hood (prepuce) which is part of labia minora anatomically that splits into a frenulum on either side of the introitus. Clitoral hood reduction is a very important surgery in this branch.
- **Vestibule**: It is almond-shaped area enclosed by Hart line laterally, external surface of the hymen medially, clitoral frenulum anteriorly and fourchette posteriorly. Fractional CO_2 laser treatment has shown significant results in treating vulval vestibulitis vulvodynia due to rigid fourchette.
- **Bartholin's glands**: They are pea-sized compound alveolar glands located slightly posterior and on either side of the vaginal orifice.

They secrete lubricating mucus from small ducts during sexual arousal. Bartholin gland cysts are common in sexually active women. It can be treated with CO_2 laser in which a skin incision is performed with focused laser beam, the capsule is opened to drain mucoid content, followed by internal vaporization of the impaired capsule.

- **Hymen**: It is a thin membrane that surrounds the opening to the vagina. Hymens can come in different shapes. Hymenoplasty is usually a simple outpatient procedure that can be done in outpatient clinic under local anaesthesia. Also called 'revirgination', it is designed to restore the hymen. It is often advertised as a 'gift' to one's partner [10]. This procedure is occasionally requested by women of certain cultural backgrounds in which premarital sex is forbidden and an intact hymen is considered evidence of virginity. Rarely imperforate hymen is also encountered leading to hematocolpos. It can be divided for free drainage of menstrual blood, secretions and sexual intercourse.

Perineoplasty—Undertaken to strengthen the pelvic floor and, in the FGCS setting, aimed at establishing penile pressure with coital thrust. This procedure is technically similar to perineal reconstruction, in which the perineal length is restored following childbirth trauma or previous surgery. It is commonly performed as part of vaginal prolapse surgery. However, even in this setting, there is no evidence that this procedure improves sexual function and, in fact, it may cause dyspareunia.

Vaginoplasty—The purpose of this procedure is vaginal creation in gender reassignment but, in the FGCS setting, it refers to tightening the vagina, which can be surgical or non-surgical—as in 'laser vaginal rejuvenation' or 'designer laser vaginoplasty'.

G-spot augmentation—G spot, also called Grafenberg spot after German scientist Ernst Grafenberg who described it; is believed to be a point 2 to 3 inches on the anterior vaginal wall, inside from the introitus. It is believed to be a confluence of several nerve endings and its stimulation is believed to lead to orgasm. Involves autologous fat or collagen transfer via injection into the pre-determined G-spot location. There is no existing scientific literature describing this procedure. Similar procedures include G-spot amplification and G-shot collagen injection into the region. Often described as a sexual and cosmetic rejuvenation procedure for the vagina using the preparation and injection of blood-derived growth factors into the G-spot, clitoris and labia.

O spot—O-spot is the space between the urethra and the vaginal wall, most distally, in the area of the periurethral glands. Injection of 4 ml of PRP at the O-spot. Fluid fills the tissue between the urethra and the vagina thus improves orgasm.

The female external genitalia varies in almost all females in shape, size and colour but despite the oestrogen-dependent anatomical variations, the functions of these structures remain the same in all women [11]. With female ageing and fall in oestrogen levels, these structures undergo atrophy and their functions also decrease (Fig. 3.2) [12, 13].

As these organs are endocrine dependent, a defect in the hormonal secretions can lead to altered anatomy and physiology which may need medical treatment (ERT, Androgens cortisol etc.), EBD treatments or even need surgical correction (FCGS) like labian fusion, clitoral hood reduction and others [10, 14] (Figs. 3.3 and 3.4).

Table 3.2 shows cosmetic genital procedures [10, 15]

Terms such as 'vaginal rejuvenation', 'designer laser vaginoplasty', 'revirgination' and 'G-shot' are commercial in nature. The consumers at whom they are targeted can then mistakenly believe such official-sounding terms refer to medically recognised procedures.

3.3 Indication and Contraindications

FCGS is non-medically indicated cosmetic surgical procedures on healthy genitalia. As these are done on normal healthy organs there is a debate that these procedures come under female genital

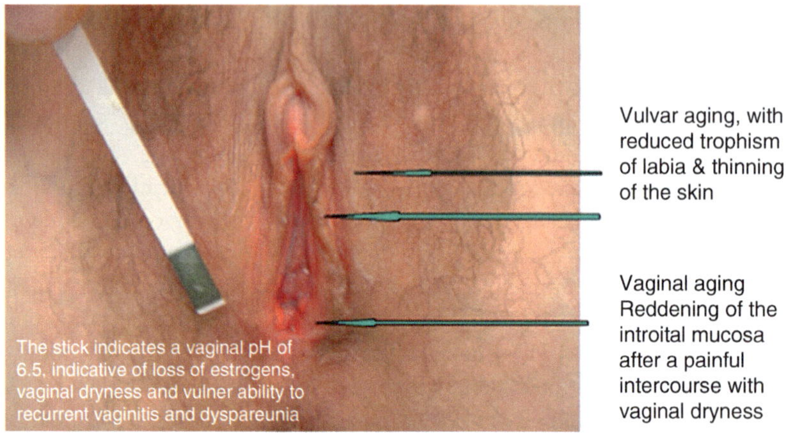

Fig. 3.2 Progressive vulvovaginal ageing after menopause

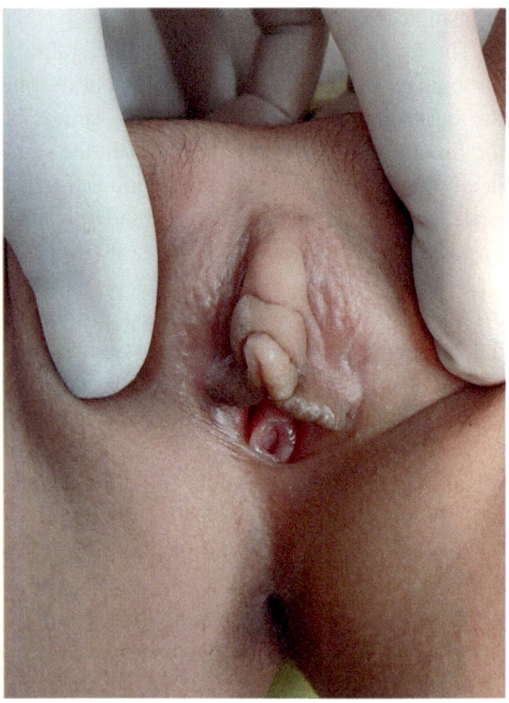

Fig. 3.3 Clitoral hood enlargement at 12 years of age

mutilation or cutting (FGM/C) as described by W.H.O.

The World Health Organization (WHO) defines FGM/C as 'all procedures involving partial or total removal of the external female genitalia or other injuries to the female genital organs for non-medical reasons'. It is generally performed on children or adolescents who are not able to provide informed consent and have no health benefits. There is some debate about whether FCGS is covered by legal definitions of FGM/C and, therefore, illegal under existing regulations. The explanation of outcome is considered mandatory in informed consent for FCGS, as for all medical procedures.

The common requests for performing FCGS are:

1. Adolescents wanting change in labial anatomy. There is only a woman's perception of what is normal because of the diverse appearances of the genitalia. Hence, many young women wish a change due to fashion trend or their visibility in tight clothes. Fashion terms such as "camel toe" and "outie" make many women in the fashion industry feel uncomfortable with genital appearances and lead to perception that the female external genitalia should be small or "barbie like Enhancement of sexual pleasure procedures like energy-based vaginal tightening vaginal rejuvenation, G-spot augmentation, orgasm shot and even bleaching.
2. Some medical indications for FCGS are:
   a. Mild degree of prolapse
   b. Cystocele
   c. Idiopathic vulva and vaginal itching
   d. GSM
   e. Dyspareunia
   f. Scar marks (Pregnancy stria/operative marks of episiotomy/C-section)
   g. Breast tightening
   h. Abdominal wall tightening

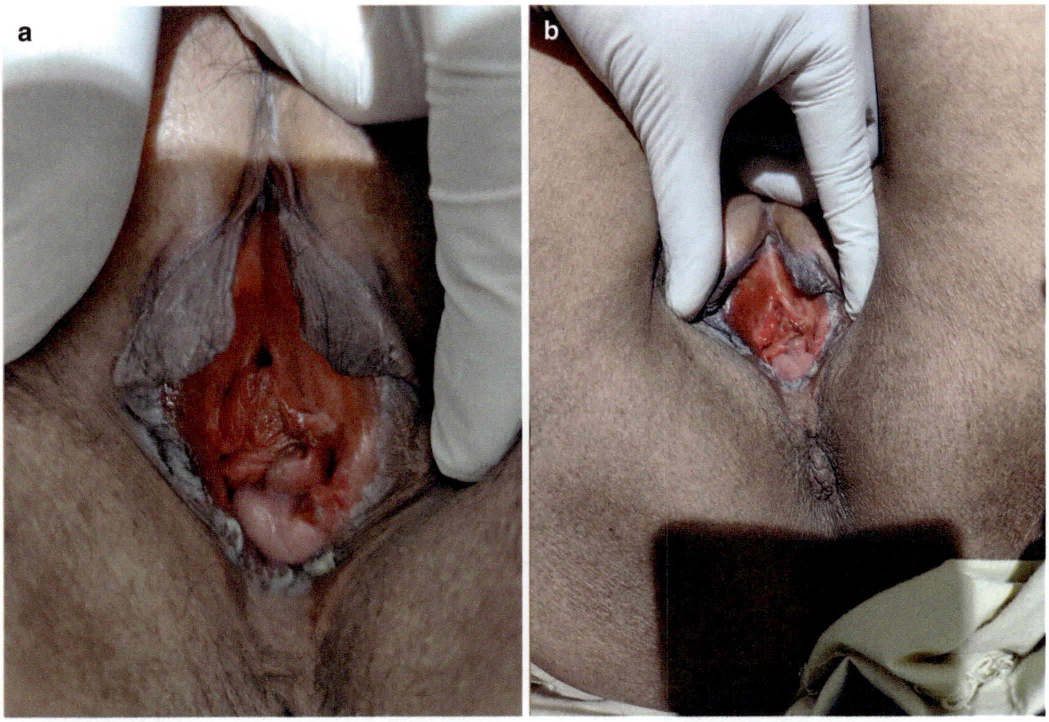

Fig. 3.4 (**a**, **b**) Posterior vaginal wall old tears

3. Fat reduction surgery on mons, lower abdomen, arms, breasts etc.
4. Fat grafting and augment

3.4 Practical Tips

1. Women's requests need to be respected.
2. Counselling is very important.
3. FCGS should be done only by trained specialists (gynecologists or plastic surgeons).
4. Always a second opinion must be documented.
5. Psychological counselling should be done.
6. All the symptoms and concerns should be discussed and documented.
7. Diagrammatically and photographically the procedures should be explained and what she wants should be documented and consented.
8. Risks and complications should be explained including non-healing, scarring etc.
9. Women undergoing these procedures should be counselled regarding all anatomy, physiology and sexual function.
10. Training in procedures and energy-based devices and equipment is essential.

3.5 Various Committee Statements

1. **Australian media code of conduct on body image**

   Australia's *Voluntary media code of conduct on body image* was designed to encourage the fashion, media and advertising industries to place greater emphasis on diversity, positive body images and a focus on health rather than body shape. In doing so, it aims to reduce young people's susceptibility to feelings of low self-esteem, eating disorders and negative body image that are associated with exposure to idealised and unrealistic images seen in the media and advertising.

Table 3.2 Cosmetic genital procedures

Type of Procedure	Purported Benefit*	Procedures Used	Reported or Potential Complications
Surgical Procedures			
Clitoral hood reduction	To improve sexual function by increasing sensitivity and allowing more direct clitoral contact	Hoodectomy Note: Often combined with labiaplasty to create labia minora symmetry and prevent clitoral hood sagging	• Scarring • Infection • Hematoma • Hypersensitivity • Damage to the glans
Labiaplasty	To eliminate unwanted tissue of the labia minora or labia majora	• Trim or edge resection • Wedge resection using a V-shaped or Y-shaped incision • Z-plasty • De-epithelialization	• Scarring • Infection • Hypersensitivity or loss of sensation • Dyspareunia • Wound dehiscence
Labia majora augmentation	To create a full, symmetric look	• Autologous fat transplantation • Injectable fillers (hyaluronic acid)	Palpable fatty cysts
Hymenoplasty	To recreate the virginal state of the hymen; has cultural roots in regions that place a value on an unmarried woman's virginity	Reconstruction of hymenal remnants, vaginal mucosal flaps, or both	Wound dehiscence
Vaginoplasty	To tighten vaginal contour and increase sexual satisfaction	• Anterior, posterior, or lateral colporrhaphy • Rugation restoration† • Energy-based devices	• Infection • Dyspareunia • Dehiscence • Fistula
Energy-Based Interventions			
Energy-based vaginal procedures†	To tighten vaginal contour and increase sexual sensation	Laser radiofrequency	• Burns • Scarring • Pain during sexual intercourse • Recurring or chronic pain
Injections			
G-spot amplification	To augment G-spot and heighten sexual satisfaction	• Autologous fat transfer • Hyaluronic acid	• Urinary tract infection • Infection

*This may not be the patient goal, but these procedures are often marketed with these outcomes.
†U.S. Food and Drug Administration. FDA warns against use of energy-based devices to perform vaginal 'rejuvenation' or vaginal cosmetic procedures: FDA safety communication. Silver Spring (MD): FDA; 2018. Available at: https://www.fda.gov/MedicalDevices/Safety/AlertsandNotices/ucm615013.htm. Retrieved August 26, 2019.

The code of conduct:

a. Discourages the use of digitally enhanced or altered pictures and suggests these digitally pictures be identified as such.
b. Encourages the use of images that represent the diversity of body shapes.
c. Encourages the considered placement of advertising on dieting, cosmetic surgery etc.
d. Discourages the 'glamourisation' of models and celebrities who are particularly underweight and instead encourages a focus on models with a healthy body shape.

2. **Guidelines for gynecological examinations and procedures**

The gynecological examination of women is a formal process and potentially intimidating to women, some of whom may have suffered various degrees of physical or sexual abuse during their lives.

Doctors should consider the information provided by women, listen and respond sensitively to their questions and concerns.

According to the Royal Australian and New Zealand College of Obstetricians and Gynecologists (RANZCOG) *Guidelines for*

Gynecological Examinations and Procedures C-Gyn 30:41.

Awareness of cultural or religious factors is essential when discussing and offering gynecological examination.

Where examination is indicated, doctors should ensure that:

a. An adequate explanation is provided about the nature of an examination and the information that it will provide.
b. The patient has the opportunity to decline examination.
c. Permission is obtained, especially for breast and/or pelvic examination.
d. Privacy is provided for disrobing.
e. Suitable cover is provided during examination, for example gown or cover sheet.
f. A chaperone is available to attend any patient undergoing physical examination when requested, irrespective of the gender of the doctor.
g. The patient must be made aware in advance of the presence of medical students and the right to decline their attendance at any examination.

With respect to examination of young women and children, see the Royal Australasian College of Physicians (RACP) policy *Genital Examinations in Girls and Young Women: A Clinical Practice Guideline*, available at https://www.racp.edu.au/docs/default-source/advocacy-library/genital-examinations-in-girls-and-youngwomen-a-clinical-practice-guideline.pdf

References

1. Committee on Gynecologic Practice, American College of Obstetricians and Gynecologists. ACOG Committee opinion no. 378.vaginal "rejuvenation" and cosmetic vaginal procedures. Obstet Gynecol. 2007;110:737–8.
2. Royal Australian and New Zealand College of Obstetricians and Gynaecologists. C-Gyn 24.Vaginal 'rejuvenation' and cosmetic vaginal procedures. Obstet Gynaecol. 2007;110:737–8.
3. American College of Obstetricians and Gynecologists. The role of the obstetrician gynecologist in cosmetic procedures.procedures [ACOG policy statement]. Washington: American College of Obstetricians and Gynecologists; 2008.
4. Royal College of Obstetricians and Gynaecologists. Hymenoplasty and labial surgery. RCOG Statement No. 6. July 2009. London: RCOG; 2009.
5. Goodman MP. Female genital cosmetic and plastic surgery: a review. J Sex Med. 2011;8:1813–25. https://doi.org/10.1111/j.1743-6109.2011.02254.x.
6. Vizintin Z, Rivera M, Fistonic I, Saracoglu F, Guimares P, Gaviria J, Garcia V, Lukac M, Tadej Perhavec LM. Novel minimally invasive VSP Er:YAG laser treatments in gynecology. LA&HA. 2012;1:46–58.
7. Julian T. Pelvic-Support defects a guide to anatomy and physiology. J Family Prac. 2002;14:3.
8. Goodman MP, PlacikOj BRH, Miklos JR, Moore RD, et al. A large multicenter outcome study of female genital plastic surgery. J Sex Med. 2010;7:1565–77.
9. Braun V. Female genital cosmetic surgery: a critical review of current knowledge and contemporary debates. J Women's Health (Larchmt). 2010;19:1393–407.
10. Goodman MP, Placik OJ, Benson RH III, Miklos RJ, Moore RD, Jason RA, et al. A large multicenter outcome study of female genital plastic surgery. J Sex Med. 2010;7:1565–77.
11. McQuillan SK, Jayasinghe Y, Grover SR. Audit of referrals for concern regarding labial appearance at the Royal Children's Hospital: 2000–2012. J Paediatr Child Health. 2018 Apr;54(4):439–42.
12. Patni R. Genitourinary syndrome of menopause. J Midlife Health. 2019 Jul-Sep;10(3):111–3.
13. Barjon K, Mahdy H. StatPearls [Internet]. Treasure Island (FL): StatPearls Publishing; May 5, 2020. Episiotomy. [PubMed]
14. Chormanski D, Muzio MR. StatPearls [Internet]. Treasure Island (FL): StatPearls Publishing; May 23, 2020. C 17 Hydroxylase Deficiency. [PubMed]
15. Goodman MP. Female cosmetic genital surgery. Obstet Gynecol. 2009;113:154–9.

Counselling Before Cosmetic Gynecology

4

Shashi Joshi

4.1 Introduction

Female genital cosmetology is the latest revolution in the field of female health. Its popularity has increased in recent years. The exact incidence is difficult to assess as many of these procedures are done in a private set-up. One reason behind the increase in prevalence of body awareness, particularly crossing the barriers of shame when discussing gynecological problems and better access to suitably qualified health professionals to address genital aesthetics.

Women favour hairless pubic area, resulting in easier visualisation of the external genitalia and making subtle irregularities more obvious (1). Increased access to female body images on Internet, TV and magazines has raised the awareness of genital appearance (but it has also skewed up the perception of what can be normal). Online promotion of normalisation of female genital aesthetics also contributes to it (2). Aggressive marketing and use of non-medical terms like designer vulva, vaginal rejuvenation, Barbie look etc. influence the demand without making realistic expectations clear (3).

With increasing popularity of cosmetic procedures, pre-procedure counselling and psychological assessment improve outcomes by assessing patient's motivations, expectations as well as identifying those that may require psychological referral (4).

Aesthetic Gynecology Includes a variety of procedures to improve genital aesthetics and enhance sexual performance (5–8):

1. SURGICAL:
   - Labia Minora plasty
   - Labia majoraplasty—enhancement or reduction
   - Clitoral hood reduction
   - Clitoroplasty
   - Vaginoplasty
   - Hymenoplasty
   - Perineoplasty
2. NON-SURGICAL LASERS—ERBIUM, YAG, CO_2 LASERS:
   - Labial whitening
   - Labial enhancement
   - Vaginoplasty
   - G-spot augmentation
   - O-spot augmentation
3. FILLERS:
   - Botulinum toxin
   - Hyaluronic acid
4. REGENERATIVE COSMETOLOGY WITH THE USE OF:
   - Fat graft
   - Platelet-rich plasma (PRP)
   - Stem cells
   - Amniotic fluid
   - Amniotic membrane

S. Joshi (✉)
City Clinic, Mahboula, Kuwait

P. Jindal et al. (eds.), *Aesthetic and Regenerative Gynecology*,
https://doi.org/10.1007/978-981-16-1743-0_4

4.2 Aesthetic and Regenerative Gynecology Counselling

It should be a priority for women seeking these procedures. In fact, counselling is as important as the procedure itself for a satisfactory outcome. Although it is not possible to define ideal aesthetic genitalia and sexual gratification, patient-specific technique based on patient-specific anatomy and applied with a realistic approach can result in high patient satisfaction.

4.2.1 Counselling Environment (Fig. 4.1)

The environment should be:

- Welcoming
- Comfortable
- Place with no distractions
- Somewhere where privacy and confidentiality can be maintained

4.2.2 Counselling Should Be

- Auditory
- Visual
- Kinesthetics

4.2.3 Auditory

- Listen to the patient in detail and understand her.
- Explain recommendations in relation to individual expectations.
- Explain the procedure in detail.
- Pre- and post-procedure care.
- Reasonable expectations.
- Give the option of alternative methods available.
- Explain inherent risks and complications.

4.2.4 Visual

Use diagrams and models to explain what is being discussed. This will help the patient to understand better. Ask them to get pictures of the result they want to achieve so that there is no difference of understanding between doctor and patient.

4.2.5 Kinesthetic

This is how the auditory and visual information given will affect a particular individual. Utilise all educational tools including imaging and models to help a patient understand the procedure and the results expected from her individual perspective.

Fig. 4.1 Counselling chamber

Make sure you take a disclaimer and documented discussions concerning realistic expectations.

4.3 Recommendations for Counselling

(A) EXPLORE THE REASON FOR THE PROCEDURE

- Discomfort because of tight clothing like body-hugging sportswear, tight jeans or G strings.
- Information on digital media may be a source for the patient to make a change in her genital organs. The free accessibility of pornographic material has made many women feel that their genitalia are inadequate. These images are often digitally modified and may convey a wrong impression.
- Physical discomfort like pain, dyspareunia, difficulty in maintaining hygiene, vaginal laxity etc. could be the reason behind it.
- Limited genital education may also play a part in seeking cosmetic procedures. The counsellor/gynecologist should play an important role in helping the woman understand her genital anatomy and respect individual variations.
- Comments directed by others at them or otherwise. They could be friends, relatives or sexual partner.
- Intimate partner abuse or sexual abuse should also be addressed.
- Grooming practices for pubic hair like waxing, depilation, shaving or laser may expose subtle irregularities (1).

(B) COMPLETE MEDICAL, GYNECOLOGICAL AND PSYCHOSEXUAL HISTORY

- Assess the degree of anxiety and concern.
- Is her concern affecting her intimate relationship, self-esteem, confidence and ability to function happily?
- Feeling unpleasant during sexual activity and how she addresses the issue.
- In the medical history, smoking and alcohol consumption can lead to compromised results. Diabetes and hypertension should be well controlled.
- In the gynecological history, pregnancy, active local infection, sexually transmitted diseases and untreated genital malignancies are a contraindication.

(C) ASSESSMENT OF MENTAL HEALTH AND SEXUAL ABUSE ISSUES

Patient's mental health and a psychological state have a bearing on the outcome. Pre-existing issues of depression, anxiety, post-traumatic stress disorder, addictions, low self-esteem and stress are likely to give poor outcomes. Body dysmorphic syndrome and marital discord should also be ruled out.

(D) COUNSELLING REGARDING GENITAL EXAMINATION

Patient should be explained that it would be a detailed examination of the genital organs to assess what she has and what can be achieved. Use diagrams and models to educate the patient objectively. She should be able to understand the procedure and the expected results.

(E) REASSURE THE PATIENT

Reassure her that the best possible will be done for her taking all her concerns into consideration. Explain in detail why a procedure has been short-listed for her. The details of what the procedure entails, including risks and complications should be explained.

(F) EFFECT OF PHYSIOLOGICAL CHANGES

Do not forget to explain that physiological changes like pregnancy, menopause and weight gain and weight loss can affect the outcome.

(G) UNINTENDED CONSEQUENCES

Possibility of inherent complications of a procedure like:

- Bleeding
- Wound infection
- Altered sensation
- Dyspareunia
- Scarring
- Potential physical or psychological risk
- Unknown delayed problems

(H) COUNSELLING ALONE MAY BE MORE BENEFICIAL IN CERTAIN CONDITIONS
- Body image distress
- Low self-esteem
- Social anxiety
- Sexual difficulties that are due to undue sexual expectations by the partner or from the partner.

(I) COUNSELLING OF BOTH THE PARTNERS

When cosmetic procedure is sought for an increase in sexual gratification, both sexual partners should be counselled. They should be explained that desire, arousal, orgasm is highly complex. They are personal experiences that do not depend on aesthetics alone. They are influenced as much by emotional, spiritual and interpersonal factors as by aesthetics.

Also, men and women view aesthetics and sexual satisfaction from different perspectives.

According to a multicentre cohort study published in SEX MEDICINE by Goodman M.P. et al., 58% of the women expected an increase in their sexual gratification and 54% expected an increase in the partner's sexual gratification irrespective of the factor quoted for the desire for genital cosmetology.

(J) COUNSELLING FOR GIRLS REQUESTING GENITAL COSMETOLOGY

Girls requesting genital cosmetic procedures before the age of 18 years should be counselled against them, irrespective of their consent. Genital maturation is not reached before the age of 18 years and hence the procedure done before that is unlikely to give the best long-term results.

Girls from 9 to 13 years of age request consideration for relief of symptoms such as rubbing, chaffing, and interference with sports. The second most common reason in this age group is the mother's perception of an abnormality in her daughter. Adolescents 15 to 17 years of age are concerned with their own appearance and have further concerns with their own sexual appearance and have further concerns that the sexual partner may find them abnormal or unattractive.

(K) POST-PROCEDURE COUNSELLING

Sometimes post-procedure support or counselling may be required if optimum result is not met or if an inherent complication of the procedure occurs.

4.4 Summary

Cosmetology is no longer only for the rich, famous, movie stars and models. It has become readily accessible to the general public. The concept of beauty has extended from having right looking cheeks to having designer private parts.

Value judgement should not be made about a person's idea of beauty. Cosmetic procedures can benefit if the patient's expectations are realistic. This can be achieved by proper pre-procedure counselling.

Genital cosmetology procedures should not be done before the age of 18 years. Always give them the opportunity to consult and reconsider again before moving ahead with a definitive surgical procedure to avoid regrets later.

References

1. Yurteri-kaplan LA, Antosh DD, Sokol AI, Park AJ, et al. Interest in cosmetic vulvar surgery and perceptions of vulvar appearance. Am J ObstetGynecol 2012;207:428. el-7.

2. Mowat H, Mc Donald K, Dobsan AS, Fisher J, Kirkman M. The contribution of online content to the promotion and normalisation of female genital cosmetic surgery: a systematic review of the literature. BMC Womens Health. 2015;Nov25;15:110.
3. Mirzabeigi MN, Jandali S, Mettel RK, Alter GJ. The nomenclature of "vaginal rejuvenation" and elective vulvovaginal plastic surgery. Aesthet Surg J. 2011;31(6):723–4.
4. Braunsteiner N, Vickers ER, Shapberg R, et al. Psychological issues for patients undergoing stem cell therapy & regenerative medicine. Open J Regen Med. 2018;7:1–17.
5. Schnatz PF, Boardman LA. Committee on Gynecologic Practice, American College of Obstetricians and Gynecologists. ACOG Committee Opinion No. 795. Vaginal "rejuvenation" and cosmetic vaginal procedures. Obstet Gynecol. 2020.
6. American College of Obstetricians and Gynecologists. The role of the obstetrician and gynecologist is cosmetic procedures, procedures [ACOG policy statement]. Washington: American College of Obstetricians and Gynecologists; 2008.
7. Vojvodic M, Lista F, Vastis P-G, Ahmad J. Luminal reduction hymenoplasty: a canadian experience with hymen restoration. Aesthet Surg J. 2018;38(7):802–6.
8. Goodman MP. Female genital cosmetic and plastic surgery: a review. J Sex Med. 2011;8:1813–25. https://doi.org/10.1111/j.1743-6109.2011.02254.x.
9. Goodman MP, Placik OJ, Benson RH III, Miklos JK, et al. A large multicenter outcome study of female genital plastic surgery. J Sex Med. 2010;7:1565–77.

5 Medico-legal Aspects in Aesthetic and Regenerative Gynecology

Geetendra Sharma and Hitesh Bhatt

5.1 Introduction

Medico-legally these procedures are different from other surgical and non-surgical treatments. In many cases, there may not be any medical indications but they are done on demand of the patient. The purpose of the treatment as the name suggests is cosmetic.

5.1.1 Who Can Do Cosmetic Surgeries? Who is a Cosmetologist?

Any Registered Medical Practitioner who is having enough experience and has taken enough training in the field of cosmetology can do a cosmetic treatment. He/She cannot write against his/her name cosmetologist or specialist or Cosmetic Gynecologist unless he/she has post-graduate or super-specialist qualification recognised by MCI. The original degree should be displayed and one can write that "Special interest in cosmetic Gynecology". There are many national and international diploma courses going on, which have no validity in the court of law, except they can be considered as training obtained.

G. Sharma (✉)
Ahmedabad, India

H. Bhatt
Mumbai, India

AIIMS has started post-Gynecology MCH is Cosmetic Gynecology. The practitioner who has done super-specialisation recognised by Medical Council of India can write specialist in Cosmetic Gynecology or Gynecological cosmetologist against his/her name.

Indian Medical Council (Professional Code of Conduct, Etiquette and Ethics) Regulation 2002, in Section 7.20—A Physician shall not claim to be a specialist unless he/she has a special qualification in that branch. Section **3.7.2**—A physician shall write his/her name and designation in full along with registration particulars in his/her prescription letterhead [1].

5.1.2 Counselling

Counselling always plays a vital role in management. There are different types of counselling.

Medical Counselling—Medical counselling should be done by a qualified medical person. In medical counselling Purpose of treatment, Procedure details, alternatives available with pros and cons of each, success rate and failure rates of treatment, risks of treatment and risks of not taking treatment should be properly explained. Patients should not be given over expectations. It is never advisable to lead the patient to a particular kind of treatment. It is always better to explain the facts in detail. One can explain by drawing sketches if possible. Ask patients to get sketches

P. Jindal et al. (eds.), *Aesthetic and Regenerative Gynecology*,
https://doi.org/10.1007/978-981-16-1743-0_5

or pictures of results they want. Video counselling is also a good option nowadays. By showing videos the facts can be explained to the patient. It is always advisable to document the counselling in the case paper. After the operation it is very difficult to prove that what was assured, and this becomes the reason for dispute. It is advisable to clearly mention in counselling, in writing, that what is assured. No guarantees or warranties should be given regarding the success of treatment.

Financial Counselling—Approximate expenditure with possibility of more expense in case of complication should be explained in detail with breakage.

Psychological Counselling—If healthcare provider feels necessary then Psychological counselling should be advised.

Consent—Consent is defined as where two or more parties agree to the same thing in the same sense that is "PARTIES AD IDEM".

Consent in Medical Context was the first time better discussed in Appeal (civil) 1949 of 2004 in the case of **Samira Kohli vs. Dr. Prabha Manchanda**, Date of judgment—16/01/2008 by the **Bench—B.N. Agrawal, P.P. Naolekar and R.V. Raveendran—Judgment delivered by Raveendran J.** [2].

Consent is to be taken for both surgical and non-surgical treatment. In case of non-surgical treatment where there is any kind of risk involved it is to be explained to the patient. Consent should contain after providing adequate information. Adequate information contains six things. Patient's own consent is a must.

1. Purpose of treatment—In Cosmetic Gynecology when there is no specific purpose and if surgery is to be done on demand for the cosmetic purposes it should be mentioned like that only. In case if there is any specific indication then an indication is to be mentioned.
2. Procedure is to explain to the patient in detail what you will be doing.
3. Alternatives available for the same are to be mentioned in detail. Suppose a patient has approached the doctor for vaginal laxity and demands surgery then in such a case it is duty of doctor to explain the alternative pelvic floor muscle exercise as one of the alternatives.
4. Success or failure rates are to be explained in detail in consent.
5. Risks or side effects are to be explained in detail. When you put something in writing the list should be exhaustive. Risks should not be explained which frightens the patient leading to refusal of treatment.
6. Risks if any for not undergoing treatment should also be explained but it should not be explained in such a way that patient becomes ready for treatment which is not required at all.

Consent in IPC—There are defense available to a doctor under the Indian Penal Code.

Section 87. Act not intended and not known to be likely to cause death or grievous hurt, done by consent. Nothing which is not intended to cause death, or grievous hurt, and which is not known by the doer to be likely to cause death or grievous hurt, is an offense by reason of any harm which it may cause, or be intended by the doer to cause, to any person, above 18 years of age, who has given consent, whether express or implied, to suffer that harm; or by reason of any harm which it may be known by the doer to be likely to cause to any such person who has consented to take the risk of that harm.

Section 88. Act not intended to cause death, done by consent in good faith for person's benefit. Nothing, which is not intended to cause death, is an offence by reason of any harm that it may cause, or be intended by the doer to cause, or be known by the doer to be likely to cause, to any person for whose benefit it is done in good faith, and who has given a consent, whether express or implied to suffer that harm, or to take the risk of that harm.

5.2 Ethical Dilemmas

There are four principles of Ethics:

Principle of Autonomy (Control by Individual)—Every person has the right to decide what she wants to undergo or what she does not want. We have to respect the autonomy of an individual. This is also known as the principle of human dignity.

Principle of Beneficence (Do Good)—We must do whatever is in good to the patient. Sometimes there is direct conflict between principle of beneficence and principle of autonomy, i.e. respecting the demand of patient.

Principle of Non-maleficence (Do no Harm)—One should not have the intention to harm. It is an obligation not to harm others. Whenever it is not possible to avoid harm them at least try to minimize harm.

Principle of Justice (Fairness)—We should be fair in treatment. There should be equality and impartiality in treatment. Your treatment should be justified.

5.2.1 Can We Do Surgery When You Think That it is Not Medically Indicated on Demand?

These situations often come in practise. If the procedure is not harmful to the patient then we can proceed with the same respecting autonomy of the patient. But when something is harmful to the patient then it becomes our duty to inform the patient about the possibility of harm in detail before proceeding with the same. We also should discuss the option of not doing the surgery at all instead of taking a risk.

5.2.2 What to Do When Everything is Normal Physiologically but Still Patient Demands Correction?

All procedures are planned, there is no medical necessity of operation, no emergency, Suppose, as per your examination patient is absolutely normal but still patient demands to undergo operation what to do?

The healthcare provider should be honest. Truth is to be explained to the patient while counselling. Counselling should be documented properly. After knowing the truth in detail if the patient still wants to undergo treatment then we have to respect the patient's autonomy. But if the procedure seems to be harmful to the patient, it should be avoided in the best interest of the patient. To refuse to treat the patient is the right of a doctor but it is advisable to document the reason for refusal.

5.2.3 Documentation

Documentation is the only evidence in the court of law to prove your case. It is said that "**Courts are not the courts of Justice but they are court of evidences**". So good document is always a good defence, poor document is a poor defence and no document is no defence. Communication, Documentation, Documentation of communication play a vital role in any medico-legal case.

Document should be Accurate in chronology, Complete, Legible and without any extraneous information. Let us see what is the difference between routine documentation and documentation in the cases of cosmetic gynecology.

In chief complaint, if there is any specific complaint then it should be mentioned in the language of the patient but if no complaint but patient want treatment not medically indicated for cosmetic purpose then it should be documented.

In examination it is advisable to do examination in the presence of one female assistant. The explanation should be done by drawing on the case paper. It is always a good idea to take pre-operative and post-operative photographs for comparison and to be attached with the records. All medically indicated pre-requisite investigations should be performed. Operative notes should be written in detail explaining which kind of instrument or laser is used during treatment. Post-operative follow-up instructions should be passed on in detail to the patient.

5.2.4 Complications

Complications per se cannot be considered negligence. If complications are explained in detail before starting the treatment that will become more acceptable. So, it is a good practise to explain complications of treatment in advance to the patient.

How to tackle complications—Disfigurement, Scarring, Haematoma, No result, and No satisfactory result. Any complication if occur should be treated as per medical norms.

5.2.5 Confidentiality

5.2.5.1 Indian Medical Council (Code of Conduct, Etiquette and Ethics) Regulation, 2002 [3]

2.2 Patience, Delicacy and Secrecy: Patience and delicacy should characterize the physician. Confidences concerning individual or domestic life entrusted by patients to a physician and defects in the disposition or character of patients observed during medical attendance should never be revealed unless their revelation is required by the laws of the State. Sometimes, however, a physician must determine whether his duty to society requires him to employ knowledge, obtained through confidence as a physician, to protect a healthy person against a communicable disease to which he is about to be exposed. In such instance, the physician should act as he/she would wish another to act towards one of his/her own family in like circumstances.

5.2.6 Disclosure of the Patient's Secret Without Valid Reason Amounts to Misconduct

7.14 The registered medical practitioner shall not disclose the secrets of a patient who have been learnt in the exercise of his/her profession except:

(i). In a court of law under orders of the Presiding Judge.
(ii). In circumstances where there is a serious and identified risk to a specific person and/or community.
(iii). Notifiable diseases.

In the case of communicable/notifiable diseases, concerned public health authorities should be informed immediately.

5.2.7 Certain Typical Medico-legal Issues in Cosmetic Gynecology

There are a few operations in Cosmetic Gynecology where there is no scale of measurement as to whether the operation is done successfully or not. Only patients can feel and tell whether any improvement is there or not and that also is subjective. In such a case if the patient claims for no improvement after treatment there is no means to know the truth.

The result of treatments like G-Spot Amplification and O-Shot® for female Sexual quality enhancement cannot be measured by any means. In such kind of treatment, patients in advance should be informed that the improvement cannot be guaranteed and as there is no measure to know the fact your claim of no improvement will not be entertained.

5.2.8 Certain Procedures Not Yet Proved: Can One Practise [4]

In many cases, doctors doing treatment is not yet proven by trials done on human beings and are not officially recognized. Laser for vaginal rejuvenation is yet not FDA approved. This should be explained to the patient and a written consent be taken that she understands it. Plasma Rich Platelet therapy is used nowadays by many doctors in different indications and treatments.

PRP is a very promising futuristic therapy. It is a vehicle to deliver a large amount of important growth factors, which are biologically active, to the injured site. Its use has increased extensively over the last decade due to advanced technology, availability of newer commercial PRP equipment, manufacturing of various PRP products in the market. It is very simple and easy to use, easily available, uses the patient own blood (autologous), potential cost-effective, and considered very safe therapy. There are many case series showing positive outcomes. But despite the promising results of several animal studies, well-controlled human studies are lacking. The research is still in its infancy. There is no consensus or protocol for the use of PRP. Even with all the limited evidence available, today PRP is becoming a very popular therapy in various fields of medicine. More research in future will clear the clouds over many questions being raised about the efficacy and evidence for PRP. PRP can be used for a long list of conditions, for example orthopaedic, neurology, musculoskeletal, cardiology, dermatology, and plastic surgery. The research on PRP is still in infancy, and there is no consensus on platelet concentration, amount of PRP, which is the best technology in preparing PRP. There are many animal studies, which are showing encouraging results, but human studies are lacking. PRP is a very promising treatment option that is nonsurgical. We need to wait for more concrete evidence to emerge to define its exact clinical role.

To conclude, we may say that there are reasonable amount of data that warrant continued research in PRP but currently, its role in clinical practise is not completely defined [4].

All readers and practitioners of the art of PRP (platelet-rich plasma), it is imperative that you be cautioned of the fact that the Indian FDA considers "preparing Platelet Concentrate" amounts to "manufacturing" of blood components (sic) and those who do contravene the provisions, as per the Drugs and Cosmetics Act 1940, read with rule 122EA of the Drugs and Cosmetics Rules, 1945 and are punishable under Section 27 of the said Act.

Some of our Plastic surgery colleagues and some Dermatologists have been served with such notices recently.

5.2.8.1 Ministry of Health and Family Welfare (Department of Health and Family Welfare) by Notification: New Delhi, the 11th March, 2020 [5]

G.S.R. 166(E). Whereas a draft of certain rules further to amend the Drugs and Cosmetics Rules, 1945, was published as required under Sub-section (1) of Section 12 and Sub-section (1) of Section 33 of the Drugs and Cosmetics Act, 1940 (23 of 1940) *vide* notification of the Government of India in the Ministry of Health and Family Welfare (Department of Health and Family Welfare) number G.S.R. 1152(E), dated on 29th November, 2018, in the Gazette of India, Extraordinary, Part II, Section 3, Sub-section (i), inviting objections and suggestions from persons likely to be affected thereby before the expiry of a period of forty-five days from the date on which the copies of the Official Gazette containing the said notification were made available to the public; And whereas copies of said Official Gazette were made available to the public on 30th November, 2018

And whereas objections and suggestions received from the public on the said rules have been considered by the Central Government; Now, therefore, in exercise of the powers conferred under Sections 12 and 33 of the Drugs and Cosmetics Act, 1940 (23 of 1940), the Central Government, after consultation with the Drugs Technical Advisory Board, hereby makes the following rules further to amend the Drugs and Cosmetics Rules, 1945, namely:

(1) These rules may be called the Drugs and Cosmetics (Second Amendment) Rules, 2020.
(2) They shall come into force on the date of their publication in the Official Gazette.

Central Government Act—Section 122 E(a) in The Drugs and Cosmetics Rules, 1945

[(a) A drug, as defined in the Act including bulk drugs substance which has not been used in the country to any significant extent under the conditions prescribed, recommended or suggested in the labelling thereof and has not been recognized as effective and safe by the licensing authority mentioned under Rule 21 for the proposed claims: Provided that the limited use, if any, has been with the permission of the licensing authority.].

Section 27 Whoever, himself or by any other person on his behalf manufactures for sale or for distribution, or sells, or stocks or exhibits or offers for sale or distributes—

(a) Any drug deemed to be adulterated under Section 17A or spurious under Section 17B or which when used by any person for or in the diagnosis, treatment, mitigation, or prevention of any disease or disorder is likely to cause his death or is likely to cause such harm on his body as would amount to grievous hurt within the meaning of Section 320 of the Indian Penal Code, solely on account of such drug being adulterated or spurious or not of standard quality, as the case may be, shall be punishable with imprisonment for a term that shall not be less than 5 years but which may extend to a term of life and with fine which shall not be less than ten thousand rupees.
(b) Any drug: (i) Deemed to be adulterated under Section 17A, but not being a drug referred to in clause (a), or (ii) Without a valid licence as required under clause (c) of Section 18, shall be punishable with imprisonment for a term which shall not be less than one year but which may extend to three years and with fine which shall not be less than five thousand rupees: Provided that the Court may, for any adequate and special reasons to be recorded in the judgment, impose a sentence of imprisonment for a term of less than one year and of fine of less than five thousand rupees.
(c) Any drug deemed to be spurious under Section 17B, but not being a drug referred to in clause (a) shall be punishable with imprisonment for a term which shall not be less than three years but which may extend to five years and with fine which shall not be less than five thousand rupees: Provided that the Court may, for any adequate and special reasons, to be recorded in the judgment, impose a sentence of imprisonment for a term of less than three years but not less than one year.
(d) Any drug, other than a drug referred to in clause (a) or clause (b) or clause (c), in contravention of any other provision of this Chapter or any rule made thereunder, shall be punishable with imprisonment for a term which shall not be less than one year but which may extend to two years and with fine: Provided that the Court may for any adequate and special reasons to be recorded in the judgment impose a sentence of imprisonment for a term of less than one year.

5.2.9 Minor Patient Coming to Take Advice: What to Do?

In minor, it is better to avoid cosmetic surgery just for beautification as minor is not competent to consent so the consent party shall always be the guardian and when minor when attains majority, may not like decision taken by the parents. However, curative cosmetic surgeries, e.g. post burns or injuries or for disease are justified on basis of necessity.

5.2.10 Insurance

It is advisable to check with the indemnity insurance company whether your indemnity policy covers the cosmetic procedures or not. If not then if you are in such practise then it is advisable to get it added by paying an extra premium if any. Every procedure is within the ambit of the Consumer Protection Act. The same thing is for the patient, most of the Mediclaim policies do not cover cosmetic surgeries. It is always better to check it with insurance provider beforehand.

5.3 Conclusion

We are in the twenty-first century in a materialistic world and most people do prefer to look better physically. So, cosmetic surgery has its place in the medical field, however, since its inception, teething problems do arise and one such problem is a medicolegal issue. It is better to consult a medicolegal consultant in such a situation.

References

1. Indian Medical Council (Professional Conduct, Etiquette and Ethics) Regulation 2002. Section—3.7.2 and Section 7.2.
2. Samira Kohli vs Dr. Prabha Manchanda & Anr. on 16 January, 2008—Author: Raveendran—Bench: B. N. Agarwal, P. P. Naolekar, R. V. Raveendran—CASE NO.: Appeal (civil) 1949 of 2004.
3. Indian Medical Council (Professional Conduct, Etiquette and ethics) Regulation, 2002—Section 2.2.
4. http://www.indianjpain.org/article.asp?issn=0970-5333;year=2017;volume=31; issue=1;spage=1;epage=6;aulast=Kothari.
5. Ministry Of Health And Family Welfare (Department of Health and Family Welfare) By Notification—New Delhi, The 11th March, 2020—G.S.R. 166(E)The Drugs and Cosmetics Rules, 1945.

Energy-Based Devices: Comparisons and Indications

6

Ohad Toledano

6.1 Introduction

Modern medicine offers a variety of treatment modalities aiming to manage a large diversity of gynecological conditions, varying from as little as topical gels or hormone-replacement therapy to invasive vaginal interventions. These feminine issues can affect women's daily lifestyle, quality of life or project physiological effects such as decreased self-confidence or impaired sexuality. Changes due to menopause, postpartum or other causes led to an emerging number of energy-based devices (EBD) for vaginal procedures that utilize ablative or thermal effects to sculpt the external vagina or to strengthen the aging vaginal wall. Available applications are mainly lasers yet other energy-based devices have also been introduced to regenerative gynecology and urogynecology, such as various types of RF (radiofrequency derives), High-intensity focused ultrasound (HIFU), Light-emitting diodes (LED) and high-intensity focused electromagnetic field (HIFEM). Karcher& Sadick [1] claim that the commonly used term of "vaginal rejuvenation" is in fact a generalized term for a wide array of gynecological aesthetic and functional procedures that aim to restore the vagina and its surrounding tissues. According to their definition, the spectrum of procedures can range from mere vaginal atrophy (VA) and dryness, treated by minimally or noninvasive strategies, to cases that require invasive intervention such as labiaplasty or vaginoplasty.This chapter aims to introduce the main available technologies offering vaginal rejuvenation and restoration, its interaction with the tissue, and compare the technological differences.

6.2 Lasers

Light amplification by stimulated emission of radiation, or LASER, was first mentioned as a theoretical concept by Albert Einstein as an atomic cascade produced by stimulated emission to create electromagnetic radiation. Yet, it took almost half a century for the first laser to be built by an American physicist named Theodore H. Maiman in 1959 [2]. Laser is defined as light, a form of an electromagnetic energy; thus it is by definition—a wave, and as such, all lasers are a part of the electromagnetic spectrum. Laser is characterized by three unique properties that differentiate it from other forms of light; monochromaticity, coherence and collimation [3]. Monochromaticity, probably the main associated characteristic of a laser, means that each laser emits light at a single wavelength, so it has only one length to its particle wave. The waves of this light also travel "in-phase" meaning in synchronization in time, therefore, are coherent. Laser

O. Toledano (✉)
Tel Aviv, Israel

P. Jindal et al. (eds.), *Aesthetic and Regenerative Gynecology*,
https://doi.org/10.1007/978-981-16-1743-0_6

waves are also collimated, meaning they travel parallel to one another and in a single direction rather than scatter like ordinary light waves. Taking all three characteristics to account, it is clearly understood why laser it defined as an ideal and precise tool for numerous indications of use. Lasers are often named or defined by their emitting wavelength as it is a crucial property of the laser, especially in the medical field, which focuses on specific chromophores to be targeted by the laser. The interaction between laser and chromophore was proposed by Anderson and Parrish at 1983, named as the theory of "selective photothermolysis" [4]. This theory revolutionized the understanding of the interaction between laser and tissue, suggesting ways in which laser can be used therapeutically. The basis of the theory claims that laser is absorbed by a defined target chromophore, with little impact on its surrounding tissue. The absorbed light is transformed from light energy into thermal energy that destroys the target in question. There are numerous chromophores that absorb light, to each its own absorption curve that defines the affinity to light for each wavelength. The main chromophores addressed in most medical disciplines are melanin, hemoglobin, and water. Absorption curves (Diagram 6.1) are important factors to consider when using laser, as different wavelengths aimed at the same chromophore will have different tissue reactions.

In practice, a desired wavelength is selected mainly according to its absorption in the main chromophore in the relevant tissue. In many procedures, in addition to the damage inflicted to the target tissue, controlled damage to the surrounding tissue is also desirable, to create other laser-tissue interactions [5] such as Vaporization (tissue ablation), protein denaturation and Hyperthermia. Tissue reaction is dependent on the temperature it is exposed to (Table 6.1), the duration of exposure, also referred to as pulse duration and the content of the tissue.

Table 6.1 Temperature-dependent laser-tissue intersections

Temperature	Molecular and tissue reactions
42–45 °C	Hyperthermia leading to protein structural changes, hydrogen bond breaking, retraction
45–50 °C	More drastical conformational changes, enzyme inactivation, changes in membrane permeabilization, oedema
50–60 °C	Coagulation, protein denaturation
~80 °C	Collagen denaturation
80–100 °C	Dehydration
> 100 °C	Boiling, steaming
100–300 °C	Vaporization, tissue ablation
> 300 °C	Carbonization

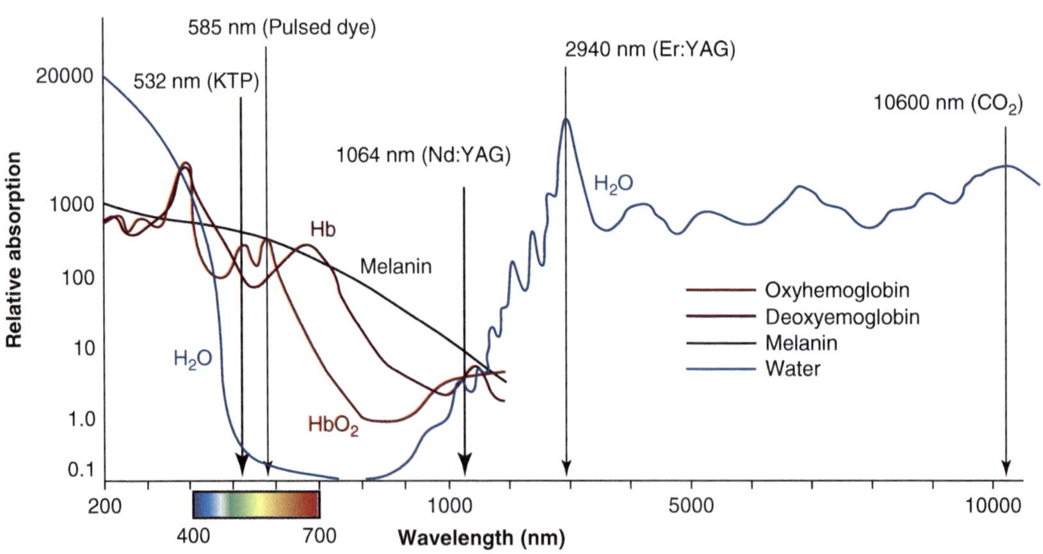

Diagram 6.1 Laser—Absorption spectra of water, hemoglobin, oxyhemoglobin, and melanin

6.2.1 Carbon Dioxide (CO_2) Laser

The CO_2 laser emits light at a 10,600 nm wavelength and has a high absorption in water (Fig. 6.1), thus causes ablation in a water-containing tissue. When CO_2 laser is used, surrounding the borders of the ablated zone, a degree of residual thermal damage is present, caused by the tissue exposure to decreased temperatures [6]. With an absorption coefficient of 800^{-1}, the CO_2 absorbed light induces mainly ablation, yet other effects such as coagulation or thermal effect are notable on the surroundings due to this conducted heat [7]. Although its main use is ablating, those other laser-tissue interactions of the CO_2 are of use and are harnessed to the success of the procedure. Coagulation formed surrounding the ablated zone causes small blood vessels to be cauterized together with the surrounding tissue, achieving hemostasis allowing a "cleaner" procedure with less complications [6]. It was also suggested that tissue exposed to hyperthermia (42–50 °C) is additionally influenced as effects of collagen shrinkage causes long-term neocollagenesis and neoelastinogenesis effects [8].

6.2.2 Erbium-Doped Yttrium Aluminum Garnet (Er: YAG) Laser

The Er:YAG is a near-infrared pulsed laser of a 2,940 nm wavelength which is of the highest absorption in the water among common practice lasers. In fact, its water absorption coefficient is 16 times higher than that of theCO_2 laser [9], making it of a superficial ablative property [7]. The difference between the Er:YAG andCO_2 lasers is in their "tissue signature" effect, as Er:YAG has a shallow penetration depth in comparison to CO_2, which allows for precise tissue ablation and minimal thermal damage to surrounding tissue. Unlike the CO_2 laser, skin-interaction of Er:YAG is mostly of ablative effect with the skin, with nearly no coagulation and minimal surrounding thermal damage, so it creates clean ablation craters and precise cuts without apparent scarring. However, it lacks sufficient coagulative margins surrounding the ablated area, thus is limited and not ideal for removal of deep or large lesions, as it cannot prevent bleeding [10].

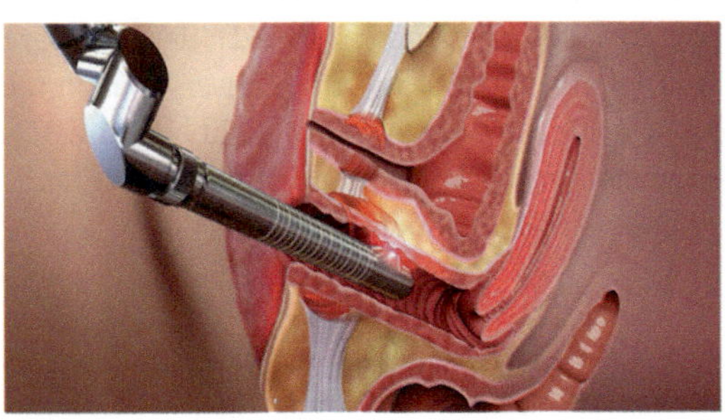

Fig. 6.1 A nonsurgical fractional laser probe emitting in the vaginal canal

6.2.3 Fractional Lasers

For external purposes or bulk removal of tissue, ablative lasers are used in a non-fractional beam profile, meaning in a full spot beam. In 2004 a new approach for laser delivery named "fractional photothermolysis" (FP) was introduced using a 1,550 nm laser as an alternative for the full non-ablation option to evaluate tissue recovery [11]. This was achieved by delivering an array of microscopic laser beams to create what the researchers named microscopic treatment zones (MTZ) of thermal injury to the skin. A few years later, this concept was applied on the ablative CO_2 & Er:YAG lasers to create MTZ in which the treatment zones were zones of ablation of only a small fraction of the skin [12]. Fractionating the ablative laser into small micron size spots allowed for rapid re-epithelialization derived from the undamaged epidermal "islands" separating the ablated MTZs, so that the ablative area was only 5–30% of the total area treated [13]. MTZ depth and width were shown to be associated with the delivered energy.

6.2.4 Lasers in Gynecology

In the second part of the twentieth century, lasers had emerged into the field of gynecology, yet it was not easily accepted [14]. As the use of older, outdated methods was found unsuitable for treating in many cases due to unwanted, yet unavoidable complications, the laser was suggested as a new alternative [15], primarily the carbon dioxide laser. Over the years, more and more researchers grew to believe that, for some indications, the use of the CO_2 laser is a safer and a more versatile option available [16]. In time, new lasers such as the Neodymium-doped yttrium aluminum garnet; (Nd:YAG, 1,064 nm) and the argon laser were proven as practical tools in gynecology [17]. Yet because of their high absorption in water, CO_2 laser and to some extent also the Er:YAG laser has become the most dominant ones in the last decades. The CO_2 laser was traditionally used for gynecological surgery and genitourinary procedures as means of ablation, vaporization, excision, incision, or coagulation of soft tissue [18]. It was adopted to the subfield of regenerative and aesthetic gynecology such as labiaplasty and removal of genital warts [19, 20], which are commonly referred to as "Vaginal rejuvenation". Both lasers are used to treat a large scope of other gynecological indications such as lichen sclerosus and more [21]. Other uses of lasers in gynecology are laser-assisted laparoscopy, hysteroscopy, and even in vitro fertilization procedures [9]. Yet it seems that most attention in recent years was drawn to the treatment of aging vagina, dryness, infections, and related stress urinary incontinence (SUI). These are some of the symptoms associated with Genitourinary syndromes of menopause (GSM), a comprehensive term embraced during the last years by the main medical societies addressing Women's Sexual Health [22]. This new definition of GSM was to replace the formally known medical terms of vulvovaginal atrophy (VVA), urogenital atrophy or atrophic vaginitis [23] as previous ones did not accurately define the medical condition and symptoms. GSM is caused due to decreased urogenital estrogen levels and is clinically described as symptoms of dryness, irritation, impaired sexual activity and other urinary disorders. All gynecological procedures aimed to treat the mentioned above indications are based on the Laser-tissue interaction, which is made possible due to the water absorption property. As previously mentioned, ablative lasers are slowly becoming tools for external ablative procedures using a full-beam, small diameter profile with precise and clean cutting abilities [24, 25] or in the fractionated beam profile [26, 27]. Fractional ablative lasers using a diffractive probe are the most common ones for vaginal rejuvenation [9, 23, 28] as they promote formation and remodeling of collagen attempting to rewind genitourinary changes such as thinning in the epithelial lining of the vagina and atrophy. The small ablative MTZs and thermal effect surrounding it stimulates collagen deposition, improves vaginal wall quality and is shown to have the capability of restoring vaginal mucosa pH levels [23, 28]. The restoration of the connective tissue, other alterations of the extracellular matrix support the effectiveness of fractional ablative lasers for the restoration of vaginal mucosa structure and related physiological tropism.

Although clinical efficacy and safety are well established in dermatology and of growing reports in gynecology, there are still some arguments yet unanswered about the best practice, ideal technology and maintenance needed [9]. In some societies, the use of lasers is still not recommended as first-line therapy due to the lack of large scope randomized controlled trials (RCTs) and long-term follow-up [29]. Nevertheless, the increasing number of publications on the use of lasers is a source of encouragement, as physicians gradually comprehend the potential of lasers as the resolution in the gynecological field. The accumulating proofs from the literature indicate that fractional ablative lasers do have histological supportive alternations and are a safe and effective option in treating symptoms of GSM in postmenopausal women [30–32] and improve wellness for those who suffer. Lasers remain the most studied and published modality of all EBDs presented in this chapter and of the widest range of indications potentially treatable.

6.3 Radiofrequency (RF)

Radiofrequency is a term aimed to describe the oscillation of electromagnetic radiation, meaning the transfer of energy in the form of (radio) waves, ranging between 3 kilohertz (kHz) to 300 gigahertz (GHz) [33]. Radiation can be either ionizing or non-ionizing, and in the case of radiofrequency electromagnetic radiation (EMR), it is non-ionizing radiation, meaning that it has insufficient energy to break chemical bonds or remove electrons thus considered as safe (adopted from ARPANSA—Australian Radiation Protection and Nuclear Safety Agency). The electromagnetic field, as depicts by its name, is described as a combination of an electric field of stationary charges and a magnetic field of moving currents that create less force than that the electric field. The electrical current formed by EMR causes oscillation and rotation of polarized molecules which in their turn exhaust heat to the surrounding tissue. The medical scope of EMR lays in the Industry, scientific and medical (ISM) part of the radio spectrum radio bands, ranging from 13 MHz to 24 GHz, and is used to induce heat is a biological tissue, which in its turn will respond to the accumulated heat and cause cellular and intracellular alternations. There are two mechanisms in which tissue heating may occur; according to the first mechanism, heat is generated due to an ionic current produced by the mobilization of charged particles in an alternating electromagnetic field. The second mechanism refers to the rotation of water molecules due to an alternating electromagnetic field [34]. In both cases, the particles interaction within the tissue leads to dissipation of the energy transferred and volumetric heating of the tissue.

6.3.1 RF-Tissue Interactions

RF can induce several phenomena on biological tissue:

A. **Ablation** is based on exposure of the tissue to high energy density so that tissue is evaporated [35]. As most of the energy is tunneled to ablation, only a small fraction is released as thermal effect to the margins of the ablated area, yet it is still more substantial than of ablative lasers. RF ablation is commonly used in cautery during surgery.
B. **Coagulation** provides hemostasis if blood vessels are present so that bleeding can be controlled during surgery. In ablation, the margins of the evaporated area are coagulated for this purpose. When applied to soft tissue, the coagulative effect induces tissue necrosis without evaporation [36].
C. **Hyperthermia** is heating of the tissue to various temperature levels that do not cause ablation or coagulation. Depending on the temperature reached and the time of exposure, different tissue reaction is seen. Physiological processes that are stimulated include fibroblasts stimulation and collagen synthesis. In higher energy levels in which the coagulation threshold is not yet reached, structural alternations in proteins are seen. In cosmetic procedures, the therapeutic temperatures are lowered to avoid epidermal and dermal damage, yet because of the use of lower temperatures, the outcomes may not be consistent, will require multiple sessions and will take longer to show results.

6.3.2 RF Devices and Configuration

RF medical applications are relatively established in fields such as electrosurgery and physiotherapy [37] utilizing tissue heating, commonly referred to as diathermy. Shortwave (SW) diathermy devices can be either continuous or pulsed waves and create deep tissue heating and superficial heating mostly localized to the skin [38]. RF devices can have different electrode configuration that delivers the current differently (Fig. 6.2), among them are monopolar, bipolar (multipolar) or unipolar, all of have been is use for cutaneous applications [9]. Penetration of monopolar is 20–25 mm, with unipolar 15–20 mm and bipolar reaching approximately 2–8 mm deep.

- *Monopolar* devices are equipped with a single electrode from which the current is released and a grounding pad to which the current passes. The grounding pad is usually of a large contact area placed outside of the treatment zone so that that high-density RF current is

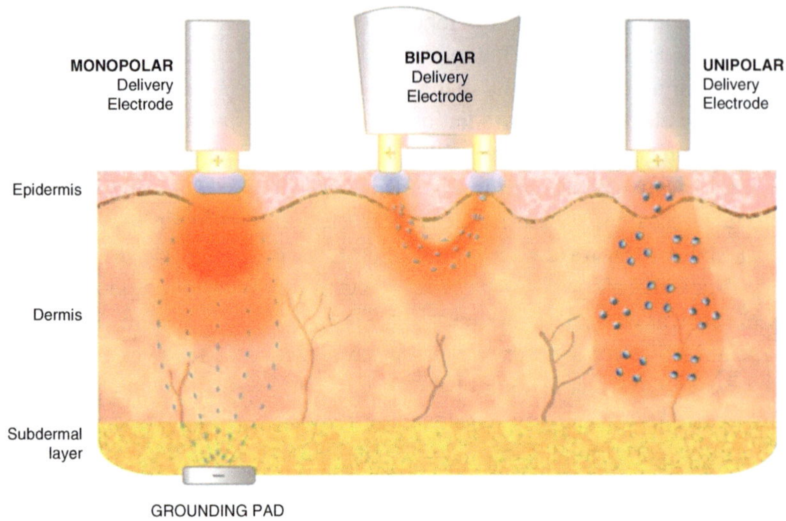

Fig. 6.2 RF electrode configurations and heating type on human skin

created near the active electrode, and the RF current diverges toward the large return electrode [33].

- *Bipolar* devices utilizes two (or more) closely placed electrodes on the applicator so that current will pass through one to the other while heating a well-defined tissue volume [34] in between the electrodes. This geometry creates a more uniform heating in the larger volume of tissue compared to monopolar. Yet, due to its electrode geometry, bipolar devices are associated with superficial tissue heating as the current density is highest along the line of the shortest distance between the electrodes. Penetration depth is a function of electrode size and the distance between them. As the distance between electrodes grows, energy density reduces dramatically to decline tissue heating.
- *Unipolar* current is applied via a single electrode with no need for a grounding pad. It is applied at frequencies higher than 13.58 MHz to create dielectric heating and induce deep tissue penetration, in which maximum heat is located at the central axis of the electrode [34].

6.3.3 Radiofrequency in Gynecology

RF has gained popularity in treating several gynecological procedures during the last years due to its thermal action, based on elevating the temperature of the treated tissue to initiate biological changes. If tissue temperature is quickly elevated beyond a certain threshold, the effect will be of evaporation and ablation of the tissue, yet if the temperature is elevated to lower levels and maintained over a relatively long period of time, a non-ablative, the thermal-dependent effect will occur. In Gynecology when applying non-ablative RF, the temperature of vaginal wall connective tissue heats up to 40–43 °C and generate tightening of the vaginal canal [39]. This can be achieved as the heating evokes several biological effects such as inflammation, fibroblasts stimulation, collagen contraction and neocollagenesis. These in their turn result in restoration of the vaginal wall, tissue shrinkage and increased moisture of the submucosa [1, 9, 40, 41]. Early findings also suggest potential neoelastinogenesis, which may contribute to vaginal canal laxity [9] and increased nerve fiber density in papillary dermis which may contribute to improved sexual activity and less emotional distress, lasting a few months post procedures [41–43]. Other than vaginal laxity, SUI [39, 44, 45] was also explored with promising efficacy and low risk of adverse events in up to one-year follow-up. Generally, these thermally related processes are triggered by a non-ablative and chromophore-independent mechanism, which makes RF considerably safe regardless to skin type. Due to moderate temperature, the use of non-ablative RF also gains benefit of high comfort and satisfaction during procedures as reported by patients yet associated with significantly longer treatment duration as compared to lasers. In the external genital area, RF was shown to improve aesthetic appearance using ablative and non-ablative procedures [41, 46, 47] such as tightening of the clitoral hood and introitus with minor improvement in labia minora pigmentation. For a nonsurgical procedure for the outer vagina, the RF may serve better than lasers, as the use of ablative lasers will result in some downtime and slight discomfort while non-ablative lasers will not penetrate as deep as RF, thus will probably have less efficacy or require more sessions.

6.4 High-Intensity Focused Ultrasound (HIFU)

Unlike laser and RF, the sound is not an electromagnetic wave but a form of acoustic energy traveling in waves. Energy is propagated by the vibration of particles through a medium (gas, liquid, or solid) in means of pressure, compression, and decompression. Audible sound lays within a frequency range of 20 Hz to 20 kHz, frequencies lower than 20 Hz are referred to as infrasound and frequencies higher than the level of human hearing are called ultrasound waves, yet all three are of the same physical properties and differ only by human hearing ability. Ultrasound is used both for diagnostic and therapeutic purposes. As a medical diagnostic technique, high-frequency sound waves provide imaging abilities inside the body with no risk of ionizing radiation. As a therapeutic tool, ultrasound waves can be used for many interactions with biological tissues, such as destruction of tumor tissues, target drugs to specific locations and more [48] which are divided into two main effects; thermal and mechanical. The thermal effect by means of ultrasound is reached since some of the energy is absorbed in the tissue and converted into heat as temperature increases faster than heat dissipation. Absorbed energy can be estimated, and tissue response predicted so that the thermal effects can be controlled by changing wave parameters such as exposure time, power, frequency, or distance. Mechanical effects are induced by high-intensity ultrasound that creates cavitation, microstreaming, and radiation force. Cavitation is a rapid collapse of a gas-filled cavity that expands during acoustic cycles, which exhaust high temperature and pressure to the surrounding tissue.

The use of HIFU for aesthetic uses (Fig. 6.3) has been fully understood and supported by publication in the last decade yet only recently introduced to gynecology. Two types of focused ultrasound are in use in the medical field; HIFU and micro-focused ultrasound

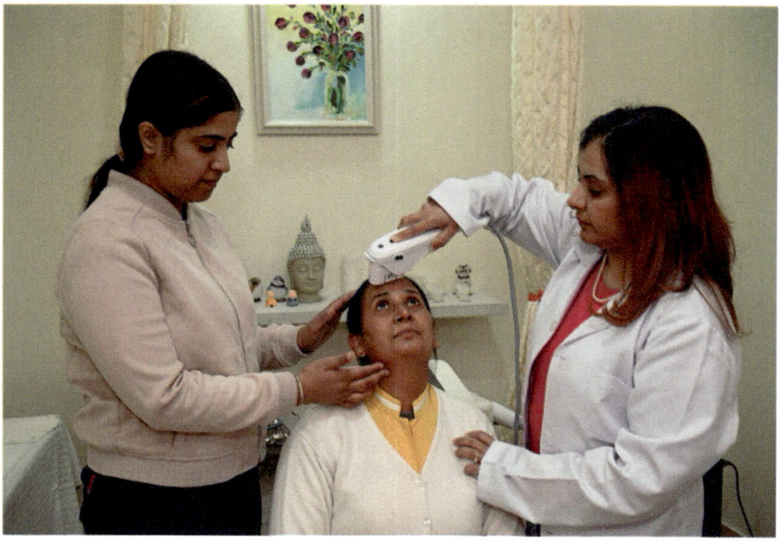

Fig. 6.3 HIFU for skin tightening. (© Jindal P. et al. 2020)

(MFU) that differ according to the energy distribution and tissue effect.

HIFU has an intense thermal effect of ultrasound with an intensity ranging from 1,000 to 20,000 times higher than that of a diagnostic ultrasound [49, 50], reaching up to 47–59 J/cm^2 at a frequency of 2 Mhz. The ultrasound focal point, achieved by focusing the soundwaves by a curved transducer, can be as little as 1 mm in diameter [48] and reach deep in tissues up to 1.8 cm in depth (Fig. 6.4) [51, 52]. The temperature at the focus reaches more than 60 °C causing coagulation and resulting in tissue necrosis [53]. Yet in addition, the compression that propagates through the tissue results in shear forces which results in frictional heating and cavitation. Apart from nonsurgical purposes such as ablating tumors, It can also be used to ablate adipose tissue for body contouring and achieve a reduction in body circumference [51, 52].

MFU utilizes lower ultrasound energy and is more superficial, reaching focal depths ranging 1.5 to 4.5 mm yet despite its lower energy, target temperature also exceeds 60 °C, producing coagulation points in the dermis and subdermal layer while sparing overlying dermal and epidermal layers [54] as no cavitation is present in upper more layers. Defused heat from coagulated foci triggers collagen denaturation and contraction, resulting in tightening of lax skin. Special transducers direct the ultrasound energy, allowing to target subcutaneous tissues, especially the superficial musculoaponeurotic system (SMAS) [55].

6.4.1 HIFU in Gynecology

Early results of in vitro experiments followed by in vivo practical testing in 2002 during a clinical trial paved the way for HIFU to be used as a nonsurgical approach for uterine fibroid treatment [56] under MR-guided HIFU (Magnetic resonance guided). The use of HIFU was found safe when coupled with MRI [53] and was granted FDA approval in 2004 after proving that patients benefit from symptomatic relief for an average of two years [48]. The thermal effect of the HIFU causes reduction in fibroid volume and in hemostasis [53, 57] that led to clinical improvement of patients symptoms, including bleeding or pain. In

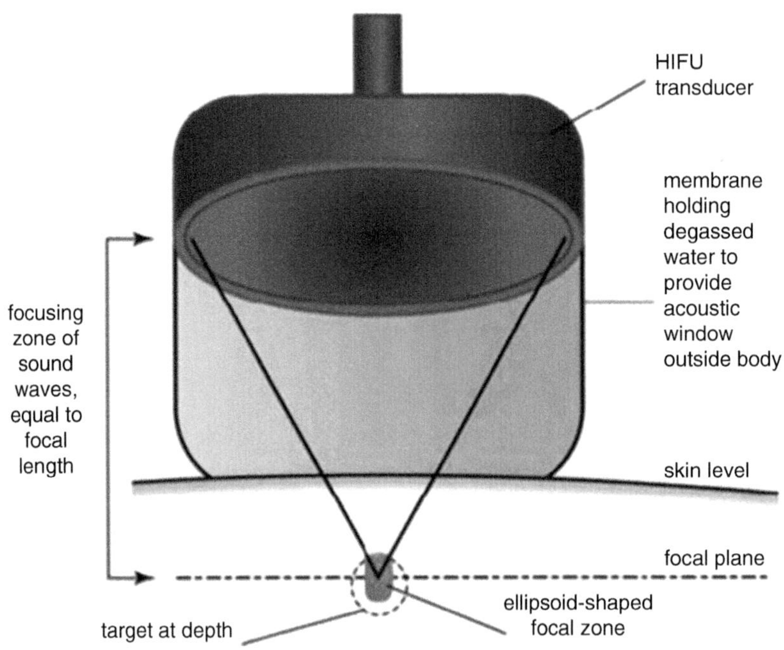

Fig. 6.4 Diagram of extracorporeal generation of HIFU focal point within a tissue target

the field of regenerative gynecology, the HIFU technology was introduces for the first time for vaginal rejuvenation in 2016 thus still lacking sufficient evidence. In vaginal procedures, the ultrasound transducer used is cylindrical narrow-shaped, plated with an emission contact plate in order to reach the vaginal interior and treat its whole length and circumference. A prospective study of GSM and SUI using HIFU in postmenopausal women by Elias et al. had demonstrated epithelial chances in mucosal histology such as stratification and increased glycogen in the superficial keratinocytes that can explain the microbiota and vaginal condition improvement [58]. Six of the patients also presented Vaginal Hyperlaxity prior to treatments, and 5 of them reported as asymptomatic in the 6-month follow-up. They concluded that the use of HIFU in treating GSM related symptoms is comparable to lasers and RF and claim that due to the HIFU unique technology it may serve better than other EBDs to reach deeper layers beyond the mucosa where SUI related issues root from and in other indications in which restoration of vaginal mucosa alone is insufficient.

6.5 High-Intensity Focused Electromagnetic Technology (HIFEM)

Electromagnetic fields are composed of both electric and magnetic fields in which electric fields are the result of electric charges, while magnetic fields are created given the movement of the electric charges [59]. The electromagnetic phenomena comprise basic principles that define its action. Electric charges attract or repel each other with a force inversely proportional to the distance between them, magnetic poles act the same and exist in pairs. An electric current may generate a circumferential magnetic field, and when induced in a loop wire, it will move towards or away the magnetic field, or the magnet will move towards or away from it (Fig. 6.5).In the case of HIFEM technology, a rapidly varying magnetic field induces an electric current to the target tissue.

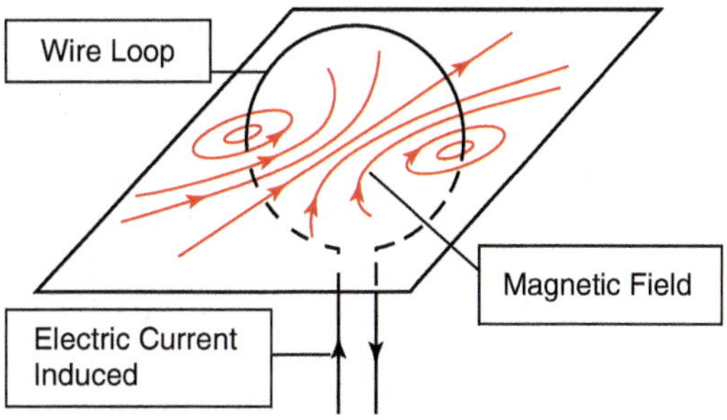

Fig. 6.5 A rapidly moving magnetic force induces an electrical current in a loop wire

HIFEM technology induces a simulative effect as the electromagnetic field passes non-invasively through the neuromuscular tissue to induce electric currents that stimulates muscle action [60, 61]. It utilizes the principles of electromagnetic induction to depolarize motor neurons, thus stimulating intense involuntary muscle contractions, bypassing the central nervous system. Such intense muscle recruitment appears to trigger an exaggerated need for energy at the cellular level.

6.5.1 HIFEM in Gynecology

When addressing SUI, the common practice to this day is the rehabilitation of pelvic floor muscles (PFMs) by means of physical exercise. PFMs are the anterior muscle, in charge of bladder and urethra contraction, middle muscle, controlling the vagina/uterus and posterior muscle for rectal function. Additional anatomical structures support these organs, such as the pelvic diagram, urogenital diaphragm and urethral/anal sphincters [62]. Both PFM and its supporting muscles are subjective to electrical and to electromagnetic stimulation. In normal functioning, PFMs prevent multiple disorders such as urinary or fecal incontinence, sexual disorders or pelvic organ prolapse. Weakness of those muscles can promote the aforementioned disorders generally referred to as pelvic floor dysfunctions (PFDs) as a result of aging or due to childbirth [63]. PFM enhancement was proven effective by mean of physical exercise [64] as training increases pelvic muscles tone, causes hypertrophy and strengthens muscle fibers. This exercise promotes the restoration of protective continence mechanisms [65]. To effectively achieve PFM re-education, hundreds of contractions are required and in a very specific manner to address the proper muscles. PFM can also be stimulated by means of electrical stimulation [66] and in recent years also by stimulation using HIFEM technology [67–69]. Both technologies deliver electrical currents to PFMs, which depolarize motoneurons to stimulate involuntary muscle contractions. Electrical stimulation utilizes direct electric charge flow that does not spare skin surface while HIFEM uses electromagnetic induction to deliver current selectively into the PFM. As magnetic field passes skin with no attenuation of energy, more energy is addressed to target tissue, thus making the electromagnetic mechanism potentially superior as it can induce deeper and more intense contractions that may result in favorable efficacy. HIFEMs' high frequency of action potentials leads not only to selective muscle contractions but also to a supramaximal muscle effect.

In a study comparing HIFEM and electrical stimulation in the treatment of pelvic floor dysfunction in postpartum women, HIFEM resulted as superior inthe maximal voluntary contraction (MVC) and by patient's subjective evaluation of decrease in symptoms [67]. In another study evaluating improvement in quality of life in a 3-month follow-up, patients reported significant improvement of their symptoms and a decrease in the use of absorbent pads. HIFEM was also shown to be a promising alternative for improvement in female sexual function by strengthening of PFM [69].

HIFEM has its drawbacks as it is still an early-stage technology with scarce support in medical literature over short follow-up periods, all with heterogenic groups, lack of control group and objective, measurable tools. Despite all, HIFEM seems like a highly promising non-invasive technology for SUI with no associated pain or discomfort. Compared to the "gold standard" PFM-targeted exercise that requires specific training and is of low compliance from patients, the HIFEM technology with its high-frequency shift of electromagnetic field can initiate thousands of accurately targeted PFM contractions during a single session.

6.6 Light Emitting Diode (LED)

Modern LED sources emit a narrow band spectrum of light, yet unlike lasers they are noncoherent. The discovery of LED dates back to 1962, which at the time was biologically ineffective [70], yet in the 90s' it was NASA researchers who developed the narrow band abilities and enabled the use for medical needs, available, and affordable [71]. LEDs are electrical components made of semiconductor chips on a reflective surface. The semiconductor is an electron-conductive material, so when electrical current passes through it, some of its electrons are excited and run freely. When the electrons return to their natural state in the atoms, energy is released and converted into emitted light. The emitted wavelength is dependent on the semiconductor within the chip, and the penetration depth of that light depends, as in lasers, on the wavelength [70]. LEDs can be either continuous or pulsed, yet photo-modulated LEDs are pulsed and usually in the blue (400–470 nm), yellow (570–590), red(630–700), or near infrared (NIR, 800–1200) spectrum reaching in depth up to 1, 2,3, or 10 mm respectively. LEDs affect cellular metabolism, alternation of collagen synthesis, increased blood flow and more, all by means of photobiomodulation and photoactivation at a cellular and molecular level. As women age or bare children, vaginal tissue can experience laxity, vaginal/urethral tissue decrease and decline in estrogen, all negative impacts for vaginal and bladder function. The rationale of using LEDs is to stimulate the synthesis of vaginal collagen and elastin to support urethrovaginal sphincter and urethra but also to promote vasodilation in the vaginal and urethral submucosa [72]. Red and NIR LEDs are theoretically the most relevant to practice in gynecology due to their action and depth of penetration, yet blue light is also recently explored as non-drug antimicrobial alternative.

6.6.1 LED in Gynecology

A preliminary study evaluated the safety and efficacy of blue light (401±5 nm) on healthy vaginal mucosa in which candidates were treated for 30 minutes and evaluated 3–4 weeks later to reveal that vaginal microflora and pH remained normal with no adverse effects [73]. The same group presented later on a case of Vulvovaginal candidiasis (VVA), a frequent infection in genitourinary tracts, treated with the same wavelength LED due to its antimicrobial properties, which were presumed to benefit [74]. In a 3-week follow-up after a single session, vaginal culture, previously confirmed for VVA, was found negative for fungus and patients symptoms resolved. Another study examined the use of 3 distinct blue spectrum wavelengths (405, 415 & 450 nm) to treat candida vaginitis, focusing on both Candida albicans and vaginal epithelial cells [75]. The survival rate of the epithelial cells was the highest during inhibiting of the Candida when the 415 nm light was in use both in experimental model simulation presented, which led researchers to argue that 415 nm has better anti-fungal outcome yet less epithelial damage compared to the others.

Intravaginal LEDs of different wavelengths were evaluated in women suffering from VA, twice a week for six weeks, using blue (415 nm, 3 minutes), red (660 nm,3 minutes) and NIR (850 nm, 3 minutes) LEDs [76]. At the one-month follow-up, all patients had an improvement in symptoms, vaginal tissue quality, Sexual Function, lubrication, and pain. Anatomical assessments presented an increase in urethra-vagina distance, vascularization, and tissue thickness, particularly for red and blue wavelengths. Researchers suggested that this treatment could be a safe option for medical centers with no other EBD or as a home-use device. Another study was conducted to evaluate efficacy in SUI and problem during intercourse using red/infrared (662/855 nm) LEDs and was concluded that multimodal vaginal toning is safe and results in improved bladder symptoms and pelvic floor muscle strength [77]

Overall, LEDs seem as a cheap, painless option when considering EBD to be used and may serve a good alternative as a complementary option used following other modalities such as lasers to speed tissue recovery or reduce related treatment side effects. Yet LED has some drawbacks as the use is of long procedure duration that requires multiple sessions, which may lead to low compliance and lack of results and satisfaction. But mainly, it seems that at this point, supporting evidence is still insufficient to determine LEDs functionality as monotherapy and its beneficial effects in gynecology require more research.

6.7 Conclusion

This chapter had presented a wide range of EBDs, all intended to better women's quality of life. Since their developments, their use in medicine has widespread into many fields and led to advancements, some imperative for numerous conditions, benefiting patients and physicians alike. The obstructions EBDs are faced with are the need for proper training and high costs but mainly the acceptance of their use and related benefits by all the medical community. Neither the less, it seems that EBDs are here to stay, and their use will grow in time as they offer low, short recovery and seem to be widely accepted by the patients. As technology continues to improve, their range of applications and efficacy will surely expand and continue providing safer profiles which will lead to a growing demand.

References

1. Karcher C, Sadick N. Vaginal rejuvenation using energy-based devices. Int J Womens Dermatol. 2016;2(3):85–8. https://doi.org/10.1016/j.ijwd.2016.05.003.
2. Maiman T. Stimulated optical radiation in ruby. Nature. 1960;187:493–4. https://doi.org/10.1038/187493a0.
3. Bogdan Allemann I, Kaufman J. Laser principles. Curr Probl Dermatol. 2011;42:7–23. https://doi.org/10.1159/000328236.
4. Anderson RR, Parrish JA. Selective photothermolysis: precise microsurgery by selective absorp-

tion of pulsed radiation. Science (New York, NY). 1983;220(4596):524–7. https://doi.org/10.1126/science.6836297.
5. Peng Q, Juzeniene A, Chen J, Svaasand LO, Warloe T, Giercksky K-E, Moan J. Lasers in medicine. Rep Prog Phys. 2008;71(5):056701. https://doi.org/10.1088/0034-4885/71/5/056701.
6. Omi T, Numano K. The role of the CO_2 laser and fractional CO_2 laser in dermatology. Laser Ther. 2014;23(1):49–60. https://doi.org/10.5978/islsm.14-RE-01.
7. Fisher JC. Photons, physiatrics, and physicians: a practical guide to understanding laser light interaction with living tissue, part I. J Clin Laser Med Surg. 1992;10(6):419–26. https://doi.org/10.1089/clm.1992.10.419.
8. Rosenberg GJ, Brito MA Jr, Aportella R, Kapoor S. Long-term histologic effects of the CO_2 laser. Plast Reconstr Surg. 1999;104(7):2239–46. https://doi.org/10.1097/00006534-199912000-00046.
9. Tadir Y, Gaspar A, Lev-Sagie A, Alexiades M, Alinsod R, Bader A, Calligaro A, Elias JA, Gambaciani M, Gaviria JE, Iglesia CB, Selih-Martinec K, Mwesigwa PL, Ogrinc UB, Salvatore S, Scollo P, Zerbinati N, Nelson JS. Light and energy based therapeutics for genitourinary syndrome of menopause: consensus and controversies. Lasers Surg Med. 2017;49(2):137–59. https://doi.org/10.1002/lsm.22637.
10. Kaufmann R, Hibst R. Pulsed 2.94-microns erbium-YAG laser skin ablationDOUBLEHYPHENexperimental results and first clinical application. Clin Exp Dermatol. 1990;15(5):389–93. https://doi.org/10.1111/j.1365-2230.1990.tb02125.x.
11. Manstein D, Herron GS, Sink RK, Tanner H, Anderson RR. Fractional photothermolysis: a new concept for cutaneous remodeling using microscopic patterns of thermal injury. Lasers Surg Med. 2004;34(5):426–38. https://doi.org/10.1002/lsm.20048.
12. Alexiades-Armenakas MR, Dover JS, Arndt KA. The spectrum of laser skin resurfacing: nonablative, fractional, and ablative laser resurfacing. J Am Acad Dermatol. 2008;58(5):719–40. https://doi.org/10.1016/j.jaad.2008.01.003.
13. Hantash BM, Bedi VP, Kapadia B, Rahman Z, Jiang K, Tanner H, Chan KF, Zachary CB. In vivo histological evaluation of a novel ablative fractional resurfacing device. Lasers Surg Med. 2007;39(2):96–107. https://doi.org/10.1002/lsm.20468.
14. Bellina JH, Fick AC, Jackson JD. Lasers in gynecology: an historical/developmental overview. Lasers Surg Med. 1985;5(1):1–22. https://doi.org/10.1002/lsm.1900050102.
15. Stafl A, Wilkinson EJ, Mattingly RF. Laser treatment of cervical and vaginal neoplasia. Am J Obstet Gynecol. 1977;128(2):128–36. https://doi.org/10.1016/0002-9378(77)90676-7.
16. Reid R, Absten GT. Lasers in gynecology: why pragmatic surgeons have not abandoned this valuable technology. Lasers Surg Med. 1995;17(3):201–301. https://doi.org/10.1002/lsm.1900170302.
17. Wallwiener D, Pollmann D, Stolz W, Rimbach S, Bastert G. Laser in gynecology: an overview. In: Bastert G, Wallwiener D, editors. Lasers in gynecology. Heidelberg: Springer; 1992. https://doi.org/10.1007/978-3-642-45683-1_1.
18. Wright VC. Laser surgery: using the carbon dioxide laser. Can Med Assoc J. 1982;126(9):1035–9.
19. Scheinfeld N, Lehman DS. An evidence-based review of medical and surgical treatments of genital warts. Dermatol Online J. 2006;12(3):5.
20. González-Isaza P, Lotti T, França K, Sanchez-Borrego R, Tórtola JE, Lotti J, Wollina U, Tchernev G, Zerbinati N. Carbon dioxide with a new pulse profile and shape: A perfect tool to perform labiaplasty for functional and cosmetic purpose. Open Access Maced J Med Sci. 2018;6(1):25–7. https://doi.org/10.3889/oamjms.2018.043.
21. Kartamaa M, Reitamo S. Treatment of lichen sclerosus with carbon dioxide laser vaporization. Br J Dermatol. 1997;136(3):356–9.
22. Portman DJ, Gass ML, Vulvovaginal Atrophy Terminology Consensus Conference Panel. Genitourinary syndrome of menopause: new terminology for vulvovaginal atrophy from the International Society for the Study of Women's Sexual Health and the North American Menopause Society. Menopause (New York, NY). 2014;21(10):1063–8. https://doi.org/10.1097/GME.0000000000000329.
23. Arroyo C. Fractional CO_2 laser treatment for vulvovaginal atrophy symptoms and vaginal rejuvenation in perimenopausal women. Int J Women's Health. 2017;9:591–5. https://doi.org/10.2147/IJWH.S136857.
24. Reid R. Superficial laser vulvectomy. I. the efficacy of extended superficial ablation for refractory and very extensive condylomas. Am J Obstet Gynecol. 1985;151(8):1047–52. https://doi.org/10.1016/0002-9378(85)90378-3.
25. Pardo J, Solà V, Ricci P, Guilloff E. Laser labioplasty of labia minora. Int J Gynaecol Obstet. 2006;93(1):38–43. https://doi.org/10.1016/j.ijgo.2006.01.002.
26. Samuels JB, Garcia MA. Treatment to external labia and vaginal canal with CO_2 laser for symptoms of vulvovaginal atrophy in postmenopausal women. Aesthet Surg J. 2019;39(1):83–93. https://doi.org/10.1093/asj/sjy087.
27. Gaspar A, Addamo G, Brandi H. Vaginal fractional CO_2 laser: a minimally invasive option for vaginal rejuvenation. Am J Cosmet Surg. 2011;28(3):156–62. https://doi.org/10.1177/074880681102800309.
28. Zerbinati N, Serati M, Origoni M, Candiani M, Iannitti T, Salvatore S, Marotta F, Calligaro A. Microscopic and ultrastructural modifications of postmenopausal atrophic vaginal mucosa after fractional carbon dioxide laser treatment. Lasers Med Sci. 2015;30(1):429–36. https://doi.org/10.1007/s10103-014-1677-2.

29. Preti M, Vieira-Baptista P, Digesu GA, Bretschneider CE, Damaser M, Demirkesen O, Heller DS, Mangir N, Marchitelli C, Mourad S, Moyal-Barracco M, Peremateu S, Tailor V, Tarcan T, De E, Stockdale CK. The clinical role of LASER for vulvar and vaginal treatments in gynecology and female urology: an ICS/ISSVD best practice consensus document. Neurourol Urodyn. 2019;38(3):1009–23. https://doi.org/10.1002/nau.23931.
30. Perino A, Calligaro A, Forlani F, Tiberio C, Cucinella G, Svelato A, Saitta S, Calagna G. Vulvo-vaginal atrophy: a new treatment modality using thermo-ablative fractional CO_2 laser. Maturitas. 2015;80(3):296–301. https://doi.org/10.1016/j.maturitas.2014.12.006.
31. Gambacciani M, Levancini M, Cervigni M. Vaginal erbium laser: the second-generation thermotherapy for the genitourinary syndrome of menopause. Climacteric. 2015;18(5):757–63. https://doi.org/10.3109/13697137.2015.1045485.
32. Salvatore S, Nappi RE, Parma M, Chionna R, Lagona F, Zerbinati N, Ferrero S, Origoni M, Candiani M, Leone Roberti Maggiore U. Sexual function after fractional microablative CO_2 laser in women with vulvovaginal atrophy. Climacteric. 2015;18(2):219–25. https://doi.org/10.3109/13697137.2014.975197.
33. Duncan D, Kreindel M. Basic radiofrequency: physics and safety and application to aesthetic medicine. In: Lapidoth M, Halachmi S, editors. Radiofrequency in cosmetic dermatology. Basel: Karger Publishers; 2015. p. 1–22.
34. Lapidoth M, Halachmi S. Radiofrequency in cosmetic dermatology. Aesthet Dermatol. Basel, Karger. 2015;2:50–61. https://doi.org/10.1159/000362767.
35. Hainer BL. Fundamentals of electrosurgery. J Am Board Fam Pract. 1991;4(6):419–26.
36. Shin MK, Park JM, Lim HK, Choi JH, Baek JH, Kim HJ, Koh JS, Lee MH. Characterization of microthermal zones induced by fractional radiofrequency using reflectance confocal microscopy: a preliminary study. Lasers Surg Med. 2013;45(8):503–8. https://doi.org/10.1002/lsm.22175.
37. Mattsson MO, Simkó M. Emerging medical applications based on non-ionizing electromagnetic fields from 0 Hz to 10 THz. Med Dev (Auckland, NZ). 2019;12:347–68. https://doi.org/10.2147/MDER.S214152.
38. Goats GC. Continuous short-wave (radio-frequency) diathermy. Br J Sports Med. 1989;23(2):123–7. https://doi.org/10.1136/bjsm.23.2.123.
39. Dillon B, Dmochowski R. Radiofrequency for the treatment of stress urinary incontinence in women. Curr Urol Rep. 2009;10(5):369–74. https://doi.org/10.1007/s11934-009-0058-z.
40. Millheiser LS, Pauls RN, Herbst SJ, Chen BH. Radiofrequency treatment of vaginal laxity after vaginal delivery: nonsurgical vaginal tightening. J Sex Med. 2010;7(9):3088–95. https://doi.org/10.1111/j.1743-6109.2010.01910.x.
41. Alinsod RM. Transcutaneous temperature controlled radiofrequency for orgasmic dysfunction. Lasers Surg Med. 2016;48(7):641–5. https://doi.org/10.1002/lsm.22537.
42. Gold M, Andriessen A, Bader A, Alinsod R, French ES, Guerette N, Kolodchenko Y, Krychman M, Murrmann S, Samuels J. Review and clinical experience exploring evidence, clinical efficacy, and safety regarding nonsurgical treatment of feminine rejuvenation. J Cosmet Dermatol. 2018;17:289–97.
43. Sekiguchi Y, Utsugisawa Y, Azekosi Y, Kinjo M, Song M, Kubota Y, Kingsberg SA, Krychman ML. Laxity of the vaginal introitus after childbirth: nonsurgical outpatient procedure for vaginal tissue restoration and improved sexual satisfaction using low-energy radiofrequency thermal therapy. J Womens Health (Larchmt). 2013;22(9):775–81. https://doi.org/10.1089/jwh.2012.4123.
44. Dmochowski RR, Avon M, Ross J, Cooper JM, Kaplan R, Love B, Kohli N, Albala D, Shingleton B. Transvaginal radio frequency treatment of the endopelvic fascia: a prospective evaluation for the treatment of genuine stress urinary incontinence. J Urol. 2003;169(3):1028–32. https://doi.org/10.1097/01.ju.0000048686.50716.ef.
45. Leibaschoff G, Izasa PG, Cardona JL, Miklos JR, Moore RD. Transcutaneous temperature controlled radiofrequency (TTCRF) for the treatment of menopausal vaginal/genitourinary symptoms. Surg Technol Int. 2016;29:149–59.
46. Caruth JC. Evaluation of the safety and efficacy of a novel radiofrequency device for vaginal treatment. Surg Technol Int. 2018;32:145–9.
47. Dayan E, Ramirez H, Theodorou S. Radiofrequency treatment of labia minora and majora: a minimally invasive approach to vulva restoration. Plast Reconstr Surg Glob Open. 2020;8(4):e2418. Published 2020 Apr 22
48. Phenix CP, Togtema M, Pichardo S, Zehbe I, Curiel L. High intensity focused ultrasound technology, its scope and applications in therapy and drug delivery. J Pharm Pharm Sci. 2014;17(1):136–53. https://doi.org/10.18433/J3ZP5F
49. ter Haar G. Intervention and therapy. Ultrasound Med Biol. 2000;26(Suppl 1):S51–4. https://doi.org/10.1016/s0301-5629(00)00164-2.
50. Zhou Y. Principles and applications of therapeutic ultrasound in healthcare. 1st ed. Boca Raton, FL: CRC Press; 2015. https://doi.org/10.1201/b19638.
51. Fatemi A, Kane MA. High-intensity focused ultrasound effectively reduces waist circumference by ablating adipose tissue from the abdomen and flanks: a retrospective case series. Aesthet Plast Surg. 2010;34(5):577–82. https://doi.org/10.1007/s00266-010-9503-0.
52. Robinson DM, Kaminer MS, Baumann L, Burns AJ, Brauer JA, Jewell M, Lupin M, Narurkar VA, Struck SK, Hledik J, Dover JS. High-intensity focused ultrasound for the reduction of subcutaneous adipose tis-

sue using multiple treatment techniques. Dermatol Surg. 2014;40(6):641–51. https://doi.org/10.1111/dsu.0000000000000022.
53. Yagel S. High-intensity focused ultrasound: a revolution in non-invasive ultrasound treatment? Ultrasound Obstet Gynecol. 2004;23(3):216–7. https://doi.org/10.1002/uog.1017.
54. Laubach HJ, Makin IR, Barthe PG, Slayton MH, Manstein D. Intense focused ultrasound: evaluation of a new treatment modality for precise microcoagulation within the skin. Dermatol Surg. 2008;34(5):727–34. https://doi.org/10.1111/j.1524-4725.2008.34196.x.
55. Fabi SG. Noninvasive skin tightening: focus on new ultrasound techniques. Clin Cosmet Investig Dermatol. 2015;8:47–52. https://doi.org/10.2147/CCID.S69118.
56. Chan AH, Fujimoto VY, Moore DE, Martin RW, Vaezy S. An image-guided high intensity focused ultrasound device for uterine fibroids treatment. Med Phys. 2002;29(11):2611–20. https://doi.org/10.1118/1.1513990.
57. Rabinovici J, Inbar Y, Revel A, Zalel Y, Gomori JM, Itzchak Y, Schiff E, Yagel S. Clinical improvement and shrinkage of uterine fibroids after thermal ablation by magnetic resonance-guided focused ultrasound surgery. Ultrasound Obstet Gynecol. 2007;30(5):771–7. https://doi.org/10.1002/uog.4099.
58. Elías MGJA, Corin G, Garcia PN, Sivo V, Nestor D, Nuñez L. Management of Vaginal Atrophy, vaginal Hyperlaxity and stress urinary incontinence with intravaginal high-intensity focused ultrasound (HIFU). Int J Obstet Gynaecol Res. 2019;6(2):735–65.
59. Alexiades M. High intensity focused electromagnetic field (HIFEM) devices in dermatology. J Drugs Dermatol. 2019;18(11):1088.
60. Yamanishi T, Yasuda K, Suda S, Ishikawa N. Effect of functional continuous magnetic stimulation on urethral closure in healthy volunteers. Urology. 1999;54(4):652–5. https://doi.org/10.1016/s0090-4295(99)00194-6.
61. Voorham-van der Zalm PJ, Pelger RC, Stiggelbout AM, Elzevier HW, Lycklama à Nijeholt, G. A. Effects of magnetic stimulation in the treatment of pelvic floor dysfunction. BJU Int. 2006;97(5):1035–8. https://doi.org/10.1111/j.1464-410X.2006.06131.x.
62. Strohbehn K. Normal pelvic floor anatomy. Obstet Gynecol Clin N Am. 1998;25(4):683–705. https://doi.org/10.1016/s0889-8545(05)70037-1.
63. Faubion SS, Shuster LT, Bharucha AE. Recognition and management of nonrelaxing pelvic floor dysfunction. Mayo Clin Proc. 2012;87(2):187–93. https://doi.org/10.1016/j.mayocp.2011.09.004.
64. Radzimińska A, Strączyńska A, Weber-Rajek M, Styczyńska H, Strojek K, Piekorz Z. The impact of pelvic floor muscle training on the quality of life of women with urinary incontinence: a systematic literature review. Clin Interv Aging. 2018;13:957–65. https://doi.org/10.2147/CIA.S160057.
65. Bø K. Pelvic floor muscle training is effective in treatment of female stress urinary incontinence, but how does it work? Int Urogynecol J Pelvic Floor Dysfunct. 2004;15(2):76–84. https://doi.org/10.1007/s00192-004-1125-0.
66. Correia GN, Pereira VS, Hirakawa HS, Driusso P. Effects of surface and intravaginal electrical stimulation in the treatment of women with stress urinary incontinence: randomized controlled trial. Eur J Obstet Gynecol Reprod Biol. 2014;173:113–8. https://doi.org/10.1016/j.ejogrb.2013.11.023.
67. Elena S, Dragana Z, Ramina S, Evgeniia A, Orazov M. Electromyographic evaluation of the pelvic muscles activity after high-intensity focused electromagnetic procedure and electrical stimulation in women with pelvic floor dysfunction. Sex Med. 2020;8(2):282–9. https://doi.org/10.1016/j.esxm.2020.01.004.
68. Samuels JB, Pezzella A, Berenholz J, Alinsod R. Safety and efficacy of a non-invasive high-intensity focused electromagnetic field (HIFEM) device for treatment of urinary incontinence and enhancement of quality of life. Lasers Surg Med. 2019;51(9):760–6. https://doi.org/10.1002/lsm.23106.
69. Hlavinka TC, Turčan P, Bader A. The use of HIFEM technology in the treatment of pelvic floor muscles as a cause of female sexual dysfunction: a multi-center pilot study. J Womens Health Care. 2019;8:1. https://doi.org/10.4172/2167-0420.1000455.
70. Opel DR, Hagstrom E, Pace AK, Sisto K, Hirano-Ali SA, Desai S, Swan J. Light-emitting diodes: A brief review and clinical experience. J Clin Aesthet Dermatol. 2015;8(6):36–44.
71. Kim WS, Calderhead RG. Is light-emitting diode phototherapy (LED-LLLT) really effective? Laser Ther. 2011;20(3):205–15. https://doi.org/10.5978/islsm.20.205.
72. Lanzafame RJ, de la Torre S, Leibaschoff GH. The rationale for Photobiomodulation therapy of vaginal tissue for treatment of genitourinary syndrome of menopause: an analysis of its mechanism of action, and current clinical outcomes. Photobiomodul Photomed Laser Surg. 2019;37(7):395–407. https://doi.org/10.1089/photob.2019.4618.
73. Pavie MC, Robatto M, Bastos M, Tozetto S, Boas AV, Vitale SG, Lordelo P. Blue light-emitting diode in healthy vaginal mucosa-a new therapeutic possibility. Lasers Med Sci. 2019;34(5):921–7. https://doi.org/10.1007/s10103-018-2678-3.
74. Robatto M, Pavie MC, Garcia I, Menezes MP, Bastos M, Leite H, Noites A, Lordelo P. Ultraviolet A/blue light-emitting diode therapy for vulvovaginal candidiasis: a case presentation. Lasers Med Sci. 2019;34(9):1819–27. https://doi.org/10.1007/s10103-019-02782-9.
75. Wang T, Dong J, Yin H, Zhang G. Blue light therapy to treat candida vaginitis with comparisons of three wavelengths: an in vitro study. Lasers Med Sci. 2020;35(6):1329–39. https://doi.org/10.1007/s10103-019-02928-9.

76. Naranjo García P, Elias JA, Parada JG, Luciañez DZ, Pinto H. Management of vaginal atrophy with intravaginal light-emitting diodes (LEDs). Int J Obstet Gynaecol. 2018;5(2):632–41.
77. de la Torre S, Miller LE. Multimodal vaginal toning for bladder symptoms and quality of life in stress urinary incontinence. Int Urogynecol J. 2017;28(8):1201–7. https://doi.org/10.1007/s00192-016-3248-5.

Ohad Toledano, is the Director of Clinical Applications and R&D in one of the largest EBD companies in the world, focusing on the development of new technologies, modalities, and applications to use. His work primarily centers on collaborating with physicians worldwide in order to create innovative protocols striving to enhance the quality of life. In his earlier work, the main focus was on clinical research and clinical trials in the field of cancer and cancer prevention, mainly colorectal and gastric, in the "Integrated Cancer Prevention Center" at Tel Aviv Medical Center, Israel. During his research, he was invited to attend the American National Cancer Institute (NCI) curriculum in cancer prevention at the FDA/NIH in Washington DC. Co-author in previously published papers on gene expression and population incidence/prevalence in inflammatory bowel diseases (IBDs), colorectal, and esophageal cancers. He holds a master's degree in medical science in the field of cellular and Developmental biology research from the Sackler Faculty of Medicine, Tel Avi University.

Laser in Aesthetic and Regenerative Gynecology: Physics, Types, Applications, Safety Profiles

7

Avitus John Raakesh Prasad, Ganesh S. Pai, and Anusha H. Pai

7.1 Introduction

The commonly used term vaginal rejuvenation is more of a reference used in practice more than a standard medical nomenclature. The aesthetics of the female genitalia have become an important concern among women over the past decade as they become aware of newer treatments and information through media. In present times women are open about their sexuality and want to alter their genital anatomy, which has undergone changes due to childbirth, hormonal changes due to aging and menopause. Whether it is for medical or cosmetic reasons, women seek out vaginal rejuvenation to gain back their self-esteem, reduce functional discomforts and difficulties, and improve sexual pleasure.

The first laser was produced by Theodore Harold Maiman at Hughes Research Laboratories in Malibu [1]. In 1960 using ruby as a lasing medium 694 nm (694.3 nm). In 1964 the Carbon Dioxide (CO_2) laser was developed by Kumar Patel at AT&T Bell Labs [2]. The concept of Fractional technology was first described by Huzaira et al. in 2003 [3]. It was further elucidated by Manstein et al. [4] a year later with the first fraxelated laser system. Fractional CO_2 was developed to give the results close to an ablative laser resurfacing clinical outcomes with less patient downtime and fewer overall adverse events. In 1926 William T. Bovie discovered that RF currents applied to a scalpel could cut and cauterize tissue in medical procedures [5].

White et al. reported the first dermatologic, aesthetic use of ultrasound as High-Intensity Focused Ultrasound HIFU in 2008 [6], and HIFU was approved by the Food and Drug Administration in 2009 for use in browlifting.

The recent technologies used in vaginal rejuvenation are mainly CO_2 fractional and erbium:yttrium-aluminum-garnet (Er:YAG) lasers, radiofrequency (RF) based devices, high-intensity focused ultrasound (HIFU) and the newer high-intensity focused electromagnetic (HIFEM) technology (Table 7.1).

Energy-based devices have proven their therapeutic efficacy in the rejuvenation of the face, neck, and décolleté, but their application in the vaginal rejuvenation is a relatively new concept. Available clinical studies in vaginal rejuvenation are currently limited, which is a hindrance to validate their efficacy. Newer modifications to existing systems with specific handpieces designed for this indication are currently available, marketed, and developed. These various technologies, their mechanism of action, and any available studies relevant to their safety and efficacy will be described in this chapter.

A. J. R. Prasad
S. P. Derma Center, Madurai, India

G. S. Pai (✉) · A. H. Pai
Derma-Care, Mangalore, India
e-mail: gspai@derma-care.in

P. Jindal et al. (eds.), *Aesthetic and Regenerative Gynecology*,
https://doi.org/10.1007/978-981-16-1743-0_7

Table 7.1 Energy-based devices for vaginal rejuvenation

	Technology	Number of treatments
Laser-based devices		
CO_2RE Intima; Syneron Candela (Wayland, MA)	Fractional CO_2 laser	3 Trx at 4 week intervals
FemiLift; Alma Lasers (Buffalo Grove, IL)	Fractional CO_2 laser	3 Trx at 4–6 week intervals
DivaTight; Quanta System (Italy)	Dual wavelength (1540 and 10,600 nm)	3 Trx at 4–6 week intervals
FemTouch; Lumenis (Israel&UK)	Fractional CO_2 laser	3 Trx at 4–6 week intervals
MonaLisa; Touch, Cynosure (Westford, MA)	Fractional CO_2 laser	3 Trx at 6 week intervals
IntimaLase; Fotona (Dallax, TX)	2940-nm nonablative Er:YAG	2 Trx at 8 week intervals
Action II Petit Lady; Lutronic (Burlington, MA)	2940-nm Er:YAG	3 Trx at 2 week intervals
Ladylift; Eufoton (Trieste, Italy)	1470-nm Diode solid state	4 Trx at 2–3 weeks intervals
Radiofrequency-based devices		
ThermiVa; ThermiAesthetics (Southlake,TX)	Temperature-controlled radiofrequency	3 Trx at 4–6 week intervals
ReVive; Viora (Jersey City, NJ)	Bipolar radiofrequency	4–6 Trx at 2–3 week intervals
Votiva; Inmode (Yokne'amIllit, Israel)	Monopolar and fractional radiofrequency	1–3 Trx at 2week intervals
Aviva; (InmodeYokne'amIllit, Israel)	Labioplasty Patented RF delivery	1Trx
Viveve System; Viveve Medical (Sunnyvale, CA)	Patented monopolar radiofrequency	1 Trx
Protégé Intima; BTL Aesthetics, (Framingham, MA)	Labioplasty Focused Rf with Ultrasound	2–4 Trx at 2–3 week intervals
ExilisUtra Femme; BTL Aesthetics (Framingham, MA)	Monopolar Rf with Ultrasound	4–6 Trx at twice a week session
Pelleve; Ellman International (Hicksville, NY)	Focused monopolar radiofrequency	3 Trx at 2–3 week intervals
Venus Fiore; VenusConcept (Toronto, CA)	Multipolar-radiofrequency with pulsed electromagnetic field	3–4Trx at 1 month intervals
Emsella; BTL Aesthetics (Framingham, MA)	High-intensity focused electromagnetic (HIFEM) technology	4–6 Trx at twice a week session
Ultrasound-based devices		
Ultravera; Hironic (Gyeonggi-do, S.Korea)	High-intensity focused ultrasound (HIFU)	1 Trx

7.2 Patient Evaluation, Consultation, and Consent

As with all treatments, a thorough history, examination, consent, and discussing the clinical outcomes and goals of treatment can help identify suitable candidates for Nonsurgical vulvovaginal rejuvenation (NVR).

Patients should be given options for alternative therapies, such as local hormone replacement therapy and lubricants, if appropriate. The procedure should be clearly explained to the patient, the costs involved and the number of treatments it would need for good clinical response and a discussion about realistic outcomes. Photographs should be taken to document pre and post-procedure changes.

Various self-reported questionnaires are available that are used to assess individual patient's perceptions of Vaginal laxity and vulvovaginal symptoms. These questionnaires act as a baseline

Table 7.2 Questionnaires Used in NVR evaluation and studies (37)

Questionnaire	Validated	Aim of questionnaire
Female Sexual Distress Scale Revised (FSDS-R)	Yes	Distress with sexual dysfunction
Female Sexual Function Index (FSFI)	Yes	Multiple domains of sexual function
Vulvovaginal Symptom Questionnaire (VSQ)	Yes	Quality of life impact from vulvovaginal symptoms, emotional, and sexual concerns
Urogenital Distress Inventory (UDI-6)	Yes	Assesses frequency of urination, urgency, and incontinence
Incontinence Impact Questionnaire (IIQ-7)	Yes	Impact of urinary leakage on quality of life
Sexual Satisfaction Questionnaire (SSQ)	No	Sexual satisfaction
Vaginal Laxity Questionnaire (VLQ)	No	Degree of patient reported vaginal laxity

to assess response with subsequent treatments and validate the clinical outcome. (Table 7.2)

7.3 Contraindications

1. Infection (e.g., candidiasis, herpes simplex) should be treated before therapy.
2. Role of any potential neoplastic lesions.
3. Prolapse ≥ grade II.
4. Active vulvar dermatoses, such as psoriasis, lichen planus and lichen sclerosis.
5. Patient on anticoagulant therapy.
6. Pregnancy.
7. Patient on Isotretinoin treatment.

7.4 Fractional Lasers

Fractional ablation which was introduced in 2004 and which involves the process of delivering multiple separate microarrays of laser beams as microscopic treatment zones as called as Micro Thermal Zone (MTZ) or Micro Ablation Zone (MAZ) that target a fraction of the skin at a time. It creates microscopic columns (islands) of treated skin surrounded by areas of untreated skin, which resulted in rapid reepithelialization. The repair mechanism occurs through the transepidermal elimination of treated necrotic skin consisting of a coagulated material containing melanin, elastin, and other dermal contents, which are called microscopic epidermal necrotic debris (MENDs) [7].

The first fractional laser introduced was a 1550 nm nonablative erbium-doped yttrium aluminum garnet (Er:YAG). Followed this the fractional technology was applied to ablative lasers such as carbon dioxide (10,600 nm), Er:YAG (2940 nm), and YSGG Yttrium Scandium Gallium Garnet (2790 nm), providing more significant clinical outcomes.

Since then, this technology has been successfully used for skin resurfacing as well as scar and burn management [8–10]. Following fractional laser, there is a greater expression of Heat Shock Protein (HSP) 70, which stimulates new formation and remodeling of collagen and elastic fibers, which resulted in tightening and textural improvement of skin [11]. Based on the efficacy and safety of the procedure and its clinical outcomes, it was hypothesized that the laser induced rejuvenation of atrophied vaginal mucosa would also restore the tissue to its premenopausal state and function [12]. Ongoing research into this new application of laser has shown to produce consistent clinical results in their patients [13–15].

7.5 Laser Tissue Interaction

The amount of energy delivered and the tissue response is dependent on four basic parameters.

1. Power/Wattage (W): Usually, 20 to 30 W is employed.
2. Time (ms or μs): Duration, in microseconds or milliseconds its the time the laser beam is

focused on the tissue. The longer duration, the more the laser remains focused on one spot, the more energy is applied to that area.

3. Spot size (mm): Smaller the spot size or beam profile, the higher the power density and vice versa. Ideally, a beam profile of 0.2 mm diameter is available in most devices. Newer devices offer 0.12 mm, which results in deeper penetration of energy and better remodeling of tissue.
4. Energy per point (mj): the energy delivered per beam on the tissue it is a value dependent on the power and duration. Power of 30 W with a duration of 1.5 ms will produce and Energy per point of 45 mj (30 W × 1.5 ms).

7.6 Histological Changes Seen with Fractional Lasers, RF, and Other Energy-Based Devices

Following a laser/RF treatment, an inflammatory and wound healing pathway in the skin is initiated that stimulates the underlying tissues to heal with both increased elastin and collagen [16].Ex vivo studies showed thickening of the mucosa, increase in fibroblasts, and improved vascularity in atrophic vaginal tissue and new collagen deposition [17]. Zerbinatiet al observed the remodeling of the vaginal thick squamous stratified epithelium with visible storage of glycogen in the epithelial cells. The active fibroblasts produced new collagen and ground substance [18]. Gaspar et al. demonstrated visible changes in all layers of the vaginal wall from vaginal biopsies from areas treated with CO_2 laser and platelet-rich plasma [19]. Premenopausal vaginal mucosa has increased water content in the tissues compared to postmenopausal mucosa, which might show a different responses to CO_2 fractional treatment as water is the main chromophore. These histological changes reflect on the postmenopausal vaginal mucosa who have undergone fractional laser by resembling a premenopausal vaginal mucosa, suggesting that lasers definitely play a role in rejuvenating the vaginal mucosa at the histologic level [20, 21].

7.7 How to Perform a Fractional Laser?

The NVR treatment is done differently for the external genitalia and the vaginal canal. The external genitalia is treated with the conventional fractional headpiece, and the vaginal canal is treated with a modified probe which can be inserted into the vaginal canal, as shown in Fig. 7.1.

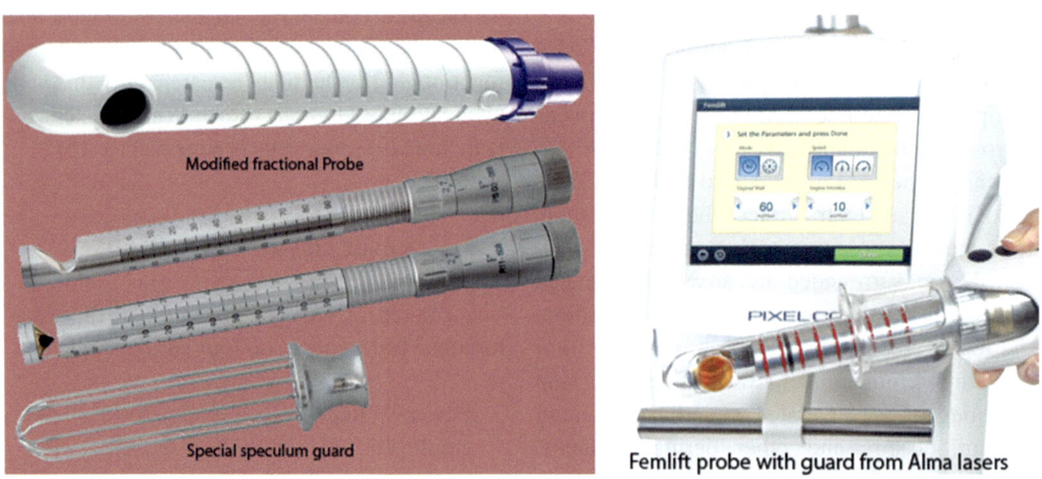

Fig. 7.1 Modified CO_2 vaginal probe with a guard speculum

For the external genitalia, fractional treatment was done under topical anesthesia 5% emulsion of 2.5% each of lidocaine and prilocaine combination, which was applied at least an hour prior to treatment. The labia majora and minora are treated with a single pass without overlapping. The most common parameters used were Energy of 30 to 45 mj, Density of 5–10%, static/normal mode preferably choosing the shape as square or hexagon was easy delivery as shown in Fig. 7.2.

For the Vaginal canal, No anesthesia is required as it is a minimally painful procedure. Application of ultrasound gel or baby oil is used for more comfortable insertion of the probe into the introitus. First, a special outer guard speculum cage which also comes as disposable with some companies, is introduced into the introitus, and then the handpiece was inserted in the center of the speculum cage up to the interior edge of the vaginal canal to the depth where the treatment has to be started. Depending on various companies, the usual depth varies from 10 to 12 cm. The probe has a marking 1 cm apart with degree marking at the base of the probe for precise delivery and better handling of the probe.

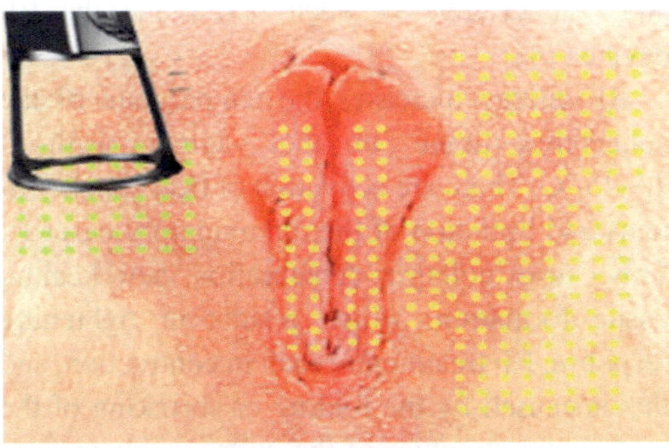

Fig. 7.2 CO_2 fractional done on the external labia

The treatment is started by giving a single pass at a depth of 10 to 12 cm, then the probe is rotated by 30 to 45°, and another pass is given. Similarly, in Fig. 7.3, the probe is rotated to full 360° treating the whole circle.

The probe is then retracted out by 1 cm, and the process is repeated throughout the whole length of the vaginal canal. Now a days fully automated user friendly robotic devices are also available in which the machine withdraws by 1 cm automatically after completion of a full rotation. The fractional beam is delivered using a specially developed laser pulse that enables fast and rapid penetration of the vaginal mucosa,

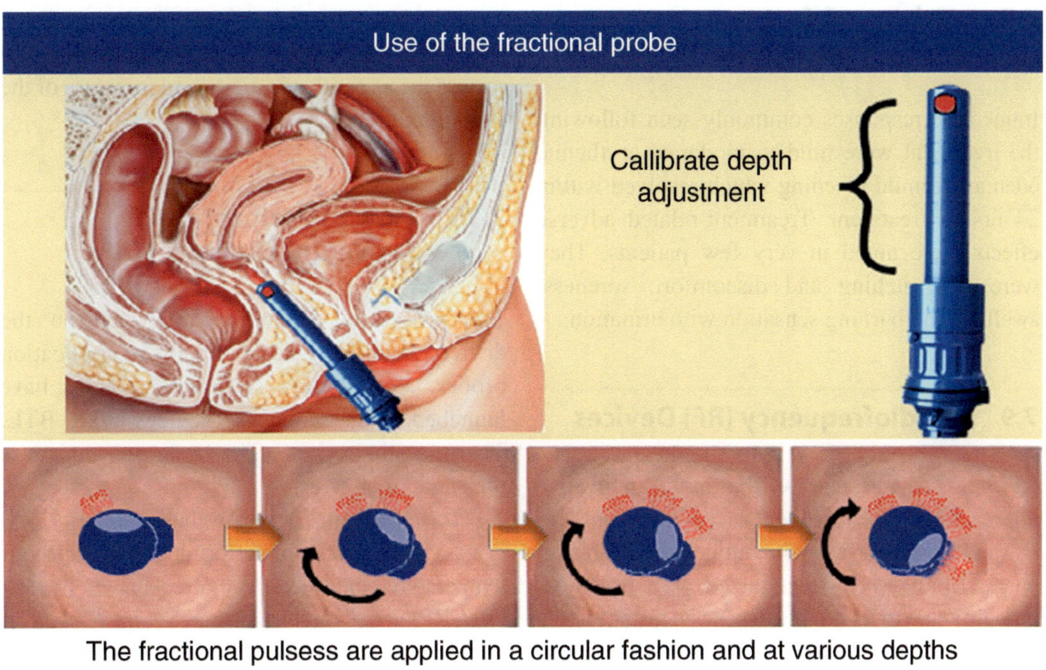

Fig. 7.3 CO_2 vaginal probe treatment of the vaginal canal

minimizing surface damage, while ensuring appropriate longer thermal effects to be diffused into the lamina propria [22]. The average parameters used were Energy of 45 to 60 mj, Density of 5–10%, static/normal mode, shape as square. Femlift from Alma Lasers (Buffalo Grove, IL) has a special calibrated probe with guard speculum and has additional parameters to control the speed of delivery of the micro beams. It also comes with a fully automatic robotic probe.

Three treatment sittings at an interval of four weeks is the protocol followed. This procedure is mostly done as an outpatient procedure and does not require systemic analgesia/anesthesia. Patients feel comfortable post-procedure and may feel a slight stinging sensation. The post-treatment recovery is quick, there may be an initial slight oozing of blood and may be a mild local inflammation, which will usually settle within 24 hrs. Patients are advised to avoid swimming or using bathtubs and intense physical exercise for 3 days.

Patient is advised to drink water regularly, avoid coital sexual activity and to use a tampon for at least seven days after treatment.

7.8 Adverse Effects and Complications

Immediate responses commonly seen following the treatment were mild to moderate erythema, edema and mild bleeding which resolved within 24 hrs of treatment. Treatment-related adverse effects were noted in very few patients. They were mild itching and discomfort, soreness, swelling and burning sensation with urination.

7.9 Radiofrequency (RF) Devices

RF has become one of the more innovative approaches in the treatment of vulvovaginal laxity and stress urinary incontinence. In recent years, RF has been recognized as a treatment modality due to its noninvasiveness, faster visible results and no downtime with the absence of adverse events [23]. The mechanism of action is based on increasing the temperature of the treated tissue to initiate a biological response. The energy produced is not absorbed by melanin, so it does not cause burns. RF devices can be unipolar, monopolar, bipolar, or multipolar. Multipolar devices can deliver the energy and produce heat quicker than bipolar and thereby reduces contact treatment time. The different devices available are illustrated in Table 7.1.

It creates an electrical field like a vortex in the tissue that causes movement of charged particles within the treated tissue and in turn generates heat. The amount of heat generated in the tissue is directly proportional to the currently employed and contact time between tissue and the RF device [24].

RF energy increases the temperature of the connective tissue in the vaginal wall to 40–45 °C. There are activation of heat shock proteins and initiation of the inflammatory cascade. This causes collagen denaturation, new healthy tissue formation, which results in tightening response. RF treatments result in contraction, and this is caused by the folding or shrinking of the triple helix structure of the collagen fibers, crea ng thicker and shorter collagen fibers.This is proposed as the mechanism of action which gives the immediate tissue tightening seen post-procedure. Over time this process helps to regenerate and restore the strength, elasticity and moisture of the vaginal mucosa [25–27].

7.10 How to Perform an RF for Vaginal Rejuvenation

The procedure differs and depends on the device, technology and variation in application probes. Now a days several companies have launched radiofrequency devices like BTL, Thermiaesthetics, etc. The basic mechanism is the same in all.

ThermiVa from ThermiAesthetics, Texas, USA) is a device designed for performing transcutaneous temperature-controlled radiofrequency therapy (TTCRF) at the vulvovaginal area (Fig. 7.4).

There is a special calibrated, slim, S-shaped treatment probe with an RF emitter on the distal

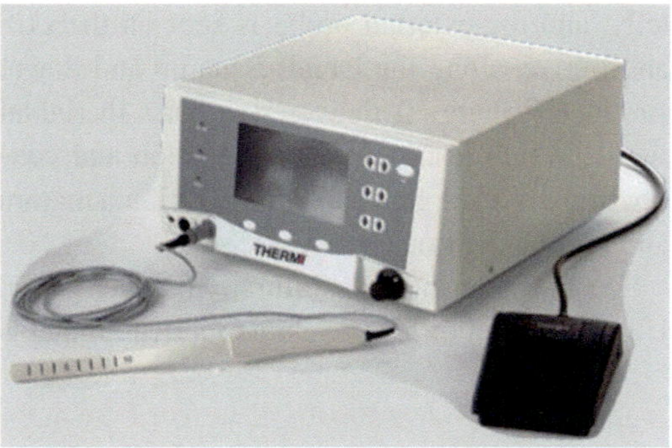

Fig. 7.4 ThermiVa RF device

Fig. 7.5 ThermiVa vaginal probe with calibration

end of the probe as shown in Fig. 7.5. This probe is applied over the external genitalia as well as inserted into the vaginal canal.

Feedback mechanisms on the probe include thermistors and thermocouples, which send temperature information back to the device itself, which modulates power output to maintain a temperature of 40–45 °C and preventing overheating.

No topical anesthesia is needed. Ultrasound gel might be used for easy mobility. The probe is moved in and out within the vaginal canal with pre-determined depth insertion delivering the RF current. The probe is rotated in a circular path while performing heating the whole vaginal canal. The treatment lasts for 30 mins and is usually done only once, but some may need 2 or 3 treatments. The effect lasts for a year in most patients.

Studies have concluded that the response following RF treatment was satisfactory. Improved texture, youthful appearance and tightening resulted in improved confidence and reduced performance anxiety [25, 28].

This procedure is noninvasive so patients can go back to normal activities, including sex, immediately post-procedure because no time is required for healing of surface tissue. ThermiVa probes are single use and hygienic and do not require cleanup.

FormaV from INMODE Israel is a specially designed handpiece and probe attachment to the multiplatform CONTURA and VOTIVA. FormaV provides non-invasive subdermal RF heating for vaginal rejuvenation. The disposable handpiece which is calibrated can be applied to the internal vaginal tissue and external vulvar tissue for laxity or labial hypertrophy. This procedure does not need anesthesia. Ultrasound gel is evenly applied to the tip of the disposable FormaV probe. Gel can also be applied in the vaginal canal if atrophy is noted. The RF energy level is set at 20 to 35, and a Cutoff temperature is set at 40 to 45 °C.

Adjust parameters according to patient tolerance and skin response and increase them gradually—higher the patient tolerance, the better the results.

Start from the full depth and treat the canal as you move out segment by segment. Apply slight pressure to ensure a full contact of the electrodes with the vaginal skin, then press the footswitch. Move the handpiece by 2 cm, then do back and forth movement along the canal and in circular movements in a 360° pattern to cover the whole area, which needs to be treated as illustrated in Fig. 7.6.

Gentle pressure over the treatment area will help to reach uniform heating. Never stop the movement in between without lifting the foot from the footswitch first, as it can cause hot spots.

Repeat the procedure by withdrawing the probe by 2 cm as you finish doing one segment of the vaginal canal. Tightening felt by both patient and operator is a typical immediate response. The procedure is done for 30 mins minimum.

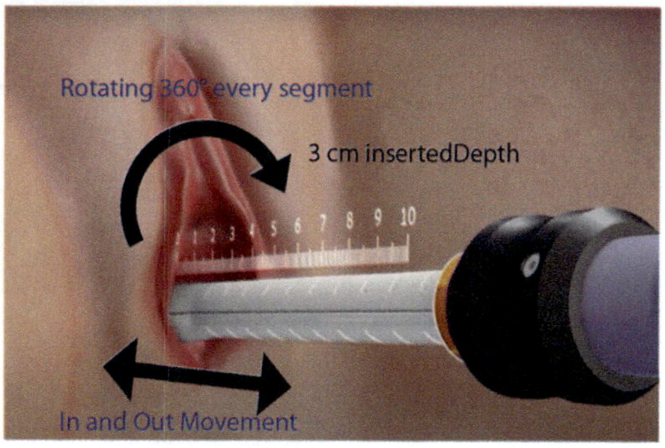

Fig. 7.6 Demonstrates the procedure

The external genitalia is treated with the same parameters. This is done by constant up and down movements on the sides of the labia minora with gentle movements for 10–20 mins. Many machines have another handpiece with a fractional RF probe which can be done on the external genitalia to give additional tightening and rejuvenation. The number of treatment sessions depends on the individual patient and is typically two–three sessions every 2–4 weeks but can vary according to patient response.

There are separate handpieces which are used to target the external genitalia. The MORPHEUS 8, which is a microneedle RF Technology, where RF is delivered through multiple specially designed needle electrodes with depth adjustable up to 4000 microns (Fig. 7.7).

Caruth et al. performed FormaV and FractoraV on Thirty women who had been diagnosed with symptoms of vaginal laxity and pelvic relaxation and concluded that the procedure is safe and effective for treating vaginal relaxation and atrophy symptoms without adverse effects [29].

The AVIVA is another unique technology attachment to the VOTIVA multiplatform, as shown in Fig. 7.8. It is a minimally invasive procedure for hemostasis and coagulation, offering a non-excisional scarless alternative to a labiaplasty.

This involves the use of RF, which is applied using a patented bipolar handpiece in which the sharp tip is inserted into the labial tissue parallelly, and the external probe is kept on the skin and moved along the labial tissue up and down. As the RF energy is delivered, it heats the labial tissue, which leads to tissue reduction and contraction due to targeted ablation. The labia majora and minora are treated separately.

Local anesthetic infiltration is given. Ultrasound gel or Hydrosoluble lubricating gel is

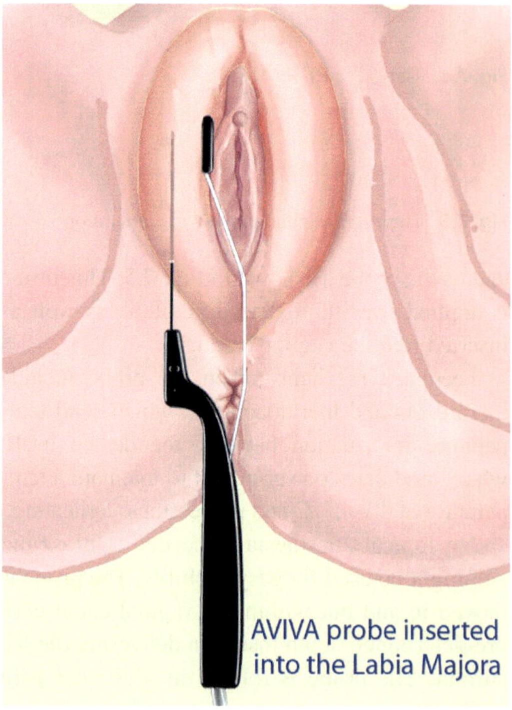

Fig. 7.8 Demonstrates the insertion of AVIVA

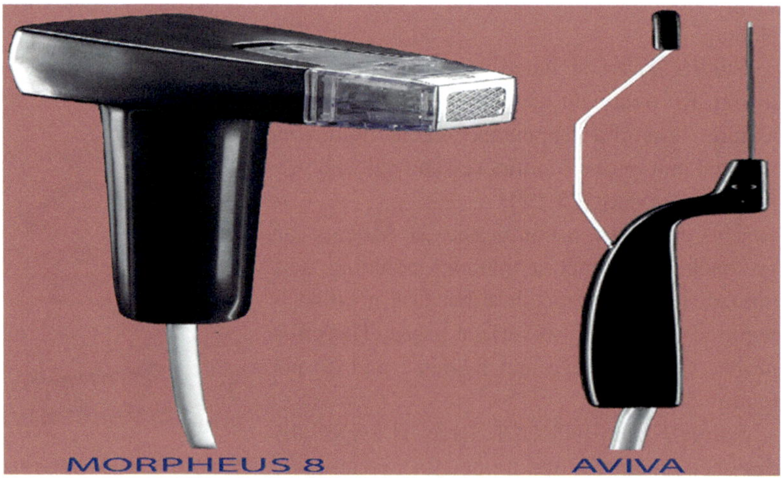

Fig. 7.7 Different probes of Morpheus 8 and AVIVA

applied over the labia to improve transduction and gliding between the 2 probes. There are two separate settings, the inner cannula probe is set with a controlled internal temperature cutoff at 60 °C, and the external probe is set at 37 °C. The inner probe is inserted into the access port and moved in an up and down motion through the tissue until the tissues reached the target temperature. Once the target temperature was reached, the procedure is continued for 1 minute and then was stopped. Then the other side of the labia majora was done. This is repeated for the labia minora also.

This technique has several benefits over the traditional surgical labiaplasty, there is no need for surgical incisions, and the patient feels little to no pain, minimal downtime, and quick results [30].

Newer devices with patented and modified technology have come into the market for last 2 years. Few of them with their technology and treatment protocol is described below.

Emsella (BTL Aesthetics Framingham, MA) uses a patented FDA approved HIFEM TM technology which specifically treats urinary incontinence as shown in Fig. 7.9. It employs High-Intensity Focused Electromagnetic technology which induces electric currents and triggers intense pelvic floor muscles contractions by targeting neuromuscular tissue. The unique effectiveness of Emsella is based on focused electromagnetic energy, penetration and targeting the precise area, and stimulation of the entire pelvic floor area. The HIFEM treats noninvasively through pelvic floor area. The patient remains fully clothed during the whole treatment. Therefore, it represents a noninvasive solution for incontinent patients.

The device consists of an electromagnetic generator connected to a treatment chair in which the patient sits comfortably. The stimulation coil is located in the seating area of the chair. A study by Samuels et al. conducted a study of 75 women who had complaints of urinary incontinence. After six treatments 61 out of 75 patients (81.33%) reported a significant reduction of their symptoms [31]. Progressive improvement was seen in most patients in subsequent visits. Patients also reported additional beneficial effects of the therapy, such as increased sexual desire and better urination control. Similar results were reported by Silantyeva et al. who reported sustained results till 1 year follow-up but cautions that relapse may occur and maintenance sessions should be considered in some patients [32].

Another newer device is the ExilisUltra Femme 360TM (BTL AestheticsFramingham, MA). It is a patented system that uses both RF and ultrasound. It has five different applicator tips to treat the external as well as internal vaginal rejuvenation. It has a built in real-time measurement of local impedance, known as Impedance

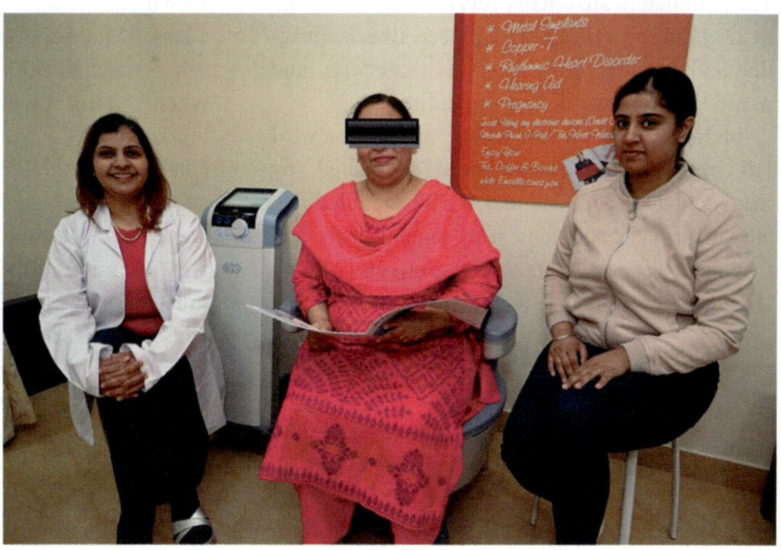

Fig. 7.9 Emsella device. In India, it was launcehed for first time at, "The Touch, Advanced obstetrics, IVF & Cosmetic Gynae Centre" Mohali by the Editor (P Jindal). Note the ease patient is in while taking therapy

Intelligence. This provides feedback from the treated area to regulate and maximize therapeutic energy delivery for improved safety and comfort. It is unique to this technology and part of the display parameters which is called Energy Flow Control (EFC™). The recommended protocol is 20 mins for both intra- and extravaginal areas. Three to four treatments are done once a week.

7.11 Ultrasound-Based Devices

Ultravera (Gyeonggi-do, S.Korea) is a modified HIFU (High-Intensity Focused Ultrasound) which has been used for skin tightening with very good response. This technology involves the use of ultrasound targeting specific depth and points in the treatment zone. There is an induction of cellular damage and volume reduction of the target area selectively created by means of coagulation by creating microthermal lesions through the accumulation of high-frequency ultrasound beams at the specific target site without any damage to the epidermis and adjacent tissue [33, 34].

There are three handpieces which is a transducer calibrated with various depths, and this is attached to a treatment handle with 360° markings to visually assist the operator while rotating the handpiece and delivering the energy precisely (Fig. 7.10).

The depths are 1.5 mm, 3.0 mm and 4.5 mm. The parameters are energy levels set between 0.2 to 2.0 J/cm^2 according to patient feedback sensation and comfort. HIFU delivers the energy and as multiple heat spots in a single line.

Initial studies on the safety and efficacy of HIFU on vaginal laxity was conducted by Sekiguchi, et al. in 2017 and 2018 concluded that all patient had improvement in vaginal laxity and treatment was safe without adverse effects [22, 35].

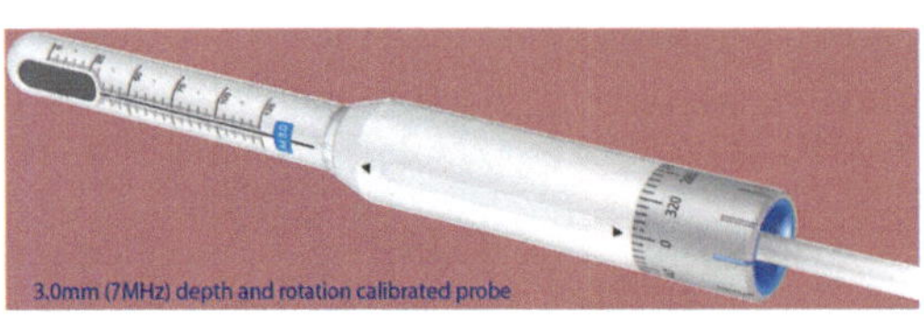

Fig. 7.10 Calibrated ultrasound transducer probe

Since there is sparing of the epidermis and its noninvasive, there is no post-procedure healing time. The patient can resume all normal activities immediately after the procedure. Treatment can be repeated after 3–4 weeks depending on the laxity. Three–Four treatments are recommended with maintenance treatment after 6–12 months.

7.12 Laser Safety

The International Electrotechnical Commission (IEC) is a global organization that prepares and publishes international standards for all electrical, electronic, and related technologies. The IEC document 60825-1 is the primary standard that outlines the safety of laser products. This is to ensure, through labels and instructions, anadequate warning to individuals of hazards associated with accessible radiation from laser products; and to reduce the possibility of injury by minimizing unnecessary accessible radiation and to give improved control of the laser radiation hazards through protective features.

Classification is based on calculations and determined by the Accessible emission limit (AEL), also incorporating viewing conditions.

- Class 1 lasers are very low risk and "safe under reasonably foreseeable use," including the use of optical instruments for intrabeam viewing.
- Class 1M lasers have wavelengths between 302.5 nm and 4000 nm and are safe except when used with optical aids (e.g., binoculars).
- Class 2 lasers do not permit human access to exposure levels beyond the Class 2 AEL for wavelengths between 400 nm and 700 nm. Any emissions outside this wavelength region must be below the Class 1 AEL.
- Class 2M lasers have wavelengths between 400 nm and 700 nm and are potentially hazardous when viewed with an optical instrument. Any emissions outside this wavelength region must be below the Class 1M AEL.
- Class 3R lasers range from 302.5 nm and 106 nm and is potentially hazardous, but the

risk is lower than that of Class 3B lasers. The accessible emission limit is within five times the Class 2 AEL for wavelengths between 400 nm and 700 nm and within five times the Class 1 AEL for wavelengths outside this region.

- Class 3B lasers are normally hazardous under direct beam viewing conditions but are normally safe when viewing diffuse reflections.
- Class 4 lasers are hazardous under both intrabeam and diffuse reflection viewing conditions. They may also cause skin injuries and are potential fire hazards.

The lasers used in dermatology are class 3 or 4. The risks can be prevented if proper safety measures are used while working with the lasers [36].

7.13 Hazards Involved Due to Laser Beam

Laser used to treat vaginal laxity, and aging atrophied skin uses a laser beam leading to the destruction of epidermis and dermis, subsequently leading to new blood flow, new collagen and elastin fiber formation through a wound healing process that results in tightening of that area [23].The vaginal laser helps to thicken the fragile vaginal lining and increase lubrication, with improvement in vaginal dryness, pain, and urinary symptoms. The laser beam can lead to potential complications such as scarring, infection, pain, and altered sexual sensation [37].

7.14 Non-Beam Hazards

Respiratory hazards—a lot of biohazardous materials are released during laser usage. Inhaled airborne contaminants can be emitted in the form of smoke or plume generated through the thermal interaction of lasers through tissue or through the accidental escape of toxic chemicals and gases from the laser itself. Over 150 chemicals have been identified in laser plumes- carbon monoxide, benzene, hydrogen cyanide, toluene are cytotoxic, genotoxic, and mutagenic.1 gm of tissue ablated with CO_2 laser is equivalent to three cigarettes smoked. The laser plumes include particulate matter more infectious than electrosurgical smoke with virus and bacteria DNA.

A study by the National Institute for Occupational Safety and Health (NIOSH) evaluated the air that operating room (OR) personnel were exposed to during laser procedures and found that detectable levels of ethanol, isopropanol, anthracene, formaldehyde, cyanide and airborne mutagenic particles were found. Inhalations of these toxic aerosols have been found to be potentially damaging to the respiratory system.

Red blood cells, bacteria, HIV, and HPV DNA have been recovered of laser plume. Q-Switched 1064 and 532 are the biggest challenge. It causes the explosion of blood and tissue fragments flying at high speed, escaping capture by smoke evacuator. After the use of the laser, the room atmosphere may take 20 mins to return to the normal concentration of particles [38].

Combustion Hazards—Electrical hazards are grouped as shock hazards and fire hazards. Most laser systems involve high potential, high current electrical supplies. The most serious accidents with lasers have been electrocutions.

Materials like cotton, towels, drapes, plastics, rubber, and sanitizers are at risk of ignition when exposed to certain lasers. Insulation, shielding, grounding, and housing of high-voltage electrical components provide adequate protection under most circumstances from electrical injury. Installation and servicing of laser equipment should always be performed by qualified personnel [39].

7.15 Laser Hazard Safety Measures

Ocular safety—eye protection is a must for the patient as well as the person operating the laser.

Protective eyewear specifically designated for the wavelength and classification of the laser in use should be worn in addition to other controls that may be in place to ensure that personnel will not be exposed to laser energy. An optical density of more than 4 at the particular wavelength of the laser used is considered safe for dermatologic lasers. Glass eyewear is heavy, reflects light, and the protective coating may get scratched. Plastic goggles, on the other hand, are lightweight, absorb light but can crack or melt at high temperatures. Each set of goggles should have a specific wavelength of rejection which should match the emission spectrum of the specific laser in use.

Patients should be made to wear eyes shields covering the eyes, thus protecting the entire periorbital area. These goggles are heatproof stainless steel with a smooth polished concave surface and anodized opaque convex surface. These eye shields can be used while the laser is performed away from the eyes. When the laser is performed near the periorbital area, control shields are preferred. Heatproof stainless steel corneal shields are used to reflect the light and do not crack like plastic eye shields. Proparacraine hydrochloride anesthetic eye drops are instilled into the cornea before insertion of the corneal shields [40].

7.16 Safety Measures Against Aerosoles

The first line of protection against the plumes is a smoke aspirator or evacuator. It should have an ultralow particulate triple filter that removes the particles up to 0.1 um in size. Charcoal filter can also be used to remove toxic chemicals in the smoke. Charcoal filter can be used to remove toxic chemicals within the smoke. The distance between the treated area and the suction tip should be 1 cm. The effectiveness drops from 90% to 50% when the distance is changed from 1 to 2 cm.

The surgical masks usually do not filter very small particles that are produced during the laser. Laser masks are made of synthetic fibers electrostatically charged. It should be replaced periodically as they become less protective after 20 mins of use. The fit and the wearing technique also affect the efficiency of the mask [38].

Other safety measures

- Protective equipment like fire extinguishers, non-flammable drapes and anodized instruments should be present in the laser room.
- Laser Manufacturers provide built in safety measures in the system to prevent the emission of the laser radiation.
- Pre-operative testing of the laser and smoke evacuator, along with using the non-flammable antiseptics, reduces the risk of complications

7.17 Conclusion

Advances have been made in the prevention, treatment and management of medical and aesthetic concerns of a women's genital area in the last 5 years. FDA has given clearance or approved laser and energy-based devices only for the treatment of serious conditions like the treatment of abnormal or pre-cancerous cervical or vaginal tissue. Though there are a large number of studies on the safety and efficacy of energy-based devices which has been published in various journals in the last 7 years, these devices has not been evaluated or confirmed by the FDA for vaginal rejuvenation. FDA has given approval for the certain device in the treatment of urinary incontinence like the Emsella from BTL Aesthetics.

A Larger, double-blinded, randomized control trials are needed to prove the reproducible results and clinical safety, including accepting these devices as standard, preventive, or first-line treatments.

These devices have given better clinical results in urinary incontinence and vaginal laxity than conventionally available medical management. These treatments will help many women to reclaim their dignity, femininity and have a normal sexual life. A good understanding of these devices, its parameter, protocol, and limitation will enable the treating physician to give satisfactory and reproducible results to their patients.

References

1. Maiman TH. Stimulated optical radiation in ruby. Nature. 1960;187(4736):493–4.
2. Patel CKN. Continuous-wave laser action on vibrational-rotational transitions of CO_2. Phys Rev. 1964;136(5A):A1187–93.
3. Huzaira M, Lodhi A, Khatri KA. Erbium:YAG laser skin resurfacing: a Pakistani experience. J Cosmet Laser Ther. 2003;5(1):43–9.

4. Manstein D, Herron GS, Sink RKH, et al. Fractional photothermolysis: a new concept for cutaneous remodeling using microscopic patterns of thermal injury. Lasers Surg Med. 2004;34(5):426–38.
5. O'Connor JL, Bloom DA, William T. Bovie and electrosurgery. Surgery. 1996;19(4):390–6.
6. Parish SJ, Nappi RE, Krychman ML, et al. Impact of vulvovaginal health on postmenopausal women: a review of surveys on symptoms of vulvovaginal atrophy. Int J Women's Health. 2013;5:437–47.
7. Gold MH. Update on fractional laser technology. The J Clin Aesthet Dermatol. 2010;3(1):42–50.
8. Rahman Z, Alam M, Dover J. Fractional laser treatment for pigmentation and texture improvement. Skin Therapy Lett. 2006;11(9):7–11.
9. Rahman Z, MacFalls H, Jiang K, et al. Fractional deep dermal ablation induces tissue tightening. Lasers Surg Med. 2009;41(2):78–86.
10. Hultman CS, Friedstat JS, Edkins RE, Cairns BA, Meyer AA. Laser resurfacing and remodeling of hypertrophic burn scars: the results of a large, prospective, before-after cohort study, with long-term follow-up. Ann Surg. 2014;260(3):519–29.
11. Xu XG, Luo YJ, Wu Y, et al. Immunohistological evaluation of skin responses after treatment using a fractional ultrapulse carbon dioxide laser on back skin. Dermatol Surg. 2011;37(8):1141–9.
12. Filippini M, Del Duca E, et al. Fractional CO_2 laser: from skin rejuvenation to vulvo-vaginal reshaping. Photomed Laser Surg. 2017;35(3):171–5.
13. Salvatore S, Nappi R, et al. 12-week treatment with fractional CO_2 laser for vulvovaginalatrophy:apilotst udy. Climacteric. 2014;17(4):363–9.
14. Salvatore S, Leone Roberti Maggiore U, et al. Microablative fractional CO_2 laser improves dyspareunia related to vulvovaginal atrophy: a pilot study. J Endometr Pelvic Pain Disord. 2014;6(3):150–6.
15. Isaza PG, Ruiz Rosas AI, VélezRizo DL. Fractional CO_2 laser treatment: a novel approach for stress urinary incontinence management in post- menopausal women. Urología Colombiana. 2017;26(1):8–11.
16. Ross EV, Domankevitz Y, Skrobal M, Anderson RR. Effects of CO_2 laser pulse duration in ablation and residual thermal damage: implications for skin resurfacing. Lasers Surg Med. 1996;19(2):123–9.
17. Salvatore S, Leone Roberti Maggiore U, Athanasiou S, et al. Histological study on the effects of microablative fractional CO_2 laser on atrophic vaginal tissue: an ex vivo study. Menopause. 2015;22(8): 845–9.
18. Zerbinati N, Serati M, Origoni M, et al. Microscopic and ultrastructural modifications of postmenopausal atrophic vaginal mucosa after fractional carbon dioxide laser treatment. Lasers Med Sci. 2015;30(1):429–36.
19. Gaspar A, Addamo G, Brandi H. Vaginal fractional CO_2 laser: a minimally invasive option for vaginal rejuvenation. Am J Cosmet Surg. 2011;28(3):156–62.
20. Salvatore S, Nappi RE, Zerbinati N, et al. A 12-week treatment with fractional CO_2 laser for vulvovaginal atrophy: a pilot study. Climacteric. 2014;17(4):363–9.
21. Salvatore S, França K, Lotti T, et al. Early regenerative modifications of human postmenopausal atrophic vaginal mucosa following fractional CO_2 laser treatment. Open Access Maced J Med Sci. 2018;6(1):6–14.
22. Qureshi AA, Tenenbaum MM, Myckatyn TM. Nonsurgical vulvovaginal rejuvenation with radiofrequency and laser devices: a literature review and comprehensive update for aesthetic surgeons. Aesthet Surg J. 2017;38(3):302–11.
23. Karcher C, Sadick N. Vaginal rejuvenation using energy-based devices. Int J Womens Dermatol. 2016;2(3):85–8.
24. Nelson AA, Beynet D, Lask GP. A novel non-invasive radiofrequency dermal heating device for skin tightening of the face and neck. J Cosmet Laser Ther. 2015;17(6):307–12.
25. Alinsod RM. Transcutaneous temperature controlled radiofrequency for orgasmic dysfunction. Lasers Surg Med. 2016;48:641–5.
26. Fistonic I, Sorta Bilajac T, et al. Short time efficacy and safety of focused monopolar radiofrequency device for labial laxity improvement—noninvasive labia tissue tightening. A prospective cohort study. Lasers Surg Med. 2016;48:254–9.
27. Lalji S, Lozanova P. Evaluation of the safety and efficacy of a monopolar nonablative radiofrequency device for the improvement of vulvo-vaginal laxity and urinary incontinence. J Cosmet Dermatol. 2017;16:230–4.
28. Magon N, Alinsod R. ThermiVa: the revolutionary technology for vulvovaginal rejuvenation and noninvasive management of female SUI. J Obstet Gynaecol India. 2016;66(4):300–2.
29. Caruth JC. Evaluation of the safety and efficacy of a novel radiofrequency device for vaginal treatment. Surg Technol Int. 2018;32:145–9.
30. Dayan E, Ramirez H, Theodorou S. Radiofrequency treatment of Labia Minora and Majora: a minimally invasive approach to Vulva restoration. Plast Reconstr Surg Glob Open. 2020;8(4):e2418.
31. Samuels JB, Alinsod R, et al. Safety and efficacy of a non-invasive high-intensity focused electromagnetic field (HIFEM) device for treatment of urinary incontinence and enhancement of quality of life. Lasers Surg Med. 2019;51:760–6.
32. Silantyeva E, Zarkovic D, et al. A comparative study on the effects of high-intensity focused electromagnetic technology and electrostimulation for the treatment of pelvic floor muscles and urinary incontinence in parous women: analysis of posttreatment data. Female Pelvic Med Reconstr Surg. 2020; https://doi.org/10.1097/SPV.0000000000000807.
33. Park H, Kim E, et al. High-intensity focused ultrasound for the treatment of wrinkles and skin laxity in seven different facial areas. Ann Dermatol. 2015;27(6):688–93.
34. Kleinerman R, Whang TB, Bard RL, Marmur ES. Ultrasound in dermatology: principles and applications. J Am Acad Dermatol. 2012;67:478–87.

35. Sekiguchi Y, Nakamura K, et al. 007 Safety and efficacy of high intensity focused ultrasound (HIFU) treatment for vagina ultra vera—preliminary results. J Sex Med. 2017;14:e353. https://doi.org/10.1016/j.jsxm.2017.04.013.
36. Andre P, Orth G, Evenou P, Guillame JC, Avril MF. Risk of papillomavirus infection incarbon dioxide laser treatment of genital lesions. J Am Acad Dermatol. 1990;22(1):131–2.
37. Buttini M, Maher C. The first published randomised controlled trial of laser treatment for vaginal atrophy raises serious questions. MJA. 2018;209(9):376–7.
38. Koska JM, Eugene J. Chemical composition of laser- tissue interaction smoke plume. J Laser Appl. 1989;3:59–63.
39. Smalley PJ. Laser safety: risks, hazards and control measures. Laser Ther. 2011;20(2):95–106.
40. Pai H. Prevention of ocular complications. In: Complications in cosmetic dermatology crafting cures. Jaypee Brothers Medical Publishers (P) Ltd; 2016. p. 178–80.

Laser in Vaginal Rejuvenation

8

Alex Bader

The Vaginal and its supportive tissues frequently remodel in response to different environmental stimuli. Pelvic Organs Prolapses are estimated to affect almost half of women above the age of 50 years [1]. Urinary incontinence (UI) is a common problem, with an estimated prevalence of 50% in the female population worldwide [2].

The evolution of medicine could not leave back the field of women's health and specifically pelvic floor organs health. With the booming of Aesthetic Gynecology, this area starts to have more interest from the surgical and non-surgical aspects. Nowadays women have been showing active interest in seeking cosmetic help to improve both the appearance and function of their genitalia. Non-invasive modalities have been introduced to correct and restore the form and functionality of the vagina and its surrounding tissues. The treatment may be aimed at improving sensitivity, hydration, and elasticity in the vulvar, introital, and vaginal tissue [3].

Discussing Non-Surgical Vaginal tissues restoration, we come across two major categories of methods. On one hand, we have the Energy-Based Devices (EBD) and on the other hand, the injectables with the most popular the Hyaluronic Acid with its both forms, cross-link and non-cross link.

While exploring the historical background of Energy-Based Devices and their applications on the female genital organs, we come across the fact that almost 5 decades ago, gynecologist and plastic surgeons pioneered the integration of lasers in their respective fields [4]. The focused CO_2 laser beam was exploited for ablation, vaporization, contraction, which helps in treating various cervical and vaginal pathologies [5]. In the 1970s, various lesions such as genital warts on the uterine cervix were treated with the CO_2 laser, which has since become a common treatment approach for genital warts with micromanipulators connected to colposcopes [6].

The Hyaluronic Acid, with its both forms, Cross-Linked HA, and Non-Cross Link HA, is gaining important ground in the field of vaginal tissue restoration. Cross-linked HA is currently the most widely used filler with a longevity of approximately 6 months [7]. Non-Cross-Linked Hyaluronic Acid (NCLHA) in its natural form is liquid, and it has been proved that it can increase the proliferative and the metabolic activity of cutaneous fibroblasts [8].

A. Bader (✉)
Bader Medical Institute of London,
London, UK

P. Jindal et al. (eds.), *Aesthetic and Regenerative Gynecology*,
https://doi.org/10.1007/978-981-16-1743-0_8

8.1 Introduction

8.1.1 Histology

The Vaginal wall is structured of four layers:

1. Superficial—stratified squamous epithelium.
2. Subepithelial—Dense connective tissue layer composite of collagen and elastin fibres.
3. Muscle layer—Smooth muscle.
4. Adventitia (composed of loose connective tissue).

The connective tissue underlying the vagina wall contains mainly fibroblasts producing components of extracellular matrix (ECM), besides fat cells and mast cells. All elements are embedded in an ECM, composed of fibrillar elements (collagen and elastic fibres) [9]. The fibrillar component (collagen and elastin fibres) contributes the most to the pelvic floor connective tissues, which helps in maintaining tone, tensile strength and provide support to all pelvic organs [10].

8.1.1.1 Collagen

There are 28 types of collagen. The collagens I, III, and V, present in the vagina and its supportive tissues, are thought to be the principal determinants of soft tissue strength.

- Collagen I fibres—high tensile strength fibres.
- Collagen III—lower tensile strenght fibres (major subtype in the vagina).
- Collagen V—lower tensile strenght fibres [11].

An increase in collagen III and V decreases the mechanical strength of connective tissue by decreasing fibre size [12]. Cross-linking of tropoelastin and procollagen to form mature functional collagen and elastin fibres is by Lysyl oxidase (LOX) family of enzymes [13].

8.1.1.2 Elastin

Elastin allows the tissue to stretch and return to its original shape without energy input. In undisturbed tissues, elastic fibres produced in the third trimester of foetal life last the rest of life [14]. In the female reproductive tract, however, elastic fibre turnover is continuous. LOX is essential for elastic fibre homeostasis [15]. Fibulin-5 acts as elastin binding protein for effective assembly of elastic fibres that helps in recovery of pelvic organ support after damage [16].

8.1.2 Tissue Remodelling and Clinical Impacts

Pelvic floor defects are visible as a bulge on anterior and/ posterior vaginal walls with resultant SUI, incomplete defecation/ faecal incontinence, chronic pelvic pain, vaginal relaxation, sexual dysfunction and social isolation. In some cases, the only complaint may be the appearance.

Factors contributing to the development of pelvic organ prolapse (POP) can be divided into genetic and acquired factors. Acquired factors include pregnancy and parity as well as myopathy and neuropathy. Obesity, smoking, pulmonary disease and constipation are examples of POP-promoting factors [17].

Where pelvic floor muscles are weakened, a decrease in pelvic connective tissue resilience related to age and menopause may facilitate progression to symptomatic POP. Increased age is a risk factor for POP. There is a 10% increased risk for each decade of life [18]. Among types of POP, cystocele has the greatest incidence [19]. The biomechanical microenvironment may be further compromised if non-resorbable polymeric meshes are used to replace tissue function [20].

Genetic predisposition may play an important role as well. When the patient has a family history of POP, the relative risk of her developing POP is also raised.

Parity is the most common predisposing factor in the development of POP. During childbirth, there occurs stretching of connective tissues, damage of neuromuscular, injury to levator ani muscle, loss of tensile strength in suspensory ligaments, which most likely contribute to the progression of POP. Other childbirth-related factors associated with prolapse include high infant birth weight, the prolonged second stage of labour and maternal age less than 25 years at first delivery [17].

Pelvic Surgery: Retropubic urethropexy or needle suspension procedures may result in more anterior deviation of the anterior vaginal wall, which alters the distribution of force on all vaginal walls. As a result, the apex and posterior vaginal wall may become prone to the development of support defects, including enterocele or rectocele. Elevated intra-abdominal pressure: Women who are labourers/factory workers, Chronic constipation, chronic obstructive pulmonary disease may cause stretch injury to the pudendal nerve. Increasing body mass index (BMI): Women with BMI >25 kg/m^2 have a threefold higher risk of having POP compared to non-obese women. Women with joint hypermobility have a significantly higher prevalence of POP, and connective tissue factor is involved.

8.1.3 Histological Changes [21]

Premenopausal vaginal mucosa is characterized by squamous stratified non-keratinized epithelium and lamina propria (thin connective tissue layer rich collagen and elastin fibres) protruding into under surface of the epithelium, with papillae rich in small blood vessels. Epithelial cells proliferate from the basal layer as they differentiate and move towards the mucosal surface and eventually shed off. Glycogen synthesized by the intermediate cells is stored in the superficial cells from where it is eventually released and utilized by Lactobacilli, maintaining the acidic pH of the vagina.

Postmenopausal vaginal mucosa is characterized by tissue atrophy, with significant thinning of epithelial layers, upward shift in vaginal pH, decreased blood flow, reduce lubrication due to decreased levels of circulating oestrogen in postmenopausal state.

8.1.4 Role of Energy-Based Devices (EBD) [21]

When absorbed in sufficient amounts, light/thermal energy can induce changes in the skin and muscle matrix. Lasers devices allow for the delivery of light to the skin in a controlled manner. The past decade has seen a surge in technological advancements energy-based devices (fractional laser and radiofrequency) that work on the vaginal wall to reverse its age-related tissue remodelling. Thermal energy delivered on the vaginal wall stimulates neovascularization, collagen formation and proliferation of glycogen-enriched epithelium leading to improved natural lubrication and urinary continence. Different wavelengths across the electromagnetic spectrum are differentially absorbed by different tissue chromophores, including haemoglobin, melanin, and water [22]. Controlled power of the energy source must be used in vaginal tissue that varies in its water content and cellular component with age and under hormonal influence, among other factors [22].

8.2 Lasers

8.2.1 Technology Parameters

Wavelength and pulse duration are the most important laser parameters that govern the effects of laser light on skin/mucosa. Fluence, irradiance and spot size are additional laser settings that influence clinical outcomes.

- **Wavelength**—Emitted light determines its depth of penetration into skin/mucosa. Long wavelengths of light penetrate more deeply into tissue than shorter wavelengths. Chromophores absorb light most effectively at different wavelengths. To exert the greatest effect, the wavelength should be near the maximum absorption of the target chromophore and should be of enough length to penetrate to the depth of the target. Because light scattering decreases with increasing wavelength, greater penetration is achieved with longer wavelengths of light. In contrast, high absorption of light by water molecules in epithelium markedly limits the depth of penetration of light emitted from mid-infrared (2940 nm erbium: yttrium aluminium garnet, Er: YAG) and far-infrared (10,600 nm carbon dioxide, CO_2) lasers.

- **Pulse Duration**—The setting is determined by the thermal relaxation time of the target chromophore.
- **Thermal Relaxation Time**—It is defined as the amount of time it takes the temperature of a target to return to the ambient temperature following heating [23]. If an object is heated for a period equal to or shorter than its thermal relaxation time, accumulated heat and resultant damage are confined to the target object alone. This selective photothermolysis dramatically reduces the risk of scarring. For example the CO_2 laser penetrates the skin to a depth of 0.05 mm, and the thermal relaxation time of this thickness of tissue is approximately 1 millisecond [24, 25].
- **Fluence**—It is a measurement of the amount of energy delivered/ unit area.
- **Irradiance—Power** describes the rate of energy delivery and is measured in Joules/ second (watts). Irradiance relates this measurement to the size of the treated area and describes the rate of energy delivery/ unit area (watts/cm^2) [26]. Irradiance is calculated from laser power output and spot size. It is a measurement of the intensity of energy delivery. Very high irradiance achieves much faster heating than low irradiance. Decreasing the pulse duration of a laser device without changing the energy results in a higher level of irradiance. Slow heating (low irradiance) coagulates tissue, while fast heating (high irradiance) can vaporize tissue.
- **Spot Size**—Spot size is the diameter of the beam of light emitted that hits the tissue's surface. The light delivered through a small spot size is more susceptible to scattering than light delivered through large spot sizes. Larger spot sizes are preferred when targeting structures in the mid or deep dermis. Spot size is less important for superficial targets.

8.2.2 Laser Technology Categories

The three major categories include continuous wave lasers, pulsed lasers, and fractionated lasers.

Continuous—It emit continuous laser beam as long as foot/ finger switch is pressed. Scanning devices are not always effective in preventing cutaneous damage. As a result, continuous wave lasers have been mostly supplanted by pulsed lasers.

Pulsed Lasers—Pulsed lasers produce a laser beam that is emitted in short pulses with long period (0.1–1 second) between pulses. Pulsed ablative CO_2 and Er: YAG lasers are used for laser skin resurfacing [24]. They are usually indicated and used for precisely ablating superficial skin lesions while leaving behind a very narrow zone of thermal damage.

Fractionated Lasers (Fig. 8.1)—Fractional photothermolysis most commonly are used for the skin resurfacing and the treatment of scars [27, 28]. They are divided into non-ablative and ablative Lasers.

- **Non-Ablative Fractionated** Lasers—thermally coagulate narrow vertical columns of the epidermis and variable portions of dermis. This is achieved with near-infrared lasers (1320, 1440, 1540, 1550, and 1927 nm). These lasers emit light that is absorbed by water in the skin to a lesser extent than light from mid and far-infrared lasers, which are used for fractional ablation. Following treatment, the thermally coagulated epidermis is rapidly sloughed off, and reepithelialization is completed within 2–3 days. The thermally coagulated dermis is gradually replaced with new collagen deposition and collagen remodeling. A series of treatments usually is necessary to remodel the entire skin surface gradually.
- **Ablative Fractionated** Laser—Creates ablative MTZs in the skin that extends into the dermis. This is achieved with mid-infrared (2940 or 2790 nm) and far-infrared (10,600 nm) lasers, which emit light that is strongly absorbed by water [29, 30]. A zone of thermally coagulated tissue also occurs around the ablated zone. With Er:YAG (2940 nm) laser, the peripheral coagulated zone is very narrow, which accounts for the frequent appearance of pinpoint bleeding during treatment. The CO_2 (10,600 nm) laser yields a thicker coagulated zone, which reduces bleeding. Reepithelialization usually occurs around four

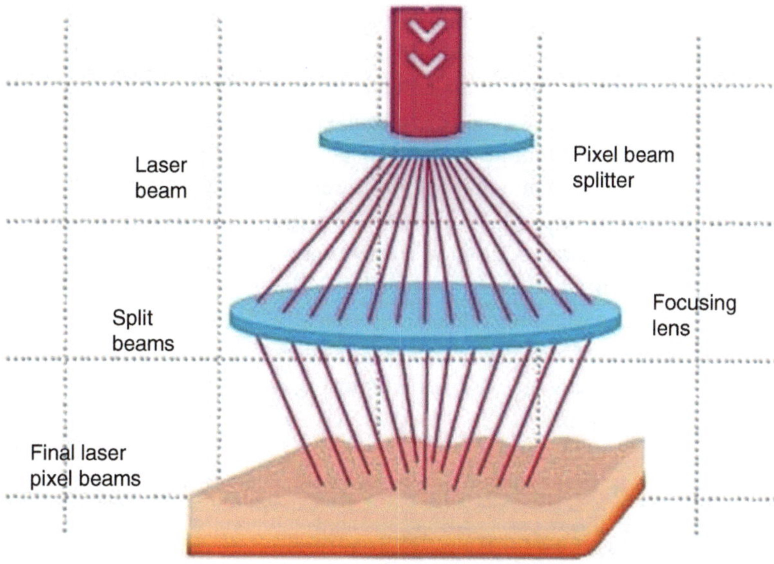

Fig. 8.1 Laser beam fractionated by the holographic lens, printing 9 × 9 (81 microbeams) on 1 × 1 cm. (Courtesy: Alma Lasers)

days after treatment with fractionated Er:YAG laser and after about seven days following fractionated CO_2 laser therapy. This relatively short time course to reepithelialization is an advantage over traditional laser resurfacing with pulsed CO_2 or Er:YAG lasers, which requires a healing period of around 7 to 10 days.

8.2.3 Clinical Applications in Vaginal Conditions

Indications

- Vaginal dryness, burning/ itching.
- Loss of elasticity and tone (laxity/ vaginal relaxation syndrome).
- Urinary incontinence.
- Thrush and bladder infections.
- Painful intercourse.
- Vaginal/vulva pain.
- Mild Prolapse.
- Reduced sensitivity and sexual arousal.
- Lichen Sclerosus.
- Vulval skin resurfacing for scars, tissue laxity or skin rejuvenation.

8.2.4 Vaginal Atrophy

For the treatment of Vaginal Atrophy, Laser Vaginal application focuses on restoring functionality by reversing the age/mechanical stress-related vaginal remodelling that results in vaginal laxity. Microablative fractional lasers, RF induces inflammatory cascade that causes superficial tissue shrinkage as well as deep stimulation of ECM layer of the submucosa [31–34]. Histologically evident alterations in postmenopausal atrophic vaginal mucosa following fractional CO_2 laser treatment has provided an understanding of the underlying molecular mechanism responsible for tissue remodelling [35–37].

8.2.4.1 Mucosal Responses Following Fractional Laser Treatment

Post laser CO_2 treatment, there occur thickening of stratified squamous epithelium, closed pack basal layer cells, increased papillary projections with blood capillaries. Shedding of superficial epithelial cells with exposure to glycogen into the vaginal lumen is seen. The Lactobacilli vaginalis dependent on glycogen from shedding of superficial epithelial cells, maintains the vaginal pH to acidic, thus preventing colonization of pathogens [38, 39]. Increased vascularity (capillaries penetrating newly formed papillae underneath the epithelium) in connective tissue supports the renewed activity of fibroblasts and provides the metabolic support for epithelial cell proliferation and differentiation (Fig. 8.2). The macroscopic imaging of the vaginal wall shows fractionated (Fig. 8.3) without any damage or bleeding. The long term efficacy is also visible macroscopically with the quality of the vaginal tissue to show improvement in terms of vascularization, rogation, and secretion (Fig. 8.4).

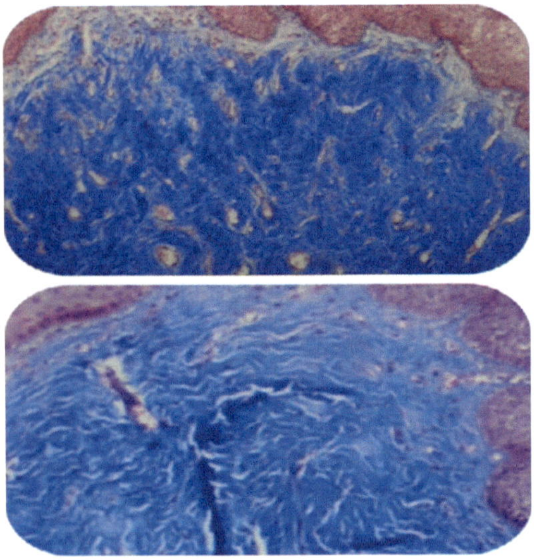

Fig. 8.2 Histological picture showing the improvement in the collagen quality 6 months after CO_2 Laser Fractional application in a patient with vaginal atrophy (Courtesy of Dr. Jorge Elias)

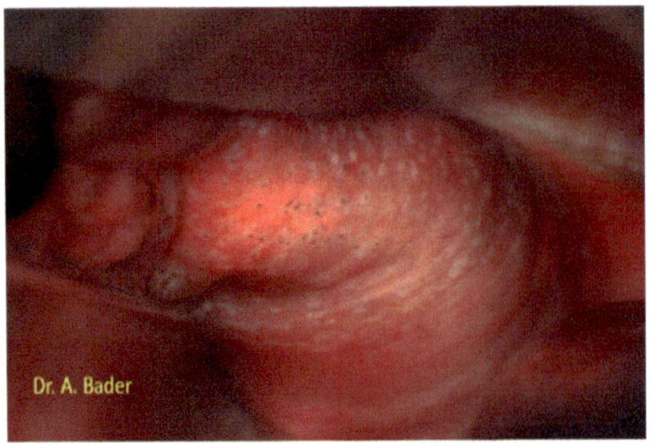

Fig. 8.3 Fractional CO_2 Laser application to the upper vaginal wall. This picture was taken immediately after the application

8.2.4.2 Biomolecular Effects of EBD Application on Vaginal Mucosa

- Thermal energy initiate stimulation of remodelling of connective tissue by fibroblasts has been correlated with activation of heat shock proteins (HSPs) response persisting at least 3 months following treatment [40]. The HSP gets activated, and other growth factors come into play to induce type 1 versus type 3 collagen [41–46].
- Inflammation cascade begins, which involves induction of cytokines, chemokines, and antimicrobial peptides that stimulate influx and activation of monocytes and neutrophils. Local increase in different **cytokines**, particularly TGF-b (stimulating collagen), basic fibroblast growth factor (bFGF), epidermal growth factor (EGF), PDGF, TNF-a, and vascular endothelial growth factor (VEGF) that results in stimulation of fibroblast activity and formation of new vessels [47].

After inflammatory phase of wound healing, dermal fibroblasts produce new collagen fibrils at 2 weeks interval and remains high for at least 5 weeks after treatment.

Average peak induction after with fractionated CO_2 laser is 8.4-fold [47]. In addition, tissue wound repair involves increased **vascularization**. Fractionated CO_2 laser treatment increase the density of CD31-positive vessels **endothelial cells** at 3 and 5 weeks after laser treatment.

Long pulse, non-ablative **modes** have been evaluated for their ability to induce deep thermal effects on the vaginal mucosa [48]. This mode generates a thermal change in deeper layers, with only minor superficial ablation (5 μm) on the surface, but with enough caloric effect to thermally alter the chromophore. Using a **large spot size** (>5 mm), deepens the heat effect in lamina propria to a depth of at least 500–1000 μm (depending on the degree of tissue hydration), inflicting the "Joule effect" (photothermal and thermochemical effect) on the vaginal mucosa. With local rise in temperature, bradykinin and histamine are released with the relaxation of precapillary arteriolar sphincters and vasodilation, an effect called "Phenomenon of Thermal Reperfusion" [48–50].

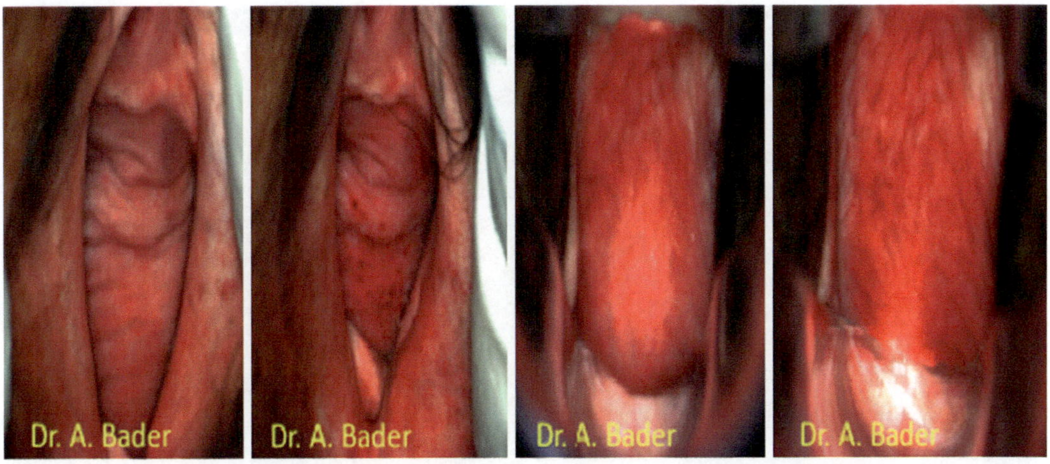

Fig. 8.4 The efficacy of Fractional CO_2 Laser applied on the atrophic anterior vaginal wall. The serial of the pictures shows the vaginal tissue before the application [1], immediately after the application [2], Three months after application [3] and six months after the application [4]

8.2.5 Stress Urinary Incontinence (SUI)

Urethral support can be improved by inducing thermal effects on suburethral tissue using energy-based devices. Numerous studies [51–57] confirm improvement in incontinence symptoms following treatment with energy-based devices (EBDs) i.e.; r:YAG laser, CO_2 laser, and RF with penetration depth of 0.5 mm inside the anterior vaginal wall, and results in 30% reduction in tissue volume. The thermally induced neocollagenases improve thickness, elasticity, and firmness of the vaginal wall and acts locally at the suburethral plexus level [52, 53].

8.2.6 Laser Tightening for Vulvar Laxity [58]

As age advances, labia majora undergoes atrophy and poor circulation, which is associated with oestrogen deficiency, following pregnancy, delivery, and significant weight loss in younger women with resultant loss of labia volume. On the other hand, protuberant labia majora can lead to functional difficulties in sexual stimulation and satisfaction, complicate hygiene maintenance, noticeable bulge while wearing sportwear and negatively impact self-confidence and self-image. Severe skin laxity is often best managed with interventions such as injectable cosmetic fillers and surgery. Surgical Labia Majora Plasty may be associated with infection, scarring, pain, dissatisfaction with the results, anatomical deformities, scarring, and hair growth toward the vaginal entrance, which can further exacerbate patient discomfort and dissatisfaction. As an alternative to surgery, dermal rejuvenation of lax labia majora using EBD is preferred with the use either of fractional laser, or RF that delivers heat gently and cause tissue remodelling by neocollagenases [58].

Procedure: A topical anaesthetic cream to be applied for 15–20 minutes prior to the application. A layer of oil is applied before introducing the probe. The device probe is then moved over the skin in circular movements ensuring the temperature does not exceed 42 °C (as indicated by the device monitor) for more than 5 seconds. Moisturizer or lubricant is applied immediately after the treatment and continued for 5 days. The patient tolerates the procedure well with no downtime. Results may be visible immediately after the session, but the continuous improvement over the 3–6 months following treatment is seen (Figs. 8.5 and 8.6), which is the estimated time for new collagen regeneration [47].

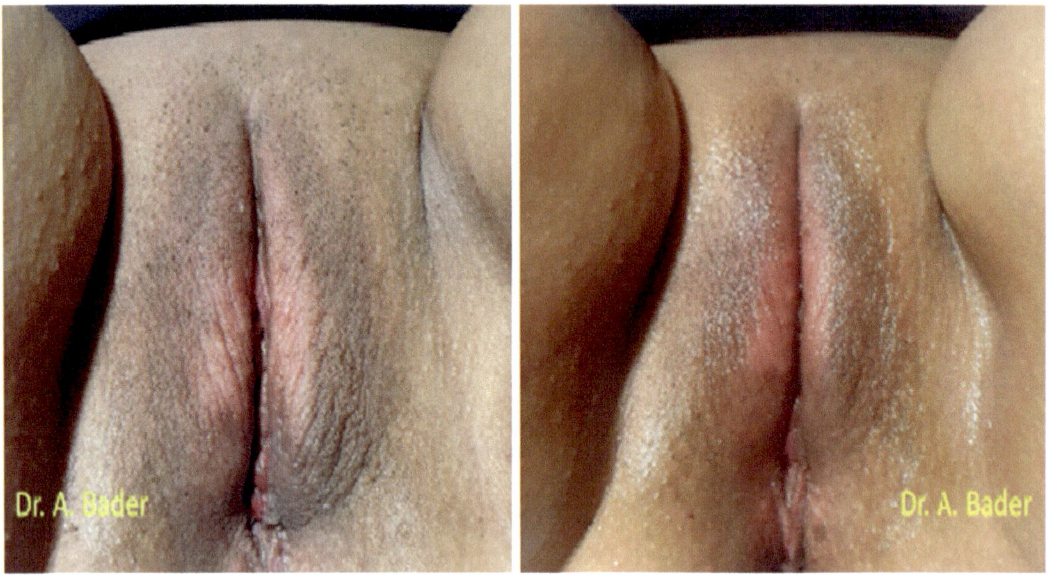

Fig. 8.5 Labia Majora Skin tightening using ablative De-Focused CO_2 Laser tip

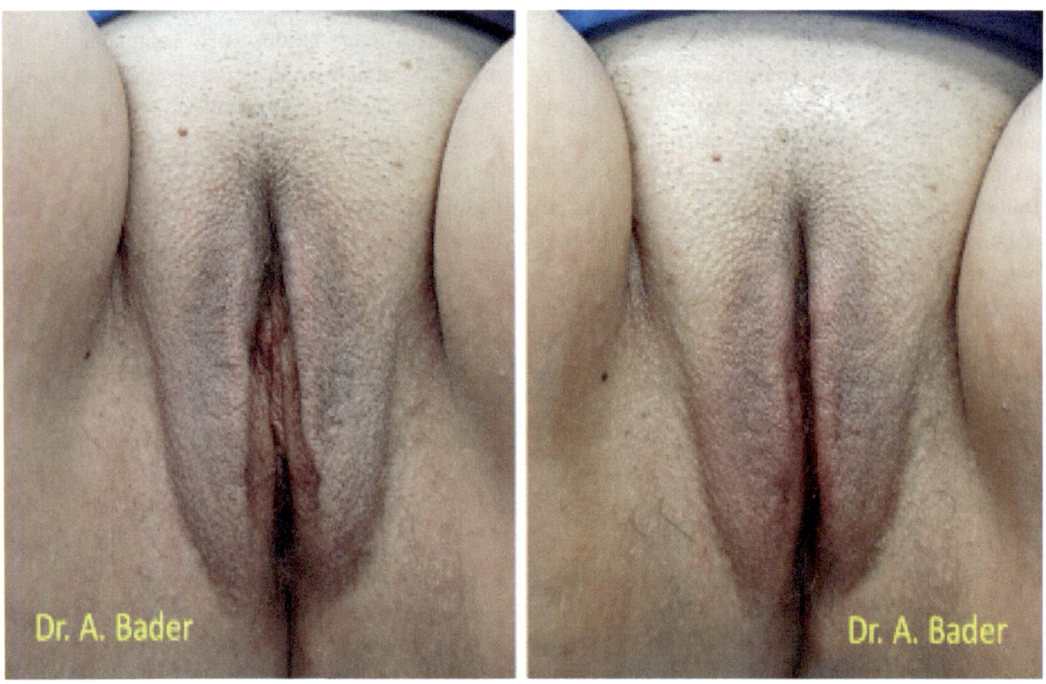

Fig. 8.6 Labia Majora Skin tightening using ablative De-Focused CO_2 Laser tip. The results usually are remarkable after one session. The technique frequently would also improve the perineal body skin with an important impact for functional enhancement alongside the aesthetic aspect

Follow-up after 6 months is advised. A maintenance treatment session every year is recommended.

8.3 Conclusion

Application of advanced energy-based devices for vaginal rejuvenation, such as minimally ablative fractional laser and radiofrequency, is currently gaining popularity by women these days due to its minimally invasive properties, less side effects and better patient tolerance.

Thermal energy generated from energy-based devices create microscopic lesions which promote wound healing subsequent new collagen and elastin fibre formation [59, 60].

Lasers are used recently been used in the cosmetic-gynecology field by clinicians that help in improving the structure, physiological function, and aesthetic look of the vagina. Recent trials and clinical studies provide evidence that the use of these non-invasive modalities seems an attractive, effective, and safe option that can be added to the toolkit of many health care professionals who are currently delivering medical aesthetic/functional treatments [61]. The review that was made by a large group of experts may lack some evidence, but we hope that it offers relevant information to medical professionals and provide define directions for their future approach.

References

1. Subak LL, Waetjen LE, van den Eeden S, Thom DH, Vittinghoff E, Brown JS. Cost of pelvic organ prolapse surgery in the United States. Obstet Gynecol. 2001;98:646–51.
2. Markland AD, Richter HE, Fwu C-W, Eggers P, Kusek JW. Prevalence and trends of urinary incontinence in adults in the United States, 2001 to 2008. J Urol. 2011;186:589–93. https://doi.org/10.1016/j.juro.2011.03.114.
3. Hutchinson-Colas J, Segal S. Genitourinary syndrome of menopause and the use of laser therapy. Maturitas. 2015;82:342–5.
4. Kaplan I, Goldman J, Ger R. The treatment of erosions of the uterine cervix by means of the CO2 laser. Obstet Gynecol. 1973;41(5):795–6.
5. Bellina JH, Polanyi TG. Management of vaginal adenosis and related cervico-vaginal disorders in DES-exposed pregnancy by means of carbon dioxide laser surgery. J Reprod Med. 1976;16:295–6.
6. Yona Tadir, MD, Adrian Gaspar, MD, Ahinoam Lev-Sagie, MD, Macrene Alexiades, MD, PhD, Red Alinsod, MD, Alex Bader, MD, Alberto Calligaro, MD, Jorge A. Elias, MD, Marco Gambaciani, MD, Jorge E. Gaviria, MD, Cheryl B. Iglesia, MD, Ksenija Selih-Martinec, MD, Patricia L. Mwesigwa, MD, Urska B. Ogrinc, MSc, MD, Stefano Salvatore, MD, Paolo Scollo, MD, Nicola Zerbinati, MD, and John Stuart Nelson, MD, PhD. Light and energy based therapeutics for genitourinary syndrome of menopause: consensus and controversies. Lasers Surg Med 49:137–159 (2017).
7. Greco RM, Iocono JA, Ehrlich HP. Hyaluronic acid stimulates human fibroblast proliferation within a collagen matrix. J Cell Physiol. 1998;177:465–73.
8. Stern R, Asari AA, Sugahara KN. Hyaluronan fragments: an information-rich system. Eur J Cell Biol. 2006;85:699–715.
9. Goh JT. Biomechanical and biochemical assessments for pelvic organ prolapse. Curr Opin Obstet Gynecol. 2003;15:391–4.
10. Alarab et al. Expression of extracellular matrix-Remodeling proteins is altered in vaginal tissue of premenopausal women with severe pelvic organ prolapse. Reprod Sci 2014, Vol. 21(6) 704–715.
11. Bailey AJ. Molecular mechanisms of ageing in connective tissues. Mech Ageing Dev. 2001;122:735–55.
12. Birk DE, Fitch JM, Babiarz JP, Doane KJ, Linsenmayer TF. Collagen fibrillogenesis in vitro: interaction of types I and V collagen regulates fibril diameter. J Cell Sci. 1990;95:649–57.
13. Shynlova O, Bortolini MAT, Alarab M. Genes responsible for vaginal extracellular matrix metabolism are modulated by women's reproductive cycle and menopause. Int Braz J Urol. March–April 2013;39(2):257–67.
14. Kerkhof MH, Hendriks L, Brölmann HAM. Changes in connective tissue in patients with pelvic organ prolapse – a review of the current literature. Int Urogynecol J. 2009;20:461–74.
15. Alperin M, Debes K, Abramowitch S, Meyn L, Moalli PA. LOXL1 deficiency negatively impacts the biomechanical properties of the mouse vagina and supportive tissues. Int Urogynecol J Pelvic Floor Dysfunct. 2008;19:977–86.
16. Nakamura T, Lozano PR, Ikeda Y, Iwanaga Y, Hinek A, Minamisawa S, et al. Fibulin-5/DANCE is essential for elastogenesis in vivo. Nature. 2002;415:171–5.
17. Ibrahim Alkatkout, Lisolette Metller, Rupinder Ruprai, Alexandros Bader (2019). Practical Manual for Laparoscopic & Hysteroscopic Gynecological Surgery(Kiel School of Gynaecological Endoscopy). ISBN 9789352701940.
18. Olsen AL, Smith VJ, Bergstrom JO, Colling JC, Clark AL. Epidemiology of surgically managed pel-

vic organ prolapse and urinary incontinence. Obstet Gynecol. 1997;89:501–6.
19. Lensen EJM, Withagen MIJ, Kluivers KB, Milani AL, Vierhout ME. Surgical treatment of pelvic organ prolapse: a historical review with emphasis on the anterior compartment. Int Urogynecol J. 2013;24:1593–602.
20. Ruiz-Zapata et al. Functional characteristics of vaginal fibroblastic cells from premenopausal women with pelvic organ prolapse. Mol Hum Reprod. 2014;20(11):1135–43.
21. Tadir et al. Light and energy based therapeutics for GSM. Lasers in surgery and medicine. Wiley Periodicals, Inc; 2017. p. 1–23.
22. Alexiades-Armenakas MR, Dover JS, Arndt KA. The spectrum of laser skin resurfacing: nonablative, fractional, and ablative laser resurfacing. J Am Acad Dermatol. 2008;58(5):719–37. quiz 738–40
23. Lev-Sagie A. Vulvar and vaginal atrophy: physiology, clinical presentation, and treatment considerations. Clin Obstet Gynecol. 2015;58(3):476–91.
24. Anderson RR, Parrish JA. Selective photothermolysis: precise microsurgery by selective absorption of pulsed radiation. Science. 1983;220:524.
25. Lowe NJ, Lask G, Griffin ME, et al. Skin resurfacing with the Ultrapulse carbon dioxide laser. Observations on 100 patients. Dermatol Surg. 1995;21:1025.
26. Teikemeier G, Goldberg DJ. Skin resurfacing with the erbium: YAG laser. Dermatol Surg. 1997;23:685.
27. Kono T, Chan HH, Groff WF, et al. Prospective direct comparison study of fractional resurfacing using different fluences and densities for skin rejuvenation in Asians. Lasers Surg Med. 2007;39:311.
28. Sherling M, Friedman PM, Adrian R, et al. Consensus recommendations on the use of an erbium-doped 1,550-nm fractionated laser and its applications in dermatologic laser surgery. Dermatol Surg. 2010;36:461.
29. Hunzeker CM, Weiss ET, Geronemus RG. Fractionated CO2 laser resurfacing: our experience with more than 2000 treatments. Aesthet Surg J. 2009;29:317.
30. Mahmoud BH, Srivastava D, Janiga JJ, et al. Safety and efficacy of erbium-doped yttrium aluminum garnet fractionated laser for treatment of acne scars in type IV to VI skin. Dermatol Surg. 2010;36:602.
31. Gaspar A, Addamo G, Brandi H. Vaginal fractional CO_2 laser: a minimally invasive option for vaginal rejuvenation. Am J Cosmetic Surg. 2011;28(3):156–62.
32. Millheiser LS, Pauls RN, Herbst SJ, Chen BH. Radiofrequency treatment of vaginal laxity after vaginal delivery: nonsurgical vaginal tightening. J Sex Med. 2010;7(9):3088–95.
33. Mulholland RS. Radio frequency energy for noninvasive and minimally invasive skin tightening. Clin Plast Surg. 2011;38(3):437–48.
34. Alinsod RM. Transcutaneous temperature controlled radiofrequency for orgasmic dysfunction. Lasers Surg Med. 2016;48(7):641–5.
35. Salvatore S, Leone Roberti Maggiore U, Athanasiou S, et al. Histological study on the effect of microablative fractional CO_2 laser on atrophic vaginal tissue: an ex vivo study. Menopause. 2015;22(8):845–9.
36. Zerbinati N, Serati M, Origoni M, et al. Microscopic and ultrastructural modifications of postmenopausal atrophis vaginal mucosa after fractional carbon dioxide laser treat-ment. Lasers Med Sci. 2015;30(1):429–36.
37. Salvatore S, Nappi RE, Zerbinati N, et al. A 12-week treatment with fractional CO2 laser for vulvovaginal atrophy: a pilot study. Climacteric. 2014;17(4):363–9.
38. Mac Bride MB, Rhodes DJ, Shuster LT. Vulvovaginal atrophy. Mayo Clin Proc. 2010;85(1):87–94.
39. Athanasiou S, Pitsouni E, Antonopoulou S, et al. The effect of microablative fractional CO_2 laser on vaginal flora of postmenopausal women. Climacteric. 2016;19(5):512–8.
40. Hantash BM, Bedi VP, Kapadia B, et al. In vivo histological evaluation of a novel ablative fractional resurfacing device. Lasers Surg Med. 2007;39:96–107.
41. Ishida Y, Nagata K. Hsp47 as a collagen-specific molecular chaperone. Methods Enzymol. 2011;499:167–82.
42. Tasab M, Batten MR, Bulleid NJ. Hsp47: a molecular chaperone that interacts with and stabilizes correctly-folded procollagen. EMBO J. 2000;19(10):2204–11.
43. Dafforn TR, Della M, Miller AD. The molecular interactions of heat shock protein 47 (Hsp47) and their implications for collagen biosynthesis. J Biol Chem. 2001;276(52):49310–9.
44. Prignano F, Campolmi P, Bonan P, et al. Fractional CO_2 laser: a novel therapeutic device upon photobiomodulation of tissue remodeling and cytokine pathway of tissue repair. Dermatol Ther. 2009;22(Suppl 1):S8–S15.
45. Nowak KC, McCormack M, Koch RJ. The effect of superpulsed carbon dioxide laser energy on keloid and normal dermal fibroblast secretion of growth factors: a serum-free study. Plast Reconstr Surg. 2000;105(6):2039–48.
46. Manolis EN, Kaklamanos IG, Spanakis N, et al. Tissue concentration of transforming growth factor b1 and basic fibroblast growth factor in skin wounds created with a CO_2 laser and scalpel: a comparative experimental study, using an animal model of skin resurfacing. Wound Repair Regen. 2007;15:252–7.
47. Orringer JS, Sachs DL, Shao Y, et al. Direct quantitative comparison of molecular responses in Photodamaged human skin to fractionated and fully ablative carbon dioxide laser resurfacing. Dermatol Surg. 2012;38:1668–77.
48. Luk€ao M, Mqjaron B, Rupnik T. Ablative and thermal effects of Er:YAG laser on human tissue. In: Waidelich W, Waidelich R, Waldschmidt J, editors. Lasers in medicine. Berlin: Springer; 1998. p. 566–72.
49. Majaron B, Plesteniak Luka e M. Thermo-mechanical laser ablation of soft biological tissue: Modeling the micro explosions. Appl Phys B Lasers Opt 1999;69:71–80.
50. Gaspar A, Brandi H, Gomez V, Luque D. Efficacy of erbium: YAG laser treatment compared to topical estriol treatment for symptoms of genitourinary syn-

drome of menopause. Lasers Surg Med. 2016; https://doi.org/10.1002/lsm.22569.
51. Fistonic N, Fistonic I, Lukanovic A, Findri-Gustek S, Sorta Bilajac Turina I, Franic D. First assessment of short term efficacy of Er:YAG laser treatment on stress urinary inconti-nence in women: prospective cohort study. Climacteric. 2015;18(Sup 1):37–42.
52. Ogrinc BU, Sencar S. Novel minimally invasive laser treatment of urinary incontinence in women. Lasers Surg Med. 2015;47:689–97.
53. Fistonic N, Fistonic I, Findri Gustek SS, et al. Minimally invasive, non-ablative Er:YAG laser treatment of stress urinary incontinence in women—a pilot study. J Laser Med Sci. 2016;31:635–43.
54. Pardo J, Sola V, Morales A. Treatment of female stress urinary incontinence with erbium YAG laser in non-ablative mode. Eur J Obstet Gynecol Reprod Biol. 2016;204:1–4.
55. Tien YW, Hsiao SM, Lee CN, Lin HH. Effects of laser procedure for female urodynamic stress incontinence on pad weight, urodynamics, and sexual function. Int Urogynecol J. 2016; [Epub ahead of print]
56. Pitsouni E, Grigoriadis T, Tsiveleka A, et al. Microablative fractional CO_2-laser therapy and the genitourinary syndrome of menopause: an observational study. Maturitas. 2016;94:131–6.
57. Leibaschoff G, Izasa PG, Cardona JL, Miklos JR, Moore RD. Transcutaneous temperature controlled radiofrequency (TTCRF) for the treatment of menopausal vaginal/genitourinary symptoms. Surg Technol Int. 2016;26(XXIX):149–59.
58. Alexandros Bader. Non Invasive Labia Majora Tightening & Skin Rejuvenation with CO_2 Laser FemiTight.
59. Kaptchuk TJ, Goldman P, Stone DA, Stason DA, Statson WB. Do medical devices have enhanced placebo effects? J Clin Epidemiol. 2000;53:786–92.
60. Gaspar A, Addamo G, Brandi H. Vaginal fractional CO_2 laser: a minimally-invasive option for vaginal rejuvenation. Am J Cosmet Surg. 2011;28:156–62.
61. Michael Gold MD1,2,3,4 | Anneke Andriessen PhD5,6 | Alexandros Bader MD7 |Red Alinsod MD8 | Elizabeth Shane French9 | Nathan Guerette MD10 | Yegor Kolodchenko MD11 | Michael Krychman MD12 | Susan Murrmann MD13 | Julene Samuels14. Review and clinical experience exploring evidence, clinical efficacy, and safety regarding nonsurgical treatment of feminine rejuvenation. JCDI. https://doi.org/10.1111/jocd.12524.

9 Other Lasers in Aesthetic and Regenerative Gynecology

Vidya Pancholia, Krishna Hari,
and Ksenija Selih Martinec

9.1 Erbium Lasers

9.1.1 Introduction and History

The aging of tissue generates laxity, and the vaginal mucosa is no exception. Almost all postmenopausal women have vaginal atrophy, dryness, dyspareunia, and other symptoms associated with menopause.

Vaginal relaxation syndrome (VRS) is a quite common medical condition described as a loss of the optimal vaginal structure and is usually associated with vaginal child delivery and natural aging. Multiple pregnancies and deliveries contribute to a worsening of VRS, as well as the onset of menopause, which causes a decline in hormone levels and vaginal atrophy. VRS results in a decrease or loss of sexual gratification.

The genitourinary syndrome of menopause (GSM) is the new definition for the variety of menopausal symptoms associated with physical changes of the vulva, vagina, and lower urinary tract and is caused by estrogen deficiency. The genitourinary syndrome of menopause (GSM), caused by estrogen deficiency, is responsible for the appearance of symptoms affecting qualify of life, such as vaginal dryness and/or dyspareunia or urinary symptoms, and affects at least 50% of postmenopausal women.

Female stress urinary incontinence (SUI) is a highly prevalent lower urinary tract dysfunction, most commonly affecting middle-aged and elderly women. Although the etiology of SUI is not fully understood, the risk factors for the condition include congenital factors, pregnancy, childbirth, hypoestrogenism, cognitive impairment, obesity, and advanced age. Its prevalence has been estimated to be as high as 40% in European countries and 50% in the USA.

There are many possible nonsurgical and surgical therapies for SUI; however, TVT (tension-free vaginal tape), until recently, has been considered as the gold standard in SUI. Initial therapy involves nonsurgical options such as behavioral changes in terms of diet reduction for overweight patients, smoking cessation, bladder training, and pelvicfloor muscle training (PFMT). Although good results can be achieved with PFMT, long-term improvement is hard to maintain due to lack of training and poor patient persistence. Mechanical devices such as pessaries, vaginal cones, and urethral obturators and electrical stimulation play an integral part in the management of these patients. Drug therapy also may reduce SUI. Surgical procedures are more effective for SUI than nonsurgical therapies, but are sometimes associated with adverse effects and complications, such as bleeding, bladder

V. Pancholia (✉) · K. Hari
Indore, India

K. S. Martinec (✉)
Kallista Medical Center, Domžale, Slovenia
e-mail: ksenija.selih@kalliste.si

P. Jindal et al. (eds.), *Aesthetic and Regenerative Gynecology*,
https://doi.org/10.1007/978-981-16-1743-0_9

perforation, urethral injury, infection, groin pain, and a sexual abstinence period of 6 weeks after surgery. It also has risk of complications of failure of sexual function, discomfort during intercourse, creating excessive obstruction of the lower urinary tract, neuritis of the obturator nerve, and postoperative restrictions. In addition, the recurrence rate for urinary incontinence after surgery ranges from 5.7% to 30–40%.

Noninvasive therapies such as electrical stimulation, radiofrequency, and laser photothermal therapy have been replacing surgical treatment as safer treatment options. Studies have demonstrated collagen remodeling effects of laser irradiation to help strengthen the pelvic floor supportive structures by heating pelvic floor tissue. Many studies have shown erbium: YAG laser therapy to be an effective and safe option for different gynecological applications, such as SUI, vaginal relaxation syndrome, and vaginal atrophy (VA).

In terms of evaluating effectiveness of laser procedures, the areas of genital rejuvenation can be broken down into three distinct areas: vaginal atrophy and laxity, stress urinary incontinence, and uterine/pelvic organ prolapse.The laser procedures can further be categorized as follows: noninvasive vs. invasive. The only noninvasive laser procedure on the market is a non-ablative fractional erbium:YAG laser (IntimaLase and IncontiLase, Fotona). All the other laser wavelengths are invasive and include fractional carbon dioxide (CO_2), fractional ablative Erbium:YAG, and fractional ablative Erbium:YAG combined with nonablative 1470 diode laser. Devices using radiofrequency are also noninvasive.

One of the first reports on the use of erbium:YAG (Er:YAG) laser technology for gynecological treatments dates back to 2000 when Dr. Claudia Pidal and colleagues reported on the use of a Fotona Fidelis erbium:YAG laser(now known as the IntimaLase and IncontiLase, Fotona) for treating vaginal tissue. The results were impressive. The treatment was effective and painless and led to the rapid development of various erbium ablative procedures, including the treatment of human papilloma virus infections, cervical ectropion, vulvar intraepithelial neoplasia, dystrophic lesions, melanosis, and many other conditions. These treatments have also obtained US FDA clearance, and since then thousands of such procedures have been performed in many countries. A high rate of success, with only minor complications, was reported when performing ablation of the lower genital tract, treatment of multifocal and multicentric lesions, excisions, and tissue coagulation. As an interesting and unexpected side effect, many patients reported that they felt a vaginal tightening effect following these erbium:YAG treatments, which resulted also in their enhanced sexual experience. This discovery initiated further research in the direction of developing a minimally invasive, nonsurgical, and nonablative erbium treatment for vaginal relaxation syndrome (VRS).

There is a large spectrum of various VRS treatment options on the market ranging from behavioral (Kegel exercises) through pharmacological therapies (hormonal, tightening creams and sprays) to various more or less invasive surgical procedures. While behavioral and pharmacological therapies are noninvasive and safe, they have limited efficacy. On the other hand, various surgical procedures promise a much better final result at the price of higher associated risks [1]. Surgical procedures require the cutting and rearrangement of vaginal and peripheral tissue in order to reduce the size of the vaginal canal. Operating on or near sensitive vaginal tissue is inherently risky and causes scarring, nerve damage, and decreased sensation. Furthermore, patients require an extended recovery period, and the procedure involves considerable pain, discomfort, risk of infection, and serious complications associated with surgery and general anesthesia, including mortality.

Vaginal relaxation syndrome (VRS) is a quite common medical condition described as a loss of the optimal vaginal structure and is usually associated with vaginal child delivery and natural aging. Multiple pregnancies and deliveries contribute to a worsening of VRS, as well as the onset of menopause, which causes a decline in hormone levels and vaginal atrophy. VRS results in a decrease or loss of sexual gratification. The first studies of the thermal effects of a Non-ablative erbium:YAG on human soft tissue using Fotona IntimaLase with SMOOTH mode were performed by Majaron and colleagues in 2000 and Drnovsek and colleagues in 2004. Based on these findings, the nonablative SMOOTH-mode erbium also began to be used on mucosal tissue. The first experiences in targeting mucosal tissue were intraorally. Application of the SMOOTH mode on the soft palate produces tissue contraction, which is an ideal noninvasive method for treating snoring and sleep apnea complications. The first trials with SMOOTH-mode erbium tightening of the vaginal canal had already begun in 2008/2009. Tightening of the vaginal canal and consequently the improvement of sexual gratification were observed.

These trials also revealed that the SMOOTH-mode erbium results in improvement of stress urinary incontinence (SUI) in women. This resulted in the development of two protocols for two new minimally invasive, nonsurgical, and nonablative procedures: IntimaLase (for VRS) and IncontiLase (for SUI). These protocols are based on heating the vaginal wall mucosa up to approximately 65 °C and include two treatment sessions with a 4–6-week interval. The time needed to execute the IntimaLase protocol is approximately 8 min and for IncontiLase around 15 min.

The wide use of erbium SMOOTH-mode technology brought further discoveries—in 2013 Bizjack-Ogrinc and Sencar reported excellent results achieved in the reduction of pelvic organ prolapse, and in the same year Gaspar presented his pioneering work on vaginal atrophy. The protocol for treatment of pelvic organ prolapse is based on the sample principle of collagen hyperthermia as are the protocols for vaginal tightening and SUI, although with a difference in the treatment intensity (increased) and the location of the major area treated (the prolapsed part of the vaginal wall). This protocol requires three to five sessions at 4–6-week intervals. The number of sessions is dependent on the severity (grade) of the prolapse. The vaginal atrophy protocol is based on a slightly different concept of milder hyperthermia, whereby the mucosa is warmed up to 45 °C, thus causing the stimulation of cell proliferation via heat shock protein activation, an increase of collagen production as well as anti-inflammatory action. This protocol consists of three sessions at intervals of 3 weeks. After immediate tissue shrinkage, the process of neocollagenesis could take up to 6 months to complete.

The vaginal laser received grade I-A level of evidence for improvement of symptoms and tropism for treatment of genitourinary syndrome of menopause (GSM).

9.1.2 Indications for Use in Gynecology (Er: YAG Smooth)

General Surgery Indications: Surgical incision/excision, vaporization and coagulation of soft tissue during any general surgery application, including tissue incision, tissue dissection, excision of lesions, complete or partial resection of internal organs, lesions, tissue ablation, and vessel coagulation.

- Specific Gynecology indications:
  - Herpes simplex
  - Endometrial adhesion
  - CIN (Cervical Intraepithelial neoplasia)
  - Cysts and Condiloma
  - Stress and mixed urinary incontinence (SUI)
  - Vaginal relaxation syndrome
  - Vulvo Vaginal Atrophy (VVA)
  - GenitoUrinary Syndrome of Menopause (GSM)
  - Pelvic Organ Prolapse (POP)
- GenitoUrinary indications:
  - Lesions on external genitilia, anus, urethra, scrotum, penis vulva
  - Polyps and familial polyps of the Colon
- Cosmetic Indications
  - Stretch marks
  - Skin resurfacing ablative and non-ablative

- Skin rejuvenation
- Treatment of scars—Acne, Post Trauma, Episiotomy
- Treatment of keloids, warts, skin tags
- Breast lifting
- Lip enhancement and refreshing
- Cheeks enhancement

9.1.3 Contra-indications (Er: YAG)

- Urinary tract infection (UTI)
- Injuries or bleeding in areas of tissue to be treated (vestibule and anterior vaginal wall)
- One of the following urinary tract abnormalities: bladder overcapacity (<300 cc), post void residual

>50 cc, spastic bladder, vesicouretral reflux, bladder stones, bladder tumors, uretral stricture, and bladder neck contracture

- Abnormal scarring
- Infection or inflammation of treatment area
- Excessive sun exposure (tanned skin)
- A history of a photosensitivity disorder or use of photosensitizing medication
- Pregnancy
- Irradiation in the region of the gonads
- Untreated Diabetes

9.1.4 Understanding the Smooth® Mode

The Smooth® mode of the laser Er: YAG Fotona® (VEL = vaginal erbium laser) is noninvasive (Fig. 9.1) because it acts by pure heating without ablation: it exploits the thermal photo effect of the laser beam on the water present at the mucosal tissue surface [2, 3]. Laser pulses, whose wavelength (2940 nm) corresponds exactly to the vibratory frequency of the water molecules (3000 Hz), lead to the rupture of these; this phenomenon generates the synthesis of reactive oxygen derivatives (ROS), which themselves induce a secretion of heat stress (shock) proteins (HSPs) at the origin of the cell repair process. The heat provided to the tissues also causes a contraction of the collagen fibers, with fibroblastic stimulation and intense neocollagenesis and angiogenesis. These changes result in an overall improvement in the tone and elasticity of the treated tissue (Fig. 9.2).

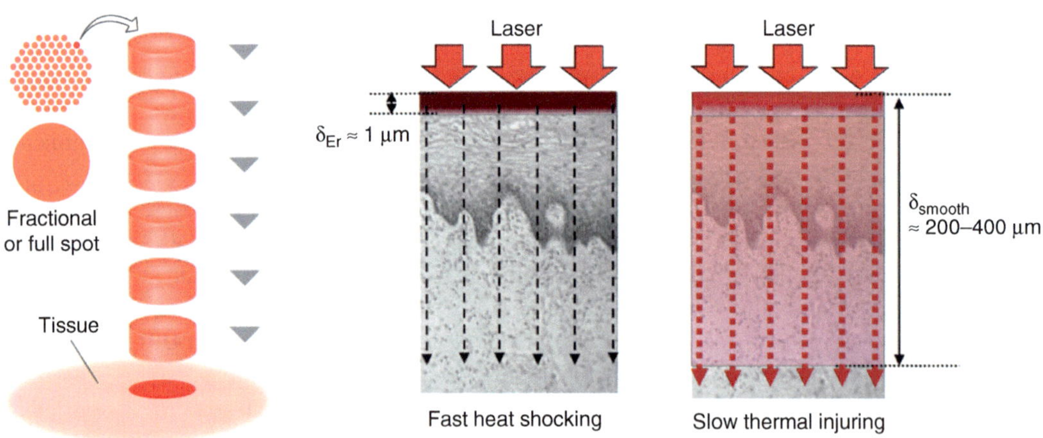

Fig. 9.1 Action of Er: YAG Smooth Laser on vaginal tissue

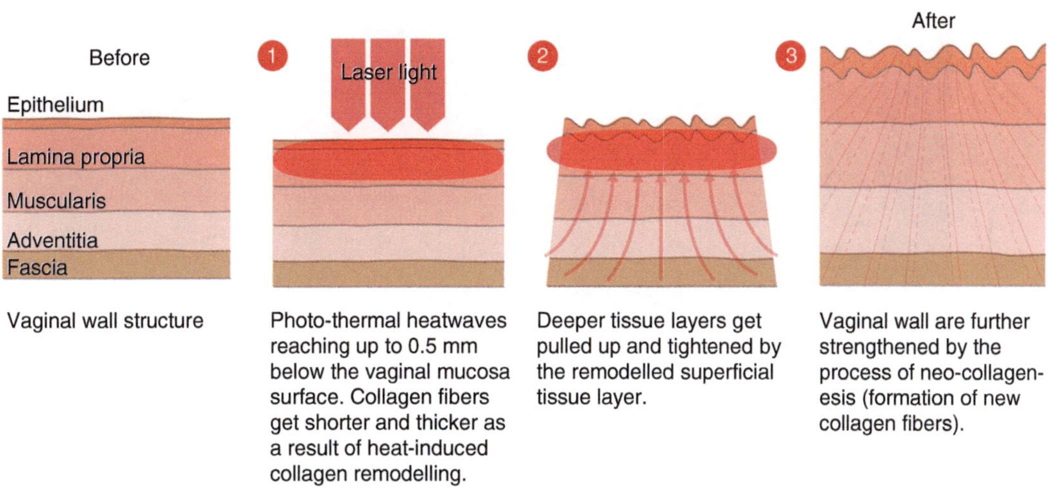

Fig. 9.2 The process of fast heat shocking of the epithelium + slow thermal stimulation of the connective tissue. Action 1: Fast heat shocking—effective and safe high temperature heat shock with superficial "triggering" activates a short exposure biochemical process. Action 2: Slow thermal injuring—involves a long exposure process and "smooth" heating of the deeper lying tissues. Direct slow thermal injury of the connective tissues results in the "conventional" stimulation of fibroblasts

The thickness of the vaginal mucosa varies but it usually measures several hundred microns. For a controlled deposit of heat at its level, an efficient and safe source of heat is required. Smooth® technology is capable of delivering effective stimulating heat over 400 μm in depth without damaging the surface of the mucosa or deeper surrounding tissues. The IntimaLase®, IncontiLase®, RenovaLase®, and ProlapLase® program techniques are based on the non-ablative Smooth® mode of Fotona®. A Smooth® pulse consists of a burst of six low-fluence laser micropulses separated by short intervals to refresh the affected surface, thereby avoiding its ablation. Ablation occurs only if the heat brought to the surface is intense and has enough duration: it is the temperature duration of exposure that determines it. The Smooth® mode allows a surface temperature of over 250 °C to be reached without creating any tissue ablation. Then, the surface heat diffuses toward the depth following a decreasing gradient allowing to obtain the optimal temperature range for neocollagenesis (between 45 and60 °C) up to 400 μm below the mucosal surface (Fig. 9.3).

It may be noted that this combination of very high temperatures and very short duration is unique to the Smooth-mode Erbium: YAG and it sets into action the regenerative process by stimulating the Heat Shock Proteins without actual injury. It is only because of the non-ablative nature of these pulses that the erbium: YAG lasers can be used in both fractional and full beam mode covering 100% of vaginal tissue for a much superior result than any ablative laser whose coverage is limited to only small fraction of the tissue since it causes micro burns and scarring in all the spots that it hits because of the longer duration for which the heat stays on the surface.

9.2 Histological Evidence

Histological examination (hematoxylin-eosin staining) of the vaginal mucosa of patients with GSM symptoms before and after Er:YAG laser treatment shows angiogenesis, increased cellularity, and restoration action at level of the extracellular matrix.

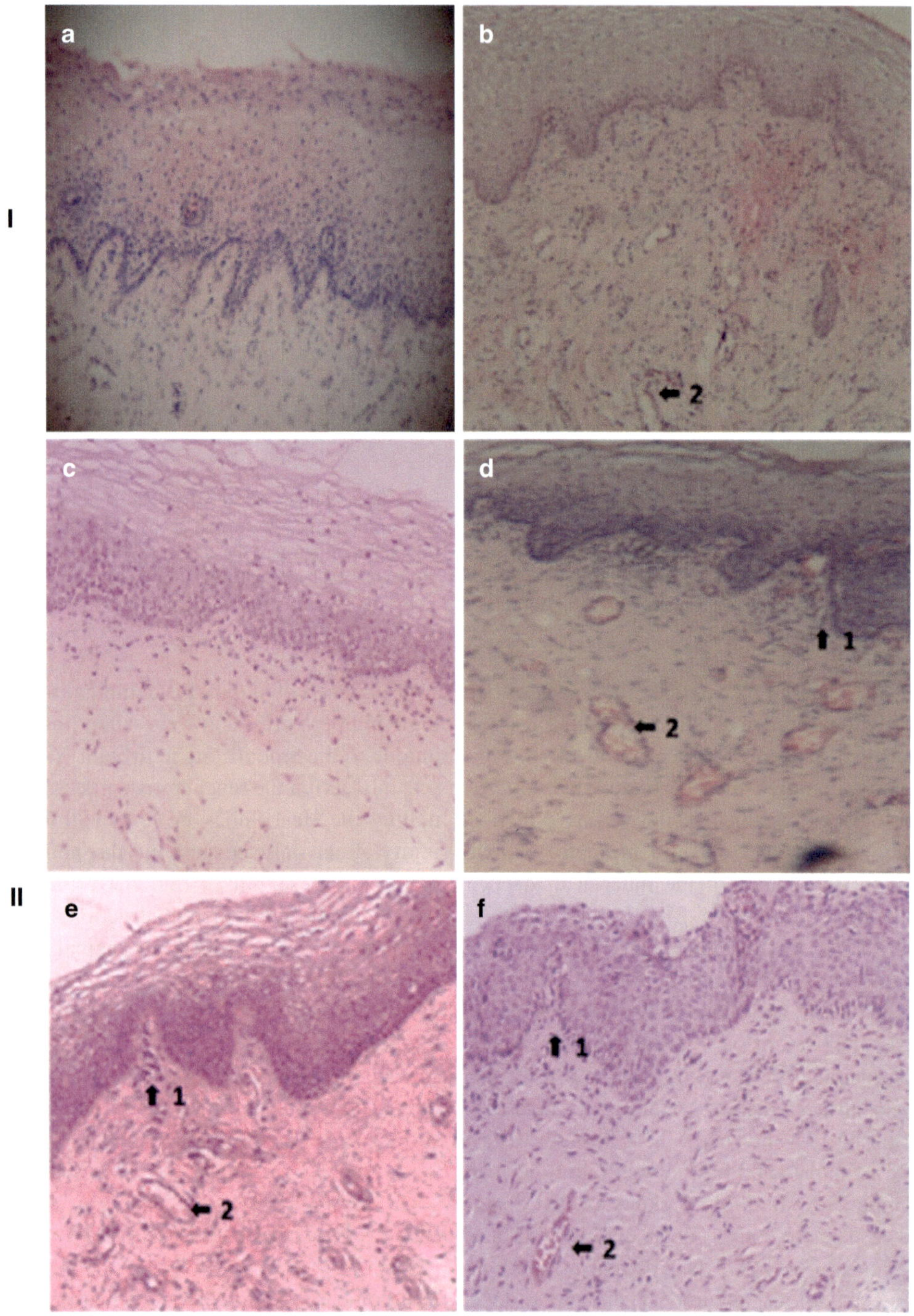

Fig. 9.3 Vaginal histology before and after Er: YAG Smooth Laser VEL [4]: reproduced with permission. (Panel I shows images before treatment (**a**) and 1 month after laser treatment (**b**). New vessel formation can be observed (2). Panel II shows images before treatment (**c**) and 3, 6, and 12 months after laser treatment (**d**–**f**, respectively). Panel I shows images before treatment (**a**) and 1 month after laser treatment (**b**). New vessel formation can be observed (2). Panel II shows images before treatment (**c**) and 3, 6, and 12 months after laser treatment (**d**–**f**, respectively). Panel III shows images at 6 months after laser treatment, displaying increased angiogenesis (**g**) and restorative reaction at the level of the extracellular matrix (**h** and **i**). Panel III shows images at 6 months after laser treatment, displaying increased angiogenesis (**g**) and restorative reaction at the level of the extracellular matrix (**h** and **i**)

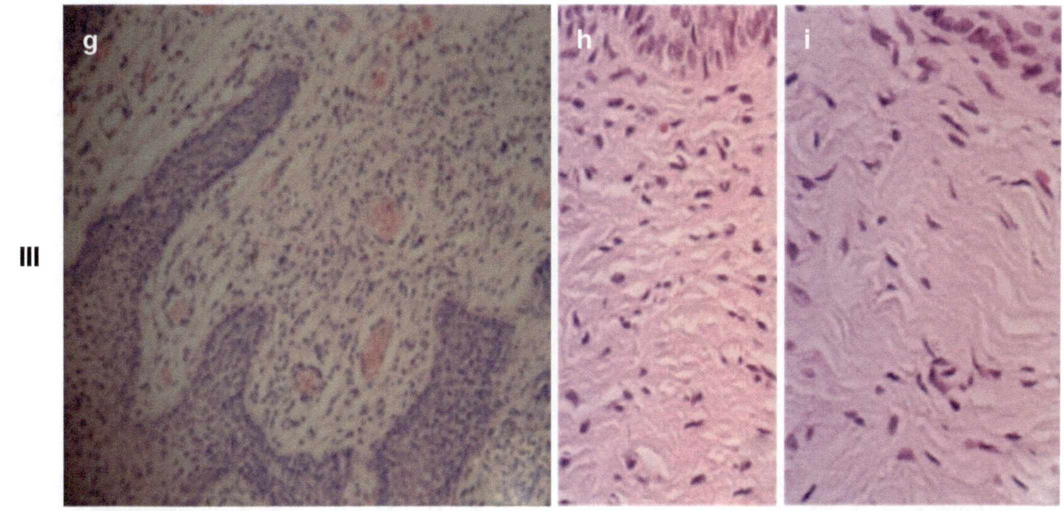

Fig. 9.3 (continued)

Laser Treatment Procedure

The use of the erbium: YAG Smooth laser for the treatment of Stress Urinary Incontinence (SUI). Vaginal Relaxation Syndrome (VRS), Vaginal Atrophy (VA), and Pelvic Organ Prolapse (POP) are a novel noninvasive non-ablative therapy based on photothermal tightening and shrinking of endopelvic fascia and pelvic floor tissue as well as neogenesis of collagen and angiogenesis which help rejuvenate and improve strength, thickness, and health of the treated tissue.

Laser energy from the 2940 nm Er:YAG Smooth laser is delivered to the desired locations of the vagina by using specialized hand pieces and accessories. A big advantage of erbium:YAG is that it can pass through glass and so a reusable glass speculum (shaped like a test tube) is used to completely isolate the patient from coming into contact with the rest of the handpiece and adapters. No anesthesia is required to be administered during procedure as the patient feels very little discomfort during the treatment—(another big advantage of the erbium: YAG treatment). The laser handpiece can be either manually inserted into the speculum or a scanner can be used to completely automate the procedure (Fig. 9.4).

When the laser is activated, it deposits thermal energy in the mucosa tissue of the walls of the vaginal canal covering 100% area with no scarring or burning unlike the ablative CO_2 or Erbium: YAG which in contrast cover only 5%–10% area and cause micro burns. Also, all accessories everything from the speculum to the hand piece adapters are autoclavable. This means that there are no expensive consumable probes or speculum like with other lasers. This greatly decreases operational expenses (Fig. 9.5).

The laser is repeatedly activated as the laser beam is gradually moved across the selected vaginal regions (depending on the indication) until the entire area is treated.

Laser Therapy for Stress Urinary Incontinence (SUI) in patients with mild to severe Stress Urinary Incontinence and Mixed Urinary Incontinence (MUI) as well as for treatments up to third degree of Pelvic Organ Prolapses involves laser shots to all the 3 regions (a,b,c) and multiple passes depending on severity and patient health.

In case of Vaginal Relaxation Syndrome or Vaginal Atrophy, laser shots are only used on two of the 3 regions (b & c) with multiple passes depending on severity and patient health.

Recovery time after an erbium: YAG treatment is very short—the patient may return to her daily routine immediately, except that it is recommended to respect standard precautions connected with relapse, such as avoiding efforts which may cause pressure to the treated area or bladder and to avoid sexual activities for one week.

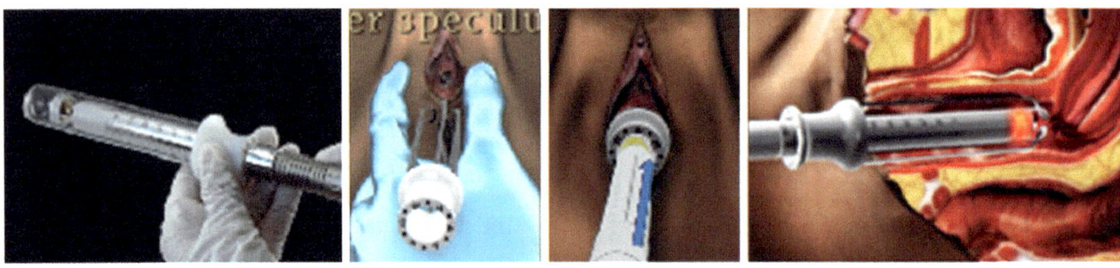

Fig. 9.4 Glass speculum, inserting speculum into patient, inserting adapter into speculum, Laser treatment on the walls of the vaginal canal

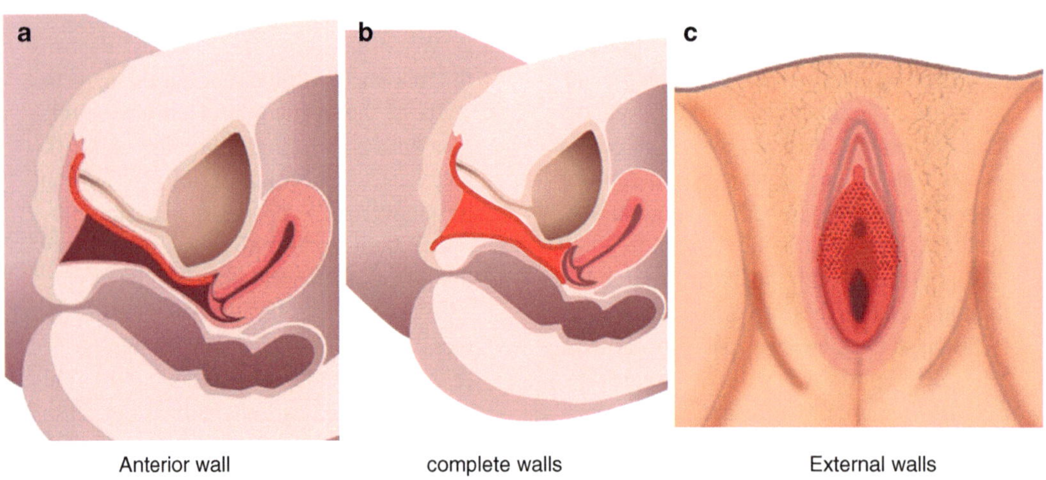

Fig. 9.5 Selected regions for laser treatment. (**a**) anterior wall, (**b**) complete walls, (**c**) external area

9.3 Conclusion

There is enough evidence now to show that the non-ablative erbium:YAG laser is an efficacious and safe new method for the treatment of vaginal laxity, stress urinary incontinence, pelvic organ prolapses, and vaginal atrophy and that it will soon become the treatment of choice for all of these indications and many more. Erbium:YAG Smooth Laser Treatment is least invasive than any other Laser or Energy based treatments or traditional surgical methods for treatment of these indications. It has a much lower complication rate, is well tolerated by patients. Moreover recovery time is very short—the patient may return to daily routine immediately.

Acknowledgment The material presented in this chapter is based on research carried out in collaboration with the EU regional Competency Center for Biomedical Engineering (www.bmecenter.com), and other research coordinated and verified by Laser and Health Academy (www.laserandhealthacademy.com).

Diode Laser in Aesthetic and Regenerative Gynecology

Ksenija Selih Martinec (Slovania)

It was in 1917 when Einstein proposed the process that makes laser possible. He theorized that electrons could be stimulated to emit light of a particular wavelength. But it took almost 40 years to make this genius idea true. 13.11.1957 acronym LASER (light amplification by stimulation of Radiation) by Gordon Gould was used for the first time. In May 1960 Theodore H. Maiman constructed the first laser. Charles J. Campbell was the first to use laser in human patient, ruby laser to destroy a retinal tumor. In 1973 CO_2 laser was first applied in gynecology for treating cervical neoplasia by Dr. Kaplan from Israel.

In September 2011 Adrian Gaspar first time published the use of fractional CO_2 laser for vaginal rejuvenation [1]. In the same year Fotona introduced Er: YAG laser for minimally invasive treatment of SUI, GSM, and relaxed vagina. In 2012 pixelated CO_2 laser entered the field. After this there was expansion of using both CO_2 and Er: YAG laser in the field of functional gynecology.

But great minds never stops. Dr. Ziv Karni, an industrial pioneer in the field of laser and energy technologies introduced Diode laser in the field of functional and regenerative gynecology.

In functional and regenerative gynecology we are treating soft tissue so the main chromophore we want to reach is water. There are three types of laser used in this field: Er: YAG CO_2 and Diode 1470, all three are absorbed in water but each with different absorption coefficient and different penetration depths. The erbium laser wavelength coincides with the strongest absorption peak of the water molecule, thus resulting in very high absorption in virtually all biological tissues [5]. CO_2 laser is 15 times less absorbed in water that Er: YAG and diode 1470 much less. So Diode 1470 laser has the deepest penetration depth. Er: YAG and CO_2 laser are ablative lasers, diode is nonablative. Ablation is not needed. Manufacturers resolved this by very long pulses in Er: YAG, CO_2 partly resolved this by pixelating laser beam.

Since Erbium and CO_2 have high absorption in tissue, deep heating is done by conduction (slow and inefficient). Diode laser penetrates deep and heat by electromagnetic radiation (quick and efficient).

With these physics characteristics, Diode 1470 emerged in the field of gynecology, proctology, EVLA-endovenous Laser Ablation, Laser Liposuction, ENT, and more.

For intravaginal treatments the wavelength is brought by 600 micron Radial Emission fiber and 5 mm Glass tube. It emits energy in a 360 degree pattern. Because the wavelength is transmitted by electromagnetic radiation, the contact with mucosa is not needed, so a very thin handpiece is used that is suitable also for very narrow vagina. The thin handpiece is inserted into vaginal canal and inside it the probe is moving. This helps to homogenously affect the vaginal wall and eliminates need to move the handpiece what results in painless treatment. The Diode 1470 wavelength combined with the controlled radial emission of the handpiece has a bio modulating effect that stimulates neocollagenesis and regenerates the epithelium and the connective tissue.

Mechanism of action: A diode laser emits an infrared light that penetrates the deep penetration and deep heating in the tissue. The treatment is non-ablative. The result is toned tissue and a thickening of the vaginal mucosa. The solid state 1470 nm vaginal laser working with sub necrotic temperatures aims to influence collagen remodelling restoring the natural composition of a pelvic floor tissue [6].

Treatment protocol: 3 sittings 3 months apart, repeat one sitting every second year. No anesthesia required.

Treatment outcome: Improvement in sexual function, dyspareunia, bladder function, vaginal sensation and lubrication.

Adverse effects: Mild warmth at vaginal introitus. No significant side effects reported.

Main indications are: GSM—genitourinary syndrome of menopause, SUI—stress urinary incontinence, recurrent vaginal infections, Postpartum rehabilitation, VVS—vulvar vestibular syndrome, and FSD—female sexual dysfunction.

Diode 1470 is used also for treating: Asherman, polyps, myomas, dismorphic uterus and T shape uterus, septum, isthmocele, retained conception products.

Bibliography

1. Gaspar A, Addamo G, Brandi H. Vaginal fractional CO_2 laser: a minimally invasive option for vaginal rejuvenation. AJCS. 2011;28 (3). https://doi.org/10.1177/074880681102800309. First Published September 1, 2011
2. Fistonić I, Štefica FG, Fistonić N. Minimally invasive laser procedure for early stages of stress urinary incontinence (SUI). J Laser Health Acad. 2012 (2012). ISSN 1855-9913.
3. Diaci J, Filipic C, Perhavec T, Lukac M. Influence of water absorption shift on ablation speed of Er:YAG and Er,Cr:YSGG dental lasers. J Laser Health Acad. 2012 (1); 67–74

References

1. Mirzabeigi MN, Jandali S, Mettel RK, Alter GJ. The nomenclature of 'vaginal rejuvenation' and elective vulvovaginal plastic surgery. Aesthet Surg J. 2011;31:723–4.
2. Jha S, Wyld L, Krishnaswamy PH. The impact of vaginal laser treatment for genitourinary syndrome of menopause in breast cancer survivors: a systematic review and meta-analysis. Clin Breast Cancer. 2019;19:e556–62.
3. Pergialiotis V, Prodromidou A, Perrea DN, Doumouchtsis SK. A systematic review on vaginal laser therapy for treating stress urinary incontinence: do we have enough evidence? Int Urogynecol J. 2017;28:1445–51.

4. Gaspar A, Brandi H. Non-ablative erbium YAG laser for the treatment of type III stress urinary incontinence (intrinsic sphincter deficiency). Lasers Med Sci. 2017;32:685–91.
5. Salvatore S, Nappi RE, Zerbinati N, Calligaro A, Ferrero S, Origoni M, et al. A 12-week treatment with fractional CO_2 laser for vulvovaginal atrophy: a pilot study. Climacteric. 2014;17:363–9.
6. US Food and Drug Administration (FDA). Statement from FDA Commissioner Scott Gottlieb, M.D., on efforts to safeguard women's health from deceptive health claims and significant risks related to devices marketed for use in medical procedures for "vaginal rejuvenation". FDA; 2018. https://www.fda.gov/NewsEvents/Newsroom/PressAnnouncements/ucm615130.htm.

Radiofrequency in Aesthetic and Regenerative Gynecology

10

Francesco Merelli and Bruno Boccioli

10.1 Introduction

10.1.1 Biological Interaction Rationale (Critical and Bibliographical Analysis)

We know that in menopause women experience many anatomical changes, thinning of vaginal epithelium, reduction of vaginal secretion, lack of vaginal fold and reduction of the turnover of collagen, low urinary tract atrophia. Furthermore after vaginal delivery we observe what we call "vaginal loose syndrome" and sometimes pelvic organ prolapse. To solve all these problems in the recent years, the use of radiofrequency became more common.

Radio frequency produces energy in the radio wave range. This energy can be used for therapeutic purposes, by changing the physical parameters of the RF generator it is possible to obtain different interactions with the biological sublayers. Therefore, the therapeutic performance of the method is related to the adequate choice of all the electro-physical variables that characterize the EM field used. Normally radio frequency is used to generate endogenous heat in the tissues which, depending on its characteristics, evokes certain therapeutic effects.

F. Merelli (✉)
Arezzo, Italy

B. Boccioli
Perugia, Italy

10.2 History

At the end of the 1960s, Frolich stated that living matter would be controlled by an electromagnetic wave system or a wave and particle phenomenon that travels at the speed of light in a vacuum [1–3].

In the 70s Popp said that there would be photons (detectable only by photomultiplier) that would act as intracellular (involving organelles and cell) and intercellular information carriers [4].

Biological systems would be EM oscillation generators with broad spectrum frequency emission from a few Hz up to 1015 HZ (ultraviolet radiation) [5, 6].

The irradiation of tissues with electromagnetic fields transmits to them an informational message through neurotransmitters and signal molecules (cytokines, growth factors, hormones, etc.) (Popp and Prigogine) [7–10].

The electromagnetic field (EMF) is characterized by the variables [5]:

- Power.
- Frequency (or wavelength [ƛ] where ƛ = c / f).

The tissue administration of electromagnetic fields determines a change in energy at the molecular, cellular, and atomic levels [1, 11].

P. Jindal et al. (eds.), *Aesthetic and Regenerative Gynecology*,
https://doi.org/10.1007/978-981-16-1743-0_10

This mechanism changes the homeodynamics of the organism by decreasing its entropy and causing energy, biological, hormonal, and immune changes [12].

The changes induced by electromagnetic fields in athermic mode can be summarized as follows [1, 3]:

- Stimulation of mitochondria with increased ATP production.
- Increase in metabolism.
- Increased synthesis of DNA and RNA.
- Activation of macrophages.
- Increase in neutrophils and fibroblasts [13].
- Stimulation of cytokines and growth factors [13].
- Regulation of cortisol production.
- Stimulation of the formation of natural opioids.
- Increased activity of the great defense system [14–17].

The organism is a unit that integrates a multiplicity of "elements." The "coordination" center of this integration is represented by the BASIC REGULATION system (Pischinger and Heine) [16, 17].

The context on which the action takes place is represented by the matrix within which everything takes place in the fractal dimension with binary characteristics (1–0/all-nothing).

Each part of our body contains information about everything else and the sum of the parts from a result superior to the individual parts.

With an RF treatment, energy is transmitted to the "target" tissues. These tissues are made up of cells and molecule; energy is transferred to the atoms of which they are made [14, 18, 19].

The energy transmitted to the atom increases the energy of the electrons which are made of. This determines an "orbit jump" of the electron itself which has acquired a higher energy charge (energy is measured in eV [electron volts]) [20].

It is known that only an electron whose position in space is not known can rotate on an orbit (Heisenberg's uncertainty principle) [21].

Having taken on an amount of energy greater than that characteristic of the orbit, the excited electron can no longer remain on the orbit of origin and transfers to an orbit characterized by greater eV.

The alternation of polarities (positive half-wave and negative half-wave), characterized by the frequency of the RF, is not able to keep the electron at the "higher" energy level, consequently the electron returns to the primitive orbit, yielding energy.

This energy will be radiated externally in the form of a photon and, more precisely of a biophoton (Pop) [7, 22–24].

From this it follows that the EM irradiation of a tissue determines the formation of biophotons on it.

The biophoton is, therefore, a quantum of "ultra-weak" light energy deriving from an orbital jump of an electron from a higher energy orbit to one characterized by a lower energy.

The biophoton has, at the same time, the characteristic of wave and particle, it propagates also in vacuum, it has no energy loss, travels at the speed of light and carries information, it has no mass or electric charge, it is subject to the force of gravity, if interacts with matter acquires the particle characteristic and its speed decreases.

A cell has the ability to emit specific electromagnetic signals since our body is "an open system of a dissipative nature," that is, capable of taking energy from the outside and transferring waste products (Prigogine) outside it [25, 26].

The state of health is determined by a state of flow that oscillates between constant values capable of maintaining the homeodynamics of the organism. When this state of flow is altered, disease is generated (Von Bertalanffy).

Specific biophotons correspond to specific information which varies from tissue to tissue and which is transmitted both through chemical intermediaries (signal molecules) and through physical means such as biophotons that always precede chemical neurotransmitters.

The physical characteristics that differentiate the biophotons giving them specificity are [27–29]:

- The shape of the wave.
- The amplitude of the wave.
- The frequency of the wave.
- The wavelength.
- The consistency of the wave.
- The redundancy of the wave.
- Wave interference (Compton effect).

Coherence means the property of a beam of electrons all in phase with each other, connected, directed towards the same target, and all using the same conduction medium [9, 10, 30].

Redundancy is the ability of the human body to implement control systems capable of correcting any damage resulting from negative information deriving from errors in the formation, transmission, reception, or decoding of the information itself.

Interference, or Compton effect, occurs when two coherent waves, overlapping each other, have the same wavelength, the same amplitude, and the same direction but are staggered over time.

There can be constructive interferences with which coupled cells divide and destructive interferences with which the opposite occurs.

According to Popp, the electromagnetic field (EMF) plays a fundamental role in the life of living beings [31].

Heisenberg says that the fundamental force on which life depends is due to ability of EM to modify the kinetic energy at the atomic and molecular level.

Again according to Popp, the cell can be compared to a "transceiver unit" that emits biophotons at the level of nuclear DNA through which all parts of the body are connected, which are felt even at a great distance when there is a variation at the level of some of them (fractal structure). The disease would be caused by a disturbance of this stimulus transmission system caused by any external or internal agent including the field electromagnetic (EMF) which, depending on its characteristics, can be both pathological and therapeutic [25, 26].

In a pathology the first alteration would be EM type where it is not yet detectable symptoms. In a second phase there would be an alteration of the chemical transmission followed by the appearance of clinical symptoms [7, 8].

The communications take place quickly because they do not only use the key/lock system, certainly very slow and uncertain, but generally prevails the EM recognition which is established between information and the membrane receptor.

The information passes through the matrix that is altered by it in its constitution and functionality [13, 32].

The extracellular fluid of the matrix is regulated by the SNV (Vegetative Nervous System), by the vascularization of the matrix and by the signal proteins [15].

At the same time the PH, hydration, permeability of capillaries and cell membranes are regulated. Sequences of biophotons pass through the matrix, consisting of a series of sub-harmonics which, together, form the "information carrier" signal [27].

The membrane receptor intercepts this signal, breaks it down into its sub-harmonics of which it uses only those corresponding to its resonance frequency. All the others are abandoned within the matrix and are not taken into consideration.

This characterizes the specificity of the information.

The membrane receptor is the glycocalyx. It is a glycoprotein formation, carrying a negative electric charge and provided with an extension (cytoskeleton) which, passing through the nuclear membrane, reaches the nucleus. The glycocalyx has a shape that resembles that of an antenna through which it resonates with the information and captures it and transmits it to the nucleus [33].

Nuclear DNA has monofilament (enzymes) or bifilament (nucleotides) molecules that are normally rolled up to save energy and to not expose their electrical charges [34, 35].

Under the photonic stimulus these structures unfold and make contact with the substrate by decoding the message and retransmitting it to the matrix using the cytoskeleton again [36].

Huge quantities of biophotons are produced at this stage which allow the message to propagate as long as the excitatory stimulus remains [27, 37].

Once the decoded information has again reached the matrix, it will be collected by a neurotransmitter that will pass it to the motor plate, then to the nerve and, finally, to the final effector.

The extracellular matrix activates the chemical part of the transmission of the message through immune antibodies, hormones, cytokines, growth factors, enzymes, neuro modulators assuming the conformation of liquid SOL that is favorable to the passage of the message [18, 38, 39].

It is a formation that incorporates various types of lipoproteins from the environment, conforming them to a hollow hyperboloid that functions as a conducting thread towards the final target.

It is a structure of very short duration that is composed at the time of the passage of the message and then immediately disappears and reforms when the next message arrives [40].

In the event of a disease state, a part of the cell membrane, more or less large depending on its severity, is not normally polarized.

In this case, the information cannot pass inside the core because the electrical conditions do not allow it and, therefore, there are great difficulties in transmitting the message [15].

The electromagnetic field (EMF), in its therapeutic variant, has the task of restoring the normal resonance between information and glycocalyx, returning the polarization of the cell membrane to normal [41].

10.3 Conclusions

The athermic part of the therapy carried out by EM radiation can be summarized as follows;

- Create photonic beams that are the basis of all the phases inherent in the message.
- Restoration of the normal polarization of the cell membrane [13, 19, 42].
- Restoration of the resonance between information and glycocalyx.
- Restoration of the normal formation, transmission, reception, and decoding of the message.
- Maintain the SOL state of the matrix [27].
- Creation of ATP at mitochondrial level.
- Increase in enzyme metabolism.
- Increase in DNA and RNA formation [43].
- Activation of macrophages.
- Stimulus and activation of fibroblasts.
- Formation of cytokines and growth factors (signal molecules).
- Formation of cortisol.
- Formation of natural opioids.
- Activation of the great defense system.

10.3.1 Action Rational in Neuro-Muscular Tissues

The causes of functional decrease of the various organs/tissues macroscopically can be attributed to:

- intrinsic or chronological factors,
- or extrinsic or environmental factors.

Depending on the typology of the aforementioned "organs/tissues," many authors have hypothesized the incidence percentages that we summarize on average:

- 30–40% for intrinsic or chronological factors
- 70–60% for extrinsic or environmental factors

In skin aging, for example, the so-called PHOTO AGING (induced by exposure to UV radiation) can represent an extrinsic factor that probably exceeds even 70% of incidence [44].

In this context, ENERGY STIMULATIONS induced by adequate radiofrequency electromagnetic fields can exert a "bio-stimulation" action on certain "target" tissues [22, 24].

The generated photons "transport" specific information to specific targets on which they can evoke increases in fibroblastic growth factors (FGF), vascular epithelial (EGF) (ESAF) and others capable of stimulating the pathological and functional anatomical recovery of systems more or less compromised by intrinsic and/or extrinsic factors.

The process is not only a local linear phenomenon, linked to a cause–effect, but it represents a non-linear, multiple connection process that interacts with very low energy levels if compatible and adequate [45, 46].

Different biological systems can be involved:

- the matrix which, with its electrically dipolar aqueous content, is sensitive to electromagnetic fields [47].
- the cell membrane which, with its receptors, is not only capable of receiving the information

wave but performs a "simulation-amplification" function of the information received [48, 49];

- all endocellular components such as microtubules, microfilaments, cytoskeleton, and microfibrils.
- It is underlined how the different cellular areas enunciated above involve all the organs in a "continuum" that reaches the whole biological system by propagating information with a fractal scheme whereby each "single" part communicates with the "whole."

In this context, microtubules represent a central element capable of receiving, storing, processing, and transmitting energy that acts as a "carrier" of information (note: all structures are ubiquitous within the cells) [50].

On these organelles, the ELECTROMAGNETIC RADIOFREQUENCY FIELDS exert their "therapeutic action" by conditioning the flow of calcium ions between the outside and the inside of the cell via the calcium channels [51].

This element is a transducer that functions as a second messenger and as a neurotransmitter [35].

As a positive ion, it has an essential function in maintaining the membrane potential through which the information reaches the nucleus for its decoding and performs the function of coenzyme in various chemical reactions [52].

In the biostimulation process, calcium stimulates the formation of fibroblasts at the muscle level, and consequently of myofibroblasts that play a fundamental role in the treatment of muscle hypotonia [53].

The striated muscle fiber is composed of a significant amount of myofibrils surrounded by a connective tissue which takes the name of sarcoplasmic reticulum whose purpose is to accumulate the calcium necessary for contraction. On the other end, myofibrils have a protein structure organized into [38, 54]:

- thin filaments composed of actin,
- thicker filaments consisting of myosin.

These two types of filaments are parallel and superimposed in a spiral pattern.

Actin fibers are anchored to other transverse protein formations called Zeta lines which take adhesion with the connective tissue that forms the endomysium.

During contraction, the actin and myosin filaments slide over each other by pulling on the zeta line which, attached to the collagen fibers of the myofasc, transmit the tension to the tendons, muscles, and finally to the bones.

Myosin filaments have a head at one end. These heads are protrusions that bind to actin, tilt, causing contraction and release, resulting in the release of muscle fiber.

At each inclination the myosin shifts, compared to the actin of about 4–10 nanometers.

Electromagnetic radiofrequency fields stimulate biological tissues through information that travels in the form of a conformational wave that is able to assemble this complex structure and, at the same time, activate the calcium channels which is the fundamental element for transmitting the message inside of the cell.

At the same time, they act at the mitochondrial level, determining the formation of that amount of ATP necessary for the whole process to take place.

In the event of contracture or muscle hypertonus, this occurs because the ionic flow of calcium is not regular. Electromagnetic radiofrequency fields tend to restore the normal functioning of the calcium channels.

In the mechanism of contraction it seems that a determining role is explicit.

By solitons, which are wave formations capable of transmitting a signal through a substrate in a stable way, even over long distances, without dispersion or corruption.

The solitons were studied by the Ukrainian Davydov and are believed to derive from the energy produced by the hydrolysis of ATP.

In the mechanism of muscle contraction it seems that solitons play a role in the rotation of myosin heads.

The calcium ions that reach the myosin heads initiate ATP hydrolysis with the formation of solitons.

The solitons move the myosin heads with a flexion movement, which in turn causes a swelling of this region causing the actin filaments anchored to the zeta line to move, thus causing contraction.

The contraction of the muscle, according to Davydov's hypothesis, arises from the kinetic energy of the solitons that propagates along the myosin filaments that roll over those of actin.

Electromagnetic radiofrequency fields can assume an important function in the mechanism of muscle contraction.

As is known, they increase the production of ATP at the mitochondrial level with the consequent passage of ATP to ADP and formation of solitons starting from the energy released by hydrolysis.

10.3.2 Rationale to Use Different Types of Electromagnetic Energy (Multifrequency, Capacitive, Resistive, Monopolar, Bipolar)

The raise of tissue temperature is the main distinguishing element of the effects of radio frequency on biological tissues.

1. TEMPERATURES between 39 and 43 degrees centigrade generate BIOSTIMULATION effects (commonly defined as "athermia") [55, 56].
2. TEMPERATURES between 63 and 75 degrees centigrade give rise to PROTEIN DENATURATION processes if thermal STRESS remains limited (time and/or surface), the open protein spontaneously returns to its biologically active native form (renaturation) [57].
3. TEMPERATURES above 90 degrees centigrade lead to a DRYING of the tissues and immediately afterwards CARBONIZATION effects occur (irreversible phenomena) [55, 56].

All the above cases are used in medical practice to obtain THERAPEUTIC effects in first and/or second intension [58].

These effects are directly related to an adequate choice of all the electro-physical variables that characterize the ELECTROMAGNETIC FIELD used [13, 59, 60].

- FREQUENCY of emission.
- TYPE OF APPLICATOR.
  - Monopolar.
  - bipolar,
- TYPE OF APPLICATOR.
  - Capacitive.
  - Resistive.

10.4 Bipolar Radiofrequency

10.4.1 Frequency

It is a characteristic element of electromagnetic fields that characterizes the number of oscillations in the unit of time (Fig. 10.1).

The unit of measurement is expressed in Hz (hertz) [1KHz = 1000 Hz, 1 MHz = 1,000,000 Hz].

In biological tissues in vivo a lower frequency corresponds to a deeper localization of "heat" vice versa a higher frequency corresponds to a more localized "heat" localization.

10.4.2 Monopolar/Bipolar Applicator

The MONOPOLAR mode involves the use of a "working" electrode and a "reference" plate which has the purpose of closing the circuit (Fig. 10.2a).

The effects on the tissues are exerted in the "portion" between the two electrodes and the increase in temperature proportionally involves all the sectors involved.

In BIPOLAR mode, the transducer implements both electrodes, the electromagnetic field is concentrated in a position very close to the electrodes themselves (Fig. 10.2b).

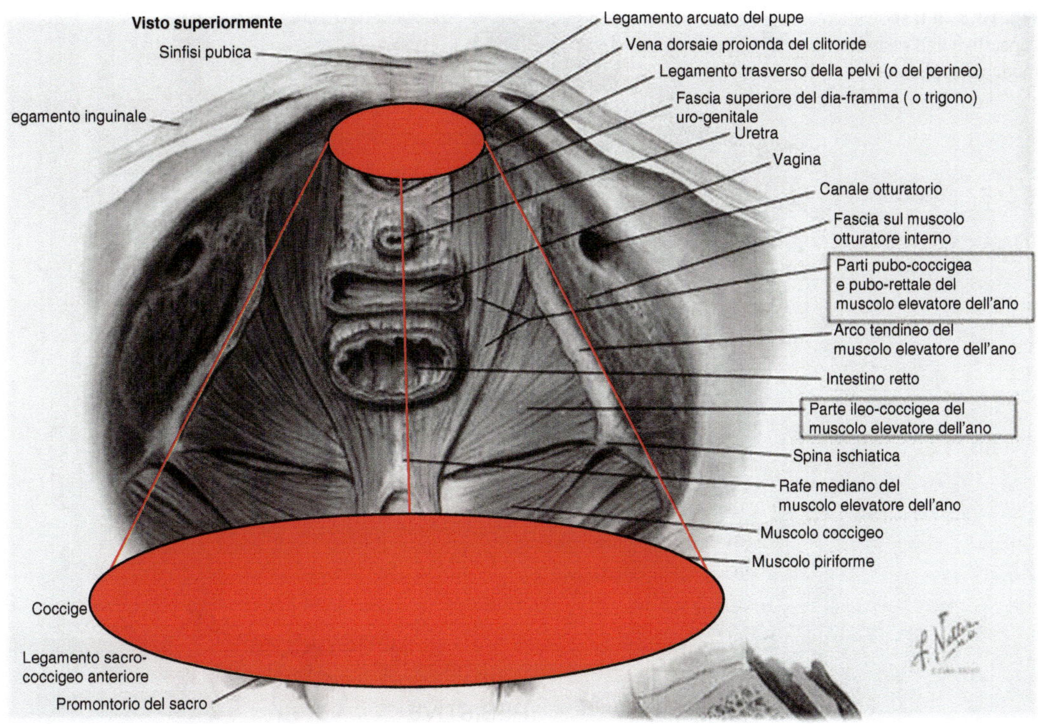

Fig. 10.1 Mechanism of monopolar radiofrequency

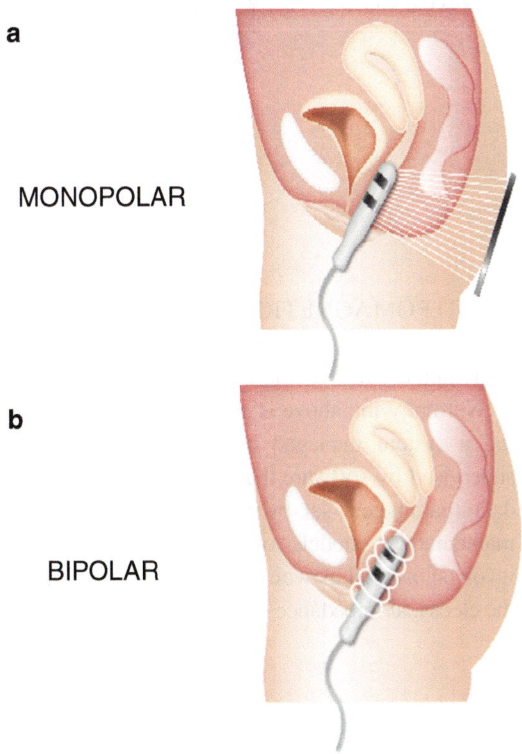

Fig. 10.2 a,b shows monopolar and bipolar radiofrequency

10.5 Monopolar Radiofrequency

10.6 Resistive/Capacitive Applicator

The CAPACITIVE system inserts a "capacitor" into the patient circuit (Fig. 10.3a).

This configuration is indicated 0 on "soft" tissues with a high content of liquids and, therefore, with low impedance.

In RESISTIVE mode (Fig. 10.3b), in the patient circuit the condenser is replaced by a "resistance," resulting in an action on high connective tissue, bone, nerve, tendon, and structures such as joint capsules, articular cartilages, fibrotic cords, keloids, and scars.

Obviously the physical parameters of the energy stimulus must be modulated (adapted) to the subject and the district being treated taking into account the "primary" pathology as well as the various interactions with any "secondary" pathologies and concomitant therapies [8, 61, 62].

Fig. 10.3 **a**,**b** shows capacitive and resistive mode of applicator

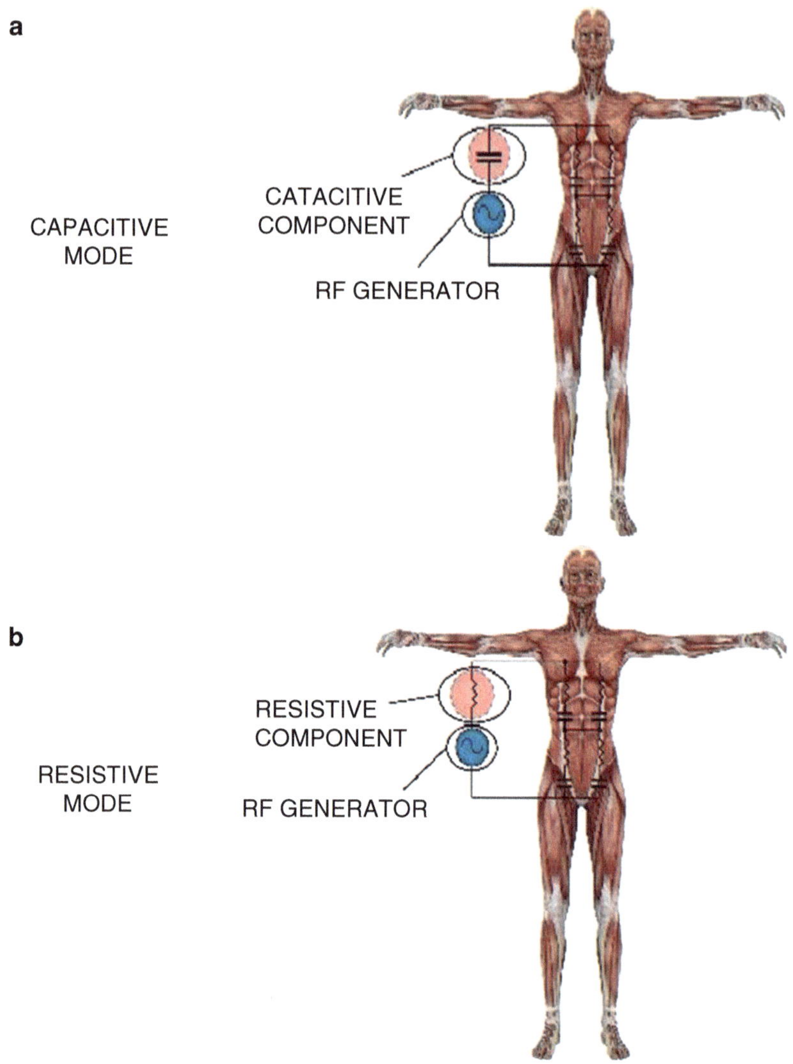

The "first" objective is to obtain the maximum therapeutic yield with the minimization of unwanted side effects.

For the purpose of correct therapy, it will also be important and necessary to objectively evaluate the electrical impedance values of the treated tissues that follow dynamic and non-definable variants.

Considering the above premises, the need for the Operator to have available all the electro-physical variables characterizing the ELECTROMAGNETIC FIELD to finalize a specific therapy for the specific electrical impedance value of the Patient.

Normally the above is achieved by implementing a system managed by microcontrollers and supervised by software that interact with the input data (parameters set by the Operator), with the mathematical models of the biological tissues involved, as well as with the dynamic variations of the electrical impedance of the tissues themselves.

In a nutshell, we tend to calculate with the maximum possible accuracy the "MINIMUM EFFECTIVE DOSE" of the characteristics of the parameters for a personalized therapy [2, 22, 25, 63–65].

10.7 Radiofrequency Treatment Protocol

- No anesthesia required.
- The device can exert its therapeutic effect on thr skin or mucosa surface at temperature ranging from 39 °C to 42 °C.
- usually RF device has been embedded safety system to prevent burns,
- Frequency used ranging from 1 MHz to 500 Hz to target deep tissues.
- Patient feels typically warming sensation at this temperature.
- One treatment takes around 12 minutes, six minutes each side.
- 4 treatments are recommended at an interval of 4–6 weeks.
- No significant side effects reported.
- Indications:

  Radiofrequency has been shown to be very helpful in treatment of GSM of menopause, loose vagina syndrome after spontaneous delivery, first and second degree genital prolapse, exerting its biological effects through several mechanisms overall the stimulation of collagen and elastin synthesis.

References

1. Adey WR. Biological effects of radiofrequency electromagnetic radiation. Loma Linda – California: Veterans Administration; 1988.
2. Alexander CM, Werb Z. Extracellular matrix degradation. In: Hay ED, editor. Cell biology of extracellular matrix. 2nd ed. New York and London: Plenum Press; 1991. p. 255–304.
3. Anemuller H. Das Grunddiat – system. Stuttgart: 2 Aufl. Hyppokrates; 1983.
4. Oberley LW. Superoxide Dismutase and Cancer. 244; Vol 2; 127–167.
5. Bergonié J. Des applications de diathermie comme ration energetique d'appoint. Comptes rendus de I'Acade.uie des Scienoes (2. decembre 1912), Band 155, S.1171.
6. Prigogine I. Vom sein zum Werden. Munchen und Zurich: Piper; 1979.
7. Bistolfi F. Campi magnetici in medicina. Torino: Minerva Medica; 1986.
8. Bistolfi F. Radiazioni non ionizzanti. Ordine, disordine e biostrutture. Minerva Medica; 1989.
9. Del Giudice E, Preparata G. Macroscopic quantum coherence. World Scientific; 1998. p. 108–29.
10. Frholich H, Kremer F. Coherent excitations in biological systems. Springer; 1983.
11. Bossy J. Basi neurobiologiche delle riflessoterapie. Milano: Masson Italia; 1977.
12. Spaggiari P, Trebbia C. Medicina quantistica. 2nd ed. Tecniche Nuove; 2005.
13. Bistolfi F. Inquadramento generale delle interazioni fra campi magnetici e materia vivente. Classificazione degli effetti. 27:249–63.
14. Heine H. Manuale di medicina biologica. Regolazione di base e matrice extracellulare. Fondamenti e sistematica – Guna editore; 1999.
15. Heine H. Manuale di medicina biologica. Regolazione di base e matrice extracellulare. 3rd ed. Guna; 2007.
16. Pischinger: Das System der Grundregulation. 7 Aufl. Haug. Verlag, Heidelberg, 1989.
17. Pischinger A. Das System der Grundregulation. Grundlagen fur eine ganzheitsbiologische theorie fur Medizin. 4 Aufl. K. F. Haug Verlag, Heidelberg, 1975.
18. Riva Sanseverino E. Interazioni ipotizzabili fra campi elettromagnetici ELF e membrane biologiche. 27:339–50.
19. Bordier H. Diathermie et Diathermothérapie. J.B. Baillière et fils 1922 – Paris.
20. Mustacchi G. Magnetoterapia. Apparecchi, tecniche di applicazione e risultati clinici. 27:375–86.
21. Kim YS. Some possible effects of static magnetic fields on cancer. TIT J Life Sci. 1976;6:11–28.
22. Bistolfi F. L'ipertermia in terapia oncologica. Piccin; 1981.
23. Bistolfi F. Radiazioni non ionizzanti. Moderni impieghi ed aspetti radio protezionistici. Torino: Edizioni Minerva Medica; 1983.
24. Bistolfi F. Campi magnetici e cancro. Torino: Edizioni Minerva Medica; 1985.
25. Gay S, Miller EJ. Collagen in the physiology and pathology of connective tissue. Stuttgart: G. Fischer Verlag; 1979.

26. Goodman R, Chizmaden Y, Shirley Henderson A: Electromagnetic fields and cells- J Cell Biochem, 51, pagg. 436–441, 1993.
27. Clegg JS. Intracellular water, metabolism and cell architecture: parte 2. In: Frohlich H, Kremer F, editors. Coherent excitations in Biological systems. Springer; 1983. p. 162–77.
28. Clegg JS. Properties and metabolism of the aqueous cytoplasm and its boundaries. Am J Phys. 1984;246:R 133–51.
29. Del Giudice E, Doglia S, Milani VG. Structures connection and electromagnetic interaction in living matter. In: Frolich H, editor. Biological coherence and response to external stimuli. Springer; 1988. p. 49–64.
30. Frholic H. Biological coherence and response to external stimuli. Springer; 1988.
31. Popp FA. Nuovi orizzonti in medicina. La teoria dei biofotoni. Tr. It. IPSA. 1985.
32. Grattarola M, Caratozzolo F, Chiabrera A. Interaction of ELF electromagnetic fields with cell membrane receptors. In: Grandolfo M, Michaelson SM, Rindi A, editors. Biological effects and dosimetry of static and ELF electromagnetic fields. Plenum Press; 1985. p. 273–94.
33. Wong H.L., Wahl S.M.: Inflammation and repair. In Sporn M.D., Roberts A.B.: Peptide growth factors and their receptors. Vol 2°, Springer Verlag, New York, Berlin 1991; Pagg 509–548.
34. Nigam VN, Altri. Glycoglycans and proteoglycans in neoplastic tissues and their role in tumor growth and metastasis. In: Varma RS, Altri, editors. Glycosaminoglycans and pro teoglycans in physiological and pathological processes of body system. Basel: Karger; 1982. p. 354–71.
35. Heines J. Regulationsmedizin - ein Ausweg aus der chronischen Krankheit naturamed 11 (1996) 30–39.
36. Iozzo RV. Cell surface heparinsulfate proteglycan and the neoplastic phenotype. J Cell Biochem. 1986;37:71–8. Family history study. Am. J. Psychiatry. 1985;142(4):441–6
37. Buselli P, Spaggiari PG. La terapia trans dermica: studio dell'efficacia di un sistema di somministrazione farmacologica con campi elettromagnetici pulsati. Europa medico – Fisica. 2004;40(Suppl. 1):3.
38. Siliprandi T. Biochimica medica, strutturale, metabolica e funzionale – Piccin. Padova: Nuova Libraria; 2005.
39. Szent Gyorgyi A. Bioelectronics and cancer; electronic desaturation of proteins in normal and cancerous tissues. 310:109–113.
40. Schlitter HE. Krebs als Folge unspezifischer primar – chronischer Funktionslahmung vegetative regulierter matrix. Natur und Ganzheitsmed. 1990;13:36–44.
41. Bistolfi F, editor. Altri.: Campi magnetici in medicina. Biologia, diagnostica, terapia. Minerva Medica; 1986.
42. Welch GR, Barry MN. Long range energy continua in the living cell: protochemical considerations. In: Frohlich H, Kremer F, editors. Coherent excitations in biological systems. Springer; 1983. p. 95–116.
43. Iozzo RV. Biology of disease. Proteoglycans: structures, function and role in neoplasia. Lab Ivest. 1985;53:337–96.
44. Yuan Z, Huan S, Pan Z. Magnetic field and tumour. IEEE Transact on magnetic. 1980;16:824–6.
45. Kittlick PD. Glycosaminoglycans. Exp Pathology Suppl. 1985;10:1–174.
46. Kraus W. Magnetfeldtherapie und magnetisch induzierte Elektrostimulation in der Orthopadie. Orthopadie. 1984;13
47. Cancedda R, Tacchetti C. Culture d'organo: crescita e differenziamento di cartilagine in vitro. Ruolo della matrice extracellulare. Lezione al 5° Corso "Culture cellulari in Oncologia" Scuola superiore di oncologia e scienze biomediche, S. Margherita.
48. Oegema TR. Junior: delayed formation of proteoglycan aggregate structures in human articular cartilage diseases states. Nature. 1980;288:583–5.
49. Xi YP, Altri. Age related changes in normal human basement membrane. Mech Age Develop. 1982;19:315–24.
50. Biologically closed electric circuits. Amer. Roentg. 1985;145:447–67.
51. Wolkowski Z.W. Interaction of non-ionizing electromagnetic radiation with living systems. Atti dell'International Symposium on Wave Therapeutics. Versailles, 19/20 Maggio 1979, 139–158.
52. Krimm S, Dwivedi AM. 407. Citato da Widder Originale in Science. 1982:216.
53. Zanolla R. Magnetoterapia vs ultrasuoni e vs laser terapia nel trattamento riabilitativo dei blocchi algici scapolo omerali dopo interventi chirurgici. Congresso Nazionale SIDSI. Fratta Terme, 30 maggio. 1 giugno 1986.
54. Belot J. A propos des dispositifs de diatermie. Arch. d' électr. méd. Nr. 316.
55. Bergonié J. La diathermie ration d'appoint. Paris mèd. 1913 Nr 5.
56. Bergonié J. La diathermie ration d'appoint. Societé francaise d'éelectrothérapie et de radiolgie médicale. Ref. Arch. d'électr. méd. Janvier 1913, Nr 349.
57. Bistolfi F. Strutture diamagnetiche anisotrope e centri paramagnetici. Basi per una magnetosensibilità cellulare differenziata. 27:273–86.
58. Bistolfi F, Olzi E, Asdente M. Targeting magnetotassico di farmaci e radio farmaci in terapia oncologica. In 26; 57–86.
59. Bergsmann O. Grundsystem, Regulation und Regulationsstorung in der praxis der rehabilitation. In: Pischinger A, editor. Das system der Grundregulation. Heidelberg: 8 Aufl.Karl F. Haug Verlag; 1990. p. 89–139.

60. Robert B, Robert L. Aging of connective tissue. General considerations. In: Robert B, Robert L, editors. Frontiers of matrix biology, vol. 1. Basel: Karger; 1973. p. 1–45.
61. Bordier H. Diathermie et Diathermothérapie. Paris: J.B. Baillière et fils.
62. Burnstock G. Autonomic neuroeffector junctions. Reflex vasodilatation of the skin. J Invest Dermatol. 1977;69:47–57.
63. Bergonié J. Des applications de diathermie comme ration énérgetique d'appoint. Cpt. Rend. Tom. 155 pag. 1171–2 decembre 1912.
64. Bistolfi F, Olzi E. Prospettive di concentrazione magnetotossica di elementi paramagnetici in terapia oncologica. Atti del Convegno: "IL progresso tecnologico in radioterapia oncologica". Varese 28–29 maggio 1984. Abstracts. p. 66–67.
65. Heines J. Regulationsmedizin ein Ausweg aus der chonischen Krankheit naturale. 1996;11:30–9.

11 Carboxytherapy in Aesthetic and Regenerative Dermatology

Karin Staab

11.1 Introduction

Carbon dioxide (CO_2) therapy refers to the transcutaneous administration of CO_2 for therapeutical purposes. Historically, CO_2 therapy started in the year 20 B.C. (discovered in France) and rediscovered in the Middle Ages. Since 1932, this was used scientifically in France at the RoyatSpas for the treatment of patients affected by obliterating arteriopathies (CO_2 was diluted in water and given transcutaneous). The use of CO_2 inside the body is not new; from the year 1914, CO_2 was used in Radiology as a contrast element for kidney studies and in laparoscopic surgery since 1924, because of its incombustible properties. The transdermal technique is still used for arterial disease, chronic wounds, and Raynaud Syndrome with good results [1–5].

In this chapter, intradermal and subcutaneous administration of CO_2 is discussed.

CO_2 administration proved to be effective not only in improving local parameters of circulation and tissue perfusion, but also in inducing a partial increase in transcutaneous PO2. Such might be due to a hypercapnia-induced rise in capillary blood flow, a drop in cutaneous oxygen consumption, or a right shift of the Oxyhemoglobin dissociation curve (Bohr effect) as shown in Fig. 11.1. As such, the effect of carbon dioxide therapy on the microcirculation, and the probability of a positive effect upon the physiological oxidative lipolytic process, led to the use of this gas in the treatment of localized adiposities. Some authors have furthermore demonstrated the coexistence of an increase in subcutaneous localization of fat with alterations in blood and lymphatic drainage [6].

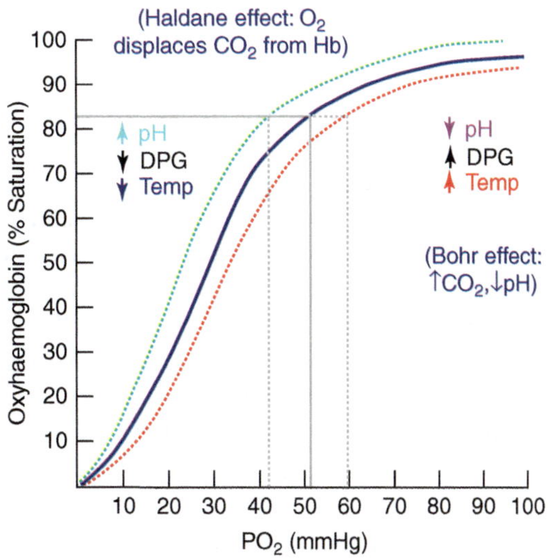

Fig. 11.1 Oxyhemoglobin dissociation curve. Bohr effect

When injected inside or under the skin, CO_2 stimulates vasodilation and neoangiogenesis. The hypercapnia induced-rise in capillary blood flow is produced, accelerating metabolic changes

K. Staab (✉)
Universidad de Chile, Santiago, Chile
e-mail: kstaab@vtr.net

P. Jindal et al. (eds.), *Aesthetic and Regenerative Gynecology*,
https://doi.org/10.1007/978-981-16-1743-0_11

and the lipolytic process. Cellular trophism is improved since interstitial CO_2 enhances tissue oxygenation by forcing oxyhemoglobin to liberate O2 (Bohr effect). There is a decrease in fibrosis [6] that can be perceived in successive treatments [1, 7]; this has a stimulating effect on the dermis [4, 8]. Histologically, the injection of CO_2 into subdermal adipose tissue has a lytic effect on adipocytes respecting vascular and nerve structures as demonstrated by Brandi et al. in 2001 and Balik et al. in 2011 [6, 9].

Berthier introduced CO_2 in the treatment of Cellulite in 2000 [10, 11]. Brandi published in 2001 the effects of Carboxytherapy in adipocytic lysis [6], and in 2004 its tensor effect [12]. In 2008, carboxytherapy was found to produce collagen synthesis [4]. There is also the mechanical undermining by the gas flow similarly like in needle subcision (e.g., in the treatment of scars, adhesions, or severe cellulite depressions). There are effects of mechanical tension on the cells (especially in rejuvenation) and pressure (especially in the treatment of cellulite or adiposities) resulting from relatively strong gas flow during the CO_2 administration. Presumably there is also some slight influence of temporary acidosis [13]. When injected intradermal, thickening of the dermis and rearrangement of collagenous fibers is shown to be increasing skin elasticity, as demonstrated with histological studies by Brandi et al. [6] shown in Figs. 11.2 and 11.3. Changes are probably due to improved microcirculation and oxygenation of the tissue [14]. This effect leads to tightening the skin and decreasing flaccidity.

Therefore, intradermal and subcutaneous administration of CO_2 (Carboxytherapy) is being used as a medical non-invasive or minimally invasive treatment for many inaesthetic conditions [10, 11].

Carboxytherapy is a safe procedure and is a good alternative in the removal and treatment of various skin and subdermal defects and inaesthetics [2]. CO_2 is drained into de blood vessels and eliminated by the lungs; while a little portion is converted into carbonic acid and is eliminated through the kidneys [10].

11.2 Treatment

Required material:

- CO_2 pressurized bottle/tube for medical use.
- Pressure regulator.
- Carboxytherapy device.
- Particle filter.
- Connection tubes.
- 30G–31G 13 mm and 4 mm long needles
- Remember what is required for intradermal/subdermal sterile technique: gloves, 95° Alcohol/2% Chlorexidine

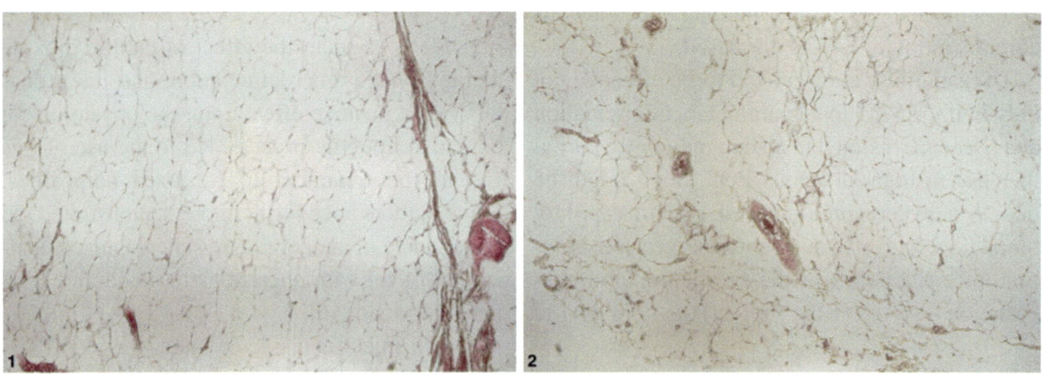

Fig. 11.2 Histology of the subcutaneous adipose tissue (from Brandi et al. [6] with permission). (1) Histological features of the subcutaneous layers before CO_2 treatment. (2) Histological appearance of the subcutaneous layers after CO_2 treatment, showing lysis of the adipocytes not involving the vascular structures

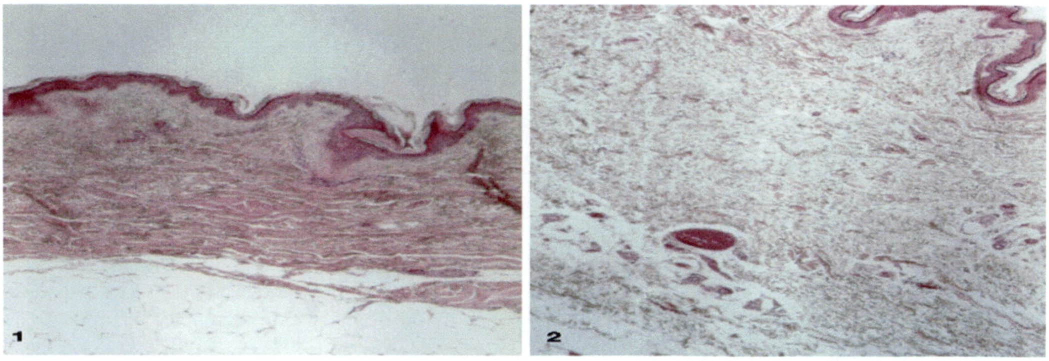

Fig. 11.3 Histological features of the Dermis (from Brandi et al. [6] with permission). (1) Histological appearance of the dermis before CO_2 treatment. (2) After CO_2 treatment, thickening of the dermis and rearrangement of collagenous fibers is shown

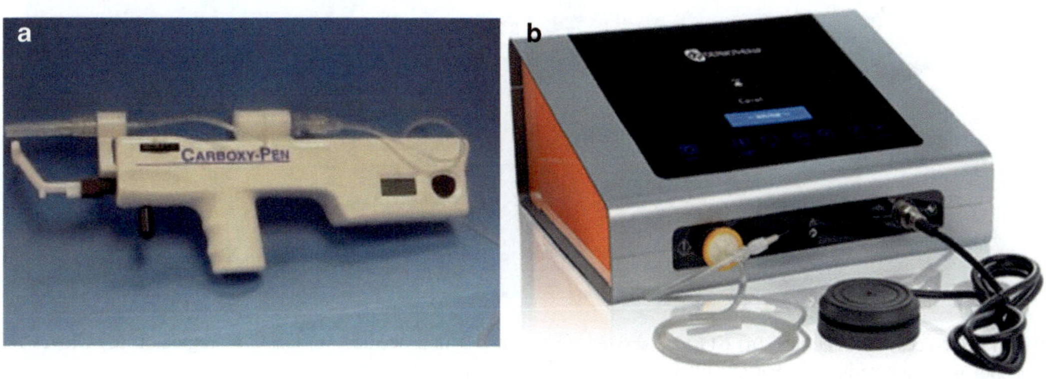

Fig. 11.4 Carboxytherapy Devices. (**a**) Carboxipen is a manual mesotherapy gun for injecting CO_2. (**b**) Coral is an automatic device

There are currently many Carboxytherapy devices available; some of them are manual or gun-like devices for mesotherapy (Fig. 11.4). For aesthetic medicine a very precise tool is required with the ability to apply very different volumes of the gas with different flows depending on the indication, state, and the compliance of the patient. The author prefers automatic devices as shown in Fig. 11.5. The CO_2 flow should be adjusted manually, as well as the gas temperature and the volume. Some devices include preloaded therapeutical programs and there are many brands and worldwide availability. Prices are affordable.

11.3 Technique

First of all, there has to be a full clinical evaluation, diagnosis, a detailed assessment of the skin, and the inaesthetics that has to be treated. As shown in Fig. 11.6, the treating physician has to define the volume to be used according to the pathology and deepness to be treated. Always warm the gas (usually the machine is automatically set to 28 °C). Then choose the right flow, according to the area and patient's personal and local sensitivity. The environmental air of the system should always be purged, so that it con-

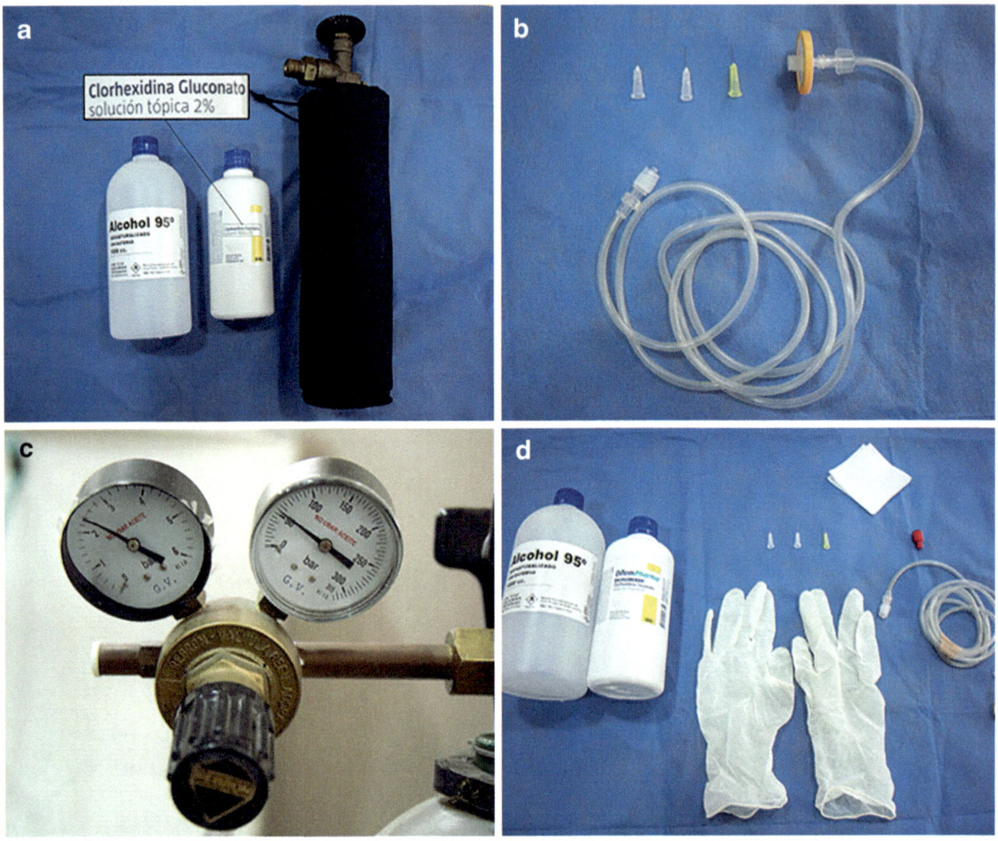

Fig. 11.5 Necessary materials. (**a**) CO_2 pressurized tube. Alcohol and Clorhexidine. (**b**) Particle filter and sterile tube for injection. 30G 1/2-inch (13 mm) needle. 31G 13 mm and 4 mm needles. (**c**) Gas flow regulator. (**d**) Gloves for asceptic technique

tains only CO_2. If the tubing system contains ambiental air, too much inflammation may occur.

Never use local anesthesia, since it is not required and may disturb the skin before the treatment. No topical anesthesia is required, neither electrical nerve stimulation (TENS) [15] or cold pads.

The needle should be introduced with a 45-degree angle into the skin with a sterile technique as shown in Fig. 11.7. For intradermal treatment, the angle should be less. The needle has to be attached through the connecting tube to the Carboxytherapy device. Only then, the treating physician is ready to push the button or pedal and start the injection of the gas.

2000 cc is the maximum dose. Higher than 2 Liters, patient may feel dizzy, because of hypercapnia. Volumes to be used depend on the pathology or inaesthetics.

It is best to do treatments every 2 days depending on the diagnosis. Most patients will get skin improvements after 2–3 weeks of treatment. When treating subdermal fat, best results are obtained in 10 weeks.

11.4 Precautions

Certain precautions should be taken when performing Carboxytherapy treatments.

- It is a medical procedure, minimally invasive and should be performed by a physician.
- Always use skin-approved antiseptics.

Fig. 11.6 Treatment settings. (1) Select mode (preset settings or manual settings). (2) Warm gas: always. (3) Set flow. (4) Select dosage system. (5) Set volume to be injected. (6) Always purge tube system. (7) Device is ready to do treatment

- For sensitive skin (face, periorbital), prefer 2% Chlorhexidine.
- Do not allow massage after the procedure for 8 h.
- Patient can return to normal activities after the procedure.
- Compression stockings or sashes are desirable.

11.5 Advantages

The treatment should be done by a trained physician; not by a non-medical professional. Furthermore, this procedure cannot be performed in a non-medical facility. It is a minimal invasive treatment, completely outpatient with fast sessions (no more than 20 minutes), and small areas can be treated without anesthesia. If done properly, there should be no side effects other than those described below, and they are transient.

One of the greatest advantages is that residual adiposities, such those left after a liposuction, can be treated and the body contour can be regularized.

The periocular area is of concern because of flaccidity, wrinkles, and dark circles. Carboxytherapy is a very good tool to treat these inaesthetics.

On the other hand, carboxytherapy devices are inexpensive and require little maintenance. CO_2 is also an inexpensive gas.

11.6 Disadvantages

Carboxytherapy treatments have 3 disadvantages. First of all, many sessions are required. Some patients may feel impatient. It is important to set real expectations on the clinical results of this minimal invasive treatment. Treating patients with very high expectations is not recommended.

Few patients dislike some of the side effects, even though they are mild, transient, and well tolerated by the great majority of patients.

Most of the devices do not require maintenance, but for some automatic devices technical support available is needed.

11.7 Side Effects

Side effects are mild, transient, and well tolerated by the vast majority of patients.

- Crackling sensation in the area of injection that lasts for 1–2 h.
- Pain at the site of injection (about 75% of the patients); it is handled by heating the gas up to 28 °C and decreasing the flow.
- Hematomas at the site of injection that may last up to 4 days and are not of aesthetic concern.

There are some clinical conditions where carboxytherapy is contraindicated as shown in Table 11.1.

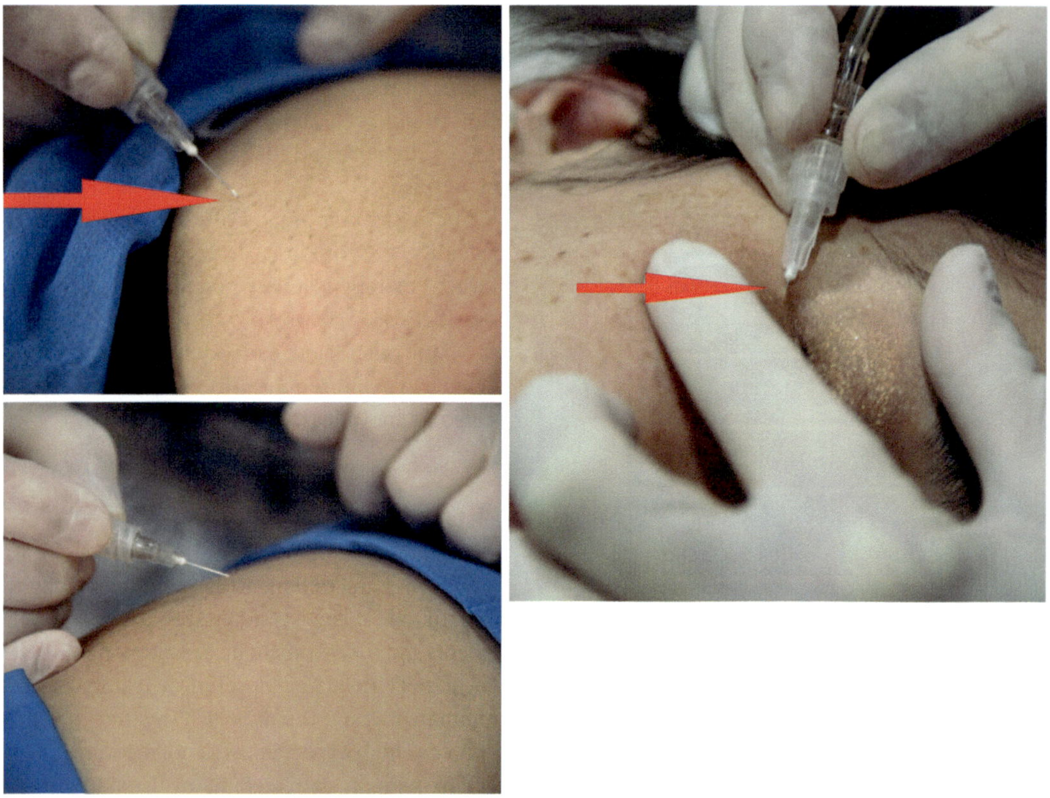

Fig. 11.7 Needle insertion techniques. For subdermal treatment, a 45° angle is best. For intradermal treatment an acute angle is required. Note that in the periocular area the tip of the needle can be seen beneath the epidermis

Table 11.1 Contraindications

• Severe respiratory insufficiency
• Severe renal failure
• Chronic congestive heart failure
• Patients treated by carboanhydrase inhibitors (e.g., acetazolamide, diclophenamide)
• Severe anemia
• Chronic liver insufficiency with decrease of plasmatic protein levels
• Gaseous gangrene (Clostridial infection)
• Pregnancy
• (Breastfeeding)
• Infection on the site of treatment
• Psychiatric disorders

11.8 Complications

Complications are rare. Some infectious complications may be possible; there is a Mycobacterium atypical infection after mesotherapy described by Herreros in 2008 [16]. Also, massive subcutaneous emphysema after carbon dioxide mesotherapy is described by Calonge et al. in 2012 [17]. There is no history about who and where the procedures were done. If 2000 ml in one session are exceeded, dizziness may appear. It is transient and just requires 30 minutes rest.

Avoiding complications is mandatory for any aesthetic treatment. The procedure has to be performed by a trained medical physician in a medical procedure ward [18].

11.9 Aesthetic Indications of Carboxytherapy

1. Skin Rejuvenation. Flaccidity (facial, neckline, or corporal).
2. Periorbital area: wrinkles or dark circles.
3. Cellulite.
4. Localized adiposities.
5. Stretch marks.

6. Chronic wounds.
7. Skin grafts.

11.9.1 Flaccidity; Skin Rejuvenation

Because of its effect on dermal collagen, CO_2 may be used in any area of the body [11, 19, 20]. With multiple punctures and low CO_2 doses separated by a few centimeters [10]. With a 31G needle and doing a little emphysema with 5–15 cc CO_2 at every puncture. Better to use lower gas flow (50–70 ml/minute) and intervals of 15 days for each session. Skin of the face, neckline, or corporal can benefit from this treatment as shown in Figs. 11.8 and 11.9.

11.9.2 Periorbital Area

In this area the greatest inaesthetics are wrinkles and flaccidity [8, 18, 21, 22]. With a 31G needle with the bezel facing up, the emphysema will be very visible as shown in Fig. 11.10. It is not necessary to use more than 80–100 cc in each periorbital area. Low flow of CO_2 is better for this area (50 ml/minute). The emphysema will be visible for hours; and in very sensitive skin, erythema might last 2 days.

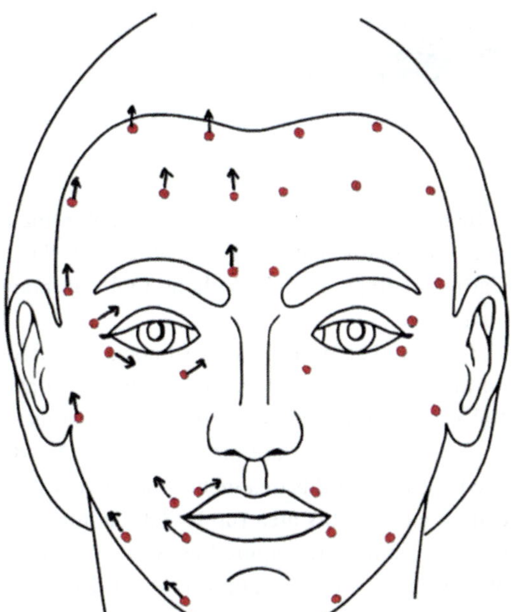

Fig. 11.8 Scheme of application of Carboxytherapy for facial rejuvenation—modification after J.C. Lopez (Sao Paulo, Brazil). It is not necessary to inject in all these points and it depends on the state of the skin, observed gas spreading, and compliance of the patient

Dark circles are of concern in this area, too. This can be of congenital cause (ethnic), but the most cases are from iron deposits due to chronic periorbital inflammation. Because of the vasodilation and increased capillary flow that occurs due to local hypercapnia, these dark circles can be treated. The mechanical effect produced by the gas flow puts pressure on the iron deposits and pushes them into the dilated capillaries. Usually good results can be achieved in 3 weeks of treatment.

11.9.3 Cellulite and Gynoid Lipodystrophy

Cellulite is an irregular alteration of the skin surface giving orange or cottage cheese appearance. It happens in 90% of women, independent of ethnicity. In advanced stages can be painful. Histologically there is fibrosis, lymphedema, and fat globules found in the subdermal tissue. This condition worsens with aging, because flaccidity of the skin and subdermal tissue [8, 23, 24].

Carboxytherapy is used among other treatments, because of its unique effect on the subdermal adipose layer [23, 25, 26]. Its infusion into the affected site produces mesolipolysis and aims to remove cellulite and improve skin texture [27, 28].

- Improves blood and lymphatic flows, which facilitates drainage of the retained liquid.
- Improves the tone of the skin, which restores elasticity and counters the typical sagging in Gynoid Lipodystrophy (GLD).
- Reduces fatty deposits and, consequently, the orange appearance of the skin.
- Fights fibrosis because it improves blood microcirculation and restores skin elasticity

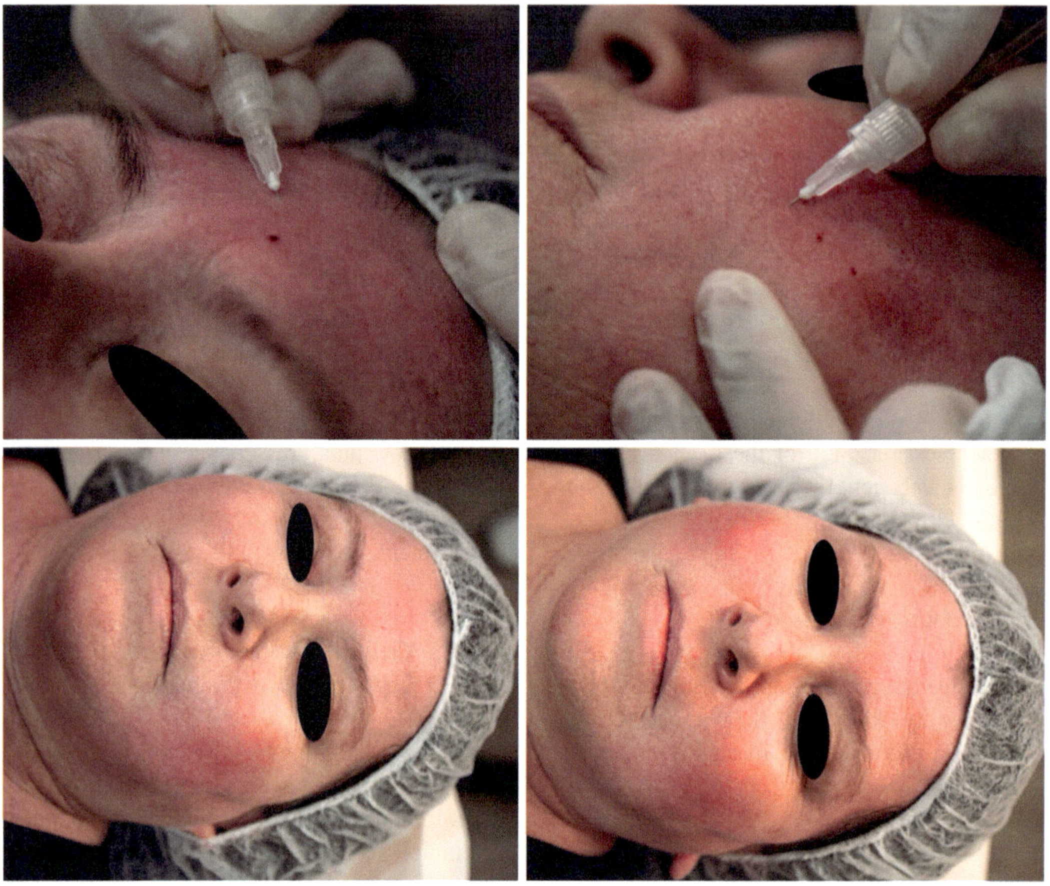

Fig. 11.9 Face treatment. 31G 4 mm needle is used. Note the deepness of the dermal injection. Use 10 to 15 ml in each puncture. Little emphysema can be seen after every puncture

(decrease of the characteristic dimples of the skin with cellulite).

Carboxytherapy is effective to treat buttocks, thighs, abdomen, and local adiposities of the knees [10, 26, 27] as shown in Fig. 11.11. The volume to be injected is higher than other treatments; 800–1000 ml each side. Gas flow may be high: 100–120 ml/minute, depending on patient tolerance.

11.9.4 Localized Adiposities

Local adiposities may be inaesthetics in any part of the body (Fig. 11.12). It affects young and old patients; and residual adiposities may remain after a liposuction surgical procedure [12]. Carboxytherapy is an excellent tool for body contouring.

Carboxytherapy produces a lipoclastic effect on fat tissue, as demonstrated by Brandi [6] and Balik [9] without effects on the blood vessels or nerves and it is a secure procedure [29]. A 30G needle is best for this treatment, since higher flow can be used in this area (up to 200 cc/min) and higher volumes are required as well; depending on the size of the adiposity 200–1000 cc. Remember to never use more than 2 Liters in one session, since the patient may get dizzy due to transient hypercapnia.

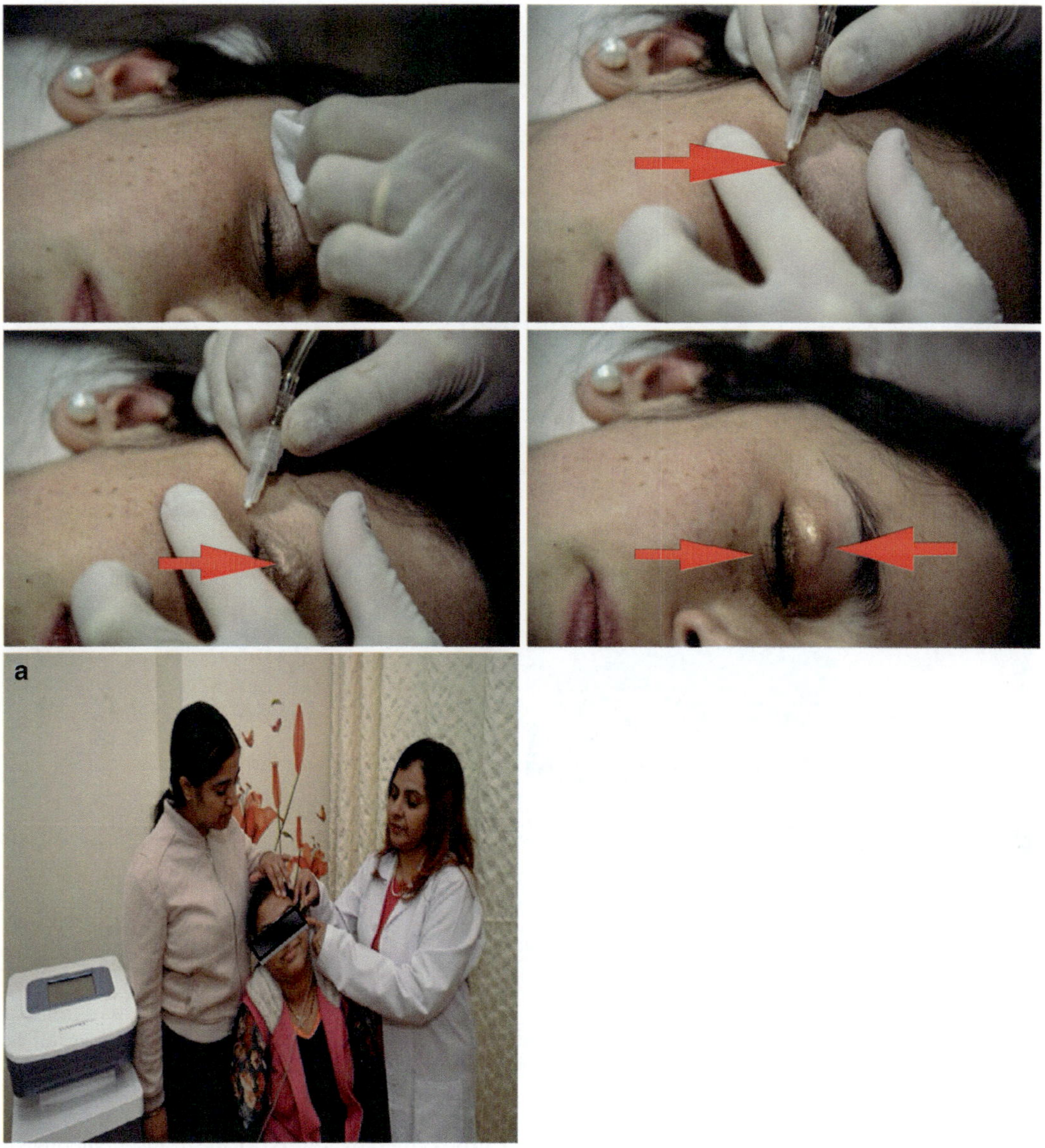

Fig. 11.10 Carboxytherapy for Periocular area. Note the deepness of the needle entrance into the dermis. The emphysema is seen in both eyelids. 50 ml were used each side. (**a**) Carboxytherapy for under eye circles © Jindal P. et al.; 2020 (with permission)

11.9.5 Stretch Marks

Stretch marks (striae distensiae) are fractures of collagen fibers within the dermis (dermal scars). Brandi et al. [6] showed thickening and an increase and rearrangement of collagen fibers in the dermis after CO_2 treatment [4, 12] and Hodeib showed an increase of fibronectin in stretch marks [30].

Better results are obtained in recent start of this condition, when stretch marks are red or pink. In more advanced stages they become pale and atrophic; better results in advanced stages are obtained with laser (resurfacing with ablative fraxel laser).

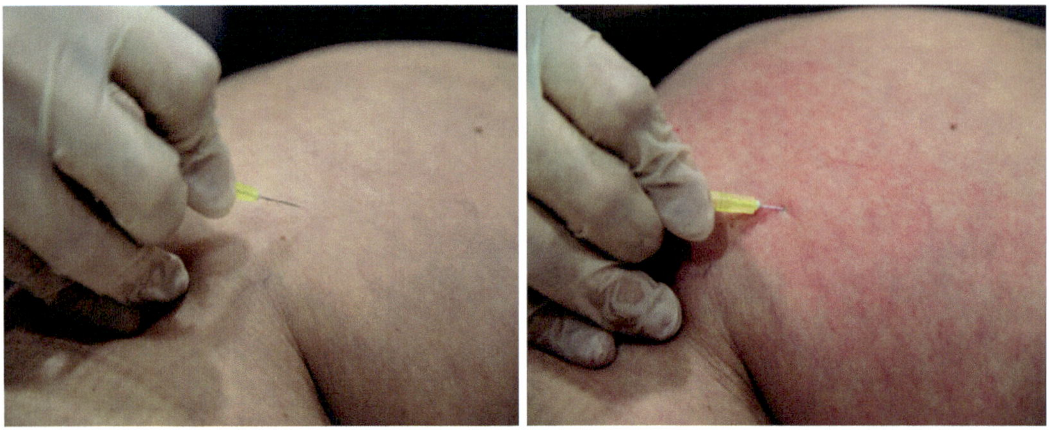

Fig. 11.11 Cellulite. Treatment of a thigh with local adipocity in the medial aspect of the knee. Note the angle of needle insertion, emphysema (200 ml), and skin erythema

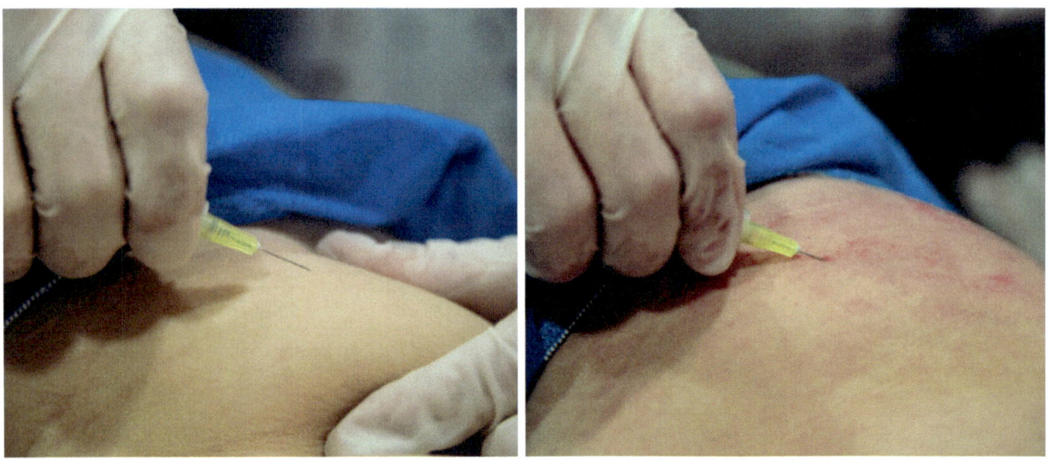

Fig. 11.12 Injection of CO_2 into a local adiposity on the flank. Note the angle of insertion of the 30G needle, the emphysema and erythema obtained with 200 ml CO_2 in the subdermal fat

Treatment should be done in each stretch mark and to its full extent. 30G or 31G needles are well tolerated using the intradermal technique. Including a low flow of 50–70 ml/min and enough volume for filling each stretch mark found in the treated area as shown in Fig. 11.13.

11.9.6 Chronic Wounds

A wound is defined as chronic when it does not heal according to the normal repair times and mechanisms (Fig. 11.14). This particular condition may be principally due to local hypoxia. Carboxytherapy refers to the transcutaneous or subcutaneous administration of CO_2 for therapeutical effects on both microcirculation and tissue oxygenation [18, 31].

Carboxytherapy administered subcutaneously may be used in addition to the routine methods of treatment for such lesions (surgical and/or chemical debridement, advanced dressings, compression, etc., according to the features of each lesion). Brandi et al. found significant improvement in progress of the lesions in terms of both healing and reduction of the injured area with Carboxytherapy [31].

For chronic wounds the same technique as for grafts is suggested.

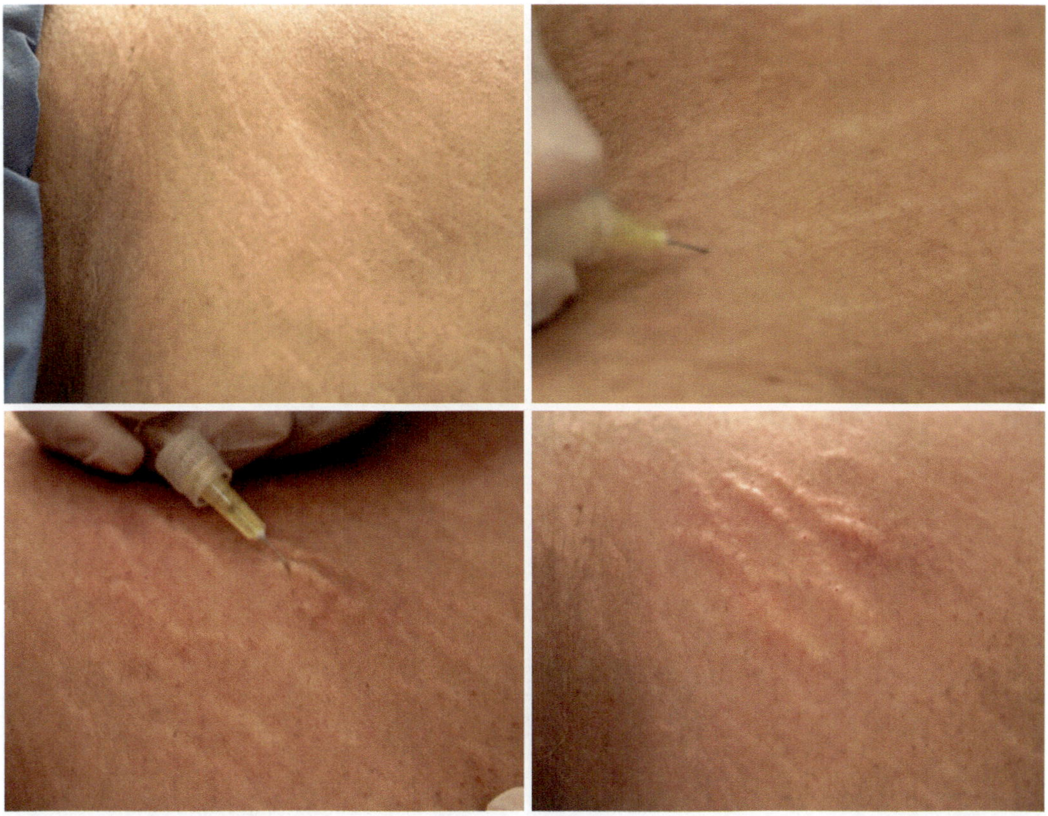

Fig. 11.13 Stretch marks treatment. Note the angle of needle punction. Each stretch mark should be punctured and filled with gas. Emphysema should appear in everyone and erythema in the whole area

11.9.7 Skin Grafts

Skin flaps and grafts procedures are widely used to reconstruct skin and soft tissue defects. Skin flap necrosis is a serious postoperative complication. Many groups work with transcutaneous CO_2 [32]. Carboxytherapy increases blood flow and tissue oxygenation [7, 15]. Nisi and Brandi concluded in their study that CO_2 injection enhances the inflammatory response of the implanted tissue and reduces the reabsorption rate. The treatment may improve the graft survival [3].

For intradermal and subdermal injections of CO_2, same technique as for flaccidity; many applications with 31G needle and low volume (10–20 cc) with low flow (50 ml/min) in the healthy tissue surrounding the graft or flap.

11.9.8 New/Other Indications

There are some other new indications for Carboxytherapy, such as different type of scars (acne, surgical), erectile dysfunction associated with microangiopathy, treatment of hair disorders; sometimes it helps to improve psoriasis, scleroderma and, paradoxically even angiectatic rosacea. In short, any skin conditions that require vasodilation [10].

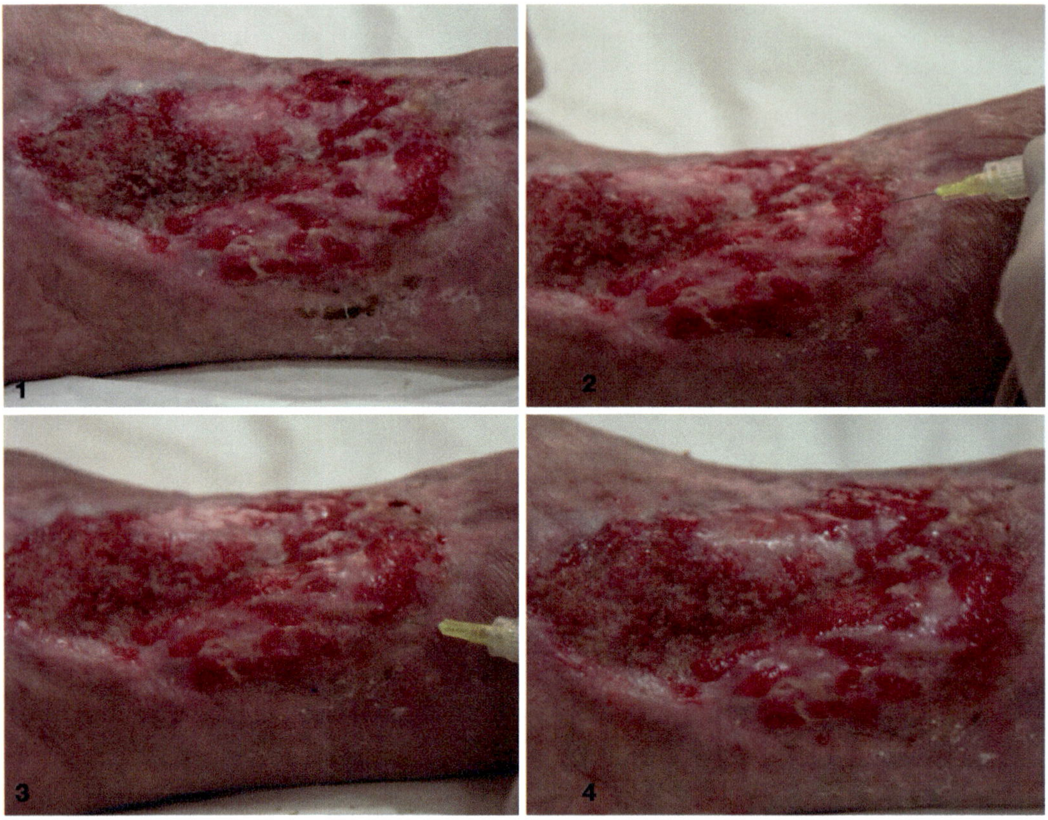

Fig. 11.14 Carboxytherapy in Venous Leg Ulcer. (1) Previous to treatment. Severe lipodermatosclerosis is seen. (2) Note the angle of the puncture using a 30 G ½ inch needle. (3) 5–10 ml in each puncture. Little emphysema is noticed from hardening of the skin. (4) Note erythema and emphysema after treatment

References

1. Gleizes E, de Goursac C. La carboxithérapie. J Med Esth et ChirDerm. Mar 2010;XXXVII(145):11–5.
2. Kołodziejczak A, et al. Is carboxytherapy a good alternative method in the removal of various skin defects? Dermatol Ther. 2018 Sep;31(5):e12699.
3. Nisi G, et al. Effect of repeated subcutaneous injections of carbon dioxide (CO_2) on inflammation linked to hypoxia in adipose tissue graft. Eur Rev Med Pharmacol Sci. 2015 Dec;19(23):4501–6.
4. Yoshitada S, et al. A novel system for transcutaneous application of carbon dioxide causing an "artificial Bohr effect" in the human body. PLoS One. 2011;6(9)
5. Hartmann BR, et al. Effects of serial percutaneous application of carbon dioxide in intermittent claudication: results of a controlled trial. Angiology. 1997;48:957.
6. Brandi C, et al. Carbon dioxide therapy in the treatment of localized adiposities: clinical study and hystopathological correlations. AesthPlast Surg. 2001;25:170–4.
7. MD Khan MH, et al. Treatment of cellulite part II. Advances and controversies. J Am Acad Dermatol March. 2010;62(3):373–84.
8. De Goursac C. Le décolleté qui se froisse. Protocole de traitement par la carboxithérapie et les LEDS. J Med Esth et ChirDerm. Sep 2012;XXXIX(155):169–72.
9. Balik O, et al. Does carbon dioxide therapy really diminish localized adiposities? Experimental study with rats. AesthPlast Surg. 2011;35:470–4.
10. Koutná N. Carboxytherapy in aesthetic medicine. Aesthetic Medicine. 2011:547–76.
11. Ordiz I. Técnicas combinadas empleadas en el Rejuvenecimiento Facial. Oviedo. 2013:106–8.
12. Brandi C, D'Aniello C. Carbon dioxide therapy: effects on skin irregularity and its use a complement to liposuction. AesthPlast Surg. 2004;28:222–5.
13. D'Arcangelo, et al. Acidosis inhibits endothelial cell apoptosis and function and induces basic fibroblast growth factor and vascular endothelial growth factor expression. Circ Res. 2000;86(3):312–8.
14. Bunyatyan ND, et al. Carboxytherapy – an innovative trend in resort medicine. VoprKurortolFizioter Lech FizKult. 2018;95(5):72–6.

15. Sadala AY, et al. Effects of transcutaneous electrical nerve stimulation on pain intensity during application of carboxytherapy in patients with cellulite: a randomized placebo-controlled trial. J Cosmet Dermatol. 2018 Dec;17(6):1175–81.
16. Herreros FOC, et al. Mesotherapy: a bibliographical review. An Bras Dermatol. 2011;86(1):96–101.
17. Calonge WM, et al. Massive subcutaneous emphysema after carbon dioxide mesotherapy. Aesthet Plast Surg 2013 Feb; 37 (1): 194–197. Epub 2013 Jan 8.
18. Fioramonti P, et al. Periorbital area rejuvenation using carbon dioxide therapy. J Cosmet Dermatol. 2012;11:223–8.
19. Seidel R, Moy R. Effect of carbon dioxide facial therapy on skin oxygenation. J Drugs Dermatol. 2015 Sep;14(9):976–80.
20. Tavares J, et al. Increase in collagen turnover induced by intradermal injection of carbon dioxide in rats. J Drug Dermatol. 2008;7(3):201–6.
21. EmanNofal MD, et al. Evaluation of carboxytherapy and platelet-rich plasma in treatment of periorbital hyperpigmentation: a comparative clinical trial. J Cosmet Dermatol. 2018:1–8.
22. Paolo F, et al. Periorbital area rejuvenation using carbon dioxide therapy. J Cosmet Dermatol. 2012 Sep;11(3):223–8.
23. Micheels P, et al. Petit essai de la carboxithérapie sur la cellulite et les vergetures. J Med Esth et ChirDerm. Mar 2012;XXXIX(153):25–33.
24. Rossi A, Katz B. A modern approach to the treatment of cellulite. Dermatol Clin. 2014;32:51–9.
25. Draelos ZD, Marenus KD. Cellulite. Etiology and purported treatment. Dermatol Surg. 1997;23:1177.
26. Pérez Atamoros FM, et al. Evidence-based treatment for gynoid lipodystrophy: a review of the recent literature. J Cosmet Dermatol. 2018 Dec;17(6):977–83.
27. Eldsouky F, Ebrahim HM. Evaluation and efficacy of carbon dioxide therapy (carboxytherapy) versus mesolipolysis in the treatment of cellulite. J Cosmet Laser Ther. 2018 Oct;20(5):307–12.
28. Ramalho Pianez L, et al. Effectiveness of carboxytherapy in the treatment of cellulite in healthy women: a pilot study. Clin Cosmet Invest Dermatol. 2016;9:183–90.
29. Brandi C, et al. Carbon dioxide: maybe not the only one but an efficient and secure gas for Treating Local Adiposities. AesthPlast Surg. 2012;36:218–9.
30. Hodeib AA, et al. Clinical and immunohistochemical comparative study of the efficacy of carboxytherapy vs platelet-rich plasma in treatment of stretch marks. J Cosmet Dermatol. 2018 Dec;17(6):1008–15.
31. Brandi C, et al. The role of carbon dioxide therapy in the treatment of chronic wounds. In Vivo. 2010 Mar–Apr;24(2):223–6.
32. Saito I, et al. Effect of local application of transcutaneous carbon dioxide on survival of random-pattern skin flaps. J PlastReconstrAesthet Surg. 2018 Nov;71(11):1644–51.

Micro-focused Ultrasound in Aesthetic and Regenerative Gynecology

12

Ting Song Lim

12.1 Introduction

The recent advancement in medicine has improved our life expectancy and prolonged functional living. With extended life expectancy, the demand for improving quality of living and physical condition, including physical appearance, has also increased.

In response to the increasing public demand for nonsurgical treatments for age-related facial skin and structural changes, various treatment modalities, including chemical peeling, resurfacing lasers, radio frequency (RF), micro-focused ultrasound (MFU), neurotoxin, dermal filler, and threads have been introduced.

Skin tightening is one of the cornerstones of skin rejuvenation and is defined as the improvement of skin laxity and crepiness of the skin. Micro-focused ultrasound (MFU) has become one of the most popular treatments to tighten skin and elevate facial structure. It generates tissue heating, leading to fibroblast stimulation and new collagen synthesis. Comparing to radio frequency (RF), which generates tissue heating at the superficial level, MFU utilizes the acoustic energy to create controlled microthermal effects at the deeper dermis and subcutaneous tissues, up to the superficial musculoaponeurotic system (SMAS) level.

The noninvasive delivery of focused ultrasound energy by MFU has enabled us to strengthen the structures that hold all soft tissues (skin, muscle, fat, etc.) to our skeletal frame of the face—the anchoring complex. The strengthening and contraction of the anchoring complex result in a gradual tightening of the skin, improved skin quality, and overall facial lifting.

12.2 Pathogenesis of Age-Related Skin Laxity Deterioration

Both intrinsic and extrinsic factors contributed to age-related skin laxity deterioration. Intrinsic factor, determined by the fundamental architecture of the skin and underlying tissues, collagen and elastin contents, anchoring complex density, and skeletal bone frame, goes through wear and tear and depletion with aging. On top of that, external factors such as ultraviolet (UV) exposures, pollutions, diet, circulation, smoking, inflammation expedite the aging process as the extrinsic factor.

At the molecular level, components of the extracellular matrix are constantly changing. Collagen is generated by fibroblasts and degraded by matrix metalloproteinases (MMPs). MMPs also deplete the elastin in our skin. UV radiation, trauma, infections, pollutions, irritation to the skin upregulates the MMPs, which results in disorganization in the

T. S. Lim (✉)
Department of Aesthetic Medicine, Clique Clinic, Kuala Lumpur, Malaysia
e-mail: drlim@cliqueclinic.com

P. Jindal et al. (eds.), *Aesthetic and Regenerative Gynecology*,
https://doi.org/10.1007/978-981-16-1743-0_12

elastic fiber network, atrophy of the fibrillin-rich microfibrils, weakening of the mechanical strength, and elasticity of the anchoring complex, leading to tissue ptosis and skin crepiness.

12.3 Mechanism of Action

When enough thermal energy is delivered to collagen, there is a structural change in the collagen polymer; causing collagen shrinkage. The amount of collagen contraction achieved depends on both temperature and duration of heating. There are no true optimal temperatures for producing collagen shrinkage, but a variety of effective combinations of these two variables: temperature and duration. The thermal injury-induced needed to be in a well-controlled manner to avoid overwhelming inflammatory response which could lead to complications such as dyschromia and scarring.

Medical ultrasound devices use acoustic waves for diagnostic imaging or medical treatments. The depth of the ultrasound energy delivery depends on its frequency: the higher the frequency the more superficial it gets and the lower the frequency the deeper it reaches. When the ultrasound energies are converged to a focused point in living tissue, molecular vibration generates heat, creating microthermal injury zones at predetermined depths while leaving the surrounding tissue unaffected. The microthermal injury causes collagen contraction, denaturation, neocollagenesis, and collagen remodelling.

Micro-focused Ultrasound (MFU) is different from RF energy in that it can be focused to target deeper tissue without affecting superficial tissues, i.e., the skin. For a RF device to achieve high temperatures, surface cooling is needed to protect the skin. MFU can target deeper tissues with better comfort and safety. The microthermal injury zones delivered by micro-focused ultrasound induce wound healing responses, involving 3 phases: inflammation, proliferation, and remodeling. Immediately after thermal injury, inflammatory response begins with the influx of neutrophils and macrophages. Viable fibroblasts start replacing the necrotic ones. Upregulation of MMPs is seen. During the proliferation phase, fibroblasts differentiate into myofibroblasts and cause tissue contraction. During the remodeling period, the connective tissues become more compact, with more horizontal alignment of collagen and elastin fibers.

12.4 Micro-focused Ultrasound (Ultherapy)

The Micro-focused Ultrasound (MFU) was approved by the US Food and Drug Administration in 2009 for noninvasive eyebrow elevation but is routinely used for panfacial and submental treatments. By using different transducers, transcutaneous energy can be delivered to target tissues at variable depths: 1.5, 3.0, and 4.5 mm, which covers the superficial dermis down to supra-SMAS and platysma level. The monitor allows direct visualization of where the energy will be delivered so that specific structures such as bone, muscles, and blood vessels can be avoided. The focal energy delivery given in predetermined "lines" results in discrete intervals between 1 and 3 mm coagulation zones that promote healing.

12.4.1 Indications

Micro-focused Ultrasound can be used to treat skin laxity issues around the eyes, the face, submental area, as well as the neck (Fig. 12.1). Other body parts such as the decolletage, arms, thighs, knees, and buttocks have also been done.

12.4.2 Treatment Time

Treatment time for the full face and neck are generally around 45 minutes to 90 minutes, depending on the operator's experience, treatment protocol, and patient's condition.

12.5 Pain and Comfort Management

With the introduction of the new protocols, the complaints of pain during the treatment have decreased and patients generally can tolerate the treatment even without any pain management. However, for some patients with lower pain threshold, the pain may limit

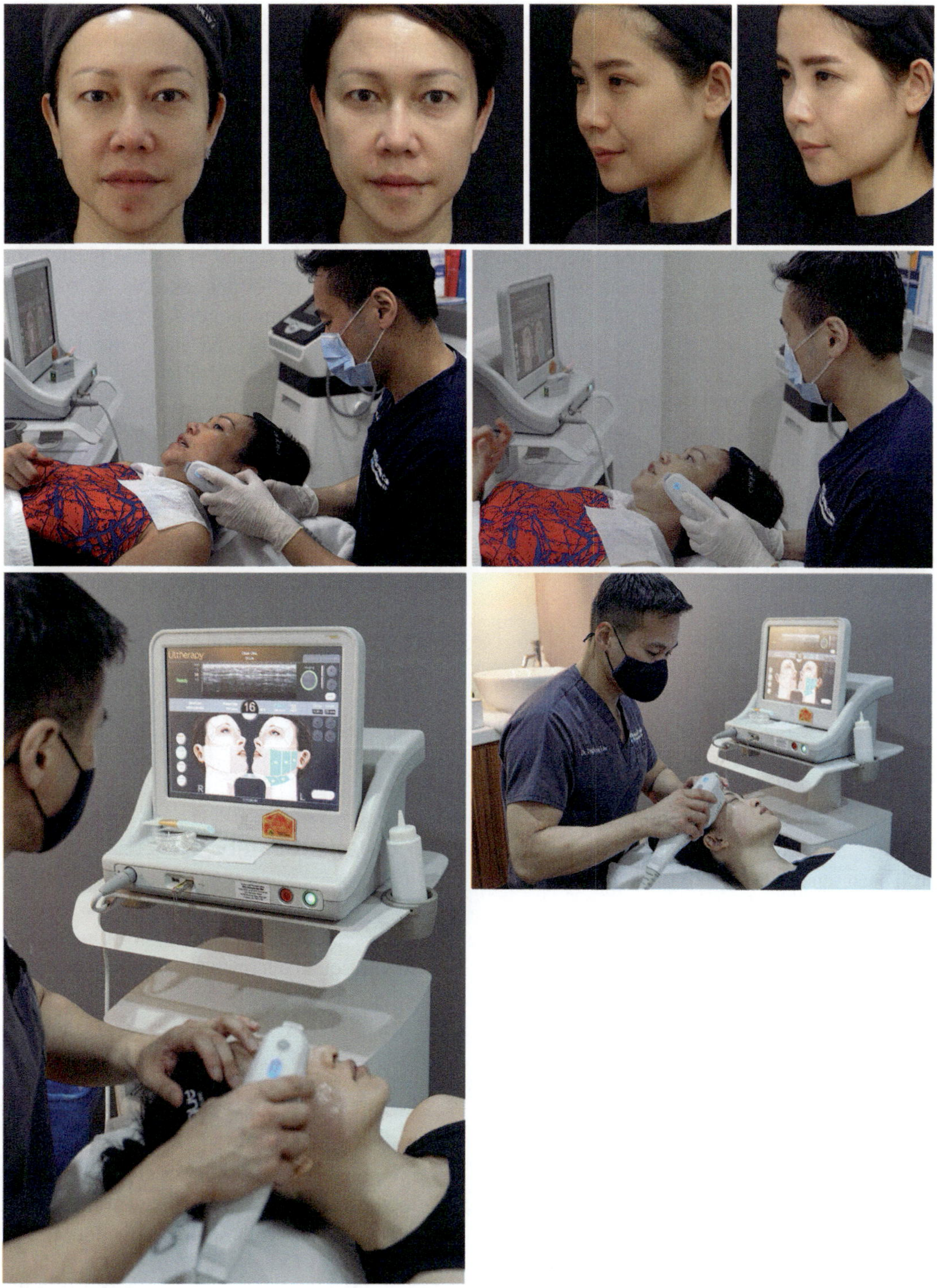

Fig. 12.1 Demonstration of MFU on face for skin tightening

the amount of energy delivered, decrease the treatment areas, and increase treatment duration times.

Topical anesthetics can provide some relief, but with minimal to moderate relief. In most practices, patients are given one or more of the following pain management prior to treatment: nitrous oxide gas, oral nonsteroidal anti-inflammatory medications, oral narcotics, oral anxiolytics, distraction techniques, cool air devices, and precooled ultrasound gels.

Also, it is found that lower energy treatment levels had less discomfort than higher energy levels without a difference in clinical outcome at 180 days, therefore there is no need to increase the energy levels in terms of trying to push for better outcomes.

12.5.1 Recovery and Down Time

Other than the mild tissue reactions (swelling, redness, etc.), micro-focused ultrasound can be considered to have no downtime and a minimal recovery period. No specific aftercare is needed.

12.5.2 Side Effects and Complications

Most patients reported only mild side effects, typically mild erythema and edema that usually resolves within 2 to7 days. However, rare complications such as transient bruising, pigmentary changes, and self-limiting neuropathic pain have also been reported.

12.5.3 Treatment Protocols

Micro-focused ultrasound is indicated for use to lift the eyebrow, tighten the lax submental and neck tissues, improve lines and wrinkles. With ultrasonic visualization of depths up to 8 mm below the surface of the skin, the operator can ensure proper coupling of the transducer to the skin, determine the appropriate depth of treatment and deliver the energy precisely at the target areas.

In the early days, the treatment protocol was set at 500 lines including both 3.0 and 4.5 mm for a full face treatment. The updated protocol recommends that 800–1200 lines are needed for a more optimal facelift. For the face and neck, it is important to concentrate the treatment on lateral forehead, temples, lateral side of the face, and neck using the 4.5 and 3.0 mm probes.

Because every patient responds to MFU treatment differently, it is important that the physician make proper assessments and come out with a patient-specific treatment plan prior to treatment.

12.5.4 Contraindications

Micro-focused ultrasound is contraindicated for use in patients with open wounds, keloids, or lesions in the treatment area, severe or cystic acne in the treatment area and active implants (e.g., pacemakers or defibrillators), or metallic implants in the treatment area.

12.5.5 Precautions

Micro-focused ultrasound should be only operated by trained personnel, preferably trained physicians with knowledge of the facial anatomy and able to read the ultrasound image correctly. It is recommended that the following areas should be avoided during treatment: thyroid gland, thyroid cartilage, trachea, major vessels, breast tissue, or breast implants.

12.5.6 Patient Safety

Micro-focused ultrasound should not be used on a patient's eyes or in a location or technique where ultrasound energy can reach the eye.

12.5.7 Results

Collagen contraction immediately after treatment may result in an initial tissue "lift." There is also mild edema that contributes to the early aesthetic improvement seen by patients. However, the heat induced collagen synthesis and remodelling can persist for up to 1 year. Although no skin tightening procedure can claim permanent results, the long-term outcomes of MFU-derived skin tightening devices is encouraging. Generally, patients are advised to repeat treatments on a yearly basis for the first 4–5 times and treatment frequency can be spaced out when the skin condition has significantly improved.

Minimal Invasive Treatment of Varicose Veins

13

Ravul Jindal, Luv Luthra, Piyush Chaudhary, Taranvir Kaur, and Shabjot Dhillon

13.1 Introduction

Varicose veins of the leg affect huge amount of population and can impact patient's day to day activities and quality of life [1].

Chronic venous disorder is related to malfunction of the leg veins leading to venous reflux [2]. Venous reflux leads to venous hypertension which leads to signs and symptoms of chronic venous insufficiency (CVI). Patients can present with dilated veins in legs which can be associated with heaviness, foot swelling, leg pain, cramps, itching, changes at the ankle and ulceration [3]. Risk factors causing CVI can be genetic, increase height, obesity, old age, female and prolonged standing.

Open surgery for varicose veins was a associated with huge morbidity and has largely been replaced in the last decade with minimally invasive techniques such as thermal and non-thermal tumescent techniques [4].

In this chapter we briefly discuss various methods on how to treat varicose veins.

R. Jindal (✉)
Fortis Hospital, Venous Association of India (VAI), Mohali, India

L. Luthra · P. Chaudhary · T. Kaur · S. Dhillon
Mohali, India

13.2 Thermal Tumescent Techniques (TTT)

Endovenous thermal ablation was started nearly two decades back and has become the gold standard treatment as compared to open surgery. All thermal techniques are associated with various advantages in terms of their minimal invasive nature, local anaesthesia, outpatient treatment and early recovery.

There are various thermal methods available with their advantages and disadvantages which we are going to discuss here.

13.2.1 Methods

- *ENDOVENOUS LASER ABLATION (EVLA)*
- *RADIOFREQUENCY ABLATION (RFA)*
- *STEAM ABLATION*
- *ENDOVENOUS MICROWAVE ABLATION (EMWA)*
- *ECHOTHERAPY USING HIGH INTENSITY FOCUSED ULTRASOUND (HIFU)*

13.2.2 Indications of Intervention

Venous reflux in long saphenous, short saphenous and perforators causing symptomatic varicose veins.

P. Jindal et al. (eds.), *Aesthetic and Regenerative Gynecology*,
https://doi.org/10.1007/978-981-16-1743-0_13

13.2.3 Contra Indications

- Superficial venous thrombosis (SVT)
- Deep venous thrombosis (DVT)
- Venous aneurysms
- Infection

13.2.4 How It Works

Thermal techniques require injection of local anaesthesia around the vein which causes compression of the vein and also provide the cushion around the vein to prevent damage from heat to the surrounding tissues.

These are minimally invasive procedures and are performed percutaneously under ultrasound guidance. The tip of the laser fibre is placed under ultrasound guidance 2 cm distal to the sapheno-femoral junction. Tumescent anaesthesia is given with 21 G needle with tumescent pump under USG guidance making sure that a halo is created around the vein.

The tumescent solution is made of Ringer lactate and local anaesthesia. The laser energy is setup according to the type of machine which is being used. If we are using 1940 nm diode laser machine, then the settings we use is 5 watts × 8 seconds. The fibre is withdrawn under ultrasound guidance delivering the required energy causing damage through heat to the water molecules in the media of the vein. This results in fibrosis over a period of time causing permanent occlusion. Post laser the residual varicosities are treated with either foam sclerotherapy or stab avulsions.

Patient is given compression therapy with stockings and discharged home after 1 h. Patient can do routine activities from next day and can also go to work. He can go to Gymnasium after one week [5].

The occlusion rates with laser and radiofrequency go up to 95–99% [6–8].

Both these techniques have nearly similar occlusion rates with minor differences in terms of pain, bruising and costing [9].

Post procedure complications are rare and can include persistent pain for few days, thrombophlebitis, bruising and very rare deep venous thrombosis and skin burns [10–12].

13.2.4.1 EVLA (Endovascular Laser Ablation)

EVLA uses a jacket tipped or bare tipped catheter to deliver laser energy in the treated segment of Long Saphenous Vein (LSV) or Short saphenous vein (SSV).

13.2.4.1.1 Principle

The heat generated damages the endothelial lining of the vein. Various laser wavelengths are used to cause endothelial damage to the vein which include 810, 940, 980, 1470 and 1940 nm. Several studies have shown that different wavelengths have different absorption qualities.

810 nm wavelength is specific for haemoglobin absorption, 940 nm has a balanced absorption for water and haemoglobin and 1470 nm is specific for water absorption. Several studies have shown that higher wavelengths cause less pain and bruising in the post-operative period.

13.2.4.1.2 Equipment

There are various companies which provide the machines varying with different wavelengths and weight and ability to travel with them. One of the smallest equipment is Biolitec Leonardo mini 1470 laser (Fig. 13.1).

This is a space-saving, economical and at the same time versatile device. This is just 900 grams light.

Latest 1940 nm laser machine is being manufactured by many companies including Biolitec and NeoLaser (NeoV1940). In today's time one should use mainly 1470 or 1940 nm as it causes less bruise and less cord formation with more comfort to the patient.

Then there is a difference in the type of fibre which we use which can change the amount of energy delivered and also the damage to the surrounding tissues. We have single or double ring fibres and the latest addition of CORONA Infinite Ring Fibre Technology from NeoLaser (Fig. 13.2). All are good and at present we really do not know if there is huge difference in the outcome with the type of fibre used.

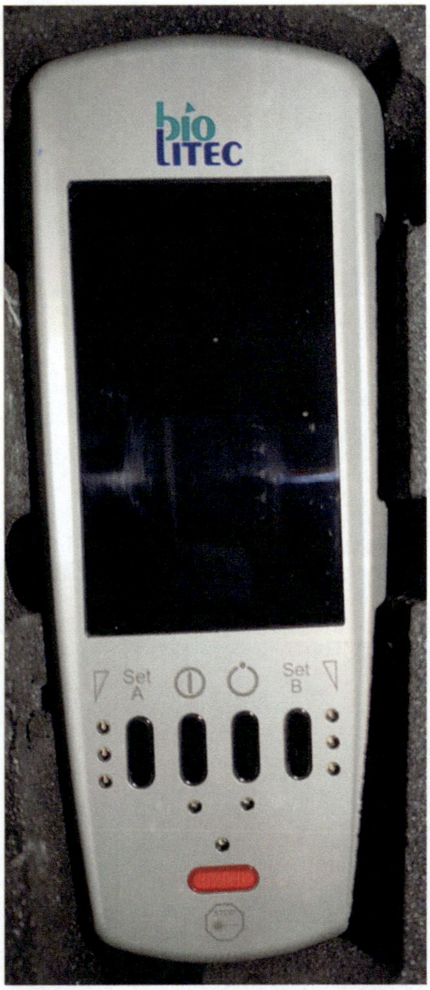

Fig. 13.1 Biolitec Leonardo mini 1470 laser

13.2.4.1.3 Procedure

After draping the patient, we select a site under USG which gives us the best access in the LSV or GSV. Usually an 18-gauge needle is used for the venepuncture. Introducer sheath is inserted in the vein (6F usually) using Seldinger technique. After placing the sheath, the laser fibre is inserted and placed around 2 cm away from Saphenofemoral Junction (SFJ) under ultrasound.

We give tumescent anaesthesia around the vein. Laser fibre is withdrawn every 1 cm on the beep from the laser machine which is fixed on a preset timing to deliver a required energy to the vein. We must make sure that we do not deliver the energy in the sheath in the last 12–15 cm of the vein and sheath has to be withdrawn out once the long black mark on the laser fibre is seen, we also should make sure that we do not deliver any laser energy around 1 cm from the entry site to avoid skin burns.

13.2.4.2 RFA (Radiofrequency Ablation)

13.2.4.2.1 Procedure

An access site is chosen with help of ultrasonography and the selected segment of the vein is punctured like previously explained for EVLA and introducer sheath is placed (usually 7F). RFA catheter is inserted (Fig. 13.3) and placed 2 cm proximal from the saphenofemoral junction. Tumescent anaesthesia is given as explained in previous section. This anaesthesia prevents tissue damage by heat as well as compresses the vein wall against the fibre.

A 21-gauge spinal needle can be used to instill the tumescent anaesthesia. Needle can be connected to a 20 ml syringe or tubing that is connected to an infusion pump which definitely helps and it is more comfortable for the patient.

Another important aspect in treatment with RFA is to put pressure over the vein segment which is being treated with RFA in order to make good contact of fibre with the vein. This is usually done with the probe.

Using RFA, vein is treated in segments of 7 cm. The duration of the treatment is 20 seconds for each segment. Usually, first segment of the vein is treated twice. If any segment of the vein is large in diameter or aneurysmatic, then again 2 treatment cycles of 20 seconds can be delivered for the effective treatment.

The parameters of RF machine are monitored during the treatment cycle. Temperature of the fibre must reach 120 degrees within 5 seconds. If not, cycle must be repeated [13].

There are black markers on the fibre which are placed at the distance of 6.5 cm. After completion of each cycle fibre is withdrawn from one marker to the next. The length of the ablating tip is 7 cm. In this way each treated segment is overlapped by the next by 0.5 cm.

The introducer sheath must be taken out before treating the last segment of the vein in order not to burn inside of the sheath.

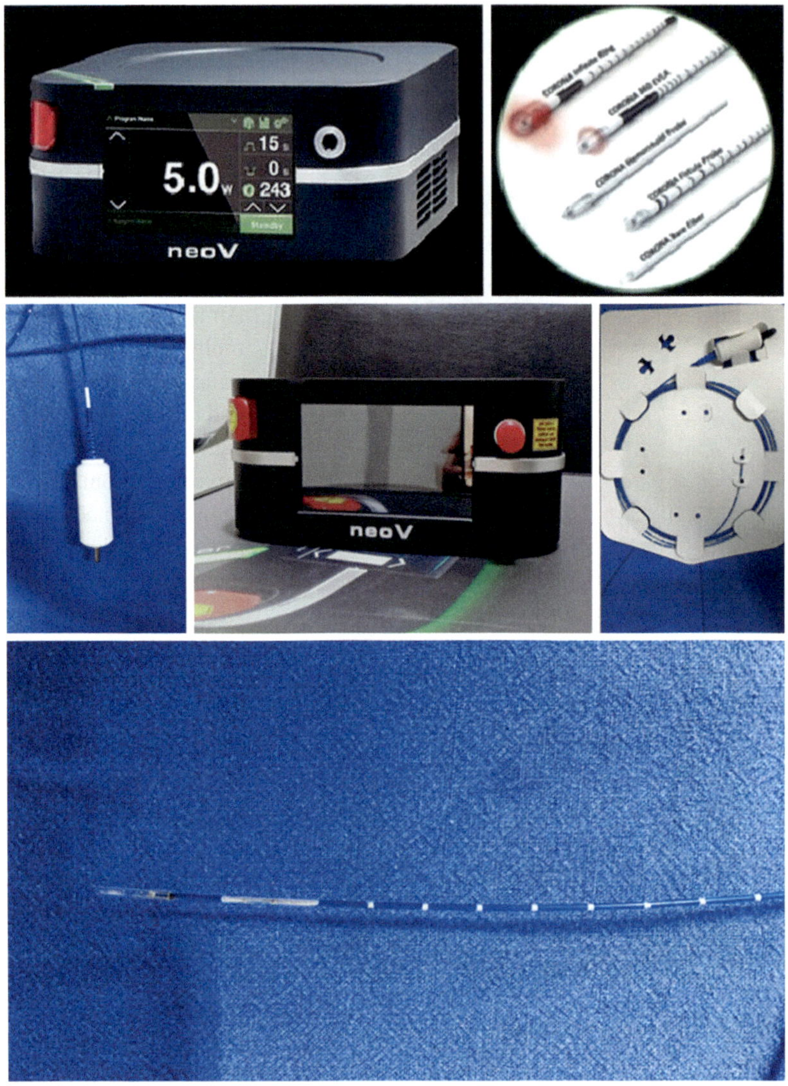

Fig. 13.2 Type of fibres

After complete treatment, occlusion can be checked with USG.

13.2.4.3 Steam Ablation

Steam ablation is the thermal ablation method done using steam (Fig. 13.4).

13.2.4.3.1 Procedure

Introducer sheath is placed in the vein using USG and the steam catheter is placed into the vein 3 cm away from the junction.

13.2.4.3.2 Principle

In this method steam is delivered from catheter to the vein wall. Heat energy is transferred to the vein wall. One pulse of steam produces 174 joules of energy which is reduced to 60 joules at the catheter tip.

Steam is produced in puffs and delivered through the catheter tip. As with the other thermal techniques, 3–4 puffs may be required for the treatment of initial segment of the vein. Vein is compressed manually and catheter pulled out sequentially.

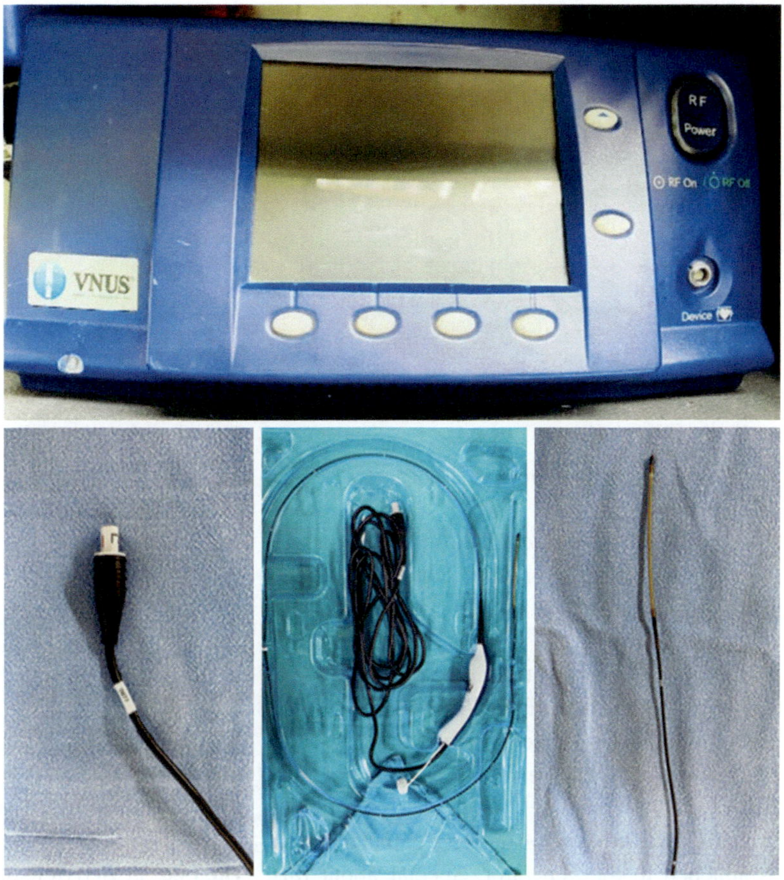

Fig. 13.3 Radiofrequency ablation

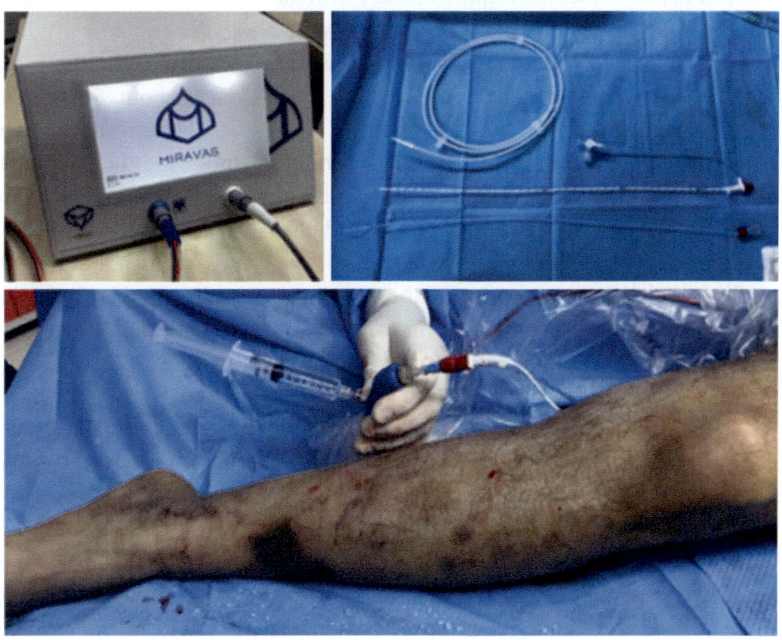

Fig. 13.4 Steam ablation

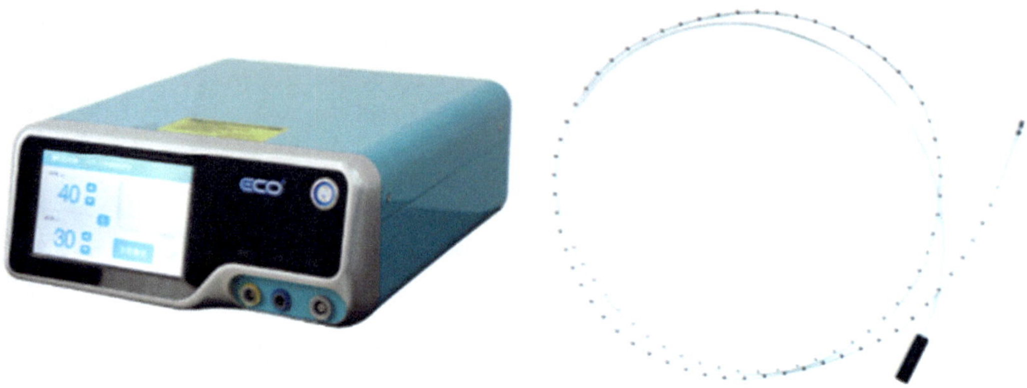

Fig. 13.5 Endovenous microwave ablation

It is still a new technique whose occlusion rate is yet to be studied in order to compare it to the other techniques.

13.2.4.4 Endovenous Microwave Ablation (EMWA)

Endovenous Microwave Ablation (EMWA) is another novel technique for the venous ablation (Fig. 13.5).

It also uses microwave delivery catheter which is attached to the generator. Special about this catheter is that it does not require contact with the vein wall for the successful ablation of the vein.

Microwave heats the blood components which in turn heats water molecules found in the cells of the vein wall.

The temperature at the fibre tip is around 80 °C which results in less chances of skin burn.

Unlike laser ablation, microwave ablation does not require protective equipment.

Since it is also a newer technique, not much data on occlusion and long-term results are available.

13.2.4.5 Echotherapy Using High Intensity Focused Ultrasound (HIFU)

In this technique High energy focused ultrasound waves are used to treat the vein (Fig. 13.6).

In this technique relatively very less amount of anaesthetic is required in comparison to other thermal techniques. Each treatment cycle produces about 900 °C temperature.

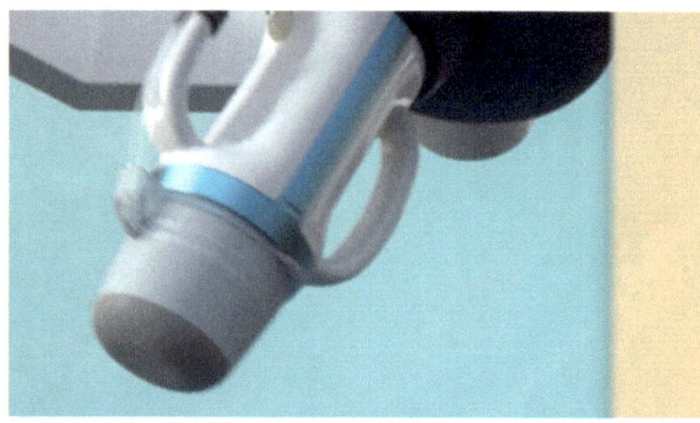

Fig. 13.6 ECHOTHERAPY using high intensity focused ultrasound

The vein can be seen in real-time which allows following the path of the vein very accurately in tortuous veins and even perforators.

Treating veins in the vicinity of an ulcer is also possible without risk of infection.

It has a big future in the treatment of the varicose veins but setup is very costly and data is limited.

13.3 Non-Thermal Non-Tumescent Technique (NTNT)

Non-Tumescent techniques do not need local anaesthesia and that really makes a big difference in the treatment of these patients. Once we do not need local anaesthesia it reduces our procedure time, need for pain killers, less bruising and also can be used in patients who are allergic to local anaesthesia and are needle phobic.

13.3.1 Methods

- *GLUE CLOSURE*
- *MECHANICO CHEMICAL ABLATION (MOCA)*
- *V-BLOCK*
- *FOAM SCLEROTHERAPY*
- *CUTANEOUS LASERS and INTENSE PULSE LIGHT*

13.3.1.1 Glue Closure

Saphenous Closure System uses n-butyl-2-cyanoacrylate-based glue closure systems. The glue as soon as it comes in contact with blood proteins polymerizes and leads to vessel occlusion after causing intense inflammatory reaction (Fig. 13.7) [14–16].

This stops the blood flow through the insufficient vein eventually leading to fibrosis of the vein causing chronic occlusion of the treated vein. The inherent risk of an over diluted CA compound is migration of the glue into deeper veins and lung tissue because of its low viscosity.

Currently, two main devices using CA are available including closure system including VenaSeal Closure System (Sapheon, Morrisville, NC, now Medtronic Medical) and VariClose Vein Sealing System [VVSS (Biolas, *Ankara, Turkey) and Invamed system].*

13.3.1.1.1 Procedure

The long saphenous vein is punctured at the most distal point of reflux under ultrasound control and 6F introducer sheath with dilator is advanced into the GSV. Glue delivering catheter tip is placed 5 cm caudal to the junction. A distance of 5 cm from the junction provides a space for ultrasound compression between the delivery catheter and the junction. It also allows propagation of

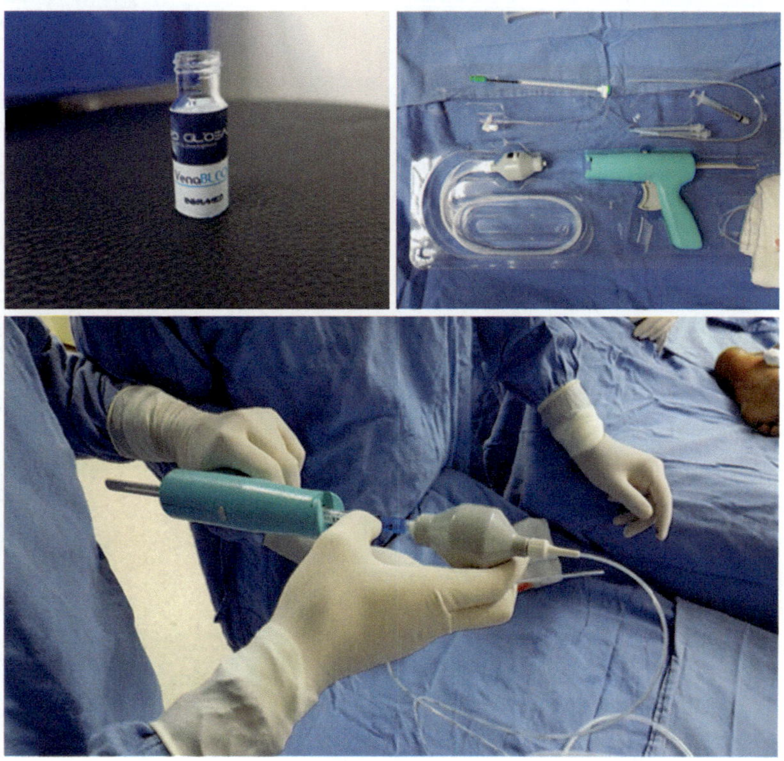

Fig. 13.7 Glue closure

glue towards the junction after injection. A 3 ml syringe filled with glue is used. The catheter is primed with the dispenser and the whole catheter, except distal 3 cm is filled with glue. Under ultrasound guidance, the glue is injected twice most centrally and then single injections are performed at 3-cm intervals on withdrawal.

13.3.1.1.2 Outcomes

The feasibility study of one of the delivery system enrolled 38 patients to evaluate the safety and effectiveness of its system for incompetent GSV [17]. Complete occlusion of the treated GSV was confirmed by duplex ultrasound in all patients except for one complete and two partial recanalization observed at 1, 3 and 6 months of follow-up, respectively.

The VeClose trial is the first randomized controlled trial comparing CA adhesives and RFA for incompetent GSV [18]. Three month closure rates were 99% for CA adhesives and 96% for RFA. Pain score during the glue procedure are mild but comparable to thermal methods (2.2 and 2.4 for CA adhesives and RFA, respectively, $P = 0.11$).

13.3.1.2 Mechanochemical Ablation (MOCA) Method

13.3.1.2.1 Mechanism

It is a type of non-thermal Endovenous ablation system. The principle of this technique is a dual injury technology. This is the first hybrid technique that works by mechanical injury of the venous wall with simultaneous sclerotherapy. MOCA is done mainly by two methods—ClariVein and Flebogrif™ (Fig. 13.8). ClariVein uses a special infusion catheter with a rotating wire tip designed for the controlled 360° dispersion of specific agents to the targeted treatment area of the vein, while Flebogrif uses catheter with metallic barbs to injure media of the venous wall followed by the chemical ablation of the vein. Sheath required for the procedure is 6F.

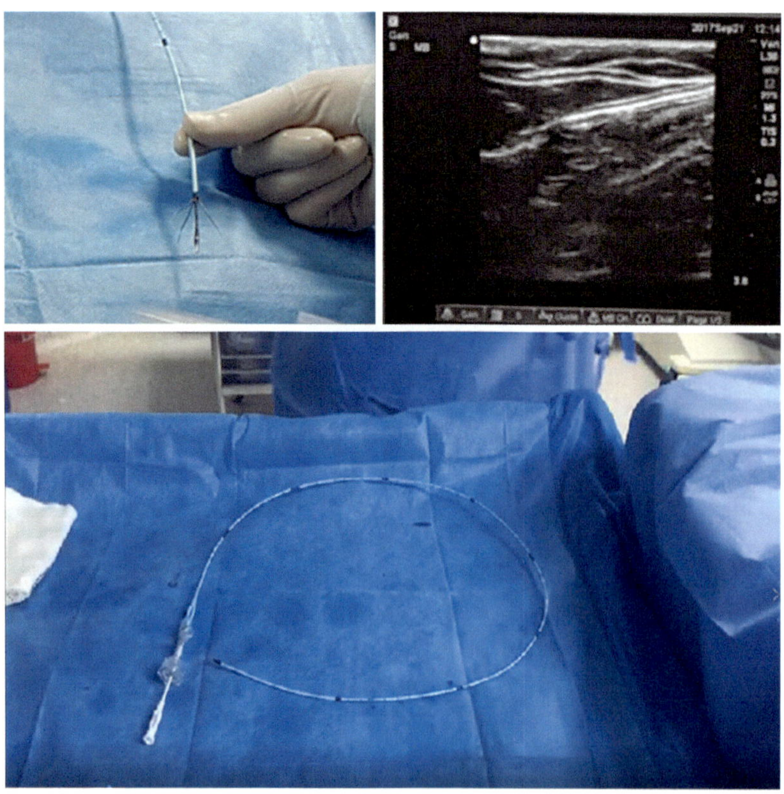

Fig. 13.8 *Flebogrif* catheter for mechanochemical ablation

13.3.1.2.2 Device Description

Flebogrif™ catheter, which has retractable metallic barbs, is passed over the guidewire and placed proximal to SF junction. The blades of the catheter were opened after its tip had been placed 2 cm below the SF junction. Guidewire is removed from the catheter. Tumescent anaesthesia not required during the procedure. Initially, 1.5 ml of sclerosant foam—Sodium tetradecyl sulphate (STD, fibrovein 3%) is injected with compression of the SF junction with ultrasonography (USG) probe. The system is withdrawn by 4 cm each time and 0.5 ml of foam is injected up to the point of entry. Compression of the vein was performed with hand or USG probe while withdrawing the catheter.

ClariVein device has two sterile, single-use components (Fig. 13.9). The catheter has one lumen to inject liquid sclerosant and a motor with 4 speeds that range from 2000 rpm to 3500 rpm. The catheter is low profile, its outer diameter 2.67 French. The internal volume of the catheter lumen is 0.7 mL for the 45-cm and 1.0 mL for the 65-cm catheter [19].

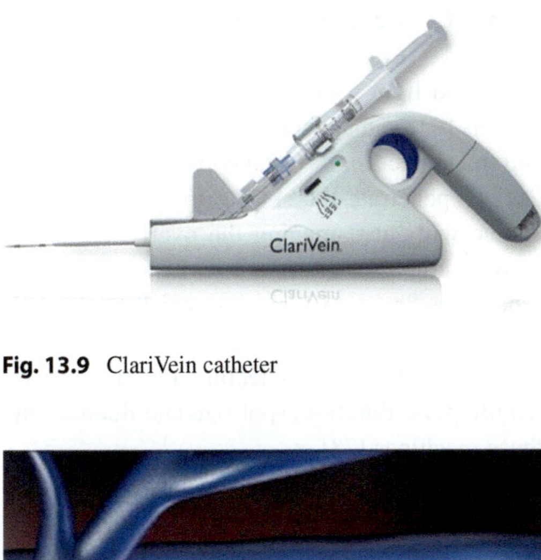

Fig. 13.9 ClariVein catheter

The treatment is started around 2 cm from the SFJ under USG guidance. As the catheter is withdrawn, the sclerosant, typically Sodium tetradecyl sulphate, is dispersed onto the vessel wall while the rotating tip injures the intima of the vein.

13.3.1.2.3 Outcomes

In a systematic review that included 1334 patients (884 women, 450 men) treated with MOCA system with a total of 1506 veins of which 1265 were GSVs and 241 were SSVs. Mean diameter was 6.41 ± 1.93 mm for GSV and 5.59 ± 1.49 mm for SSV. Mean length of the treated GSV and SSV was 38.4 ± 10.5 cm and 21.7 ± 7.30 cm, respectively.

Around *twenty* studies have reported on technical success and complications. Studies with both catheters have been published. Technical success rates with these procedures varies between 90–97% though long term follow up shows occlusion rated between 85–90%. There are randomized controlled trials that also compared the degree of pain after MOCA and RFA and showed that MOCA has less pain during the procedure but in post-operative period the pain is similar in both groups [20].

13.3.1.3 V-Block

V-Block is the newest NTNT technology (Fig. 13.10). In this method we place an occlusion device at the SFJ and deliver liquid sclerosant through a dual syringe system. The occlusion device is like a nitinol filter covered by a thin membrane of polytetrafluoroethylene. The dual

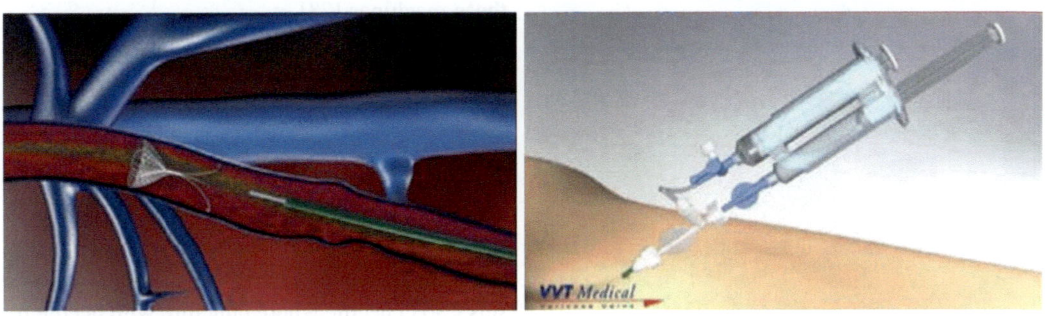

Fig. 13.10 V-Block

syringe system used helps in aspirating blood and simultaneously injecting liquid sclerosant [21].

13.3.1.3.1 Outcomes

Ralf R. Kolvenbach et al. 2019 showed in 51 Patients with varicose veins complete occlusion during the 7-day postprocedural examination. No complication of the procedure was noted with improvement in quality of life. There was a significant drop in follow-up of patients but after 3 years the occlusion rate dropped significantly to 77%.

Considering the low long-term occlusion rates and deploying a foreign body made this technique inferior to the available methods.

13.3.1.4 Foam Sclerotherapy

It is a chemical ablation technique which causes intense inflammation of the veins by injecting sodium tetradecyl sulphate (STS), polidocanol, sodium morrhuate, glycerine or hypertonic saline (Fig. 13.11).

These sclerosants cause irreversible damage of the endothelium of the veins and thus fibrosis. They can be injected as solution in various concentrations depending on the size of the vein or as a foam. The foam is made by mixing air and sclerosant in ratio of 4:1 by Tessari method [22, 23].

This is the most versatile method and can be used to treat any veins including LSV, SSV, perforators or reticular or spider veins.

Sclerotherapy has its advantages of being cheap, does not require local anaesthesia, can be done as outpatient, can be repeated and is associated with quick recovery. It should be avoided in situations like allergy to sclerosant, patent foramen ovale, infections or in cases of acute DVT.

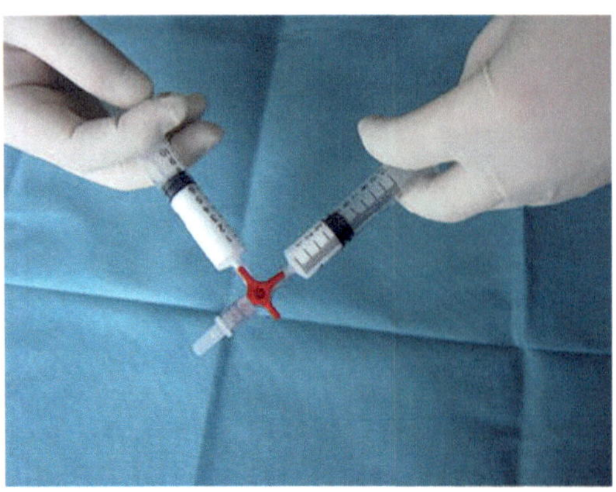

Fig. 13.11 Foam sclerotherapy

It is associated with good cosmetic results in 82% and can cause up to 90% improvement in symptoms. The drawback is poor closer rates in treatment of longer trunk veins of around 70–80% [13].

Sclerotherapy can be associated with skin necrosis and thrombophlebitis as most common complications. Most of these complications happen due to poor technique and knowledge. Other adverse events could be allergic reactions, visual disturbances, hyperpigmentation and deep venous thrombosis [13].

13.3.1.5 Polidocanol Endovenous Microfoam

This is a mixture of polidocanol and gas (mixture of oxygen and carbon dioxide with only trace (0.01–0.08%) nitrogen) present in a canister that produces uniform foam (Fig. 13.12). This is given similar to other agents [24].

13.3.1.6 Cutaneous Lasers and Intense Pulse Light (IPL)

Laser and light source technology are emerging options for the management of small spider veins and telangiectasias. Cutaneous lasers and IPL devices penetrate the skin to treat the vessel of choice without damaging the overlying skin or surrounding tissues (Fig. 13.13) [25]. These modalities are used for veins which are smaller than 3 mm in diameter [26, 27]. The results of the treatment depend on selecting the right wavelength, pulse duration, spot size and fluence with these machines [28].

If vein is deeper we use longer wavelength and if superficial we use shorter wavelength. Pulse duration is ideally matched to the diameter of the vein [29, 30]; larger diameter veins require a longer pulse duration. A large spot size, preferably between 3–8 mm, is optimal to maximize penetration and endothelial destruction [30].

This techniques are specially useful for vessels <30 G needle where puncture with needle for sclerosants could be difficult [31]. This technique

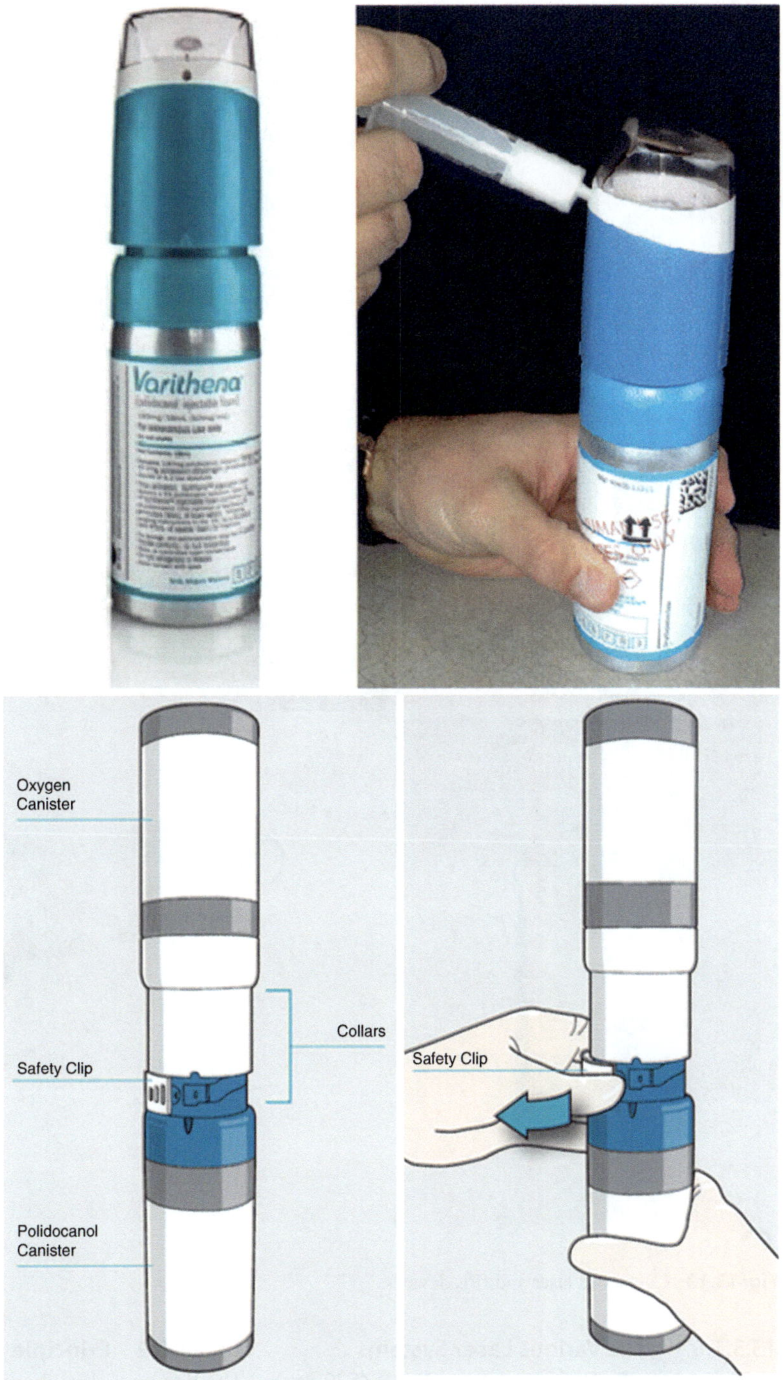

Fig. 13.12 Polidocanol Endovenous microfoam

is also useful for Telangiectatic matting, sclerotherapy resistant vessels and Patients who are afraid of needles.

Cutaneous lasers are not to be used in Patients with tanned skin, patients who have Photosensitivity disorders and who are prone to Hypertrophic and Keloid formation [30].

There are many different technologies available with varying wavelengths to damage the vessels [33].

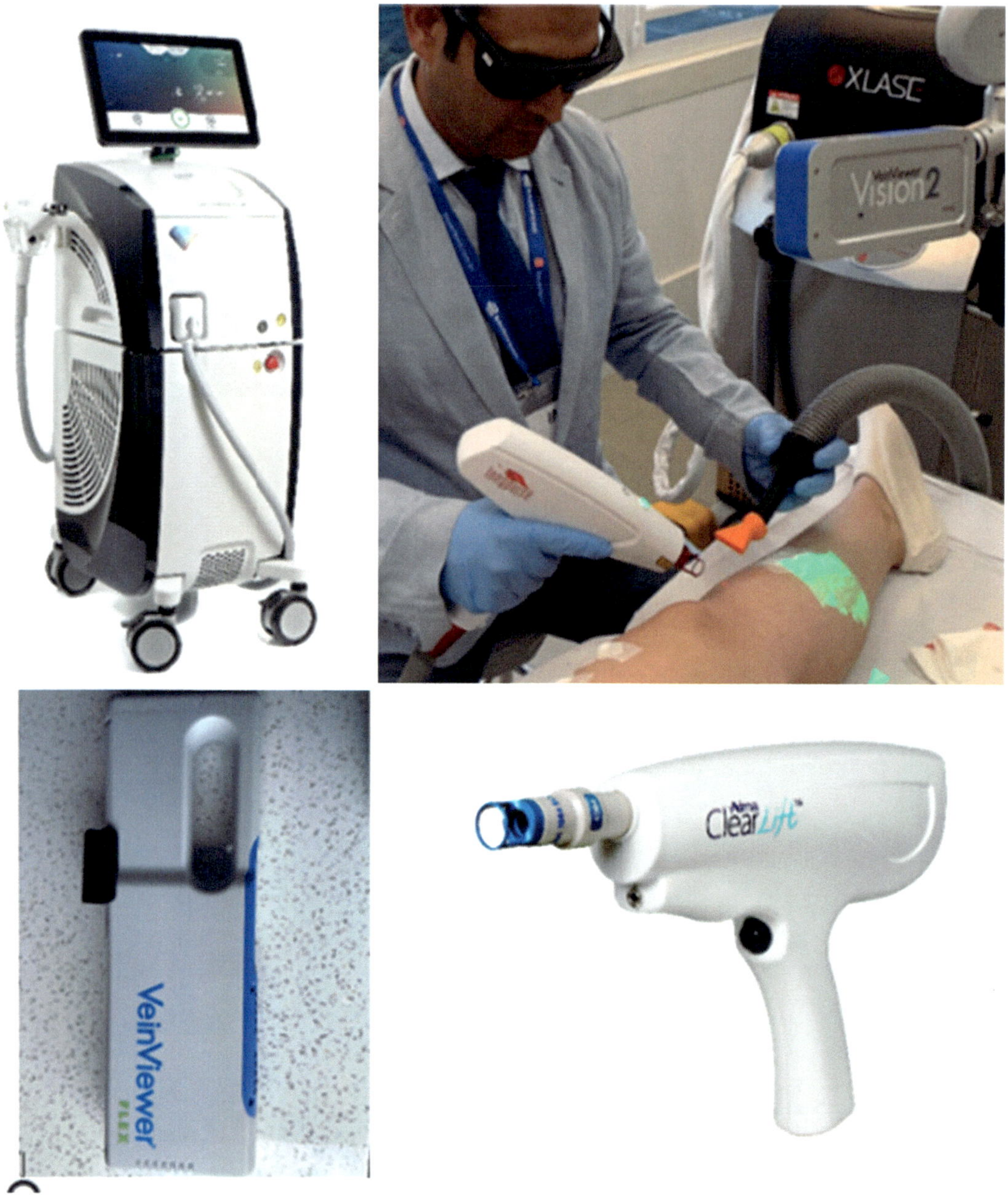

Fig. 13.13 Cutaneous laser and IPL devices

13.3.1.6.1 The Various Laser Systems

- Potassium-Titanyl-phosphate (KTP) (532 nm)
- Pulsed dye (585 to 605 nm)
- Alexandrite (755 nm)
- Diode Laser (810 nm)
- Neodymium:Yttrium aluminium-Garnet (Nd: YAG) (1064 nm)

13.3.1.6.2 Principle

The haemoglobin shows a preferential absorption of the laser light and once the Hb absorbs energy that is converted into thermal energy and lead to occlusion of the vessel after causing the heat damage [29, 32, 33].

13.3.1.6.3 Lasers

If we want to treat telangiectasias <1.0 mm, *then* KTP lasers with a short wavelength of 532 nm with a long pulse duration and low fluence is the best option [34].

Pulsed dye lasers use one of the shortest wavelengths and are likewise recommended for veins smaller than 1.0 mm because a laser with this wavelength also has a relatively short pulse duration [35].

Alexandrite lasers is used to treat larger and deeper veins with a wavelength of 755 nm. Combined with midrange fluence (60 to 70 J/cm2), a short pulse duration (3 msec), and a large spot size (8 mm), an alexandrite laser can treat veins up to 3 mm in diameter [35].

Diode lasers having wavelength of 810-nm have shown good results in veins 3 mm or smaller with use of a very high fluence (210 to 336 J/cm2) with a 50-msec pulse duration and a smaller spot size of 3 mm [36].

The longest-wavelength laser (1064 nm) is system of choice for treating both telangiectasias and reticular veins [30]. This wavelength demonstrates minimal interaction with melanin [37], allowing it to treat darker skin phenotypes effectively [30]. For vessels less than 1.0 mm, a fluence of 400 to 600 J/cm2 combined with a pulse width of 30 to 50 msec and a spot size of 1.5 mm has shown excellent results [38]. With slight modification of the treatment parameters for vessels 1 to 3 mm, success can be achieved with a lower fluence (250–370 J/cm2), similar pulse width (50–60 msec), and twice the spot size (3 mm) [38].

IPL (500–1200 nm) devices use a different mechanism; they emit a spectrum of light between 500 and 1200 nm instead of a single wavelength [30]. This technology provides longer-duration pulses, sequential rapid pulsing and longer wavelengths than laser devices do [29]. A wavelength cut off filter may be used to accommodate skin colour; light skin benefits from a 550-nm and darker skin from either a 570- or 590-nm wavelength [29, 39]. Weiss et al. [40] found greater efficacy when treating both smaller and larger veins with a 12-msec pulse duration in two consecutive pulses, separated by a 20- to 30-msec delay, a fluence of 70 J/cm2 and a 570- to 590-nm cut off filter.

13.3.1.6.4 Procedure

The operating surgeon and the assisting staff including the patient should wear protective eyewear before the device is powered on. Skin cooling is recommended during and after the procedure to prevent further thermal damage to tissues [30]. Once prepared, the handpiece is positioned over the target vein while avoiding pressure great enough to compress the vein. Laser pulses are then delivered one at a time, spaced 1 to 2 mm apart, and overlap of treatment areas is avoided. This process is repeated until the whole length of the vein has undergone laser. If instantaneous changes are not observed after treatment, re-treatment should be delayed at least 5 minutes [39]. Treated areas should be allotted time to cool before a second pass is performed [30].

13.3.1.6.5 *Complications* [30, 41, 42]

Transient urticaria: develops in most patients after treatment, but it typically resolves within a few hours.

- Hyperpigmentation
- Hypopigmentation
- Telangiectatic matting
- Thrombosis
- Epidermal damage
- Purpura and pain

13.4 Conventional Surgery

It has been the treatment of choice for venous insufficiency for more than 50 years. It includes Stripping, Ligation and avulsion of the affected veins.

The traditional open surgery for varicose veins is associated with need for hospitalization, general anaesthesia and postsurgical complications.

Up to half of the patients have recurrence after conventional surgery within 5 years [32, 43, 44] of surgery.

The reason behind recurrence is neovascularization or incomplete treatment.

Various surgical techniques for varicose veins include Ligation, axial stripping and phlebectomies.

13.4.1 High Ligation

The procedure is done under regional anaesthesia. The SF junction is marked preoperatively with duplex for better precision of incision and cosmetic results. Usually there are 6 tributaries draining the GSV, but anatomic variations are seen in few patients. All these tributaries are individually identified and should be ligated independently (Fig. 13.14). Flush ligation of varicose veins has a high recurrence rate so it is always better to combine it with stripping of GSV.

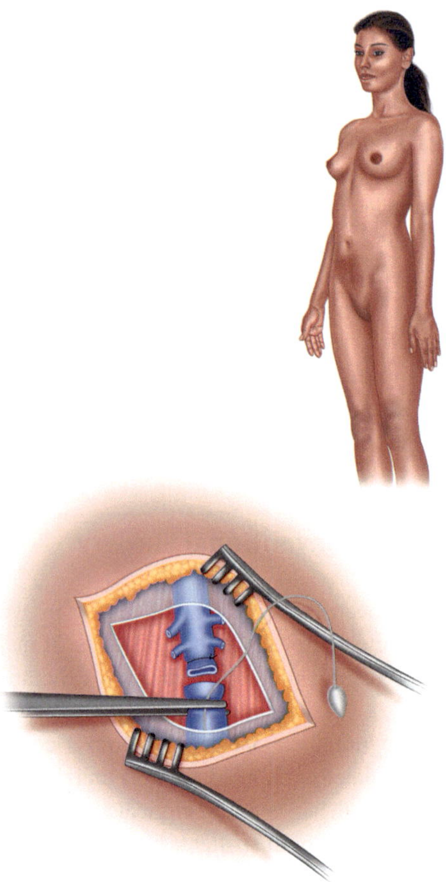

Fig. 13.14 High ligation

13.4.2 Stripping of GSV

SFJ is dissected and flush ligation is performed. Then a stripper is passed distally into the vein through a longitudinal venotomy incision and recovered at the knee through transverse incision (Fig. 13.15). The GSV is stripped in a downward direction. The head size of the stripper should be selected so that tissue injury and bruising are less but, on the other hand, entire vein and its tributaries are recovered.

13.4.3 Stripping Under Tumescent Anaesthesia

This is stripping of the GSV under Tumescent anaesthesia which requires infusion of it into the saphenous sheath under ultrasound guidance.

This anaesthesia allows compression of the tributaries resulting in rapid haemostasis ultimately leading to less pain and bruising.

13.4.4 Saphenous Sparing Operations

- CHIVA (Cure conservatrice et Hémodynamique de l'InsuffisanceVeineuseenAmbulatoire) (Fig. 13.16)
- ASVAL (Ambulatory Selective Varices Ablation)

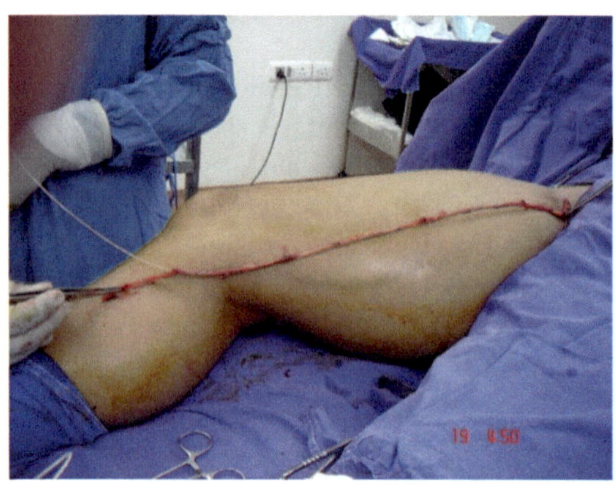

Fig. 13.15 GSV stripping

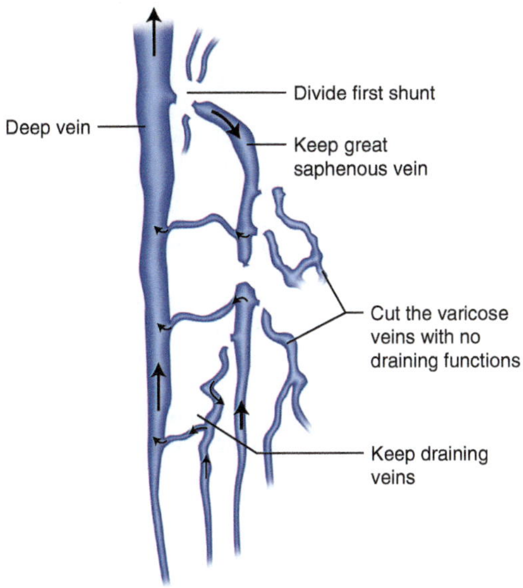

Fig. 13.16 CHIVA

13.4.4.1 CHIVA

The detailed duplex ultrasound study determines the starting point of the reflux, the entry point and where it re-enters. The aim is to promote normal superficial to deep flow through re-entry perforating veins during muscular diastole. Finding competence of the terminal and preterminal valves of the saphenous guides the decision of whether it should be ligated. The tributaries are assessed for competence (flow into saphenous) or incompetence (flow into tributaries and associated varicosities). Incompetent tributaries are ligated and the varicosities removed.

13.4.4.2 ASVAL

This procedure is done under local and tumescent anaesthesia with removal of the venous varicose reservoir (multiple varicosities) through multiple stab incisions (average, 30), generally using Muller phlebectomy hooks. This technique has poor cosmetic results and is associated with high incidence of recurrence.

Newer surgical methods include Cure conservatrice et Hémodynamique de l'InsuffisanceVein euseenAmbulatoire (CHIVA) and ambulatory phlebectomy.

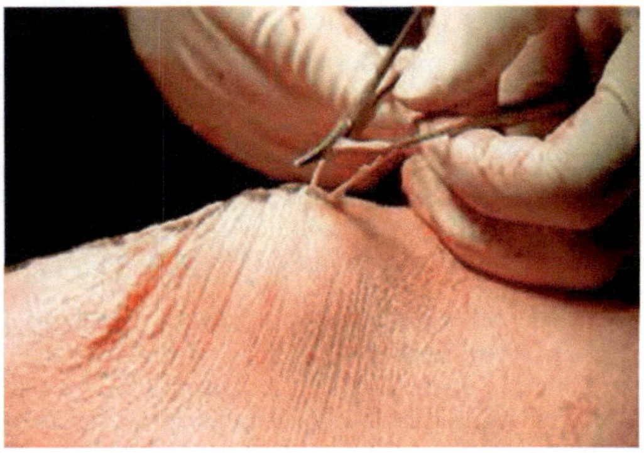

Fig. 13.17 Ambulatory phlebectomy

Ambulatory phlebectomy is a simple procedure where we remove tributaries under local anaesthesia using small stab incisions. CHIVA is, on the other hand, a shunt ligation technique.

13.4.5 Ambulatory Phlebectomy

In most of the cases ligation and/or stripping or laser ablation reduces the axial reflux. What to do with the remaining varicosities which if not treated will drain via alternate pathways and remain both symptomatic and cause cosmetic issues.

Phlebectomies can be done under general, regional anaesthesia (Fig. 13.17). If performed in isolation, then local anaesthesia is a preferred choice. Multiple small stab incisions are given (1–3 mm) and the veins are avulsed.

13.5 Summary

The management of varicose vein can be divided into open surgery including high ligation and stripping, thermal/tumescent (TT), and NTNT technologies. The complication rates are relatively low for all of these. Thermal/tumescent procedures such as RFA and EVLT have shown very good long-term results but do need tumescence anaesthesia which causes some patient discomfort and also require more expertise.

The newer NTNT procedures do not require the tumescent anaesthesia and are less painful but has poorer results than the thermal techniques.

References

1. Callam MJ. Epidemiology of varicose veins. Br J Surg. 1994;81:167–73.
2. Khilnani NM, Grassi CJ, Kundu S, et al. Cardiovascular interventional radiological Society of Europe, American College of Phlebology, and Society of Interventional Radiology Standards of practice committees. Multi-society consensus quality improvement guidelines for the treatment of lower-extremity superficial venous insufficiency with endovenous thermal ablation from the Society of Interventional Radiology, cardiovascular interventional radiological Society of Europe, American College of Phlebology and Canadian Interventional Radiology Association. J VascIntervRadiol. 2010;21(1):14–31.
3. Meissner MH, Gloviczki P, Bergan J, et al. Primary chronic venous disorders. J VascSurg. 2007;46(Suppl S):54S–67S.
4. Gloviczki P, Comerota AJ, Dalsing MC. Et al; Society for Vascular Surgery; American venous forum. The care of patients with varicose veins and associated chronic venous diseases: clinical practice guidelines of the Society for Vascular Surgery and the American venous forum. J VascSurg. 2011;53(5 Suppl):2S–48S.
5. Rasmussen LH, Lawaetz M, Bjoern L, Vennits B, Blemings A, Eklof B. Randomized clinical trial comparing endovenous laser ablation, radiofrequency ablation, foam sclerotherapy and surgical stripping for great saphenous varicose veins. Br J Surg. 2011;98(8):1079–87.
6. van den Bos R, Arends L, Kockaert M, Neumann M, Nijsten T. Endovenous therapies of lower extremity varicosities: a meta-analysis. J VascSurg. 2009;49(1):230–9.
7. Gibson KD, Ferris BL, Polissar N, Neradilek B, Pepper D. Endovenous laser treatment of the small [corrected] saphenous vein: efficacy and complications. J VascSurg. 2007;45(4):795–801. discussion 801–803
8. Lawrence PF, Chandra A, Wu M, et al. Classification of proximal endovenous closure levels and treatment algorithm. J VascSurg. 2010;52(2):388–93.
9. Rathbun S, Norris A, Morrison N, et al. Performance of endovenous foam sclerotherapy in the USA for the treatment of venous disorders: ACP/SVM/AVF/SIR quality improvement guidelines. Phlebology. 2014;29(2):76–82.
10. Morrison N, Gibson K, McEnroe S, et al. Randomized trial comparing cyanoacrylate embolization and radiofrequency ablation for incompetent great saphenous veins (VeClose). J VascSurg. 2015;61(4):985–94.
11. Rasmussen L, Lawaetz M, Bjoern L, et al. Randomized clinical trial comparing endovenous laser ablation and stripping of the great saphenous vein with clinical and duplex outcome after 5 years. J Vasc Surg. 2013;58(2):421–6.
12. Jose IA, Robert JM, Rod R, et al. Cyanoacrylate adhesive for the closure of truncal veins: 60-day swine model results. VascEndovasc Surg. 2011;45(7):631–5.
13. Sadick N, et al. Laser treatment of telangiectasias and reticular veins. In: Bergan J, editor. The vein book. Boston: Elsevier; 2007.
14. Almeida II, Javier JJ, Mackay EG, et al. Two-year follow-up of first human use of cyanoacrylate adhesive for treatment of saphenous vein incompetence. Phlebology. 2015;30(6):397–404.
15. Morrison N, Gibson K, McEnroe S, et al. Randomized trial comparing cyanoacrylate embolization and radiofrequency ablation for incompetent great saphenous veins (VeClose). J Vasc Surg. 2015;61(4):985–94.
16. Mueller RL, Raines JK. ClarWein mechanochemical ablation: background and procedural details. VascEndovascular Surg. 2013;47(3):195–205.
17. Bootun R, Lane TR, Dharmarajah B, et al. Intra-procedural pain score in a randomised controlled trial comparing mechanochemical ablation to radio frequency ablation: the multicentre Venefit versus clan-vein for varicose veins trial. Phlebology. 2016;31(1):61–5.
18. Cavezzi A, Tessari L. Foam sclerotherapy techniques: different gases and methods of preparation, catheter versus direct injection. Phlebology. 2009;24(6):247–51.
19. Tessari L, Cavezzi A, Frullini A. Preliminary experience with a new sclerosing foam in the treatment of varicose veins. Dermatol Surg. 2001;27(1):58–60.
20. Eckmann DM, Kobayashi S, Li M. Microvascular embolization following polidocanol microfoam sclerosant administration. Dermatol Surg. 2005;31(6):636–43.
21. Fournier N, et al. Treatment of leg telangiectasias with a 532 nm KTP laser in multi-pulse model. Dermatol Surg. 2002;28:564–71.
22. Anderson RR, et al. Mechanisms of selective vascular changes caused by dye lasers. Lasers Surg Med. 1983;3:211–5.
23. Goldman MP, et al. Sclerotherapy treatment of varicose and telangiectatic leg veins. 3rd ed. St Louis: CV Mosby; 2001.
24. Lupton J, et al. Clinical comparison of sclerotherapy versus long-pulsed Nd:YAG laser treatment for lower extremity telangiectasias. Dermatol Surg. 2002;28:694–7.
25. Van Rij AM, Jiang P, Solomon C, Christie RA, Hill GB. Recurrence after varicose vein surgery: a prospective long-term clinical study with duplex ultrasound scanning and air plethysmography. J VascSurg. 2003;38(5):935–43.
26. Babilas P, et al. Selective photothermolysis of blood vessels following flashlamp-pumped pulsed dye laser irradiation: in vivo results and mathemati-

cal modeling are in agreement. J Invest Dermatol. 2005;125:343–52.
27. Goldberg D. Evaluation of KTP/532nm laser with StarPulse for treatment of leg telangiectasia. Available at http://www.iridex.com/pdf/ktpevaluation.pdf; 2008. Accessed April 24.
28. Goldman MP, et al. Pulse-dye laser treatment of leg telangiectasia: with and without simultaneous sclerotherapy. J Dermatol Surg. 1990;16:338–44.
29. Weiss RA, et al. Early clinical results with a multiple synchronized pulse 1064 nm laser for leg telangiectasias and reticular veins. Dermatol Surg. 1999;25:399–402.
30. Sadick NS. Laser treatment with a 1064-nm laser for lower extremity class I-III veins employing variable spots and pulse width parameters. Dermatol Surg. 2003;29:916–9.
31. Weiss RA, et al. Non-coherent filtered flashlamp intense pulsed light source for leg telangiectasias: long pulse durations for improved results. Lasers Surg Med. 1998;10(Suppl):40.
32. Carradice D, Mekako AI, Mazari FA, Samuel N, Hatfield J, Chetter IC. Randomized clinical trial of endovenous laser ablation compared with conventional surgery for great saphenous varicose veins. Br J Surg. 2011;98(4):501–10.
33. Eremia S, et al. A side-by-side comparative study of 1064 nm Nd AG, 810 nm diode and 755 nm alexandrite lasers for treatment of 0.3-3 mm leg veins. Dermatol Surg. 2002;28:224–30.
34. Rogachefsky AS, et al. Nd AG laser (1064 nm) irradiation for lower extremity telangiectases and small reticular veins: efficacy as measured by vessel color and size. Dermatol Surg. 2002;28:220–3.
35. Royle JP. Recurrent varicose veins. World J Surg. 1986;10(6):944–53.
36. Blomgren L, Johansson G, Dahlberg A, Kerman A, et al. Recurrent varicose veins: incidence, risk factors and groin anatomy. Eur J VascEndovascSurg. 2004;27(3):269–74.
37. Alexander CJ. The epidemiology of varicose veins. Med J Aust. 1972;1(5):215–8.
38. Proebstle TM, Alm J, Dimitri S, et al. The European multicenter cohort study on cyanoacrylate embolization of refluxing great saphenous veins. I VascSurgVenousLymphatDisord. 2015;3(1):2–7.
39. Todd KL 3rd, Wright D, Orfe E. The durability of Polidocanol endogenous microfoam treatment effect on varicose vein symptoms and appearance in patients with Saphenofemoral junction incompetence: one-year results from the VANISH. Study J VascSurg Venous LymphatDisord. 2014;2(1):12.
40. Sadick NS. Laser treatment of leg veins. Skin Therapy Lett. 2004;9.
41. Trelles MA, et al. The 800-nm diode laser in the treatment of leg veins: assessment at 6 months. J Am Acad Dermatol. 2006;54:282–9.
42. Weiss RA, et al. Vein diagnosis and treatment: a comprehensive approach. New York: McGraw-Hill; 2001. p. 211–21.
43. Mundy L, Merlin TL, Fitridge RA, Hiller JE. Systematic review of endovenous laser treatment for varicose veins. Br J Surg. 2005;92(10):1189–94.
44. Disselhoff BC, der Kinderen DJ, Kelder JC, Moll FL. Randomized clinical trial comparing endovenous laser with cryostripping for great saphenous varicose veins. Br J Surg. 2008;95(10):1232–8.

Chapter on Testosterone Therapy

14

Carolyn DeLucia

Carolyn DeLucia went to medical school at New York Medical College followed by her internship and residency at Long Island Jewish Medical Center in New York. After several years of Obstetrics and Gynecology she has evolved her practice to gynecology and female sexual health. With her interest growing in preventative techniques she opened ViVa Rejuvenation center in Hillsborough New Jersey and the Secret Orchid in NYC. She is the author of best seller ULTIMATE INTIMACY, the Revolutionary Science of Female Sexual Health.

She is a faculty member for the Cellular Medicine Association, the American Aesthetic Association, the International Society of Regenerative Medicine, and the Intimate Wellness Institute.

14.1 Introduction

Testosterone is thought of as male hormone. However it is critically important for women as well. Testosterone or androgens are released from two locations in the female body. 80% of testosterone is created by the ovary and 20% is released by the adrenal gland. It is not valued as much as it should be by the medical community. The most critical factor is that there is not an accepted normal range of testosterone in women worldwide. The impact of testosterone on women tends to improve muscle growth and diminish fat. It also helps with mood and a sense of well-being. It is critical in libido, sexual sensation and response. As a woman goes through life this is critical. The impact on social interaction is greatly underestimated. Replacement of testosterone in women has been used for over 30 years. The methods used to replace testosterone have varied from oral synthetic forms to bio identical forms that come in topical gels, injections, troches, and pellets. The history and development of these options will be explored in this chapter [1, 3, 6, 23].

C. DeLucia (✉)
Hillsborough, NJ, USA

P. Jindal et al. (eds.), *Aesthetic and Regenerative Gynecology*,
https://doi.org/10.1007/978-981-16-1743-0_14

The mainly accepted use of testosterone (T) is for hypoactive sexual desire disorder (HSDD).Common causes of low libido in menopausal women include menopausal symptoms including insomnia, urinary incontinence, breast cancer or another major medical problem, weight gain, poor body image, genitourinary syndrome of menopause (GSM) with dyspareunia, fatigue, stress, aging, relationship duration issues or conflict, and a partner's sexual problems. These areas should be treated as best as possible before using testosterone because unresolved psychological factors or relationship problems are not solved by hormone therapy alone [4, 22, 24–26].

14.2 Basic Physiology of Testosterone Production, Function, and Metabolism

As mentioned above, androgens are produced in the female body in the ovary and the adrenal gland. The ovary secretes testosterone as well as DHEAS and androstenedione. The adrenal gland secrets DHEAS and androstenedione that gets converted into testosterone. It is also important to mention that estrone can be converted into testosterone in the peripheral tissues. The adrenal gland androgens are inactive but get converted at a slow rate in the periphery to testosterone. Only five percent of these hormones are converted to testosterone. The circulating testosterone gets metabolized to and excreted as androstenedione and etiocholanolone and passed in the urine. Circulating testosterone is bound to Sex Hormone Binding Globulin (SHBG) 85% of the time and is considered inactive when bound. The active form of testosterone occurs at the cellular level where it is converted to DHT, 5 alpha-dihydrotesterone. There are several conditions that can cause elevated testosterone but we will be focusing on replacing testosterone in cases of deficiency that happens with aging and ovarian failure [5, 22, 31–34, 36].

14.2.1 Testosterone Excess

Most often patients present with irregular menstrual cycles and hirsutism. They are evaluated for endogenous androgen excess. These conditions are of either ovarian or adrenal origin. The Poly Cystic ovarian syndrome (PCOS) is the most common condition. It occurs in reproductive age women with prevalence of 4–10%. It is characterized by irregular periods, male pattern hair growth, and acne. Upon pelvic ultrasound the ovaries seem to have multiple small follicles rather than a dominant follicle. The appearance may be described as a string of pearls. The adrenal excess may be congenital adrenal hyperplasia and on lab values there is an elevation on DHEAS, 17-OH Progesterone, and testosterone. The physical findings are similar in both conditions. They are both associated with increased risk for metabolic syndrome and respond to treatment aimed at correcting insulin resistance. However the exogenous increase in testosterone levels achieved by testosterone treatment does have the same affect [27, 29].

14.2.2 Testosterone Deficiency

If a woman loses ovarian function for any reason whether it be surgical removal, natural menopause, or premature ovarian failure; then circulating testosterone will be very low. Symptoms of deficiency include emotional liability, disturbance of sleep, loss of libido, loss of sensation, muscle weakness, fatigue, breast soreness, lack of focus and memory. These have devastating effects on everyday function.

Measuring testosterone is a challenge in both men and women. The levels of most hormones in the body vary from hour to hour, day to day. The best way to get an idea of testosterone is to measure total testosterone taking into consideration bound and free testosterone. Measuring SHBG is helpful in a way to evaluate non-responders and to help assess if the levels are being bonded and excreted without having efficient end organ function.

Table 14.1 Testosterone treatment options for menopausal women

Topical compounded testosterone
1% Testosterone compounded cream or gel
Apply 0.5 g topically to thigh or low abdomen daily
Monitor testosterone level approximately 2 weeks after receiving a new tube or jar of compounded cream or gel to confirm that the testosterone level remains within the normal range for reproductive-aged women
Dose-adjusted topical testosterone product approved for men
1% Testosterone gel (Testim)
Apply 3–4 drops topically to thigh or low abdomen daily (warm gel slightly before use)
Re-cap tube tightly after use
Reduce dose as needed to ensure 1 tube lasts for 10 days
Monitor testosterone level approximately every 3–6 months to confirm testosterone level remains within the normal range for reproductive-aged women
Testosterone formulations not recommended
Due to risk of supraphysiologic dosing
• Testosterone injections
• Testosterone implants or subcutaneous pellets
Due to adverse effects associated with oral administration
• Oral methyltestosterone

Unfortunately serum testosterone levels are variable, inconsistent, and vary person to person and even within the same person never directly reflecting end organ response or symptom relief. This will be relevant when we speak about dosing. Testosterone levels should not be tested as an efficacy measure, as there is no testosterone level that will assure a satisfactory sex life [34, 35, 37–39].

14.2.3 Testosterone Replacement Methods

Replacing Testosterone when found to be deficient and symptomatic can be done using topical gels, oral troches, or injections or pellets. Treatment with compounded bioidentical creams or gels is very common as shown in Table 14.1. The delivery method does have weaknesses due to batch to batch variance in dosing and absorption changes daily as well. Controlling or regulating levels is challenging using gels or cream for these reasons. Another issue with topical testosterone is transference. It can be transferred to a partner, a baby, or a pet.

A study of intramuscular (IM) testosterone administration in seventy one women estrogen-treated, hysterectomized, aged 45 to 70 years suggests that significant increases in sexual desire and activity with testosterone occur only with supraphysiologic dosing. Women were randomized to receive weekly IM injections of placebo or testosterone (3 mg, 6.25 mg, 12.5 mg, or 25 mg) for 24 weeks. Total testosterone levels ranged from19 ng/dL with placebo to as high as 210 ng/dL with 25-mg IM testosterone, with the highest doses of testosterone resulting in hormone concentrations approaching the lower limit of normal for men.

Table 14.2 Historical dosing testosterone pellet implants, relief of symptoms: end organ response, no serious adverse side effects

Author	Year	Dose	Indications
Greenblatt	1949	75, 150, 225mg (1-3, 75 mg testosterone pt implants)	• Menopausal syndrome • Prevent uterine bleeding caused by estrogens • Dysmenorrheic patients with endometriosis or small fibroids • Fibromyomata • Nocturia • Increased libido desired • Palliative measure: carcinoma of the breast • Addison's disease (Adrenal fatigue)
Jones	2004	50-150 mg most common 100 mg	Reduced the incidence of breast cancer • Increased bone density • Memory and concentration • Sex drive, libido • Strength
Gambrell/Natrajan	2006	75 , 150 (most common), 225 mg	• To Treat symptons of T deficiency • No increase in the incidence of breast cancer with estradiol (E2) 50-75 mg implant (with T) • No adverse affects on lipids or lipoproteins • Mean Serum T levels 80-262 ng/dL • Implanted 3-6 m up to 29 y fu
Natrajan	2002	75, 150 (most common), 225 mg	• No Increase in recurrence of breast cancer (with or without estrogen therapy)
Tutera	Submitted for publication	75-225 mg	• 967 women treated for 10 years • 1 case of breast cancer at year 2
Burger	1984	100mg. Levels peaked at one month 101-283 ng/dL and declined by month 5 (38-106 ng/dL; average 71.4 ng/dL) at which time symptoms returned and pellets were reinserted	Un-supplemented ranges • No symptoms of testosterone excess • No change in lipid profile • One patient with mild increase in facial hair • Symptoms improved • Libido • Sexual satisfaction • Fatigue • Memory/concentration
Thom	1981	Dose: 100 mg, 200 mg • T levels mo. 1 100 mg 143 ng/dL (range 60-171 ng/dL) • T levels mo. 1 200 mg 174 ng/dL (range 148-326 ng/dL) • No change in FSH, did increase E2(not ss)	100 mg worked as well as 200 mg for libido
Brincat	1984	100mg	• Lethargy • Depression • Loss of libido
Brincat	1987	100 (with E2 50)	Increase collagen
Cardoza	1984	100mg	Headache, insomnia, hot flushes, bone pain, palpitations, dyparunia, libido, irritability, memory, concentration, depression, lethargy, urethral syndrome • Re-implant as early as 3 months (3-12) • Acne 2%, Increase facial hair 20% • No other symptoms of excess T • Return of Symptoms when levels in upper normal range
Barlow	1986	100 mg (with E2 50)	
Montgomery	1987	T 100 mg (E2 50)	Improved: anxiety, somatic disturbances, Depression

Table 14.2 (continued)

Dow	1983	T 100 (E2 50)	Psychological complaints, somatic complaints, vasomotor Sxs, libido, response, dyspareunia
Studd	1990	100 mg (with E2 75)	• Increased bone density
Garnett	1991	Testosterone 100 mg q 6 mos. (with E2 50-75) up to 24 Years (mean5.2y) • Dose thought to be insignificant for bone density • Testosterone levels at re-implantation .91-1.51 nmol/L 11.4-169 ng/dL	• Libido • Mood • Depression
Saavas	1992	T 100 (with E2 75) increased bmd	
Davis	1995	50mg(with E50)	Increased bmd and sexuality
Davis	2000	T 50 (E2 50)	
Seed	2000	T 50 (with E2 50)	not attenuate benefits of E2 on LDL TG
Sands	1997	T 100 (E2 50)	not accentuate benefits of E2 on lipid metabolism
Worboys	2000	T 50	Improved blood flow, brachial artery vaso-dilation
Loeser	1941		• T implants to treat breast cancer • Prevented recurrence • Improvement/control in metastatic disease
		500-700 mg implant. Most common 600 mg (every 6 Months)	• Symptoms of excess (hoarse voice, enlarged Clitoris, increased hair) Treats breast cancer • Re-implantation at return of mets then every 6 months • Only hoarse voice remained/continued with q 6 Month implantation
		1500-1850 mg	use in metastatic disease Obvious improved health' from bedridden state

Significant increases compared with placebo for libido and frequency of sexual activity were seen only in women assigned to the 25-mg testosterone group. Sexual encounters per week increased by approximately 0.5 in women treated with placebo and lower testosterone doses compared with an increase of 2.7 per week in women assigned to the highest testosterone dose [13, 17, 19, 35].

Pellets have been around for 40 years and are a very consistent delivery system that is gaining popularity worldwide (Table 14.2). They offer a compounded bioidentical option delivering very consistent levels of hormone evenly over several months as the body gradually degrades them [10–18, 20, 21, 28, 30, 31, 33].

14.3 Pellet Insertion

With the use of trocars the insertion technique is extremely simple. The location of choice is in the adipose tissue overlying the gluteal muscles. The area is cleaned using an antiseptic liquid. The local is injected using a 22 gauge one and half inch needle at a thirty degree angle. Approximately 10 cc of Lidocaine 1% with epinephrine is used injected along the expected path of the trocar. The 3 mm skin incision is made using an 11 blade which is the diameter of the trocar. The trocar has a window in which to place the sterile pellet and a blunt ended introducer is used to slide the pellets into the adipose tissue. The trocar is then removed and steri-strips and bandage are applied. It is not necessary to stitch the incision. It is recommended to keep the bandage dry for two days and allow the steri-strips to fall off. The complication rate of infection or pellet expulsion is very low [33, 34].

14.4 Conclusion

Testosterone is an essential and critical hormone in both men and women. With appropriate levels it is an anti-aging factor improving the quality of life. Women should be aware of the influences of bioidentical hormone therapy. They should be given the option to optimize their aging health. Physician should share the therapy as an option where it is clinically necessary [2, 7, 8, 10, 9, 14, 27].

References

1. The North American Menopause Society Hormone Therapy Position Statement Advisory Panel. The 2017 hormone therapy position statement of The North American Menopause Society. Menopause. 2017;24:728–53.
2. Simon JA, Nappi RE, Kingsberg SA, et al. Clarifying Vaginal Atrophy's Impact on Sex and Relationships (CLOSER) survey: emotional and physical impact of vaginal discomfort on North American postmenopausal women and their partners. Menopause. 2014;21:137–42.
3. The North American Menopause Society. Management of symptomatic vulvovaginal atrophy: 2013 position statement of The North American Menopause Society. Menopause. 2013;20:888–902.
4. Shifren J, Monz B, Russo P, et al. Sexual problems and distress in United States women: prevalence and correlates. Obstet Gynecol. 2008;112:970–8.
5. Davison S, Bell R, Donath S, et al. Androgen levels in adult females: changes with age, menopause, and oophorectomy. J Clin Endocrinol Metab. 2005;90:3847–53.
6. Shifren JL, Desindes S, McIlwain M, et al. A randomized, open-label, crossover study comparing the effects of oral versus transdermal estrogen therapy on serum androgens, thyroid hormones, and adrenal hormones in naturally menopausal women. Menopause. 2007;14:985–94.
7. Davis SR, Davison SL, Donath S, et al. Circulating androgen levels and self-reported sexual function in women. JAMA. 2005;294:91–6.
8. Wahlin-Jacobsen S, Pedersen AT, Kristensen E, et al. Is there a correlation between androgens and sexual desire in women? J Sex Med. 2015;12:358–73.
9. Randolph JF Jr, Zheng H, Avis NE, et al. Masturbation frequency and sexual function domains are associated with serum reproductive hormone levels across the menopausal transition. J Clin Endocrinol Metab. 2015;100:258–66.
10. Dennerstein L, Lehert P, Burger H. The relative effects of hormones and relationship factors on sexual function of women through the natural menopausal transition. Fertil Steril. 2005;84:174–80.
11. Shifren JL, Davis SR. Androgens in postmenopausal women: a review. Menopause. 2017;24:970–9.
12. Sherwin BB, Gelfand MM, Brender W. Androgen enhances sexual motivation in females: a prospective, crossover study of sex steroid administration in the surgical menopause. Psychosom Med. 1985;47:339–51.
13. Huang G, Basaria S, Travison TG, et al. Testosterone dose response relationships in hysterectomized women with or without oophorectomy: effects on sexual function, body composition, muscle performance and physical function in a randomized trial. Menopause. 2014;21:612–23.
14. Shifren JL, Braunstein GD, Simon JA, et al. Transdermal testosterone treatment in women with impaired sexual function after oophorectomy. N Engl J Med. 2000;343:682–8.
15. Simon J, Braunstein G, Nachtigall L, et al. Testosterone patch increases sexual activity and desire in surgically menopausal women with hypoactive sexual desire disorder. J Clin Endocrinol Metab. 2005;90:5226–33.
16. Shifren JL, Davis SR, Moreau M, et al. Testosterone patch for the treatment of hypoactive sexual desire disorder in naturally menopausal women: results from the INTIMATE NM1 study. Menopause. 2006;13:770–9.
17. Davis SR, Moreau M, Kroll R, APHRODITE Study Team, et al. Testosterone for low libido in postmenopausal women not taking estrogen. N Engl J Med. 2008;359:2005–17.
18. Kingsberg S, Shifren J, Wekselman K, et al. Evaluation of the clinical relevance of benefits associated with transdermal testosterone treatment in postmenopausal women with hypoactive sexual desire disorder. J Sex Med. 2007;4:1001–8.
19. Wierman ME, Arlt W, Basson R, et al. Androgen therapy in women: a reappraisal: an Endocrine Society clinical practice guideline. J Clin Endocrinol Metab. 2014;99:3489–510.
20. Snabes M, Zborowski J, Simes S. Libigel (testosterone gel) does not differentiate from placebo therapy in the treatment of hypoactive sexual desire in postmenopausal women (abstract). J Sex Med. 2012;9(suppl 3):171.
21. Lobo RA, Rosen RC, Yang HM, et al. Comparative effects of oral esterified estrogens with and without methyltestosterone on endocrine profiles and dimensions of sexual function in postmenopausal women with hypoactive sexual desire. Fertil Steril. 2003;79:1341–52.
22. Somboonporn W, Davis S, Seif M, et al. Testosterone for peri and postmenopausal women. Cochrane Database Syst Rev. 2005;19:CD004509.
23. American College of Obstetricians and Gynecologists Committee on Gynecologic Practice and American Society for Reproductive Medicine. Committee opinion 532: compounded bioidentical menopausal hormone therapy. Obstet Gynecol. 2012;2(1):411–5.
24. Fooladi E, Reuter SE, Bell RJ, et al. Pharmacokinetics of a transdermal testosterone cream in healthy postmenopausal women. Menopause. 2015;22:44–9.
25. Braunstein GD, Sundwall DA, Katz M, et al. Safety and efficacy of a testosterone patch for the treatment of

hypoactive sexual desire disorder in surgically menopausal women: a randomized, placebo-controlled trial. Arch Intern Med. 2005;165:1582–9.
26. Greenblatt RB, Oettinger M, Bohler CS. Estrogen-androgen levels in aging men and women: therapeutic considerations. J Am Geriatr Soc. 1976 Apr;24(4):173–178. https://doi.org/10.1111/j.1532-5415.1976.tb04294.x. PMID: 1254880.
27. Nieschlag E, Behre HM, Bouchard P, Corrales JJ, Jones TH, Stalla GK, Webb SM, Wu FC. Testosterone replacement therapy: current trends and future directions. Hum Reprod Update. 2004 Sep-Oct;10(5):409–419. https://doi.org/10.1093/humupd/dmh035. Epub 2004 Aug 5.PMID: 15297434.
28. Gambrell RD Jr, Natrajan PK. Moderate dosage estrogen–androgen therapy improves continuation rates in postmenopausal women: impact of the WHI reports. Climacteric. 2006;9:224–233. Received 06 Jul 2005, Accepted 13 Feb 2006, Published online: 26 Aug 2009.
29. Natrajan PK, Gambrell RD Jr. Estrogen replacement therapy in patients with early breast cancer. Am J Obstet Gynecol. 2002 Aug;187(2):289–294. https://doi.org/10.1067/mob.2002.125999; discussion 294–5.
30. Burger HG, Hailes J, Menelaus M, Nelson J, Hudson B, Balazs N. The management of persistent menopausal symptoms with oestradiol-testosterone implants: clinical, lipid and hormonal results. Maturitas. 1984 Dec;6(4):351–358. https://doi.org/10.1016/0378-5122(84)90008-2. PMID: 6442755.
31. Studd JW, Thom MH. Ovarian failure and ageing. Clin Endocrinol Metab 1981 Mar;10(1):89–113. https://doi.org/10.1016/s0300-595x(81)80040-0. PMID: 6784985.
32. Brincat M, Magos A, Studd JW, Cardozo LD, O'Dowd T, Wardle PJ, Cooper D. Subcutaneous hormone implants for the control of climacteric symptoms. A prospective study. Lancet. 1984 Jan 7;1(8367):16–18. https://doi.org/10.1016/s0140-6736(84)90183-1. PMID: 6140343.
33. Montgomery JC, Appleby L, Brincat M, Versi E, Tapp A, Fenwick PB, Studd JW. Effect of oestrogen and testosterone implants on psychological disorders in the climacteric. Lancet 1987 Feb 7;1(8528):297–299. https://doi.org/10.1016/s0140-6736(87)92026-5. PMID: 2880114.
34. Roberts AD, Al Azzawi F, Leggate I, Hart DM, Barlow DH, Abdalla HI. Long-term hormone implant therapy—hormonal and clinical effects. Obstet Gynecol 1986 Mar;67(3):321–325. PMID: 3945443.
35. Dow MGT, Hart DM, Forrest CA. Hormonal treatments of sexual unresponsiveness in postmenopausal women: a comparative study. Br J Obstet Gynaecol. 1983;90(4):361–6.
36. Studd J, Savvas M, Waston N, Garnett T, Fogelman I, Cooper D. The relationship between plasma estradiol and the increase in bone density in postmenopausal women after treatment with subcutaneous hormone implants. Am J Obstet Gynecol. 1990 Nov;163(5 Pt 1):1474–1479. https://doi.org/10.1016/0002-9378(90)90608-a. 2240090.
37. Garnett T, Studd J, Watson N, Savvas M. A cross-sectional study of the effects of long-term percutaneous hormone replacement therapy on bone density. Obstet Gynecol. 1991;78(6):1002–7.
38. Davis SR, McCloud P, Strauss BJ, Burger H. Testosterone enhances estradiol's effects on postmenopausal bone density and sexuality. Maturitas. 1995;21(3):227–36. https://doi.org/10.1016/0378-5122(94)00898-h.
39. Shifren JL. Testosterone for midlife women: the hormone of desire? Menopause. 2015;22:1147–9.

15 Laser Hair Removal

A. B. Prabhu

15.1 Introduction

Laser stands for Light Amplification by Stimulated Emission of Radiation, has brought in an era of precision and power to many a field of science and led to its advancement. This technology and terminology has been around for nearly 70 years. The term Laser was coined by Gordon Gould in 1957, and within the next decade or so it found immense applications in communication, spectroscopy, cutting and welding industry, and also in medicine.

The initial medical use of lasers was restricted to the field of ophthalmology, but under the entrepreneurship and leadership of Dr. Leon Goldman a dermatologist the evolution of lasers in medical field began to take place.

In the past half century, Lasers have been used to treat epidermis and dermis but because of its unwanted side effects soon fell into disrepute. But with advent of time and technological inventions like CO2 laser, interest began to develop again in it.

Laser technology is safe now with the application of the "Theory of Selective Photothermolysis" by Rox Anderson. Since then Lasers has been used to treat various pigmentatory lesions, vascular pathology, unwanted hairs, and issues related to aging.

Lasers are now becoming an integral part of any cosmetic practice in the world.

In gynecology the role of lasers was limited to ablation only or during colposcopy procedure. With an increase in demand, due to awareness about one's own sexuality as well as an increase in longevity, there is a new and potentially challenging field, in use of lasers in gynecology.

15.1.1 Laser and Light Assisted Hair Removal

The discovery of Lasers in the use of hair removal was by serendipity. It was noticed that when a Q switched laser was used to remove a tattoo also caused some reduction of hair in that region.

Because of lack of complete knowledge and rapid commercialization, laser hair removal became quite popular in the late 90s but soon fell into disrepute due to lack of results. With better understanding of this technology and applying the theory of selective photothermolysis, the efficiency and the outcome of results have improved significantly.

Unwanted hair or excessive hair growth makes an individual lose their confidence and can have a negative impact in their psyche. In this present era, having a smooth hairless appearance not

A. B. Prabhu (✉)
Plastic, Cosmetic & Reconstructive Surgery, Fortis Hospital, Mohali, India

P. Jindal et al. (eds.), *Aesthetic and Regenerative Gynecology*,
https://doi.org/10.1007/978-981-16-1743-0_15

only makes one feel good but also makes them socially acceptable.

Hair removal methods have been traditionally by plucking, waxing, use of depilatory cream, shaving, etc. These major draw backs of these methods are that they have to be done frequently, time consuming, and can cause quite a few irritations to skin like cuts or boils and can also cause pigmentations. Besides these techniques are temporary and have to be repeated. Although a majority of people opt for these above methods. Nowadays significant number of them look for a permanent or long-lasting solution. A permanent hair removal is a fallacy and the correct term to be used is permanent hair reduction.

Before we delve further, certain terms and the principles of hair growth needs to be thoroughly understood.

The reader is guided to relevant anatomical textbooks for relevant structural details of hair. One need to know about the phases of development of hair growth cycle consists of anagen, catagen and telogen.

- Anagen: phase of active hair growth
- Catagen: transitional phase of hair growth
- Telogen: resting phase of hair growth

In the human body the length of hair cycle varies from location to location. Hair in the scalp is usually in a longer anagen phase as compared to the limbs. It is imperative to understand that only during the period of active anagen phase, the hair bulb area containing maximum of melanin lies in close proximity to the bulge area of the hair structure as compared to other phases of hair cycle. Hence when a laser is applied to a hair root in the anagen phase, it causes sufficient damage to effect a good hair removal. Counseling the same to the patient about this and also explaining that hair in the surrounding areas which were in the catagen or telogen phase can revert to anagen phase, puts the mind at rest of achieving optimal results, which will happen eventually.

Selective photothermolysis is laser light of selective wavelength and sufficient fluence, with pulse duration shorter than the cooling time of the target chromophore. Lasers affect the target chromophore, either melanin or hemoglobin or water and the light energy gets converted to heat and destroys the target chromophore by either photochemical, photomechanical, or photothermal destruction. Cooling of the epidermis and the surrounding skin prevents any damage and decreases the unwanted side effects.

Laser assisted hair removal causes majority of hairs reduction, which do not grow back for a longer period of time. A typical laser therapy can nearly complete but temporary hair loss.

15.1.2 Principle of Hair Removal

When a sharp, focused monochromatic, light of a particular wave length, hits a target chromophore, the effect it causes leads to photothermal destruction, photomechanical or photochemical destruction.

15.1.3 Photothermal Destruction

Melanin is an ubiquitous pigment which is predominantly found in the hair or epidermis. Even in the hair it is distributed in the hair bulge area during anagen phase, parts of hair shaft, and outer root sheath. Essentially the bulge area, which contains the stem cells for hair regrowth needs to be ablated for permanent hair reduction.

This bulge area is present close to the bulb during anagen phase, hence laser therapy is effective during anagen phase. For effective photothermal destruction, certain laser parameters like wavelength of laser, fluence, pulse duration, spot size, and cooling need to considered.

15.1.4 Wavelength

We need to understand that longer the wavelength, deeper the penetration, the wavelength of ruby, alexandrite, and diode is from 600 to 800 nanometers which is the effective range of absorption for targeted chromophore melanin of hair roots.

So a wavelength ranging from 600–1000 nanometers is in the ideal optical window which penetrates the skin up to a length of 2–4 mm where the targeted hair root, with melanin is situated.

15.1.5 Pulse Duration

Defined as the amount of time the tissue is exposed to laser light. Whenever a laser beam hits a tissue, heat is produced and dissipated.

Heat in smaller tissues will dissipate quicker than in other complex structure. For example, if the targeted chromophore is melanin, which is present in epidermis as well as hair roots, because of increased concentration of melanin in hair roots. Laser will be specifically absorbed by the melanin in hair roots, which is, converted to heat, and leads to destruction of the hair roots. This basic principle of laser is called selective photothermolysis.

15.1.6 Fluence

It is associated with safety and pain tolerance. By using the highest fluence one can cause effective destruction of the targeted chromophore but at the same time it causes pain, burns sometimes blistering even leading to scarring, especially in skin types IV–VI. To reduce this risk one has to use optimal fluence.

15.1.7 Spot Size

Larger the spot size, deeper the penetration of laser. Effective and successful outcomes are possible if the laser penetrates an ideal depth of 5–10 mm where the targeted chromophore is located. But the corollary is also true that smaller the spot size then lesser is the pain and discomfort. This is one of the important factors in determining patient's compliance. Also as sessions of laser therapy progresses, the hair roots become thinner and superficial. Hence one can decrease the spot size and increase the fluence to acquire good results.

15.2 Indications for Hair Removal

It is mostly for cosmesis and also associated with medical conditions.

Treatment of hirsutism in females, which is associated very commonly with polycystic ovarian disease, needs thorough evaluation and treatment while concurrently undergoing Laser

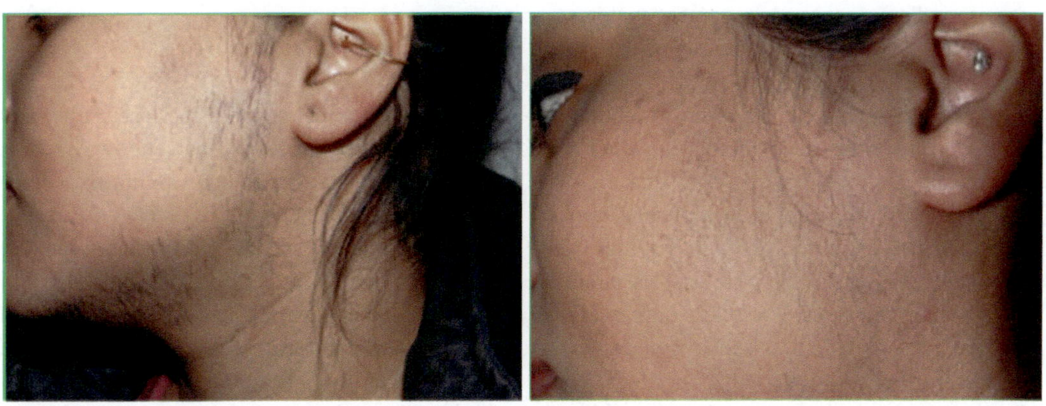

Fig. 15.1 Laser hair reduction before and after

therapy in order to optimize the results (Fig. 15.1). The other conditions which require treatment by laser are Pseudofolliculitis barbae, Pilonidal sinus, and sometimes Hidradenitis suppurativa.

15.2.1 Contraindications

- Patients having unrealistic or exaggerated expectations.
- Patients having skin conditions like SLE or other medical conditions that warrant a specialist clearance for the same.
- Patient having keloidal tendencies.
- Any patient suffering from other skin conditions like Psoriasis and Vitiligo.

15.2.2 Laser Therapy

Lasers should be handled after proper training in a clean and safe environment so as to enable best safety and positive outcomes.

The laser theater should be designed keeping fire safety, electrical safety, proper lighting, ventilation, and smoke evacuation. Besides it should have good eye protection for the patient as well as the operator. The theater should be having non-reflective surfaces, unlimited power supply, asepsis, and emergency management systems.

Most of the lasers procedures are office procedures. Which most of the time are with or without anesthesia, so any procedure room which has all the safety guidelines in place can be used as a laser theater?

Many laser equipment are simple and easy to use, a trained operator, who is having basic knowledge of laser physics and laser tissue interaction will be able to give excellent outcomes and good patient satisfaction. It is also imperative that the operator is able to recognize any local complications early, prevent as much as possible any untoward incidents, and also manage the complications reasonably well will go a long way in running a successful enterprise.

The safe motto should be not to use laser unless it is absolutely indicated. Good positive outcomes are expected, though variation in this general rule does occur.

15.3 Procedure

The various factors to be kept in mind before starting a laser therapy session are:

- A good and detailed history regarding anatomical distribution of hair, the density and thickness of hair, frequency of hair growth.
- Menstrual history, acneiform or any other skin eruptions, or any skin infections like herpes need to be considered and appropriately treated.
- If required the services of a qualified endocrinologist should be sought to deal with the recalcitrant cases of PCOS.
- A thorough physical examination of the area which is undergoing lasing needs to be done.
- It is also imperative to take a consent after explaining the hair growth cycles, the number of sessions, and cost of therapy.
- Early recognition and possibly prevention of any post laser complications is imperative.

15.3.1 During Procedure

After taking the patient into the laser suite, any unwanted personnel should be sent away which includes overzealous parents wanting to accompany their wards.

The patient should lie comfortably on the operating couch and the area to be lased should be exposed. It is imperative that the patient should be warm and comfortable in the presence of the operator even though the laser suites are generally keep in cool temperatures so as to enable the machines to function optimally.

Since the patient has been instructed not to use any methods of hair removal prior to the procedure, the part should be prepped on table. Generally a female matron or a technician with a pleasant and friendly personality should make the patient comfortable and make process easy. We generally use a clipper with a disposable head to be used on each patient individually. Though recommended we generally avoid the razor so as not to cause any cuts or breaks in the skin and also many of the new clients undergoing laser therapy and have not got anything done prior to the therapy are not comfortable with razors.

Usually the area to be treated are circled or marked with a red or white pencil so as to ensure optimal lasing which leads to acceptable results.

Eye protection is of paramount importance so as to prevent inadvertent lasing of the retina leading to blindness, a major reason to keep unwanted personnel away from the area of operations. Every operator and assistants should wear the safety goggles of correct optical density recommended by the company for reasons of safety.

While prepping the parts for laser, any metallic objects should be ensured is not in the field, and if any make up or sunblock cream should be washed and removed.

Once laser is started it is advisable to be as gentle and slow during the procedure ensuring the area being treated is covered in cooling gel and also cooled adequately with the chilled tip so as to prevent any burns or dyspigmentation.

A word of caution about lasers on darker skin, though deemed safe patients with type IV–VI have higher pain and greater chances of sustaining skin burns. The way to counter this problem is to anesthetize the area to be treated adequately with topical application of local anesthesia at least 45 to 60 minutes prior to the procedure and also ensure that the area is cooled adequately. It helps to have the laser settings in lower fluence or spot size along with pulse stacking to acheive optimal results.

It is prudent to give a test patch on the skin near the treatment area to check whether the patient is able to tolerate the laser and is not in discomfort. Prior to lasing the patient is informed about the shots and in case of a new patient it is helpful to explain it to them that the sensation of laser on the skin is similar to a tap or like a rubber band snapping against the skin.

Once the procedure is done the end point is usually an erythematous zone which is mildly swollen and roots of hair popping out. There is also a smell of burnt hair which indicates that the therapy has been optimal.

Thread Lift in Aesthetic and Regenerative Gynecology

16

Biplav Agarwal and Poonam Mishra

16.1 Introduction

The desire to lift a sagging face has been ever increasing, and so has the interest in various non-surgical and minimally invasive modalities to help do so. Thread lift for facial rejuvenation is one such common and popular modality that we use these days.

In the late 1990s, PDO threads were initially designed to lift the ptotic facial tissues, and now different variations such as mono, screw and cog threads have been developed. These threads are placed along certain vectors and are pulled to lift the sagging skin.

As the technology and the techniques pertaining to thread lift have been advancing, so is the faith and widespread use of it by dermatologists, plastic surgeons and aesthetic physicians all over the world. There is a lot of literature that is now available which suggests the effectiveness, consistency in results and safety of use of thread lift in aesthetic practice.

Due to advancement in technology, newer variations of threads in terms of designs, materials, techniques of application and different clinical indications of use of threads lift, along with the significantly lower incidence of complications associated with these, have come into being.

Thread lifts are performed primarily as antiaging treatment modalities to tighten skin, with their secondary effect to contour and shape the face. Also, the evolution of thread lifting has progressed so much that aside from the popularly known and used PDO (Polydioxanone) threads, the market has now availed new threads such as PLLA (Poly-I-lactic acid) and PCL (Polycaprolactone). And thread lifting is now being applied to other body areas like arms, thighs, abdomen, and even buttocks.

16.2 Facial Aging

Before we discuss further threads and their application it is imperative to understand about facial aging and how it reflects a combination of skeletal and connective tissue changes. Some signs of aging include drooping of the zygomatic malar region, prominence, and deepening of nasolabial line, the appearance of marionette line, poor definition of mandibular margin, on the upper face wrinkles frown lines as well as wrinkles on the forehead start to become more prominent. Another major factor apart from chronological aging is gravity, which causes ptosis of facial soft tissue like the malar fat pad, thereby, creating hollowness of the mid face and infra-orbital area.

B. Agarwal (✉)
Gurgaon, India

P. Mishra
Lucknow, India

P. Jindal et al. (eds.), *Aesthetic and Regenerative Gynecology*,
https://doi.org/10.1007/978-981-16-1743-0_16

Hence, the vectors of facial rejuvenation are placed vertically.

16.3 Types of Threads

The types of thread can be divided on the basis of:

1. Mode of Absorption (Table 16.1)
2. Texture: Presence of Barbs (Table 16.2)
3. Length (Table 16.3)

Table 16.1 Types of thread based on the mode of absorption

Absorbable threads	Nonabsorbable thread
1. Polydioxanone thread (PDO)—get completely dissolved and absorbed in about 6 months, but they help to stimulate the production and synthesis of collagen in the skin for up to 12 months. 2. Poly-I-lactic acid threads (Silhouette Soft thread)—This material has a lifetime of two years or more, and once dissolved, the PLLA breaks down into ordinary harmless substances like lactate, glucose, carbon dioxide and water.PLLA can function as a volumizer and is able to stimulate the production of Type 1 and Type 3 collagen. 3. Polycaprolactone (PCL): Can last up to more than 2 years because the chemical bonds and structure in PCL are stronger and more complex, hence it takes a much longer time to be dissolved completely. PCL is also known to be very potent at stimulating collagen production as compared to PDO or PLLA, and even after the thread has dissolved, collagen production can last up to one year.	Natural (surgical steel, silk, cotton, and linen) and synthetic nonabsorbable ones (nylon, polypropylene, and polybutester). Example APTOS thread, Contour thread, Silhouette lift thread, Woffles thread (Polypropylene) Polypropylene is a type of material that the body cannot absorb.

Table 16.2 Types of threads based on texture

Barbed threads	Non-barbed threads (smooth thread)
(1). Bi-directional thread (Long suture) is inserted into a hollow needle and then placed in the treated area.	1. Monofilament Plain: These thread types are not used for lifting, but for their skin tightening effects by improving and stimulating collagen formation around the thread. They are placed in a mesh-like fashion, usually in the neck area, forehead, and under eyes. Examples include Miracu plain thread and TR lift thread.
(2). Uni-directional barbed threads (Long sutures) are designed to be anchored to a fixed structure, such as the deep temporal fascia.	2. Monofilament Screw or Spiral or Tornado threads. They are often in the form of single or double threads intertwined together around the inserting needle. These threads have a stronger effect over the mono threads and are usually used for general face-lifting. Examples include K2 screw lifting and T Screw lifting thread.
(3). Cogged Threads (Short sutures): Cog threads are essentially mono threads but with barbs designed to hook to the underside of the skin. These barbs are usually either cut or molded as part of the thread and purposed as a support structure to lift sagging tissues. Examples include PDO Uni-directional cogged thread, PDO Bi-directional cogged thread, and PDO Multi-directional cogged thread	

Table 16.3 Types of thread based on length

Short suture	Long suture
Any thread shorter than 90 mm in length	Any thread longer than 90 mm in length.
PDO and all its thread types usually are short in length	Barbed threads tend to be longer in length and can be used across the face, neck, chest, and even parts of the body

16.4 Polydioxanone (PDO Threads)

These are by far the most common threads used in practice, and their effects can last between 6 to 12 months. Though, these threads get dissolved and absorbed by the skin after 6 months, and they help to stimulate the production and synthesis of collagen in the skin for up to 12 months. There are various ways by which PDO threads function to lift and tighten the skin.

(1) PDO has been shown to be effective in improving pores and fine wrinkles by stimulating collagen formation in the dermal matrix and increasing thickness of the papillary dermis.
(2) PDO threads, when placed in the subcutaneous layer, induce tissue changes such as are gathering of PMNs, including eosinophils making granulation tissues around the thread after insertion. It has been observed in some studies that there is an abundant of newly made collagenous connective tissue in the formed granulation tissues.
(3) Some studies have also shown that these newly made collagenous connective tissues converge into the pre-existing fibrous connective tissue creating a "merging effect." Through this effect, it causes an inflammatory reaction to the surrounding area where the thread is inserted, and by the effect of mechanotransduction, it will spread as waves to the surrounding tissues.
(4) In the granulation tissues newly made near the threads, fibroblast, and myofibroblasts have also been reportedly observed, which can effectively improve the elasticity of the skin in the area of the procedure.
(5) Lastly, fat cells have also been observed to have been denatured by the granulation tissues in the area of treatment, proving that PDO threads, if used correctly, can be very effective in face shaping, contouring, and rejuvenation procedures.

16.5 How Can We Achieve Facial and Body Contour Restoration?

- Open surgical technique where the parallel with excising excess skin volume should be added and fat pads repositioned.
- As a minimally invasive technique involving the placement of surgical threads in the subcutaneous layer known as "thread lift."

16.6 Concept

- Thread lift is a procedure where specially designed or ordinary surgical sutures are passed into the subcutaneous tissues where they relay on the trabecular connective tissue network that runs between SMAS and the skin in order to elevate and reposition fat pads affected by ptosis and consequently have a "contour restoring effect." The aim is to tighten the subcutaneous tissues as well as the overlying skin.

16.7 Two Stage Effect

- Immediate lifting effect due to the mechanical properties of the thread
- Lifting effect due to the fibrosis formed by the thread "the secondary vector of pull"

16.8 History of Thread Lifting

- First Generation—Catgut (Plain and Chromic)
- Second Generation—PGA (Polyglycolide)
- Third Generation—PDO (Polydioxanone)

16.9 Introduction of PDO

16.9.1 PDO (Polydioxanone) as Shown in Fig. 16.1

- It is the first synthetic suture.
- Is an absorbable monofilament thread (6~8 months).
- Has longest tenacity sustainability among biodegradable sutures (Table 16.4).

Table 16.4 Thread types and its absorption period

Thread types	Absortion period
Plain Catcut	~ 70 days
Chromic Catgut	~ 90 days
PGA	60 ~ 90 days
PGLA	60 ~ 80 days
PGCL	80 ~ 120 days
PDO	180 ~ 240 days

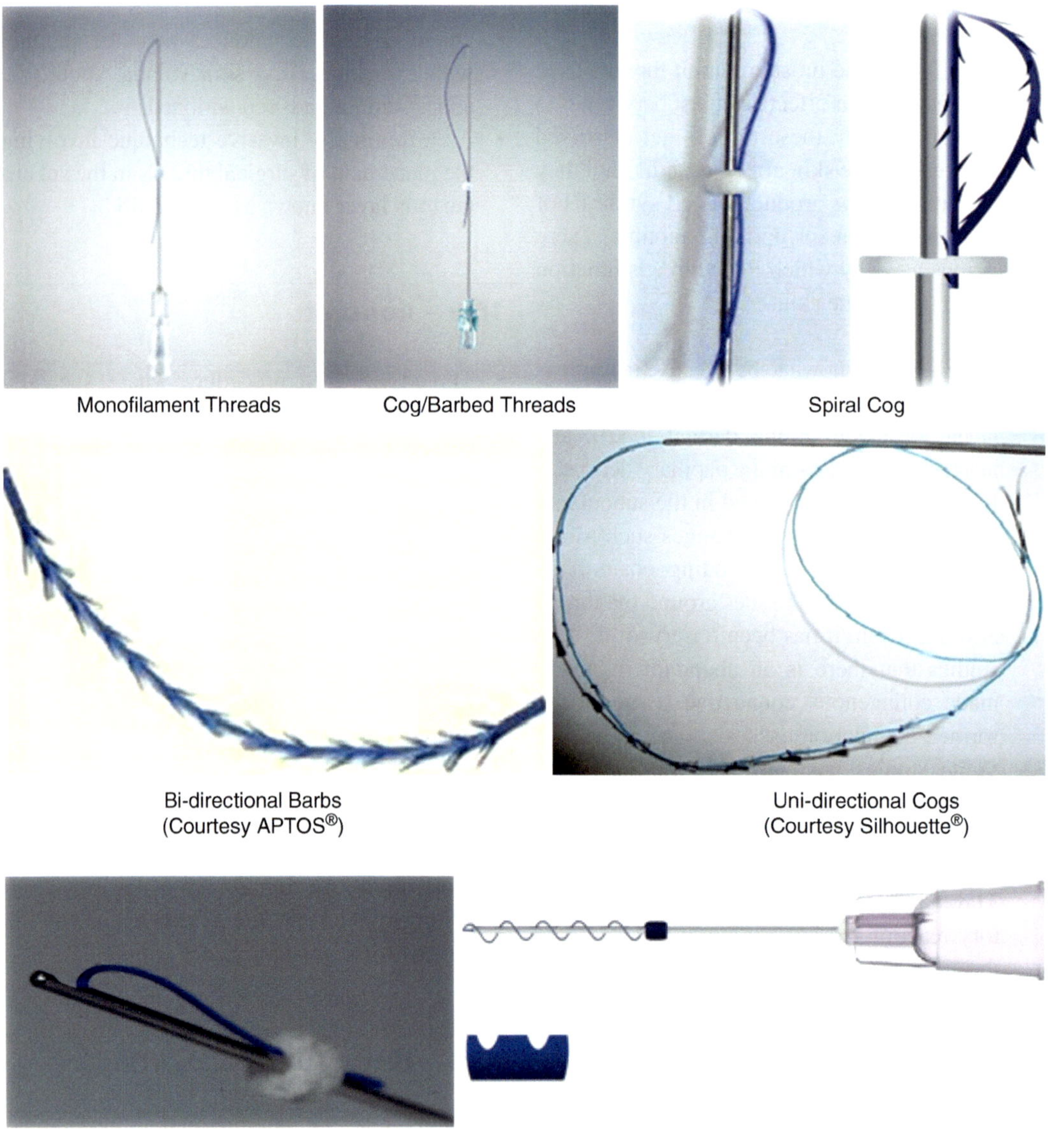

Fig. 16.1 Different types of thread

16.9.2 PDO (Polydioxanone)

- *Very high tensile strength*
- *Excellent pliability*
- *Smooth tissue passage*

16.10 Types of Thread (Fig. 16.1)

16.10.1 Vaginal Threads

Most women have problems with vaginal looseness and stretching the vagina. Vaginal tightening is a procedure that treats this problem. Today's modern technologies are becoming popular. In this article, we discuss one of these technologies, which uses the threads to tighten the vagina.

Vaginal tightening or vaginoplasty is a procedure to tighten the vaginal muscles or tissues.

Vaginal tightening is usually performed in the woman whose vaginal muscles and tissues have become loose, stretched, or torn after childbirth, weight changes or due to age. This article provides a new method of vaginal tightening, which is called, thread vaginal tightening. It is a procedure of tightening and lifting the posterior perineum.

During aging, the vagina may be felt, so most women want to turn it back into primary form.

The procedure is usually performed on an outpatient basis under general or local anesthesia. Your surgeon makes a small incision in the perineum (area between the anus and the vaginal opening) and tightens the vaginal canal length or removes excess skin and tissues.In the past, doctors use threads to lift the skin of the face and body, but, today and with using this new method, doctors can lift the perineum. Absorbable threads provide a safe method of tightening the vagina. This is an outpatient procedure with no anesthesia and bleeding.

16.10.2 Procedure of Tightening

Canical form of tightening the vagina and perineal without any anesthesia and bleeding during the surgery. A secondary effect, which is more effective than the previous procedure. In this procedure, the vagina turns back to its primary form by collagen. There is a fibrotic response that increases the level of collagen. This is a constant procedure to absorb the threads.

16.10.3 The Uses of Vaginal Threads

- Vaginal rejuvenation
- Vaginal tightening
- Vaginal looseness recovery
- Perineal reconstruction after childbirth
- Vaginal lifting
- Perineal reconstruction after epithosomy

16.10.4 Benefits of this Procedure

- Short recovery
- No anesthesia is needed
- No bleeding
- A non-invasive method
- Pelvic floor recovery
- Improved the physical and medical condition

16.10.5 Complication of Vaginal Looseness

- Reducing pleasurable intercourse.
- Creating vaginal leakage.
- Increasing vaginal microbes.
- Before performing, women should be checked by a gynecologist, and the infections should be treated. Pap-smear is an important examination.
- After the surgery, partners should not have intercourse for at least 6 months.

16.10.6 Contraindications

- Pregnant women
- Single girls
- Those with PID infection

16.10.7 Functions of PDO

1. PDO Threads generate collagen by foreign body reaction.
2. There is increased microcirculation.
3. Tissue adhesion and anchoring effect is seen around the foreign material.
4. Increased metabolism in the localized tissue.
5. Decrease in cellulite due to increased lipolysis.
6. They ease muscle tension → Muscle relaxation → muscle volume decreased.
7. Lifting effect.
8. Increased skin tightening: improved fine wrinkles and deep wrinkles.
9. Improves skin tone and gives a whitening effect.

16.10.8 Effects (Post-Procedure)

- First day: Mechanical trauma → Necrosis, hemorrhage, and disintegration of fatty tissue
- Three–seven days: Polymorphonuclear neutrophils (PMNs), macrophage-like and fusiform or fibroblast-like cells
- At 1 week: Proliferation of myofibroblasts and formation of new blood capillaries
- After 1 week: Formation of a fibrous capsule with abundant and dense collagenous fibers

16.10.9 Mechanotransduction

The ability of the cells to sense, process, and respond to mechanical stimuli is utilized in threads procedures, as shown in Fig. 16.2.

It is an important regulator of physiologic function that has been found to play a role in regulating gene expression, protein synthesis, cell differentiation, tissue growth and most recently, the pathophysiology of disease

16.10.10 Sequen.ce of Events

- Signal transduction
- Cytoskeleton change
- Immediate contracture

16.10.11 Effect at the Tissue Level

- Adipocyte disintegration
- Fibroblast proliferation (Fig. 16.3a)
- New blood capillary formation (Fig. 16.3b)
- Collagen matrix formation
- Increased eosinophils (which may mean increased histamine release); eosinophils induce fibrosis.
- Wound fibrosis, continuous with the dermal to skeletal, muscular connective tissue> mechanical signal.

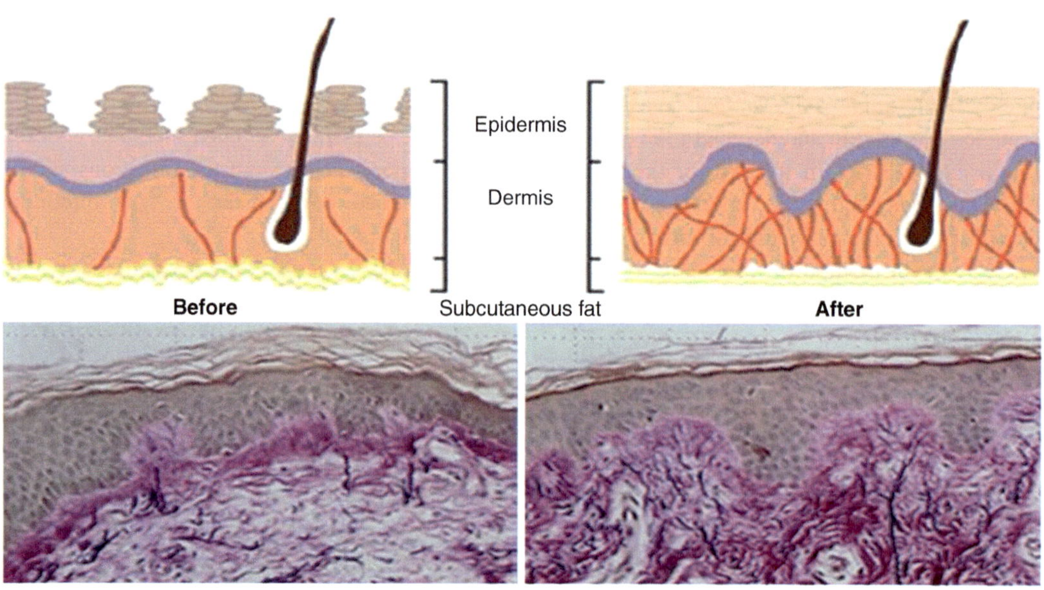

Fig. 16.2 Effect—Mechanotransduction

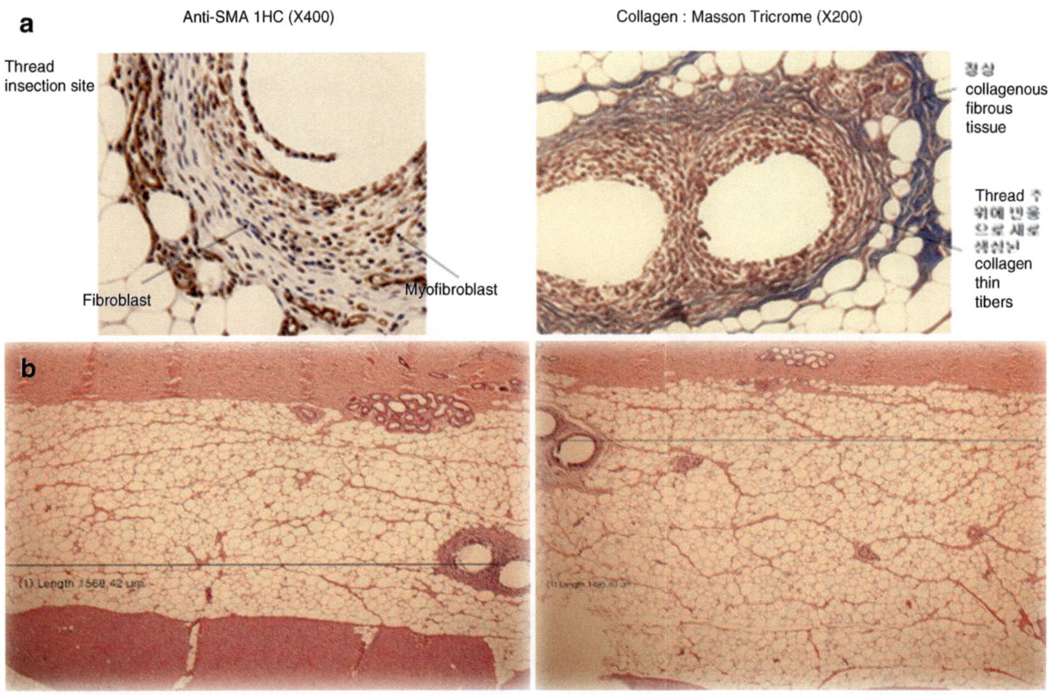

Fig. 16.3 (**a**) Effect at the tissue level—fibrotic network. (**b**) Effect at the tissue level—Increased microcirculation

16.11 Applications (Fig. 16.4)

- **Face**—forehead, mid face (cheek), jaw line, nose
- **Neck**—tightening, lifting
- **Specific winkles**—glabella, crow's feet, naso-labial folds, fine wrinkles
- **Breast**—firming, shaping
- **Hip up**
- Obesity—arm, abdomen, flank, etc.

16.11.1 How to Perform the Procedure

- **Thread insertion is an invasive procedure and needs skill.**
- **Under correction is always preferred over the overcorrection**.

Threads are injected into the skin and sometimes the muscle tightens the skin. Site to be selected in sitting position or in dorsal position taking into account the effect of gravity on the sagging part. It has a powerful dual effect.

- The first is to give an immediate lifting effect.
- The second is to stimulate bodies own collagen, elastin and hyaluronic acid.)

16.11.2 Procedure Protocol

- Consent is must.
- Tetanus Toxoid should be covered.
- Investigations needed are Blood counts, Coagulation profile, and viral markers.
- Site should be cleaned properly, taking care of asepsis.
- Drawing the VECTOR is very important for COG THREADS—Point A is the entry point, and Point B is the end point or Anchoring point.
- MESH is formed with MONO THREADS, and specific points (wrinkled and sagging) are selected.

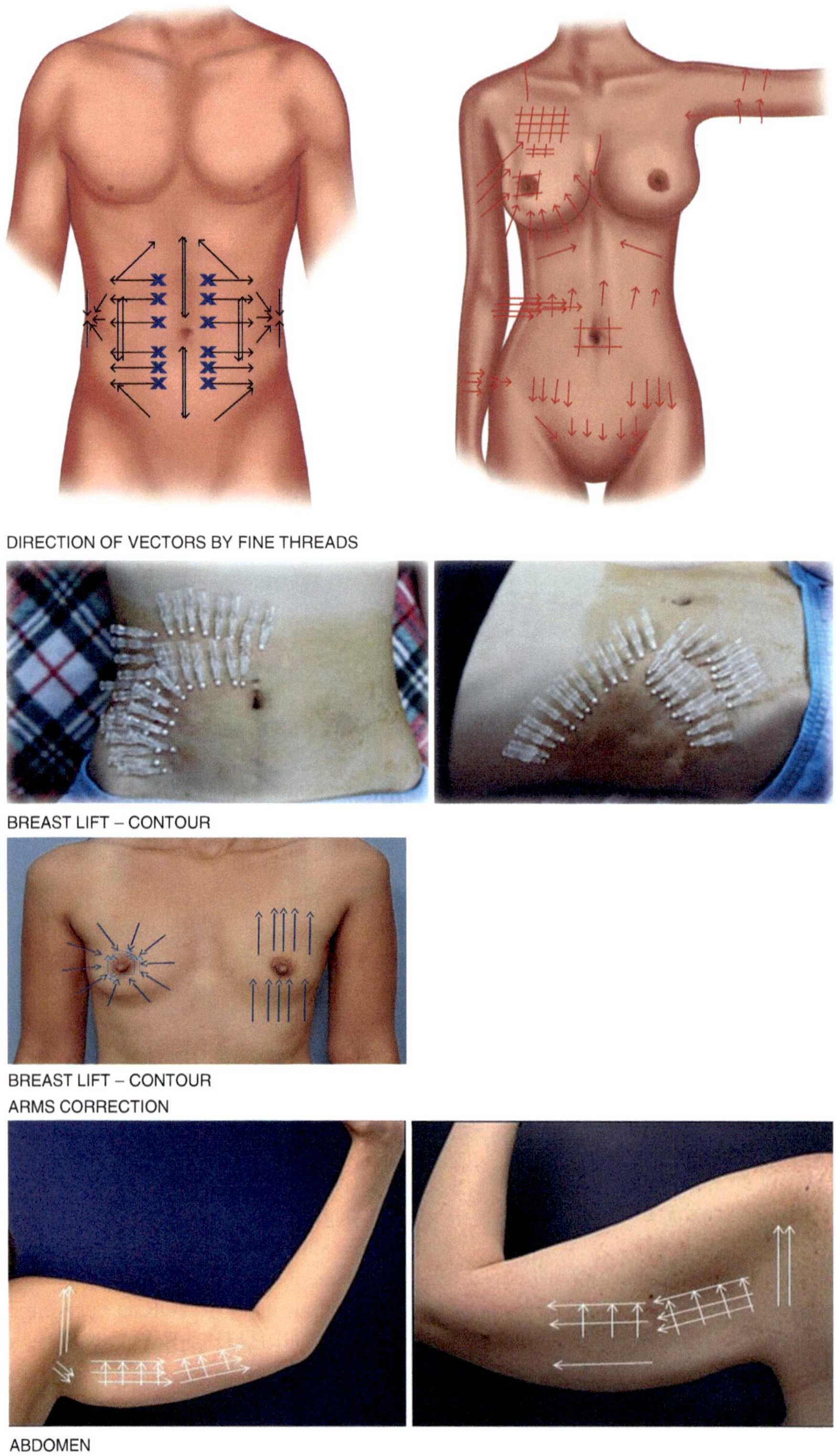

Fig. 16.4 Applications of thread lift from head to toe

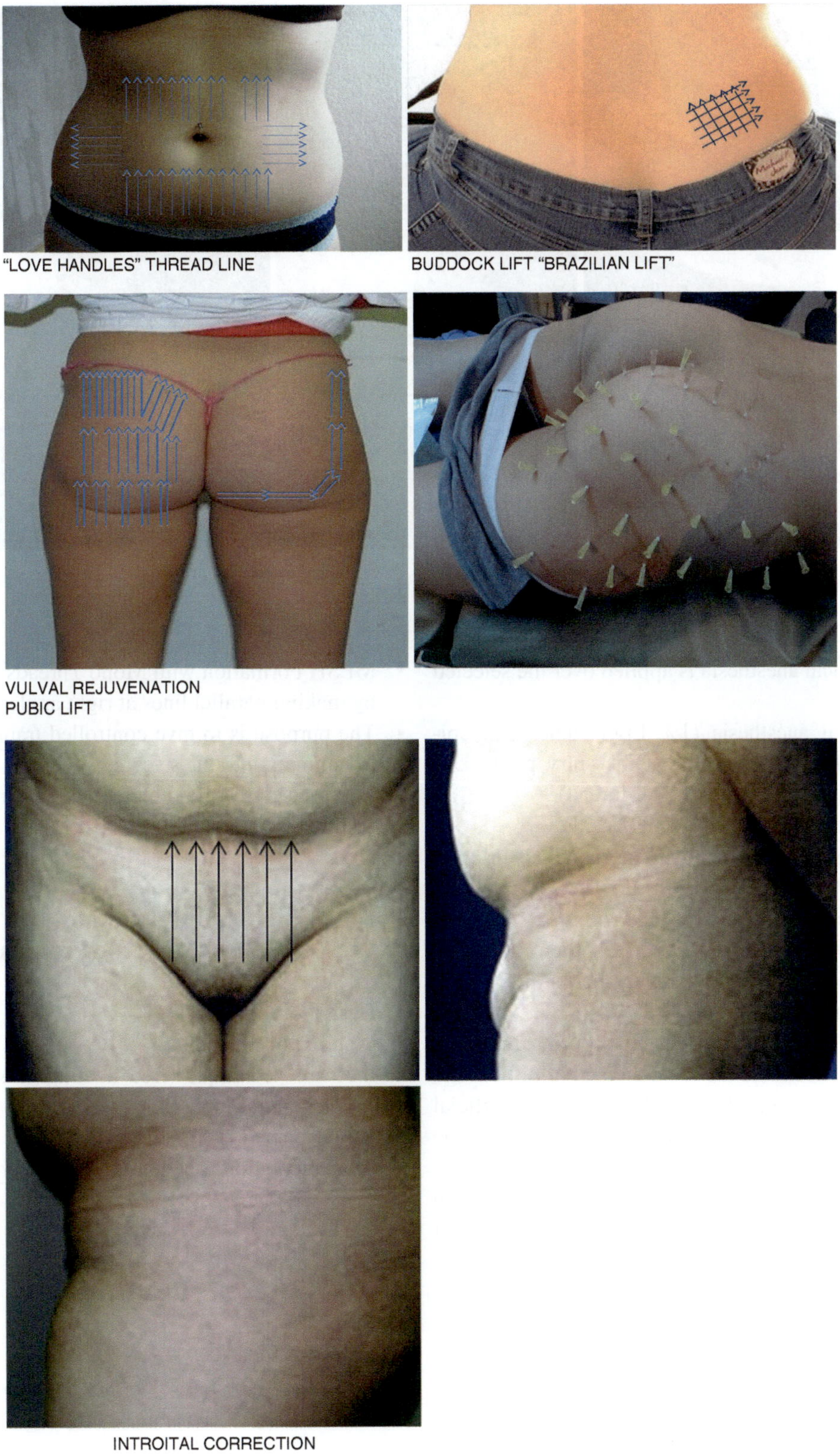

Fig. 16.4 (continued)

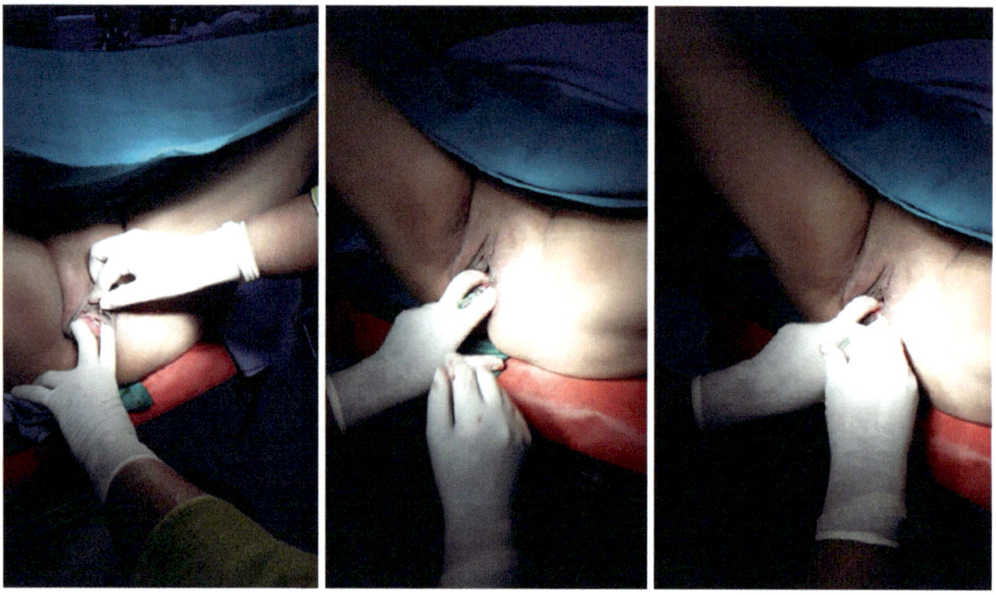

Fig. 16.4 (continued)

- Topical anesthesia is applied over the selected part.
- Local anesthesia (1% Lignocaine with epinephrine) is instilled at the entry point. Use less local (0.7–0.9 ml per thread).
- The port is made by 18 G needle.
- Cog thread mounted on the cannula is inserted at 90 degree to the skin surface.
- The hypodermis is reached once the give way feeling is achieved, and then the direction of the cannula should be parallel to the skin.
- The vector path is followed till the end point, which should be overshooted a little. The thread passes through the SMAS (Superficial Muscular Aponeurotic System) layer of the skin.
- The cannula should be rotated a little and then drawn back pulling and sliding the skin over it a little. Cog thread immediately gets entangled and cannot be removed once inserted.
- Lift the thread from each side before cutting.
- Cut enough the ends of the threads.
- Be careful with the previous thread while inserting.
- MESH Formation with Mono Threads is done by making parallel lines at right angles.
- The purpose is to give controlled trauma and inflammation to the selected site.

The procedure is completed with an immediate pulling effect visible.

16.11.3 Factors Determining the Success of the Procedure

- Design of the thread
- Material of the thread
- Insertion technique (free floating or suspension technique)
- Patient selection
- Implementation of combination techniques when indicated

16.11.4 Aftercare Instructions

- Ice packs for 1 day
- No massage for 3–4 weeks/special low-impact pillow

16.11.5 Complications

- Bruises
- Edema
- Pain
- Sense of muscle contraction/skin pulling

16.11.5.1 Issues from Technical Problem (Improper Procedure)

- No change
- Asymmetry
- Visible threads
- Dimpling

16.11.5.2 Generate Elasticity and Contour Changed

- **Initial result**: Grasping effect
- **Proper result**: Visible after 3~4 weeks
- **Sustainability**: PDO remains at least 6 months, dissolves slowly afterwards. Artificially synthesized collagen lasts up to 2 years.
- **Whitenening Effect**: Due to blood circulation and collagen regeneration, not by the absence of pigments.
- **Reduction of** Cellulite **and Fat**: Effects similar to the reduction of cellulite and fat from fat cytolysis, improved microcirculation, and increased metabolism

16.12 Conclusions

- "Thread lift" bridged the gap between injectables and traditional surgery.
- Combining different threads and different techniques, as well as both anchoring and free free floating threads, gives the most dramatic results.
- For the right patient, this procedure can give results similar to open surgery.

Chemical Peels and Vulval Whitening in Aesthetic and Regenerative Gynecology

17

Vivek Malhotra and Purva Pande

17.1 Introduction

Chemical peels form an integral part of dermatology and aesthetic medicine practice. The history of chemical peeling dates back to the era of ancient Egyptians (around 1500 BC) when Egyptian women used to apply sour milk to their faces. Legend has it that Cleopatra used to bathe in sour donkey milk for her skin brightening and beautification. Sour milk contains lactic acid (an alpha-hydroxy acid), which has the property of rejuvenating the skin. In modern times, Dr. Ferdinand von Hebra, an Austrian dermatologist, is credited with popularising the use of various exfoliative substances to treat conditions like freckles and melasma [1]. The early 1900s saw the increasing use of phenol peels by physicians, and the research in the field paved the way for the future of chemical peeling in dermatology. In the 1980s, further researches led to the elaboration in the uses of alpha-hydroxy acids and subsequently other compounds as chemical peeling agents [2, 3].

V. Malhotra (✉)
Beau Visage Skin Clinic, Chandigarh, India

P. Pande
Dr. Purva's Skin Clinic, Panchkula, Haryana, India

17.1.1 Scope of This chapter

Awareness and demand for procedures for improving the aesthetic appearance of the genital areas in females are on the rise. It is a relatively newer domain, being explored under varied names like "genital beautification," "feminine intimate rejuvenation," etc. Surgical procedures for the same have been projected as simple and complication-free. These not only aim at improving or restoring aesthetic appearance but also claimed to increase self-esteem and enhance sexual pleasure for both women and their partners [4]. But there is a large proportion of patients unwilling to go under the knife, and there is an ever-increasing demand for non-surgical procedures, being termed as "non-surgical feminine rejuvenation." Feminine genital brightening is one such field that is gaining popularity, and chemical peels are being used as easily available, relatively safer option with gratifying results. Although there is a growing area of interest about using chemical peels for feminine genital brightening and rejuvenation, published data about this indication is lacking. Chemical peels have been traditionally used in a dermatologist's clinic mainly on the face, extending to other visible areas like neck, décolletage, upper back, dorsum of hands, etc. Experience of this procedure in other areas especially softer, sun shielded skins like body folds, genital regions, etc. is quite limited. This chapter deals with the basics of

P. Jindal et al. (eds.), *Aesthetic and Regenerative Gynecology*,
https://doi.org/10.1007/978-981-16-1743-0_17

chemical peeling as well as the applications of various peeling agents, with reference to the facial skin only. The basic principles of chemical peeling should hold true for areas like the genital region as well, keeping in mind the anatomical differences of the skin in these areas and also the fact that these areas are prone to more complications due to occlusion. So the approach to chemical peeling in these areas should be cautious and conservative.

17.1.2 Definition

Chemical peeling involves the application of a keratolytic or caustic chemical agent to instil a controlled injury/ablation till a specific depth of the skin, which leads to regeneration of the skin by means of controlled wound healing and subsequent improvement in the skin texture and appearance [5]. The process of peeling has to be performed in a controlled manner so as to facilitate exfoliation without significant side effects [2].

17.2 Classification of Peels

Chemical peels can be classified as per the active chemical ingredient or according to their depth of penetration into the skin [2, 6].

Chemical Classification: Based on the active ingredient/ peeling agent, the chemical peels can be broadly classified as detailed in Table 17.1.

Histological Classification: Based on the depth of action of the chemical peeling agent, the peels can be divided as represented in Table 17.2.

17.3 Mechanism of Action

To understand the chemical peels, their mechanism of action and their effect on the skin, it is imperative to have a basic knowledge of the histology of the skin and the basic wound healing mechanisms. The human skin comprises the upper epidermis and lower dermis followed by the subcutaneous tissue. The dermis has two sub-layers;

Table 17.1 Chemical Classification of Peels

1.	Alpha-hydroxy acids and monocarboxylic acids: Glycolic acid, Lactic acid, Mandelic acid, Tartaric acid, Malic acid, Phytic acid.
2.	Beta hydroxy acids: Salicylic acid, Citric acid
3.	Trichloroacetic acid (TCA)
4.	Alpha ketoacids: Pyruvic acid
5.	Retinoic acid, Retinol
6.	Resorcinol
7.	Phenol
8.	Combination peels: Jessner's solution, modified Jessner's solution, Brody's peel, Coleman's solution, Monheit's peel

Table 17.2 Histological Classification of chemical peels [2]

Depth of penetration		Peeling agent	Conditions
Superficial	Stratum corneum to papillary dermis (upto 60 μm)	Alpha-hydroxy acids (glycolic acid 20–70%, lactic acid, malic acid, pyruvic acid, tartaric acid); beta hydroxy acids (salicylic acid 10–30%), carbon dioxide snow, Jessner's solution, lipohydroxy acid, resorcinol, retinoic acid, TCA <20%,	Mild photoageing (actinic keratoses, fine lines, roughness, solar lentigines, yellow stains), acne vulgaris, mild acne scarring, pigmentary disorders (melasma, mild dyschromia, post-inflammatory hyperpigmentation)
Medium	Papillary dermis to upper reticular dermis (upto 450 μm)	Unna paste, 5-fluorouracil, TCA 35–50%, TCA 35% + solid CO_2, TCA 35% + glycolic acid 70%, TCA 35% + Jessner's solution	Mild-to-moderate photoageing, actinic keratoses, fine lines, rhytides, solar lentigines, pigmentary disorders, (melasma, mild-to-moderate dyschromia) seborrhoeic keratosis, superficial atrophic scars
Deep	Mid-reticular dermis to (upto 600 μm)	Baker–Gordon, TCA >50%	Severe photoageing (advanced rhytides), pigmentary disorders, premalignant skin tumours, scars

Co_2, Carbon dioxide; TCA Trichloroacetic acid

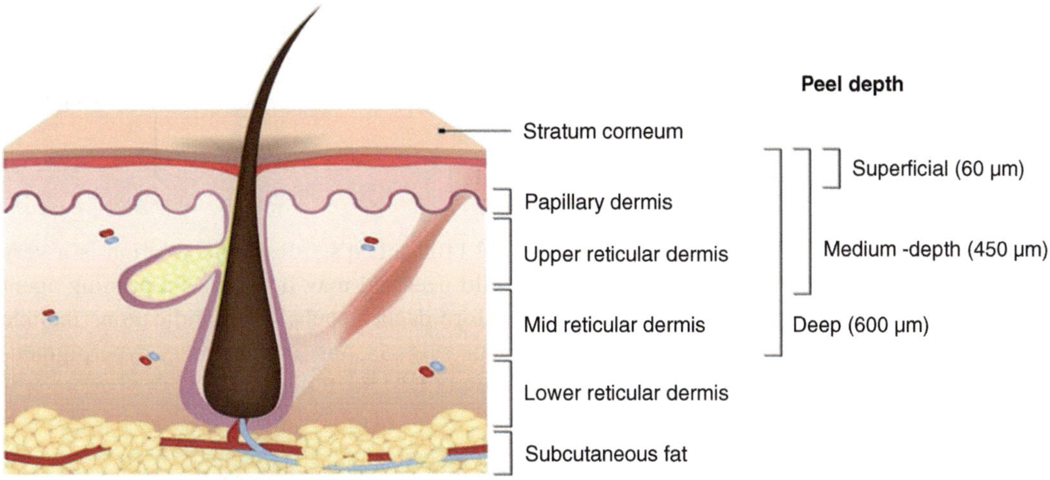

Fig. 17.1 Layers of human skin and depth of chemical peels action [2]

the upper papillary dermis and the lower reticular dermis (Fig. 17.1). The wound healing mechanism induced by the peel depends on the depth of the action of the peel (Fig. 17.1; Table 17.3).

Superficial chemical peels penetrate the epidermis, dermo-epidermal junction down till the papillary dermis [4, 7]. Their effect occurs through breakdown of corneosomes, increased epidermal enzyme activity leading to epidermolysis and exfoliation and prevention of excessive keratinization of the pilosebaceous unit [8]. This increases dermal collagen and epidermal thickness, and melanin gets more uniformly distributed [5]. Exfoliation is mild and may occur up to 7 days [3].

Medium-depth peels target the region from the papillary dermis to the upper reticular dermis. They cause precipitation of proteins, thus causing coagulative necrosis of the cells in these dermal layers. This leads to uniformity in the structure of papillary dermis. Cutaneous injury is relatively more than superficial peels, and exfoliation may be visible for up to 10–14 days [2].

Deep peels go down till the mid-reticular dermis. These peels damage the epidermal keratin along with dermal proteins, thus causing injury up to mid-dermis [5]. Wound healing occurs in its different phases, including scab formation and re-epithelization, which may take approximately 2 weeks to complete. Post-peel erythema can persist for 2–3 months [2].

It is possible to achieve greater peel depths and subsequently desired results without complications when properly done [9]. The depth of penetration of the peeling agent and the outcome is dependent on several factors (Table 17.4).

17.4 Common Superficial Peeling Agents

17.4.1 Glycolic Acid

Glycolic acid is an alpha-hydroxy acid derived from fruits and has exfoliative, anti-inflammatory properties and some activity against *Propionibacterium acnes*. It causes thickening of epidermis and dermis with deposition of collagen and mucopolysaccharides along with melanin dispersion. Glycolic acid is available in concentrations ranging from 20% to 70%. It has been found to be useful in conditions like acne, melasma and other facial melanosis, skin rejuvenation, etc. [10–12].

Table 17.3 Wound healing and chemical peels

1) Coagulation and inflammation	6) **Superficial and Medium Peels**	7) **Deep Peels**
2) Re-epitheliazation		
3) Granulation tissue formation		
4) Angiogenesis		
5) Collagen and matrix remodelling		

Table 17.4 Factors affecting peel depth and outcome [7]

1. Nature of peeling agent and its concentration
2. Type and thickness of skin
3. Anatomic location of area for peeling
4. Pre peel priming
5. Degreasing method
6. Application technique and peel procedure
7. Number of coats
8. Post-peel procedures and applications

17.4.2 Salicylic Acid

Salicylic acid (SA) is 2-hydroxybenzoic acid, derived from the willow tree and is a lipophilic agent with keratolytic and anti-inflammatory properties. It has a stimulant effect on the dermal fibroblasts. The commercially available concentrations are 5% to 30%. SA is used for acne, photodamaged skin, melasma, post-inflammatory hyperpigmentation, etc. [7, 13, 14].

17.4.3 Lactic Acid

Lactic acid is an alpha-hydroxy acid. Sour milk and its products are common sources. It is found to be useful in wrinkles, superficial hyperpigmentation, UV-induced damage, etc. [15].

17.4.4 Mandelic Acid

It is an aromatic alpha-hydroxy acid. Mandelic acid has demonstrated efficacy in superficial pigmentary changes and acne vulgaris [15].

17.4.5 Jessner's Solution

It is a combination of salicylic acid (14 g), resorcinol (14 g), lactic acid 85% (14 g) and ethanol (100 mL q.s.). Jessner's solution is a very mild peel and may be used as a priming agent before deeper peels. Other indications include acne vulgaris, post-inflammatory hyperpigmentation [2].

17.5 Medium-Depth Peels

17.5.1 Trichloroacetic Acid

TCA is a trichlorinated carbonic acid and occurs in crystals. It causes denaturation of epidermal proteins and damages the upper dermis. A concentration of 35–50% TCA forms a medium-depth peel. It is used for lentigines, melasma, rhytides, dyspigmentation, superficial scars [2].

17.6 Deep Peels

Baker–Gordon peel and TCA >50% are common deep peels. Baker–Gordon peel consists of 88% phenol (3 mL), croton oil (3 drops), hexachlorophene soap (8 drops), and distilled water (2 mL). These peels are used for severe photodamage, deep scars, dyspigmentation, premalignant skin lesions like actinic keratoses, etc [2].

17.7 Indications and Contradications of Chemical Peels

17.7.1 Indications

Chemical peeling is used for a variety of dermatological indications, both cosmetic as well as therapeutic (Table 17.5) [2, 15].

1. Pigmentation disorders: Melasma, lentigines, freckles, facial melanosis like lichen

planus pigmentosus, post-inflammatory hyperpigmentation
2. Cosmetic indications: Facial rejuvenation, rhytides, photodamage, scars
3. Inflammatory disorders: Acne, rosacea
4. Premalignant cutaneous lesions: Actinic keratoses

17.7.2 Contraindications [2, 7]

17.7.2.1 Absolute

- Active infections
- Known hypersensitivity to peeling agent
- Open ulcers/wounds

17.7.2.2 Relative

- Unrealistic expectations of the patient
- Inability to have adequate sun protection
- Keloidal tendency
- Pre-existing dermatoses like atopic dermatitis
- Pregnancy
- Lactation

17.8 Procedure of Peeling

17.8.1 General Steps

1. Assessment of the Patient:
   (a) *History:* Detailed history should be taken about the cutaneous disease, comorbidities, hypersensitivity to any agent/ drug, keloidal tendency, patient's expectations, photosensitivity and prior radiation exposure.
   (b) *Examination:* Thorough examination of the cutaneous lesions along with the estimation of Fitzpatrick skin type, presence of any keloid, active infection, etc. Objective assessment of indications like acne, photoageing should be done using proper scales/ scoring system.
2. Counselling and Consent: A written informed consent should be taken for the procedure as well as for photographic documentation. The patient's apprehensions and doubts should be addressed. An explanation should be given about the procedure, its expected outcome and possible complications.
3. Photography: Photographic documentation of the lesion/ indication should be done after taking proper consent. Photographs should be taken at baseline, before each peeling session and at the follow-ups.

17.8.2 Pre-Peel Steps

Priming is an important pre-peel step. The purpose is to prepare the skin for a peeling procedure, enhance the penetration of the peeling agent as well as wound healing and decrease the chances of complications. It should be started at least 2–3 weeks before the peel procedure. Usual agents used for priming are topical retinoids, tyrosinase inhibitors like hydroquinone and low concentrations of alpha-hydroxy acids like citric acid, maleic acid, tartaric acid.

Additionally, the patient should be instructed to avoid waxing, threading, topical retinoids, facials, etc. for at least a week before the peel.

17.8.3 Peel

Patient is asked to wash the skin gently with soap and water. The patient is asked to lie supine with head lifted at an angle of 45°. The required area is thoroughly cleaned with 70% alcohol, followed by cleansing with acetone-soaked gauze. This is done to de-grease the area and dissolve the excess sebum, if present, and enhance the penetration of the peeling agent. The peeling agent is applied with the help of a sterile cotton swab or gauze or brush.

17.8.3.1 Termination of Peel

Timely termination of peel is an essential step. Any delay in termination can lead to greater penetration and thus, leading to significant complications. To terminate, neutralization is done with 10% sodium bicarbonate solution or distilled water-soaked gauze pads. The end point for peel termination may vary with the agent used and the depth of injury it causes. In case of glycolic acid, the peel is kept for

Table 17.5 Indications and contraindications for peeling procedures [16]

Treatment	Peeling agent	Peeling level		
		Superficial	Medium-deep	Deep
Indications		Photoageing: Roughness, yellow stains fine lines, keratosis, solar lentigines Pigmentary disorders: Melasma, Post-inflammatory	Photoageing Fine lines wrinkles Pigmentary disorders, Superficial atrophic scars	Severe photoageing Pigmentary disorders scars
Contraindications	Absolute	Pregnant/nursing patients Active herpes simplex (cold sores) Fitzpatrick skin types V–VI	Pregnant/nursing patients Active herpes simplex (cold sores) Fitzpatrick skin types V–VI Questionable patient compliance	Pregnant/nursing patients Active herpes simplex (cold sores) Fitzpatrick skin types IV–VI Questionable patient compliance
	Relative	Cold sores: 4–6 weeks after healing Botulinum toxin: 1–2 weeks after collagen injections: 2 weeks before or after Facial surgery: 6 weeks after oedema Laser: 8 weeks after electrolysis and dying: 7 days before or after Waxing, depilatories: 3 weeks after Isotretinoin treatment: 6 months after	Regular sun exposure heavy cigarette smoking Inactive but recurring herpes infections Oral oestrogen intake history of hypertrophic scarring Connective tissue disorders advanced AIDS stages Isotretinoin treatment: 6 months after	Isotretinoin treatment: 6 months after

TCA, trichloroacetic acid; AHA, alfa-hydroxy acids; BHA, beta-hydroxy acids; LHA, lipo-hydroxy acid

3–5 minutes or till the appearance of erythema, whichever is earlier. In case of salicylic acid, the peel is terminated when there is a uniform light white coat of pseudo-frost (due to crystallization). TCA is neutralized when there is dense white frosting (due to coagulation of proteins).

17.8.4 Post-Peel Steps

Adequate post-peel care is very important to prevent complications. Ice packs can be given immediately after the procedure to reduce the burning and erythema. Patient should be advised to apply a broad-spectrum sunscreen of SPF >30 before going out of the clinic. Strict sun protection should be advised with the application of sunscreen 3-hourly daily along with protective clothing. Emollients should be applied frequently. The patient should be made aware of the visible desquamation in the subsequent days and should be advised against picking the skin.

Mild soaps should be used, and irritants like Dettol should be strictly avoided. Advise should be given to avoid the use of any topicals other than the emollient and sunscreen. Facials and related procedures should be avoided for at least a week after the peel.

17.8.5 Complications

Chemical peeling is a relatively safe procedure if done by an experienced clinician. Complications depend on a number of factors, including depth of peel, application technique, pre- and post-peel care. The risk of complications can be reduced by taking appropriate pre- and post-peel steps, including patient selection and counselling. Complications are more commonly observed in darker skin types. At times, the complications may occur inevitably, and that depends mainly on the depth of the peel. Various complications are summarised in Table 17.6.

Table 17.6 Complications of chemical peeling

Early:
1. Persistent erythema
2. Burning
3 Pain
4. Scabbing
5. Pruritus
6. Oedema
7. Blistering
8. Secondary infection
Delayed:
1. Post-inflammatory hyperpigmentation
2. Post-inflammatory hypopigmentation
3. Allergic contact dermatitis
4. Acneiform eruptions
5. Scarring (atrophic, hypertrophic, keloid)
6. Systemic toxicity in case of deep peel like phenol peel

17.9 Emerging Concepts in Chemical Peels

17.9.1 Newer Peels

A variety of newer superficial chemical peels are gaining popularity for various dermatological and aesthetic indications. These peels have shown efficacy in few published studies, but substantial data is still lacking. Many such newer peels have been commercially available over time. They have been formulated and introduced to achieve better outcomes with lesser side effects and downtime as compared to the traditional peels. These may be single peeling agents, or as combinations, where-in different agents are combined in lower concentrations (Table 17.7). These may even have additives like antioxidants and humectants to improve their tolerance and decrease the irritant potential of the active chemicals. Various active agents in these peels are tretinoin (acne, hyperpigmentation, acne scars), arginine (for periorbital melanosis), pyruvic acid (for acne, melasma), ferulic acid (for facial rejuvenation), kojic acid (for hyperpigmentation) as well as peels derived from fruit sources like maleic acid, tartaric acid or citric acid. The fruit-derived peels can be used for mild superficial exfoliation or for priming before other deeper peels. Most of these peels are proprietary products of different companies and usually patented. Many of these are available commercially as kits with different chemicals being applied sequentially in 2–3 steps, as per the instructions provided by the company.

17.9.2 Combination Peels

A combination peel refers to a formulation wherein multiple chemical agents are combined in relatively lower concentrations. These usually consist of two or more superficial peeling agents combined in a single formulation. These agents act synergistically in such combinations, thereby increasing the overall efficacy of the peel and at the same time reducing the chances of side effects seen with individual agents when used in higher concentrations alone [18]. These peels have a relatively wider range of action, possess a good safety profile and are also easier to use when compared to their individual components used in higher strengths. Many of these are commercial preparations available as proprietary products from various companies. Some of the available formulations are:

1. Jessner's Peel: 14% Salicylic acid, 14% Lactic acid and 14% Resorcinol
2. Coleman's peel: 70% Glycolic acid and 35% TCA
3. Monheit's peel: Jessner's solution plus 35% TCA
4. Azelac M peel (Mediderma): 18% Salicylic acid, 15% Azelaic acid and 10% Mandelic acid
5. Vedasol SM peel (Cipla): 20% Salicylic acid and 10% Mandelic acid
6. Melanostop peel (Mesoestetic): 20% Azelaic acid, 10% Resorcinol and 6% Phytic acid
7. Claze peel (Cipla): 20% Lactic acid, 20% Arginine, 5% Kojic acid, 5% Urea, 2% Citric acid and 2% Arbutin

Table 17.7 Newer Chemical Peels [17]

Tretinoin
Pyruvic acid
Ferulic acid
Kojic acid
Azelaic acid
Mandelic acid
Arginine
Polyhydroxy acid
Phytic acid
Citric acid
Fruit peels
Enzyme peels
Black peels
Obagi blue peels
Dermamelan peel
Cosmelan peel
Milk peel
Mela peel
Cosmo peel
Mask peel
Bio C

8. Sesglicopeel K peel (Mediderma): 33% Glycolic acid, 9% Lactic acid, 7% Kojic acid, 5% Citric acid and 3% Salicylic acid
9. Nomelan A peel (Rejsol): 15% TCA, 8% Phenol, 2% Glycolic acid, 5% Ascorbic acid, 2% Salicylic acid, 20% Mandelic acid, 2% Ferulic acid, 2% Phytic acid and 2% Retinol.
10. Retises CT Yellow peel (Mediderma): packaged as a sachet containing 4% Retinol and 1% Retinyl propionate, along with an ampule of 10% Ascorbyl glucoside and 5% niacinamide

17.9.3 Party Peels

This is a relatively new concept, catering more so to the cosmetic needs of the patients rather than being therapeutic. These may be known and promoted by various names like "office peels," 'instant glow," "office time peels," "lunch time peels," etc. [19]. They are widely practiced but hardly find mention in published literature. These are very superficial chemical peels, either a single large molecule chemical or a combination of lower concentrations of multiple superficial peels. They have the advantage of no or minimal visible peeling and almost nil downtime. The patients can straight away resume their daily activities and are safe for any skin type. Owing to their safety profile, these can be good candidates as peeling agents for feminine genital brightening.

The list of peels being used as party peels includes:

1. Lactic acid
2. Phytic acid
3. Citric acid
4. Ascorbic acid
5. Polyhydroxy acids
6. Arginine
7. Ferulic acid
8. Kojic acid
9. Various proprietary combination peels

17.10 Combination Treatments [2]

Chemical peels can be combined with other aesthetic procedures particularly for facial rejuvenation, photoageing, melasma and acne scarring. Chemical peel acts synergistically with other modalities to enhance collagen remodelling and rejuvenate the skin. Various modalities that can be combined are:

17.10.1 Microdermabrasion

Superficial ablation of skin with the help of microcrystals.

17.10.2 Microneedling

Also known as Dermaroller, it involves the use of a device with small attached needles to create micro-pores in the skin leading to the release of various growth factors and collagen stimulation.

17.10.3 Lasers

Chemical peels can be combined with various ablative lasers like fractional carbon dioxide laser for resurfacing and rejuvenation of skin.

17.11 Conclusion

Female genital beautification is a relatively newer, but a fast upcoming field. It is more of an on-demand procedure. As the awareness about this is increasing, or rather being increased, it is imperative for the physicians to be well prepared. Since documented data are lacking in the field, it comes down to the physician's own experience and comfort level with any intended procedure in the area. Chemical peeling is a proven tool for quick and relatively safer treatment of a wide range of clinical and aesthetic indications. It is useful in the treatment of acne, pigmentary disorders like melasma, photoageing, rhytides, etc. It is a common place procedure in any dermatologist's practice. But the application in the genital area, and that too by non-dermatologists not formally trained for the purpose, it can be a different ball game. Appropriate patient selection, counselling and precise technique are essential to achieve excellent results of chemical peeling along with avoidance of complications. A thorough theoretical and practical knowledge about chemical peels is imperative to achieve desirable peel results.

References

1. Hebra F, Kaposi M. On diseases of the skin. London: New Sydenham Society; 1874. p. 22–3.
2. O'Connor AA, Lowe PM, Shumack S, Lim AC. Chemical peels: a review of current practice. Australas J Dermatol. 2018 Aug;59(3):171–81.
3. Smith Victoria. An overview of chemical peels [Internet]. The PMFA Journal. [cited 2019 Oct 3]. Available from https://www.thepmfajournal.com/features/post/an-overview-of-chemical-peels
4. Vieira-Baptista P, Almeida G, Bogliatto F, et al. International Society for the Study of vulvovaginal Diease recommendation regarding female cosmetic genital surgery. J Low Genit Tract Dis. 2018;22:415–34.
5. Rendon MI, Berson DS, Cohen JL, Roberts WE, Starker I, Wang B. Evidence and considerations in the application of chemical peels in skin disorders and aesthetic resurfacing. J Clin Aesthetic Dermatol. 2010 Jul;3(7):32–43.
6. Castillo DE, Keri JE. Chemical peels in the treatment of acne: patient selection and perspectives. Clin Cosmet Investig Dermatol. 2018;11:365–72.
7. Zakopoulou N, Kontochristopoulos G. Superficial chemical peels. J Cosmet Dermatol. 2006 Sep;5(3):246–53.
8. Dréno B, Fischer TC, Perosino E, Poli F, Viera MS, Rendon MI, et al. Expert opinion: efficacy of superficial chemical peels in active acne management – what can we learn from the literature today? Evidence-based recommendations. J Eur Acad Dermatol Venereol JEADV. 2011 Jun;25(6):695–704.
9. Al-Talib H, Al-Khateeb A, Hameed A, Murugaiah C. Efficacy and safety of superficial chemical peeling in treatment of active acne vulgaris. An Bras Dermatol. 2017 Apr;92(2):212–6.
10. Sharad J. Glycolic acid peel therapy – a current review. Clin Cosmet Investig Dermatol. 2013 Nov 11;6:281–8.
11. Salam A, Dadzie OE, Galadari H. Chemical peeling in ethnic skin: an update. Br J Dermatol. 2013 Oct;169(Suppl 3):82–90.
12. Takenaka Y, Hayashi N, Takeda M, Ashikaga S, Kawashima M. Glycolic acid chemical peeling improves inflammatory acne eruptions through its inhibitory and bactericidal effects on Propionibacterium acnes. J Dermatol. 2012 Apr;39(4):350–4.
13. Arif T. Salicylic acid as a peeling agent: a comprehensive review. Clin Cosmet Investig Dermatol. 2015;8:455–61.
14. Imayama S, Ueda S, Isoda M. Histologic changes in the skin of hairless mice following peeling with salicylic acid. Arch Dermatol. 2000 Nov;136(11):1390–5.
15. Soleymani T, Lanoue J, Rahman Z. A practical approach to chemical peels: a review of fundamentals and step-by-step algorithmic protocol for treatment. J Clin Aesthetic Dermatol. 2018 Aug;11(8):21–8.
16. Fischer TC, Perosino E, Poli F, et al. Chemical peels in aesthetic dermatology: an update. J Eur Acad Dermatol Venreol. 2010;24:281–92.
17. Sachdev M, Shilpakar R. Proprietary peels. In: Sarkar R, Roberts WE, editors. Chemical peels a global perspective. 1st ed. New Delhi: Jaypee Brothers Medical Publishers; 2019. p. 192–202.
18. Sarkar R, Mandlewala ZK. Combination peels. In: Sarkar R, Roberts WE, editors. Chemical peels a global perspective. 1st ed. New Delhi: Jaypee Brothers Medical Publishers; 2019. p. 180–91.
19. Gupta A. Party peels. In: Sarkar R, Roberts WE, editors. Chemical peels a global perspective. 1st ed. New Delhi: Jaypee Brothers Medical Publishers; 2019. p. 252–8.

Scars in Aesthetic and Regenerative Gynecology

18

Rashmi Sarkar and Preethi B. Nayak

18.1 Introduction

A scar is an undesirable outcome of wound healing. Whenever tissue is injured by trauma, inflammation or surgery, various cells and mediators get activated and try to repair this wound. When only the epidermal layer of the skin is injured, it heals with mild or no scar. When the dermal layer is injured, the tissues cannot reproduce back the initial, pre-injury structure, leading to reorganization of fibrous tissue resulting in scar formation [1, 2].

There are various invasive and non-invasive treatment options that are available now for the treatment and prevention of these scars. The patient's risk of developing a scar and their level of concern about the scar are the factors considered before deciding the scar management measures [3].

The scars are broadly classified depending on the net loss or gain of collagen into

1. Atrophic scars
2. Hypertrophic scars and keloids

The scars most commonly seen in women are those due to acne, and in gynecology are striae or stretch marks.

R. Sarkar (✉)
Lady Hardinge Medical College, New Delhi, India

P. B. Nayak
Cutis Academy of Cutaneous Sciences, Bengaluru, Karnataka, India

18.2 Management of Acne Scars [4–7]

Acne scars can be emotionally or psychologically distressing, and with time these scars may worsen due to normal ageing or photodamage. There is a wide array of treatment modalities for acne and acne scars. The treatment of each acne patient is different, and it has to be individualized. The physician should make it clear at the beginning about the unpredictability of acne scar treatment and the need for multiple approaches, multiple sessions and long treatment plans. Prevention of acne scars by early intervention is the most effective approach. The various treatment options available for the treatment of acne scars are dependent on the type and severity of acne scars. The various types of acne scars (Fig. 18.1)are scars, which show loss of collagen—ice pick scars, rolling scars, box scars, atrophic scars, stellate scars, follicular elastolysis and crateriform scars, scars which show increased collagen—hypertrophic scar and keloid.

The procedures for acne scars can also be divided into as shown in Table 18.1:

1. Resurfacing Procedures

These help in the removal of layers of skin from top to down, injury to the dermis while resurfacing induce neocollagenesis and dermal remodelling.

P. Jindal et al. (eds.), *Aesthetic and Regenerative Gynecology*,
https://doi.org/10.1007/978-981-16-1743-0_18

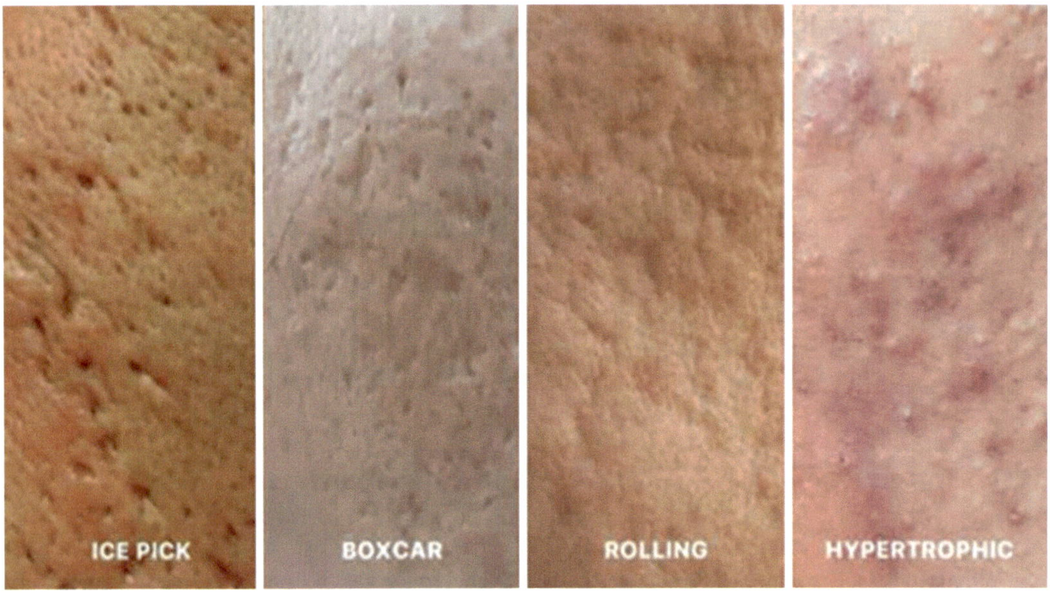

Fig. 18.1 Type of acne scars © Copyright Skin Renewal 2017

Table 18.1 Broad Categorization of Management in Acne Scars

Medical	Topical—Retinoids, steroids Intralesional—Triamcinolone acetonide, bleomycin, 5-fluorouracil
Procedures	Microdermabrasion, dermabrasion, platelet-rich plasma, chemical peels, electro-dissection, cryotherapy, radiation, radiofrequency, fillers
Light and lasers	Ablative, non-ablative, fractional, IPL
Surgical	Punch techniques, subcision, elliptical techniques, autologous fat transfer, autologous dermal grafting

Example—Chemical peels—full face, CROSS technique, Dermabrasion—Laser resurfacing (ablative, non-ablative, fractional).

2. Lifting Procedures

This attempt to uplift the base of the deep scar towards the surface resulting in smooth skin.

Example—Subcision, Fillers—directly under the scar, autologous fat transfer, volumizing, Punch flotation.

3. Excisional Techniques

It is the complete excision of the scar.

Example—Punch excision, Punch grafting, Elliptical excision.

4. Miscellaneous

Example—Thread lift, skin needling, combination techniques.

Various procedures done for acne are

1. Chemical Reconstruction of Skin Scars (CROSS)—Trichloroacetic acid (50%/90%) is applied to the base of the scar. It is done once a month for four–six sessions, especially for ice pick scars.
2. Subcision—It is mainly done for rolling or depressed scars. A tri-bevel needle/ 18-gauge needle is probed under the lesion into the deep dermis and with back and forth, and in fan-like motion, the fibrous bands are released. Post-operative hematoma is common after the procedure.
3. Chemical peels—Medium depth (glycolic 70%, trichloroacetic acid 35–50%) and deep peels (phenol 88%, trichloroacetic acid >50%) are commonly used.
4. Dermabrasion—Manual or handheld motor dermabraders are used with wire brushes, diamond fraises and serrated wheels as end pieces.

5. Microdermabrasion—Aluminium oxide crystals are used with pressure application, and vacuum removal is usually used.
6. Microneedling—It is popularly called as dermaroller, a procedure in which skin needling devices with many small surgical needles are used to create microscopic channels through the skin leading to activation of wound healing cascade causing new tissue production and remodelling.
7. Cryosurgery—It is used to cause superficial peeling by epidermal necrosis leading to desquamation. Cryoslush, cryopeel and cryoroller are used.
8. Tissue augmentation—It can be done by various methods and by using various filler materials; body fat is used for augmentation in autologous fat transfer, dermis is used for scar reconstruction in autologous dermal grafting and fillers like hyaluronic acid and other natural or synthetic materials are used commonly in tissue augmentation with quicker results.
9. Punch techniques—Punch excision—punch biopsy is done up to subcutaneous level for deep boxcar scar. Punch elevation is done for shallow boxcar scars, when the base of the scar appears normal.
10. Elliptical excision techniques—a scar requiring a punch larger than 3.5 mm is usually excised by the ellipse. The ellipse is planned along the relaxed skin tension line.
11. Lasers—Ablative lasers like CO2, erbium YAG, Non-ablative lasers like pulsed dye laser, 1064 Nd: YAG, 1540 erbium glass, 1450 diode laser, fractional ablative and non-ablative lasers can be used for the treatment of acne scars
12. Light and energy-based devices—Intense pulsed light, radiofrequency and plasma devices work in the management of acne scars.

The recommendation is based on the type of scars as follows

- Rolling scars—subcision, fractional laser therapy, filler injection directly under the scars, skin needling.
- Ice pick scars—CROSS chemical peels, punch excision.
- Boxcar scars—punch elevation, focal dermabrasion, punch excision, elliptical excision, CROSS chemical peels, skin needling.
- Fibrotic scars—excisional techniques.
- Macular scars—fractional ablative/ non-ablative laser, non-ablative laser for pigment, skin needling.
- Hypertrophic scar and keloids—intralesional therapy, cryosurgery, ablative lasers, radiofrequency, surgery.

Having an approach to efficiently evaluate and develop an appropriate treatment plan for acne scar patients will increase the chances of patient satisfaction. Setting the appropriate expectations and goals for improvement is imperative during the initial consultation. Selecting the most appropriate procedures for each lesion type will increase the chance of success.

18.3 Management of Striae/Stretch Marks [8–11]

Stretch marks or striae distensae are the most common dermal scars and they are the most common connective tissue changes in pregnancy. There are various treatment options for striae, but prevention is better than cure works very well here. There are various topical treatments that help in the prevention of striae; olive oil, almond oil, cocoa butter, hyaluronic acid, tocopherol and *Centella asciatica*.

The different treatment modalities for the treatment of stretch marks are

1. Topicals—tretinoin, silicone gel
2. Procedures—Microdermabrasion, dermaroller, platelet-rich plasma, chemical peels
3. Lasers—fractional CO2, pulsed dye laser, Nd: YAG, Fractional Er: glass, 589 nm, 1319 nm
4. Energy-based devices—Excimer light, ultraviolet light, microneedling radiofrequency, carboxytherapy

- *Silicon Gel*: It helps in increasing collagen and melanin levels in striae alba with daily application
- *Procedures*
  - *Platelet-rich plasma*: It is done alone or in combination. Combination therapy has shown the requirement of the lesser session compared to when used alone. It induces the collagen and fibroblast synthesis.
  - Microdermabrasion: It causes induction of epidermal signal transduction pathways, along with remodelling of dermal matrix leading to improvement of striae alba and striae rubra.
  - Chemical peels: Superficial peels like glycolic acid and trichloroacetic acid are very effective in striae distensae.
- *Lasers*
  - Fractional non-ablative lasers: Fractional erbium: glass lasers with 1540, 1550 and 1565 nm wavelengths are usually used in striae distansae of milder form.
  - Ablative lasers: Fractional lasers of both ablative and non-ablative varieties are used for treatment of striae. Ablative fractional lasers like CO2 though show good results but are associated with longer downtime and pain. Ablative fractional Er: YAG lasers can also be used effectively but are associated with hyperpigmentation as seen with CO2 laser.
  - *Pulsed Dye Laser*: The dilated blood vessels of striae rubra are the targets here, but chances of post-inflammatory hyperpigmentation are high.
  - *Nd: YAG Laser*: This is an ideal laser both in striae rubra and striae alba, as it acts on vascular chromophore and increases dermal collagen.
- *Energy-Based Devices*

Ultraviolet light and excimer light have been used in stria alba, with good benefits but might not be long lasting and would need maintenance.

There are various other scars like post-traumatic scar, post-burn scar which can occur. The treatment of these scars depends on the type of lesion, and the modality of treatment used will be in similar lines as those used for the above conditions.

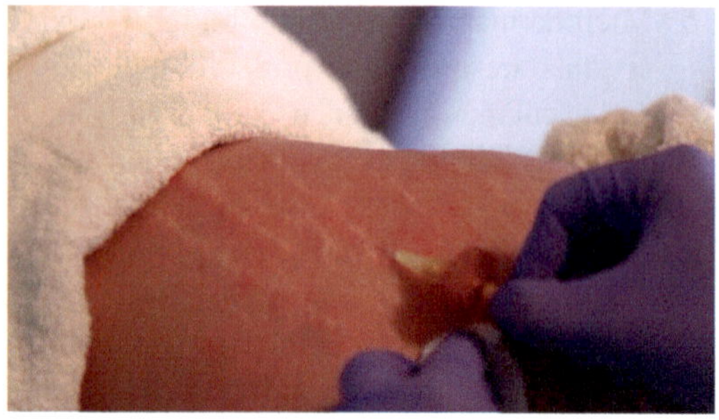

Fig. 18.2 Carboxytherapy for stretch marks

Carboxytherapy in stretch marks: There is an increase in collagen fibres in dermis after CO_2 infusion, as shown in Fig. 18.2. More information has been discussed in Chap. 16.

18.4 Conclusion

A scar, once formed, is very difficult to treat and cure completely. Patients have to be counselled about this from the beginning, and expectation from the treatment outcome should be realistic and practical. Patients' counselling is as important as treatment of scars. A good scar outcome is obtained by using multimodality treatment approaches, with multistep planning and adequate knowledge about the scar evolution.

References

1. Pape SA. The management of scars. J Wound Care. 1993 Nov 2;2(6):354–60.
2. Thomas K, Critchley P. Management of scars. Surgery (Oxford). 2006 Jan 1;24(1):18–20.
3. Meaume S, Le Pillouer-Prost A, Richert B, Roseeuw D, Vadoud J. Management of scars: updated practical guidelines and use of silicones. Eur J Dermatol. 2014 Jul 1;24(4):435–43.
4. Fife D. Practical evaluation and management of atrophic acne scars: tips for the general dermatologist. J Clin Aesthet Dermatol. 2011 Aug;4(8):50.
5. Garg S, Baveja S. Combination therapy in the management of atrophic acne scars. J Cutan Aesthet Surg. 2014 Jan;7(1):18.
6. Kumar P, Rani V, Chaudhary SS, Shubham K. Subcision combined with microneedling in the management of atrophic acne scars. IOSR Journal of Dental and Medical Sciences. 2020;19(4);26–32.

7. Abraham A. Overview of acne scaring. In: Venkataram M, editor. ACS (I) textbook on Cutaneous & Aesthetic Surgery: two volume set. JP Medical Ltd; 2017 Jul 17.
8. Hague A, Bayat A. Therapeutic targets in the management of striae distensae: a systematic review. J Am Acad Dermatol. 2017 Sep 1;77(3):559–68.
9. Wollina U, Goldman A. Management of stretch marks (with a focus on striae rubrae). J Cutan Aesthet Surg. 2017 Jul;10(3):124.
10. Chandrashekar BS, Shenoy C. Stretch marks. In: Venkataram M. ACS (I) textbook on Cutaneous & Aesthetic Surgery: two volume set. JP Medical Ltd; 2017 Jul 17.
11. Ud-Din S, McGeorge D, Bayat A. Topical management of striae distensae (stretch marks): prevention and therapy of striae rubrae and albae. J Eur Acad Dermatol Venereol. 2016 Feb;30(2):211–22.

19 Hypo and Hyperpigmentary Disorders of Vulva

Surajit Gorai and Koushik Lahiri

Learning Points/Key Points

1. As vulval anatomy is complex, a fusion of ecto, endo, and mesoderm, diagnosing a vulval disease demand a careful examination of the private parts, and this needs a blend of social skill, scientific mind, clear-cut knowledge, and earned faith with the subject.
2. Among the hyperpigmentary disorders, malignancies like carcinoma in situ, melanoma to be kept in mind, and physiologic or post-inflammatory pigmentation to be distinguished from to avoid unnecessary interventions.
3. Contact dermatitis from many hygiene products is possible and needs to be enquired about.
4. Many common drugs like sulphonamides, NSAIDS, etc., commonly used by females during menstrual pain, may cause Fixed Drug Reaction (FDE). It may cause acute inflammation of the vulva, occurs at the same site with repeated exposure that heals with hyperpigmentation.
5. Vitiligo and Lichen Simplex et Atrophicus (LSA) are two main Hypopigmentary disorders. These two must be distinguished from each other as LSA can cause erosion, dyspareunia, atrophy of the vulva and it is a premalignant condition.

19.1 Introduction

The vulva is a significant organ that was left unattended for a long time only to look at it carefully until it is diseased. Recent changes in women empowerment, instrumentation, and scientific advancement, drawn attention not only towards early diagnosis and treatment of female genital disease but also to its cosmetic enhancement. Diagnosing a vulval disease demand a careful examination of the private parts, and this needs a blend of social skill, scientific mind, clear-cut knowledge, and earned faith with the subject. All vulvar diseases are out of the scope of this chapter, and hence our discussion will be limited to pigmentary disorders of female external genitalia. 1 out of 10 women face pigmented lesions on their vulva though most of them are physiological and benign, yet knowledge about some treatable diseases is essential to avoid unnecessary cosmetic intervention and avoidance of disfigurement or severe disease upon delaying exact treatment [1].

We will divide the disorders of pigmentations into two broad categories:

1. Disorders of hyperpigmentation and 2. Disorders of hypopigmentation.

S. Gorai · K. Lahiri (✉)
Department of Dermatology, Apollo Hospitals, Kolkata, India

P. Jindal et al. (eds.), *Aesthetic and Regenerative Gynecology*,
https://doi.org/10.1007/978-981-16-1743-0_19

19.2 Disorders of Hyperpigmentation

Like all other areas of skin, the vulva has its special appearance and pigmentation. Knowledge of normal vulval pigmentation and its deviation is the key to identify abnormal vulval pigmentation. Developmentally vulva is a structure that is a junction point of three embryonic structures e. gecto, meso, and endoderm. Also, vulval skin gradually transforms into resilient vaginal mucosa and its color changes. So all the skin and mucosal pigmentary disorders that may happen there will be discussed in brief for understanding and applying its knowledge in cosmetic vulval dermatology.

19.3 Physiological Pigmentation

Vulval anatomy and physiology are dynamic, through the different stages of a woman's life it changes itself, e.g. Childhood, puberty, adulthood, pregnancy, peri, and post-menopausal; so for the color of the vulva. Hormonal factors, lifestyle, ethnic, and cultural issues also affect in a significant manner. Also, different parts of the external genitalia are of having different colors. With darker skin types color of the vulva becomes darker. Physiologic hyperpigmentation is normal and is accentuated at the posterior introitus, the tips of the labia minora, and the perianal skin [2]. Sometimes, there is also hyperpigmentation of the hair-bearing labia majora.

The proximal, medial thighs sometimes exhibit uniform hyperpigmentation that fades to the color of the non-modified mucous membrane. These lesions are asymptomatic, macular, and symmetric, without scale or change in texture from normal skin. Hormonal changes in the body like in puberty, pregnancy, or in menopause this physiologic pigmentation may be accentuated and must not be confused with the disease process [3]. But a clinician should always be vigilant enough to consider the other possible clue from history or general survey to catch and refer the patient with congenital adrenal hyperplasia, Addison's disease, or Cushing's disease that can result in hyperpigmentation of similar nature.

19.4 Lichen Simplex Chronicus

Pruritus vulvae is fairly common and most of the time neglected. Evaluation of its exact cause is not always done, be it a lack of patients' urge to seek care or a casual approach to evaluate it from the physician's end [4]. On top of that, self-medication, repeated application lead to temporary relief followed by aggravation. Chronic itching over that sensitive area leads to thickening, rugosity, rough, pigmented skin of labia majora that is collectively known as lichenification.

This is a telltale sign of chronic eczema that is termed as lichen simplex chronicus [5]. The primary cause may be widely different, ranging from psychogenic issues, fungal, or subtle bacterial infection, neurogenic and endogenous eczema, like atopic dermatitis or contact dermatitis. Repeated itching leads to subclinical inflammation and release of various chemical mediators of itch, and patients continue scratching, leading to the release of more chemicals. Thus it becomes an itch-scratch cycle that gets repeated [5]. There is hyperkeratosis, spongiosis, acanthosis, and chronic dermal infiltrate histopathologically. The thick, leathery skin of LSC is not healthy enough to perform all physiological functions of the skin, like protection from infection. Thus superimposed bacterial or fungal infection sometimes makes the situation worse. To treat this condition effectively, we need to evaluate the condition by a dermatologist, and after taking care of the secondary infection, the main aim would be essential to break the itch-scratch cycle [6]. Elimination of any possible contact allergen, long-term antihistamines, repeated application of emollient and intermittent short-term corticosteroid cream tackles the issue most of the time [7].

19.5 Lichen Planus (LP)

LP is an autoimmune inflammatory mucocutaneous disorder without any specific cause of autoimmunity known till now. It has multiple clinical variants, but only a few are found affecting genitalia. Erosive, classical, and hypertrophic variants are most commonly seen over these areas (Fig. 19.1) [8]. Classically, an LP lesion is defined as an itchy plaque with violaceous color, mainly affecting the flexural body parts of keratinized skin. In the mucosal aspect though erosive variety predominates that presents as whitish lacy plaques inside labia minora, vestibulum, and vagina. The central part of the lesion may show erosion that becomes very distinctive in older lesions. A typical violaceous tint is seen on the periphery of the lesion most of the time. Vulval lesions are usually associated with the buccal mucosal lesion [9]. Patient complains of pain, intense itching, dyspareunia, burning & irritation.

A specific syndrome called vulvovaginal-gingival syndrome presents with the severe burning sensation of the vulva and oral mucosa with LP lesions [10].

Treatment of LP is mainly with topical or short course systemic steroids.

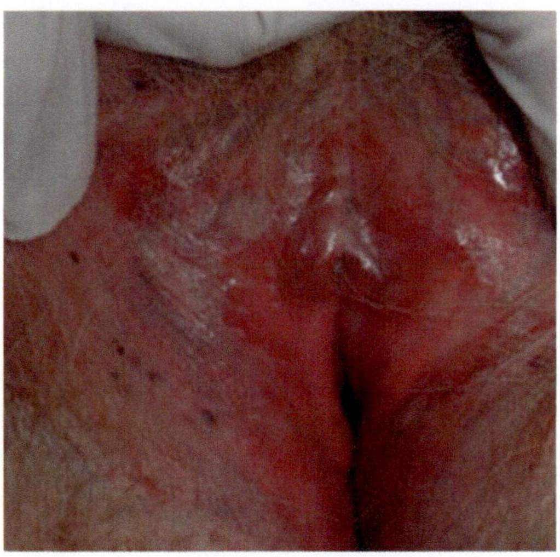

Fig. 19.1 Lichen planus © Libby Edwards, MD

19.6 Fixed Drug Eruption (FDE)

FDE is an adverse cutaneous drug reaction that usually happens on the same site, with repeated exposure to the offending drug. One of the most common sites for it is genitalia in both men and women. Common drugs being NSAIDS & Mefenamic acid, women fall prey to it to reduce menstrual pain or headache. Few antibiotics like Doxycycline, Sulfonamides, and Cephalosporin groups are the usual culprit [11]. Pseudoephedrine is also notorious to produce lesions without any pigmentation. Few drugs typically cause a type of FDE called SDRIFE, Symmetric Drug-Related Intertrigenous and Flexural Exanthema or Baboon syndrome. In this type of lesion buttocks, upper inner thighs become red, like a baboon, that may extend up to labia, and it happens in a previously sensitized individual with mercury, nickel, ampicillin, potassium iodide, etc. [12] It is still an illusion, why it happens on the same site. One theory is, on exposure to the drug, there is a production of memory T cells specific to the affected site. Their homage to the same site and causing an immunological reaction on re-exposure to the drug is the most accepted explanation [13].

Identification and stoppage of the offending drug is the main therapy with a reduction of existing inflammation by steroids [14].

19.7 Contact Dermatitis (CD)

CD is an inflammatory reaction due to an external agent that may cause irritants or allergic reactions. The irritant is a substance that causes itching, redness, and edema, even exposure to it for the first time due to its direct toxic effect (Fig. 19.2). Whereas in case of allergic CD, clinical effects are seen only when the substance is used repeatedly. It happens due to sensitization while using for the first time and types IV/delayed hypersensitivity reaction later [15]. There may be a combination of the two, usually when patient use some other over-the-counter products or

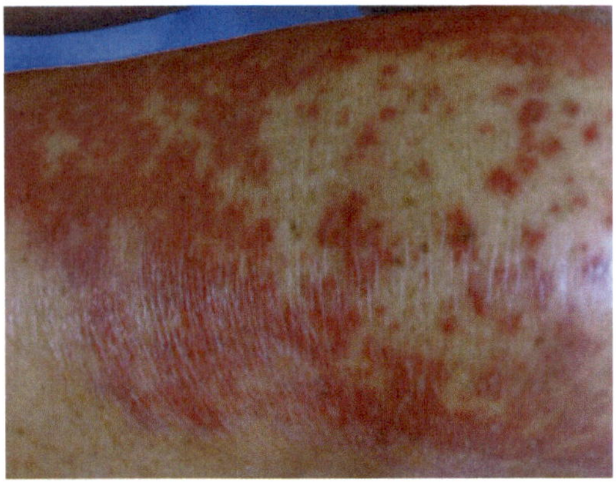

Fig. 19.2 Contact dermatitis © 2021 DermNet New Zealand Trust

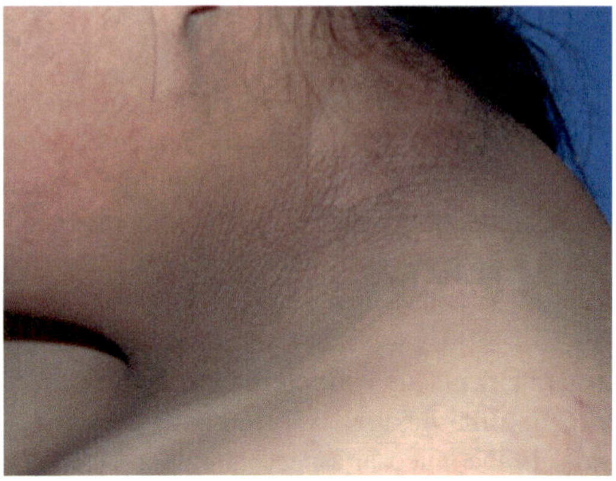

Fig. 19.3 Acanthosis Nigricans © 2021 DermNet New Zealand Trust

medicines to alleviate the symptoms of one type of CD. The exact incidence worldwide is hard to find out as data is not available from all countries. Few reported data shows its occurrence from 10–30% among patients visits for genital lesions [16]. A rising trend is striking in recent years, maybe due to using genital hygiene products or an increase in reporting. Common products blamed to cause CD are soaps, menstrual pads, panty liners, toilet paper, diapers, fabric detergents, fabric softeners, feminine sprays, cosmetics, spermicides, pessaries, and condoms. Medications used to treat genital dermatoses may also cause CD, like corticosteroid, benzocaine, antibiotic, or antifungal creams, hormonal, or vitamin preparation [17]. Sign symptoms can be divided into acute subacute and chronic dermatitis or eczema. Itching, burning, erythema edema is maximum in acute, and it gradually becomes less in subacute while in chronic, it tends toward Lichen simplex chronicus, described earlier [18]. The mainstay of therapy is always the identification and removal of the allergen. Then antihistamines and low to mid potent corticosteroids and bland emollients will do the rest [19].

19.8 Post-Inflammatory Pigmentation (PIH)

PIH, as the name suggests, happens after some prior insight event such as trauma, lichen planus. Contact dermatitis (CD) or other primary skin disorders. Lesions are different than physiologic pigmentation, being better demarcated, having no specific size or shape [20]. There may be itching if the underlying disease is still active. Histopathology shows the presence of melanin in the dermis and basal cell degeneration along with features of the primary disease [21]. Treatment is with various depigmenting agents like hydroquinone, arbutin, kojic acid, etc. and modern lasers like Qs ND: Yag [22].

19.9 Acanthosis Nigricans (AN)

AN is a pure form of epidermal keratin layer proliferation leading to pigmented, velvety, rugose appearance of skin mainly on axilla, groin, etc. (Fig. 19.3). It can also be found in vulva with the same appearance [23]. It i's thought to happen due to insulin resistance and implies an impending metabolic disaster. Insulin resistance and increase in the production of insulin-like growth factor 1 is responsible for keratinocyte proliferation. Diabetes, dyslipidemia, obesity, hypothyroidism, insulin hormone-producing tumor; all may the cause of it [24]. The sudden appearance of AN with marked hyperplasia of keratinocytes and in atypical sites like palm, the sole may indicate underlying malignancy, mainly adenocarcinoma [25]. Usually, skin tags or acrochordons are associated with AN. Treatment depends on the correction of the underlying condition and keratolytic agents topically [26].

19.10 Pigmented Condyloma Acuminate/Anogenital Wart

Condyloma acuminate/anogenital wart is HPV-related verrucous growth involving the genital area. It is contagious and often the result of sexual interaction with an infected person. Usually, its skin color, flat-topped/papular/fleshy mass, but in females of darker skin type, it may be pigmented [27]. Many times, these are indistinguishable from seborrheic keratosis, and it should be evaluated to differentiate from bowenoid papulosis/VIN. Diagnosis often clinical and treatment is removal with electro or cryosurgery [28].

19.11 Vulval Lentiginosis

Lentigines are the most common vulval melanocytic lesion accounting for more than 65% of all pigmented lesions. Those are macular (flat), small (<5 mm), brown to black, roundish lesions with an irregular border. It usually occurs in labia minora, medial border of majora, introitus & perineum, as shown in Fig. 19.4 [29]. These benign lesions always to be evaluated keeping in mind ABCDE rule, any Asymmetry in shape, disruption, or irregularity in Border, variation in Color, Diameter more than 6 mm, Evolution in terms of size, shape, and color is indicative of malignancy and to be biopsied [30]. Dermoscopy is now evolved as a non-invasive tool. Dermoscopic features are a structureless, parallel, reticular, or ringlike pattern [31]. Histopathologically, it shows mild melanocytic hyperplasia without atypia and pigmentation confined to the basal layer of squamous epithelium [32]. No treatment is needed if not cosmetically troublesome. But there is an increased chance of vulval intraepithelial neoplasia or melanoma, so evaluation is necessary. Q switched Nd: Yag laser may be used to treat if needed [33].

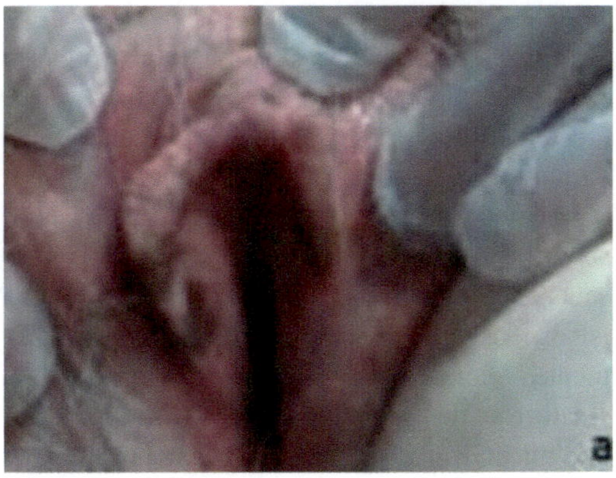

Fig. 19.4 Vulvar Melanosis © 2018 R Giménez-García

19.12 Genodermatosis Characterized by Lentigines

Lentigines at some sites like labia, vermilion border of lips, palm, sole, or other mucosae like conjunctiva usually provide a signal of an underlying syndrome associated with it. Peutz-Jeghers syndrome (PJS) is one of the most commonly known. Many other disorders associated with lentigines can mimic PJS like Carney Complex (CNC), Laugier-Hunziker syndrome (LHS), Ruvalcaba-Myhre-Smith, Bannayan-Zonnana syndrome (BRRS), Cowden disease (CD), and LEOPARD/Noonan syndrome (Table 19.1) [34]. Most of these syndromes are inherited in an autosomal dominant. They also have a relatively high rate of de novo cases and predispose to a variety of neoplasms [35]. Protein kinase A (PKA), Ras-MAP kinase, and the mammalian target of rapamycin (mTOR) are the downstream signaling pathways involved in these disorders [36]. Mutations affecting these signaling pathways may include conditions ranging from benign lentigines to aggressive malignancies. An algorithm for the approach of patients presenting with the three most prevalent syndromes associated with lentigines is presented in Table 19.1.

19.13 Benign Melanocytic Nevi

Benign melanocytic nevi are common vulval lesions ranging from 1–3% of all genital dermatosis. It may be macular (flat) or papule/plaque type (elevated), brown to black in color, symmetric with a well-defined border. Size is usually less than 6 mm [29]. Histologically, they are identified

Table 19.1 The Three most prevalent syndromes associated with lentigines

C/F	PJ	Carney complex	Pten syndrome Ruvalcaba-Myhre-Smith or Bannayan-Zonnanasyndrome (BRRS), and Cowden disease (CD)
MUCOCUTANEOUS FEATURES	Genital Lentigines Freckling of lips, conjunctiva, inner/outer canthi	Freckling/Lentiginosis of inner canthi and genital mucosa Blue nevi Cutaneous Myxoma Melanocytic Schwannoma	Skin tags Acral keratosis Small warty papules Mucosal pigmentation
ASSOCIATED NEOPLASMS	Hamartomatous polyp in GI tract	Cardiac, breast, testis, thyroid neoplasm Cardiac myxoma Association with MEN syndrome	Breast, thyroid, endometrium cancer

with classical nevus cells, those are melanocytes with more cuboidal in shape and arranged in clusters. They are clonal, retains pigment in the cytoplasm, usually start from the dermo-epidermal junction (junctional nevus). Gradually they bulge toward the epidermis without dermal involvement (epidermal nevus) or with only dermal involvement (dermal nevus) or both (compound nevus) [37].

19.14 Melanoma

Melanoma is a malignancy that carries a bad prognosis due to its metastatic potential, so to detect early, we need to be vigilant in all pigmented lesions of the vulva. Vulvar melanoma is a rare disease, less than 1% when we consider all types of malignancies in females [38]. But comes in second, after squamous cell carcinoma, accounting for all genital malignancy. Histological types, in order of incidence, are mucosal lentiginous (27–57%), nodular (22–28%), unclassified (12–16%), and superficial spreading (4–56%) [39].

The ABCDE rule for any pigmented lesions of the vulva should always be kept in mind as the lesion on the mucosal surface may be confused with seborrheic keratosis, flat wart, benign nevi, etc.

- A—Asymmetry in shape.
- B—Border irregularity.
- C—Color variation.
- D—Diameter > 10 mm.
- E—Elevation from the surface.

Diagnosis can be made now with dermoscopy but always need to be confirmed with biopsy and histopathology. With dermoscopy also focus on ABCD rule, and under D we look for more differential features like irregular pigment network, irregular dots, and globules, blue-white veils etc. [40]. These dermoscopic features now can be evaluated through algorithms artificially within a very short time and with high sensitivity and specificity (Gorai S, one of the authors, himself has developed one, with his team) [41].

19.15 Pigmented Basal Cell Carcinoma

Basal cell carcinoma is a low-grade neoplasm with low potential for metastasis, but it can be locally eroding, even bone-deep. It has many variants and is not common for the vulva. But pigmented variant may be confused with a nevus, seborrheic keratosis, or condyloma. BCC represents only 5% of all vulval malignancies, and that is also usually the ulcerative variant. So pigmented variant is very rare in occurrence. A small, roundish, indurated ulcer with pigmentation in and around is the clinical presentation [42]. Dermoscopy shows arborizing vessels, large blue-gray ovoid nests, ulceration, maple-leaf-like areas, spoke wheel areas, or multiple blue-gray dots/globule [43]. Histopathology shows basaloid cells with peripheral palisading of lesional nuclei, specialized stroma, clefting artifact between epithelium and the stroma, arranged in nets or clusters [44].

19.16 Disorders of Hypopigmentation

Two common disorders associated with hypopigmentation that involve the vulva are vitiligo and lichen sclerosus et atrophicus (LSA). Both are clinically confusing when it starts, but with careful observation, we can differentiate the two. This is important for the various reason, one being the chance of malignancy in LSA; others are morbidity associated with LSA in the form of erosion, dyspareunia, etc. In vitiligo, there is no other complication except for psychological issues associated with the sexual relationship. For all the caregiver who deals with genital problems, knowledge of skin changes are necessary to refer to an expert in proper time as well as to avoid unnecessary intervention. These two disorders will be discussed in detail with clinical features and management.

19.17 Vitiligo

Vitiligo is a disorder of depigmentation without any other epidermal changes. Autoimmunity against melanocytes is thought to be responsible for it. Melanocyte destruction leading to depigmentation of skin is difficult to explain with autoimmunity only; cytoskeleton instability, oxidant stress, drug-induced direct toxic effect are also in discussion for a long time. Genetic susceptibility and environmental factors both in combination are thought responsible for its pathogenesis [45]. Clinically, it manifested as asymptomatic depigmentation of any body part (Fig. 19.5). Genital involvement is fairly common, and it is confusing with genital **Lichen sclerosus et atrophicus** (LSA), discussed later. Differentiating point is, in

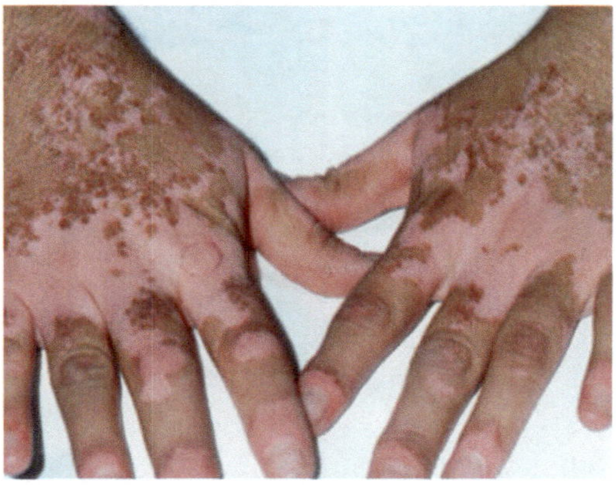

Fig. 19.5 Vitiligo © 2021 DermNet New Zealand Trust

vitiligo, there are no secondary cutaneous changes like atrophy, fissuring, or erosion as in LSA. Pubic hairs may also be white sometimes. There may be an association with other autoimmune diseases like thyroid disease, pernicious anemia, etc. Treatment is NBUVB, Topical steroid, Tacrolimus, and surgical modalities as grafting [46].

19.18 Lichen Sclerosus Et Atrophicus (LSA)

LSA is a chronic inflammatory skin disorder affecting the vulva. It can also affect male genitalia but in less proportion. There is a bimodal peak of incidence, one in pre-menarchal girls and the other is post-menopausal women. Extra genital lesions may also be there [47]. The cause of this disorder is still an enigma, blaming autoimmunity in most of cases. High level of circulating autoantibodies and co-presence with other autoimmune disorders like autoimmune thyroid disease, pernicious anemia, etc. This inflammatory condition needs early diagnosis and treatment as there is a definite chance of malignancy, vulvar carcinoma ranging from 4–6% [48].

Clinically it usually involves labia minora, inner portion of majora, interlabial area, perineum, perianal region, and often forms a figure of eight-shaped involvement (Fig. 19.6). Porcelain white atrophic patch that gradually increases and often causes fissuring of interlabial

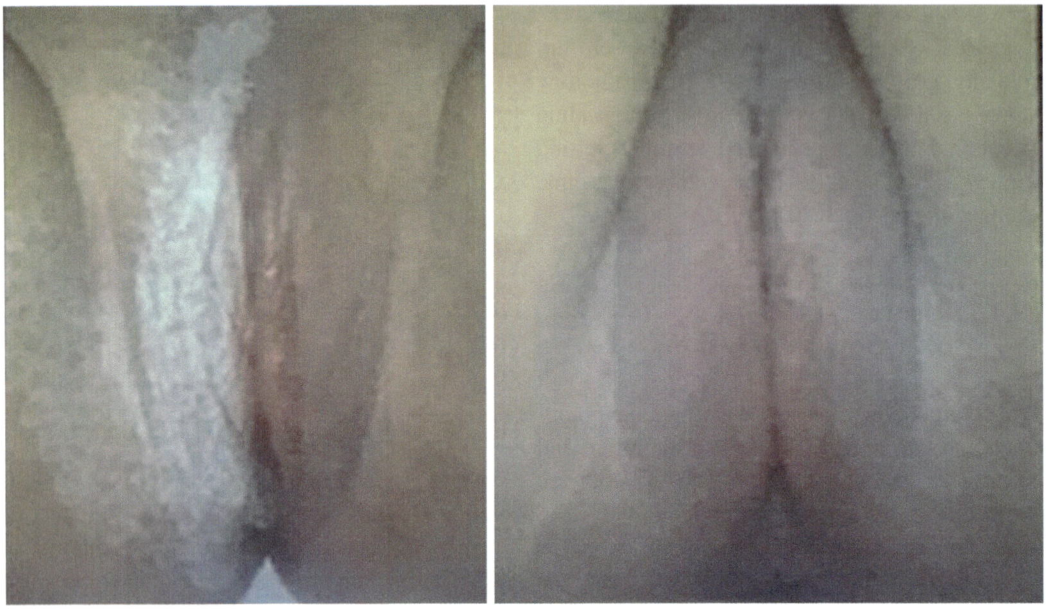

Fig. 19.6 Lichen sclerosus

sulcus and depigmentation that needs to be differentiated with vitiligo. Patients present with chronic vulvar pain, dyspareunia (though at this age, patient is usually not active sexually), itching, discharge, and secondary infection in advanced stages. The long-standing disease causes atrophy of minora and other parts of external genitalia [49].

After having an in-depth history of complaints, medical, menstrual, sexual, and treatment done so far the next most important step is proper clinical examination. Here the most important from the dermatology point of view is inspection; positioning, privacy, illumination, and attendant are necessary. Best done in lithotomy position with a headlamp, sequentially from outer surface including perianal region, groin to labia majora, and then to proceed to introitus and minora and finally to intravaginal examination. Most of the time, we can reach a diagnosis with inspection aided by palpation, but definitely to take help of dermoscopy and sometimes a small punch biopsy becomes necessary. Basic understanding of the common pigmentary changes and their anomaly is quite helpful and rewarding.

References

1. Rock B, Hood AF, Rock JA. Prospective study of vulvar nevi. J Am Acad Dermatol. 1990;22(1):104e6.
2. Friedrich E, Burch K, Bahr J. The vulvar clinic: an eight-year appraisal. Am J ObstetGynecol. 1979;135:1036e40.
3. Lynch PJ. Pigmented disorders. In: Edwards L, Lynch PJ, editors. Genital dermatology atlas. 2nd ed. Philadelphia: Lippincott Williams & Wilkins; 2010. p. 170e85.
4. Ball SB, Wojnarowska F. Vulvar dermatoses: lichen sclerosus, lichen planus, and vulval dermatitis/lichen simplex chronicus. In Seminars in cutaneous medicine and surgery. 1998;17(3):182–8.
5. McKay M. Vulvodynia versus pruritus vulvae. Clin Obstet Gynecol. 1985;28(1):123–33.
6. Carlson JA, Lamb PM, Malfetano J, Ambros RA, Mihm MC. Clinicopathologic comparison of vulvar and extragenital lichen sclerosus: histologic variants, evolving lesions, and etiology of 141 cases. Mod Pathol. 1998;11(9):844–54.
7. Stewart KMA. Clinical care of vulvar pruritus, with emphasis on one common cause, lichen simplex chronicus. Dermatol Clin. 2010;28(4):669–80.
8. Thorstensen KA, Birenbaum DL. Recognition and management of vulvar dermatologic conditions: lichen sclerosus, lichen planus, and lichen simplex chronicus. J Midwifery Womens Health. 2012;57(3):260–75.

9. Moyal-Barracco M, Edwards L. Diagnosis and therapy of anogenital lichen planus. Dermatol Ther. 2004;17:38–46.
10. Ridley CM. Chronic erosive vulval disease. Clin Exp Dermatol. 1990;15:245–52.
11. Fischer G. Vulvar fixed drug eruption. J Reprod Med. 2007;52:81–6.
12. Arnold AW, Häusermann P, Bach S, Bircher AJ. Recurrent flexural exanthema (SDRIFE or baboon syndrome) after administration of two different iodinated radio contrast media. Dermatology. 2007;214(1):89–93.
13. Ozdemir S, Kepil F, Erkekol O, Aydin O, Celik GE, Misirligil Z. Fixed drug eruption due to meloxicam. J Investig Allergol Clin Immunol. 2011;21(5):419.
14. Wain EM, Neill S. Fixed drug eruption of the vulva secondary to fluconazole. Clinical and Experimental Dermatology: Viewpoints in dermatology. 2008;33(6):784–5.
15. Margesson LJ. Contact dermatitis of the vulva. Dermatol Ther. 2004;17:20–7.
16. Crone AM, et al. Aetiological factors in vulvar dermatitis. J Eur Acad Dermatol Venereol. 2000;14:181–6.
17. Brenan JA, et al. Evaluation of patch testing in patients with chronic vulvar symptoms. Australas J Dermatol. 1996;37:40–3. Margesson LJ. Contact dermatitis of the vulva. Dermatol Ther2004;17:20–7. https://doi.org/10.1111/j.1396-0296.2004.04003.x.
18. Haverhoek E, Reid C, Gordon L, et al. Prospective study of patch testing in patients with vulval pruritus. Australas J Dermatol. 2008;49:80–5. https://doi.org/10.1111/j.1440-0960.2008.00442.x.
19. Bhate K, Landeck L, Gonzalez E, Neumann K, Schalock PC. Genital contact dermatitis: a retrospective analysis. Dermatitis. 2010;21(6):317–20.
20. Edwards L. Pigmented vulvar lesions. Dermatol Ther. 2010;23(5):449–57.
21. Moyal-Barracco M, Wendling J. Vulvar dermatosis. Best Practice & Research Clinical Obstetrics &Gynaecology. 2014;28(7):946–58.
22. Venkatesan A. Pigmented lesions of the vulva. Dermatol Clin. 2010;28(4):795–805.
23. Grasinger CC, Wild RA, Parker IJ. Vulvar acanthosis nigricans: a marker for insulin resistance in hirsute women. Fertil Steril. 1993;59(3):583–6.
24. Puri N. A study of pathogenesis of acanthosis nigricans and its clinical implications. Indian J Dermatol. 2011;56(6):678.
25. Mikhail GR, FachnieD'Anne M, Drukker BH, Farah R, Allen HM. Generalized malignant acanthosis nigricans. Arch Dermatol. 1979;115(2):201–2.
26. Kellett HA, Collier A, Taylor R, Sawers JSA, Benton C, Doig A, Baird D, Clarke BF. Hyperandrogenism, insulin resistance, acanthosis nigricans, and systemic lupus erythematosis associated with insulin receptor antibodies. Metabolism. 1988;37(7):656–9.
27. Rock B. Pigmented lesions of the vulva. Dermatol Clin. 1992;10(2):361–70.
28. Alam M, Stiller M. Direct medical costs for surgical and medical treatment of condylomata acuminata. Arch Dermatol. 2001;137(3):337–41.
29. Rock B, Hood AF, Rock JA. Prospective study of vulvar nevi. J Am Acad Dermatol. 1990;22(1):104–6.
30. El Shabrawi-Caelen L, Soyer HP, Schaeppi H, Cerroni L, Schirren CG, Rudolph C, Kerl H. Genital lentigines and melanocytic nevi with superimposed lichen sclerosus: a diagnostic challenge. J Am Acad Dermatol. 2004;50(5):690–4.
31. Mannone F. Vincenzo De Giorgi, Antonio Cattaneo, Daniela Massi, Angelina De Magnis, and Paolo Carli. "Dermoscopic features of mucosal melanosis.". Dermatol Surg. 2004;30(8):1118–23.
32. Barnhill RL, Albert LS, Shama SK, Goldenhersh MA, Rhodes AR, Sober AJ. Genital lentiginosis: a clinical and histopathologic study. J Am Acad Dermatol. 1990;22(3):453–60.
33. Nogita T, Mitsuhashi Y, Takeo C, Tsuboi R. Removal of facial and labial lentigines in dyschromatosis universalis hereditaria with a Q-switched alexandrite laser. J Am Acad Dermatol. 2011;65(2):e61–3.
34. Erkek E, SeldaHizel CS, Erkek AB, Tombakoglu M, Onder Bozdogan SU, Akarsu C. Clinical and histopathological findings in Bannayan-Riley-Ruvalcaba syndrome. J Am Acad Dermatol. 2005;53(4):639–43.
35. Lodish MB, Stratakis CA. The differential diagnosis of familial lentiginosis syndromes. Familial Cancer. 2011;10(3):481–90.
36. Blume-Jensen P, Hunter T. Oncogenic kinase signalling. Nature. 2001;411(6835):355–65.
37. Murzaku EC, Penn LA, Hale CS, Pomeranz MK, Polsky D. Vulvar nevi, melanosis, and melanoma: an epidemiologic, clinical, and histopathologic review. J Am Acad Dermatol. 2014;71(6):1241–9.
38. Sturgeon SR, Brinton LA, Devesa SS, Kurman RJ. In situ and invasive vulvar cancer incidence trends. Am J Obstet Gynecol. 1992;166:1482–5.
39. Trone JC, Guy JB, Mery B, LangrandEscure J, Lahmar R, Moncharmont C, et al. Melanomas of the female genital tract: state of the art. Bull Cancer. 2014;101:102–6.
40. Rigel DS, Friedman RJ, Kopf AW, Polsky D. ABCDE—an evolving concept in the early detection of melanoma. Arch Dermatol. 2005;141(8):1032–4.
41. Chatterjee S. Debangshu Dey, Sugata Munshi, and SurajitGorai. "Extraction of features from cross correlation in space and frequency domains for classification of skin lesions.". Biomedical Signal Processing and Control. 2019;53:101581.
42. Kara M. E. Colgecen, and Erdogan Nilsen Yildirim. "Vulvar basal cell carcinoma.". Indian J Pathol Microbiol. 2012;55(4):583.
43. Rambhia K, Shah V, Singh R. Dermoscopy of pigmented basal cell carcinoma. Pigment International. 2018;5(2):123–4.
44. Deepadarshan K, Mallikarjun M, Abdu Noshin N. Pigmented basal cell carcinoma: a clinical

variant, report of two cases. J Clin Diagn Res. 2013;7(12):3010.
45. Sarhan D, Mohammed GFA, Gomaa AHA, Eyada MMK. Female genital dialogues: female genital self-image, sexual dysfunction, and quality of life in patients with vitiligo with and without genital affection. J Sex Marital Ther. 2016;42(3):267–76.
46. Parsad D. Mucosal vitiligo. In: Vitiligo. Berlin, Heidelberg: Springer; 2010. p. 57–9.
47. Nair PA. Vulvar lichen sclerosus et atrophicus. Journal of mid-life health. 2017;8(2):55.
48. Ridley CM. Lichen sclerosus et atrophicus. Arch Dermatol. 1987;123(4):457–60.
49. Barker LP, Gross P. Lichen sclerosus et atrophicus of the female genitalia: a clinical study and diagnostic guide. Arch Dermatol. 1962;85(3):362–73.

Breast Lifting and Reduction in Aesthetic and Regenerative Gynecology

20

Mohan Thomas and James D'silva

20.1 Introduction

Breasts undergo significant changes over a woman's lifetime. Many women have noticed changes in their breasts that are different from their usual cyclical changes at some stage in their lives. Usually, these changes are not cause for concern but some changes in the breast may make a woman feel less confident or incomplete.

Young women usually have dense breasts because of the predominance of glandular tissue which hypertrophies during breastfeeding. This thickness may be felt as a lump or a mass of firm tissue. When women grow older, the glandular tissue shrinks and fat takes its place and so by the time most women achieve menopause their breasts are completely soft due to which normal lumps may become more visible as well as palpable.

When women gain or lose weight, they might feel that their breasts have changed which is also possible due to the presence of genetic factors. Deflation of breasts due to weight loss and post breastfeeding is akin to having loose clothes. This is part of the "Deflation syndrome." Women who are well endowed have to fight the vagaries of gravity which causes continuous strain and pain on the neck and shoulders.

The breast in a female is one of those rare organs that anatomically have no bony or cartilaginous supports to keep it in place while gravity is continuously acting on it. Soft fibrous structures suspend it to the chest wall and its adult shape is influenced by the genetic predisposition and the bone, cartilage, and muscle elements that surround the breast's base. A "Perfect Breast" for a woman involves having the right proportion of

The original version of this chapter was revised. The correction to this chapter can be found at https://doi.org/10.1007/978-981-16-1743-0_34

M. Thomas (✉) · J. D'silva
The Cosmetic Surgery Institute, Mumbai, India

P. Jindal et al. (eds.), *Aesthetic and Regenerative Gynecology*, https://doi.org/10.1007/978-981-16-1743-0_20

fat, glandular tissue, connective tissue in a skin envelope that holds everything in place on a chest which is proportionate to her height, weight, and lower body, making it look conical and perky [1].

Breast Reductions and **Breast uplifts** (also known as mastopexy) are different types in the same procedure spectrum. The principal aim of both these procedures is to reposition the droopy breast having a low nipple and a large areola into a well-positioned, uplifted, and perky breast with the nipple pointing forward and outward. Breast lift surgery involves repositioning the nipple–areola complex on a vascular pedicle along with reshaping the breast tissue and tailoring the excess skin without reducing the breast volume. It may be combined with an implant to add volume, firmness and enlarge the breast which is called an augmentation mastopexy. A breast reduction on the other hand includes moving the nipple on its vascular pedicle which includes the residual breast tissue left behind after adequate excision of the breast tissue has been undertaken.

Excess of the breast tissue, fat deposits and skin either naturally or due to pathologies can lead to breast enlargement and drooping, also called "breast ptosis." Breast tissue is further affected by aging, gravity, and smoking which cause glandular and skin stretching primarily in the lower pole and the peri-areolar area which wears the brunt of the excess weight.

Etiology of changes in the breast:

- Pregnancy and Breastfeeding—Pregnancy is considered to be directly involved and is a major risk factor in the development of breast ptosis, so much so that the degree of drooping is directly proportional to the number of pregnancies. Breastfeeding may not cause a worsening of the ptosis but is responsible for the flattening of the upper pole [2].
- Genetic predisposition—The breast weight and the degree of laxity in the ligaments suspending the breast (Cooper's ligaments) depend on the person's genes [3].
- Hormonal changes over each menstrual cycle and menopause—Estrogen levels have a direct correlation to the amount of breast volume and a reduction in their levels can make the breasts smaller and less full than they used to be. Menopause also causes shrinkage in the breast tissue which along with loss of elasticity and denaturation of the connective tissue make them look deflated [3].
- Weight loss or gain—Since the breast is composed of a large percentage of fat, changes in body weight of more than 20 kgs will cause drooping of the breasts. Surprisingly this does not happen when ladies lose weight after pregnancy [2].
- Aging—Good breast support if not used at the time of breast enlargement is associated with having ptosis. Aging also has a direct association due to the reduced hormones causing reduced breast volume [3].
- Smoking—Experimental and clinical studies have shown that smoking causes elastoses by causing degenerative changes in the dermal connective tissue [4].

20.1.1 Surgical Anatomy of the Breast

20.1.1.1 Anatomically the Woman's Breast Extends From

Superiorly—the 2nd rib

Inferiorly—the level of the 6th rib

Medially—Middle of the sternum

Laterally—Usually the midaxillary line but may extend to the edge of the Latissimus dorsi muscle

Base—lying over the Pectoralis major muscle and can slide easily on the pectoral fascia (Fig. 20.1).

The NAC (nipple–areola complex) is usually situated at the center of the mound when the person lies down flat, while in the standing position it is usually situated in the mid-clavicular axis at the level of the inframammary crease (4th or 5th intercostal space).

The NAC contains small Montgomery tubercles which contain the Montgomery glands that are embryologically placed between the sweat and mammary glands and are capable of secreting milk and are small (1–2-mm-diameter) raised papules. The NAC also contains many sensory nerve endings, smooth muscle, and an abundant lymphatic system called the subareolar or Sappey plexus [5].

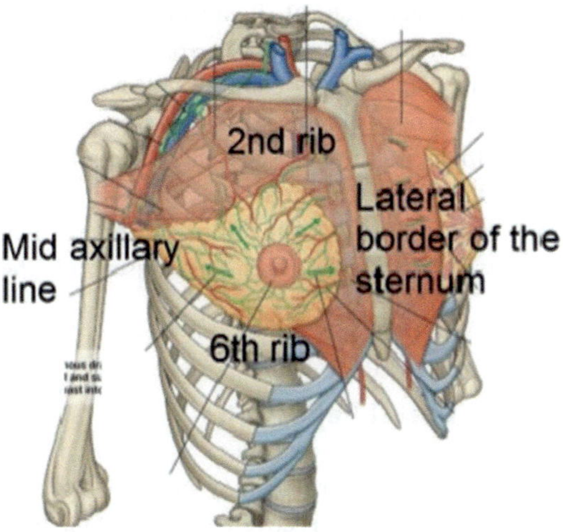

Fig. 20.1 Showing the anatomical extent of the Breast

The mammary gland forms a cone with its base at the chest wall and its apex at the nipple (Fig. 20.2).

The breast is not a homogeneous structure and contains various types of tissues (Fig. 20.2), fat (becomes predominant during weight gain), and glandular tissue (which supports lactation).

Each breast is composed of 15–20 lobes which are present in a radial manner around the nipple and are further divided into smaller sections, called lobules. Tiny "bulbs" present at the end of the lobule produce milk. Small tubes called ducts that connect these structures carry milk to the nipples. Each nipple has up to 20 ducts bringing milk to the surface. The space between the lobules, lobes, and ducts is filled with fat.

The size of the breast varies substantially and can be 500 gms or less in a small to moderate breast and about 750–1000 gms in a large heavy breast. The ratio of connective tissue to fat content determines the firmness of the breast and this

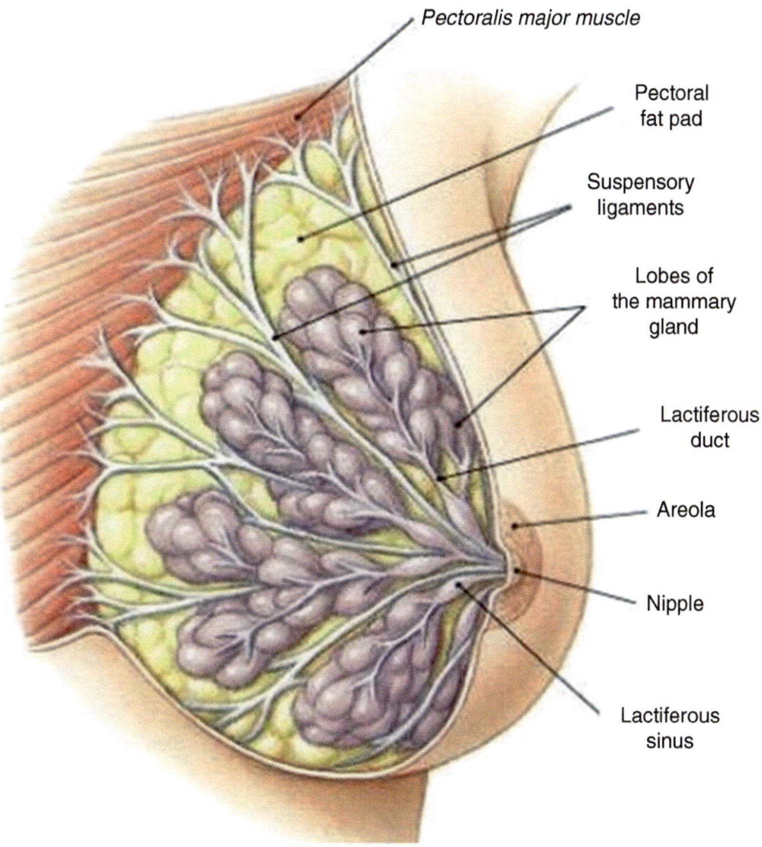

Fig. 20.2 The conical shape of the breast

ratio varies genetically as well as depends on the age of the person as well as the hormonal levels. The changes in the size and shape of the breast also change during the different periods of the menstrual cycle, pregnancy, after weaning, and during menopause [6].

20.1.2 The Breast Suspensions

The facial suspensory system of the breast is composed of a deep and superficial fascia. The superficial fascia divides into two layers: deep and superficial layers when the breast is growing during puberty. 0.5–2.5 cm thickness of subcutaneous fat separates the skin from the superficial layer. The breast grows primarily between the superficial and the deep fascia and is traversed by thin ligaments called the suspensory ligaments or Cooper's ligaments (Fig. 20.3) which connect the two. The ligaments run obliquely from the skin to the deep pectoral fascia and contain collagen fiber bundles with interspersed fibroblasts which are packed closely and oriented in a parallel fashion to provide for structural stability [6]. Aging causes a reduction of ligament elasticity by ~40%, and tensile strength by ~65%.

Regular upper body exercise will not prevent the breast from drooping since it has loose attachments to the pectoral muscles and is anatomically part of the skin to which it has strong ligamentous attachments [7]. Therefore, the breast may "fall with the skin" rather than "rise with the muscle."

20.1.3 Breast Reduction

Breast reduction surgery may be the only cosmetic procedure that is covered as part of medical

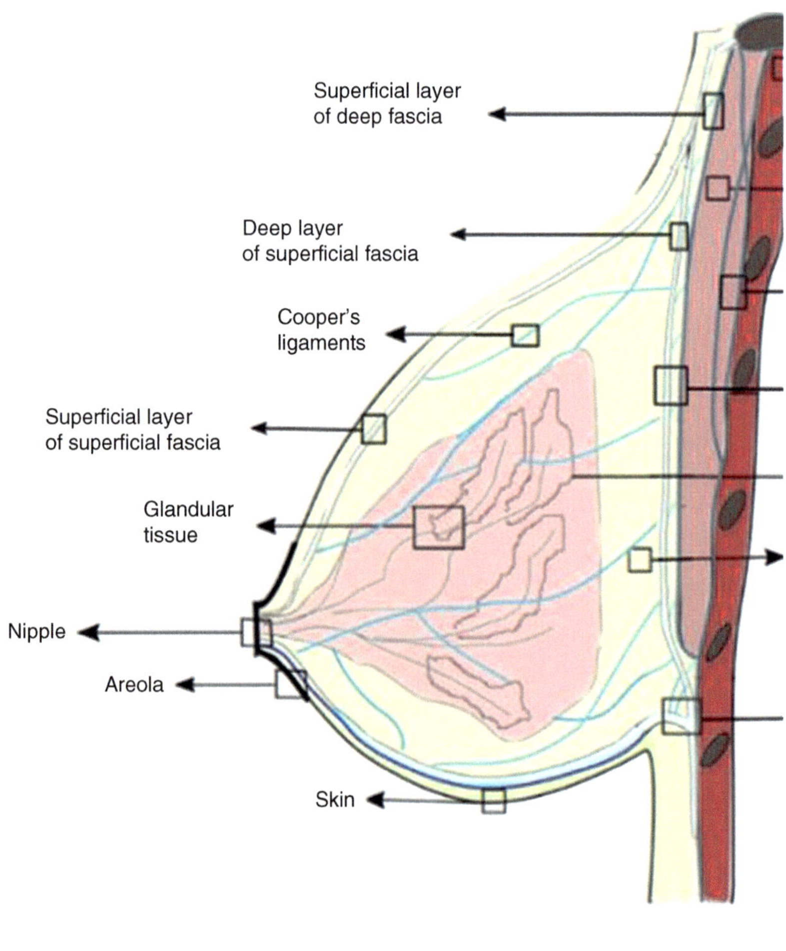

Fig. 20.3 Suspensory ligaments of the breast

insurance in many countries as there are usually medical issues such as pain in the neck and back in association with excess skin, fat, and parenchyma with an overall enlarged breast shape. By surgically manipulating the breast volume along with precise planning of the skin incision and excision, an aesthetically pleasing breast shape can be created that matches the reduced size of the breast. Breast reduction surgery is a combination of skin and parenchymal resection taking care to maintain the blood supply of the NAC on a pedicle to allow movement of the NAC to the right level.

20.1.3.1 Tenets of an Ideal Breast Reduction

These were first presented by Biesenberger in 1928 and has stood the test of time [8]-

1. The breast should look natural as well as lifted to a youthful proportion with other parts of the body.
2. Both breasts should be symmetrical
3. The nipple–areola complex should be appropriately relocated.
4. The blood supply to nipple and areola should be preserved.
5. Breast function should be preserved if possible.
6. All scars should be below the areola and not visible through normal clothes.
7. It should be possible to correct all deformities with this technique.
8. It should be a single-stage procedure.

20.1.3.2 Preoperative Evaluation

1. Complaints—

   Patients usually present with all or some of the following concerns—neck and back pain, headaches, formation of grooves in the shoulder due to bra straps and traction on the spinal ligaments, sub-mammary intertrigo due to the presence of moisture and sweat with fungal infection and paresthesias in the hands and fingers due to traction on the brachial plexus. More often, there is a general discomfort about the upper body due to the heavy weight dragging down from the shoulders for many years.

   History should include documentation of the number of children born, whether breast-fed and how long since it has been stopped?

   Past History should elicit details about prior breast surgery as it will influence the planning of skin incision as well as technique of glandular excision.

   Medical conditions such as diabetes which can influence the viability of the nipple–areola complex (NAC), pedicle or skin flaps should be recorded so that patient fully understands the risks involved preoperatively.
2. Examination—the patient is first taken in a gown and her weight checked. In the standing position, the following are examined and measured:
   (a) Measurements—The first measurement to be taken to check the symmetry for the position of NAC is the distance between the Nipple and the sternal notch on both sides. The distance of the NAC from the midline should also be recorded along with the diameter of the NAC and the length of the nipple. All these measurements have to be taken in the standing position.
   (b) Base of the breast—the breast diameter is measured along with its footprint. For breasts that have very wide bases, liposuction on the breast extensions can be considered along with narrowing of the breast.
   (c) Fullness in the Upper pole—Concavity in the upper pole is assessed. Its presence may necessitate suspension sutures to hold the tissue in the upper pole. A small implant or fat grafting may also be considered.
   (d) Skin texture and the lower pole—The extent of breast ptosis along with the distance of the NAC to inframammary crease are taken on both sides. Patient is checked for fungal infections and venous congestion in the lower pole. Any previous scar has to be noted and marked.

(e) Finally, check for asymmetry of the breast along with the consistency of the breast tissue to assess whether it is fatty or glandular. Glandular breasts may be challenging to dissect along with chances of increased bleeding.

20.1.4 Patient Profiles Coming for Breast Reduction [9]

1. Young Adults—These are girls who have suddenly developed very large breasts during puberty and are not in proportion to their body called "Giant Virginal Hypertrophy." These enlargements are due to the hormonal influence and may continue to grow even after breast reduction has been undertaken. They are unable to wear the right clothes nor able to participate in physical activities. Blood Prolactin levels have to be checked.
2. Women after childbearing—These patients have had significant enlargement during lactation and now are concerned about heavy droopy breasts due to lactation and also want fullness in the empty upper pole of the breast.
3. Women after menopause—These patients have always had large breasts but present late due to the severity of symptoms of large and heavy breasts compounded by the effects of aging. These patients usually request dramatic reductions in cup size.

20.1.5 Operative Technique

The surgical technique varies from Surgeon to surgeon. Some surgeons use a modification of the same technique for all sizes and shapes of the breast while others choose the technique based on the extent of breast ptosis, amount of breast tissue to be reduced, and the amount of skin excess.

The different techniques described for breast reduction are primarily based on two variables namely the technique of glandular resection which is different from the skin incision and excision. Even though many surgeons can vouch for the "one technique" approach serving them for many years, there is no denying the fact that familiarity with other less invasive techniques allows customization based on evidence for better outcomes with minimal or no complications.

Any breast reduction surgery is essentially composed of four components with each component targeting a specific surgical aim:

1. NAC (Nipple–Areola complex) and its blood supply—All breast reduction procedures require the movement of NAC to a higher level as they are at a lower level. This can be undertaken only by creating a vascular pedicle that preserves the blood supply and possibly nerve supply of NAC.

   The vessels supplying these pedicles are not named and are not specifically identified during the procedure but rather assumed based on previous experience.

   Many different pedicles have thus been described as seen in Fig. 20.4.

(a) Inferior pedicle technique—The blood supply of the NAC from the Inferior pedicle is robust and also provides good sensation with the possibility of breastfeeding. This technique has now replaced the vertical bipedicle technique [10, 11].

(b) Superior pedicle—provides very good circulation but is difficult to inset due to tissue bulk and has to be thinned [12].

(c) Central pedicle—The Central pedicle does not have a dermal bridge and the blood supply is from the perforating branches of the intercostal arteries. Since most of the milk ducts are left intact, the chances of breastfeeding are higher at the same time shear injury to the base of the breast can cause loss of NAC [13].

(d) Lateral pedicle—This pedicle even though has a good vascular supply from the lateral thoracic artery perforators and is easy to inset is seldom used [14].

(e) Medial pedicle—is the most common technique used these days for mild to moderate breast reduction due to the realization that this provides good NAC sensation, strong blood supply, and can be inset easily [15].

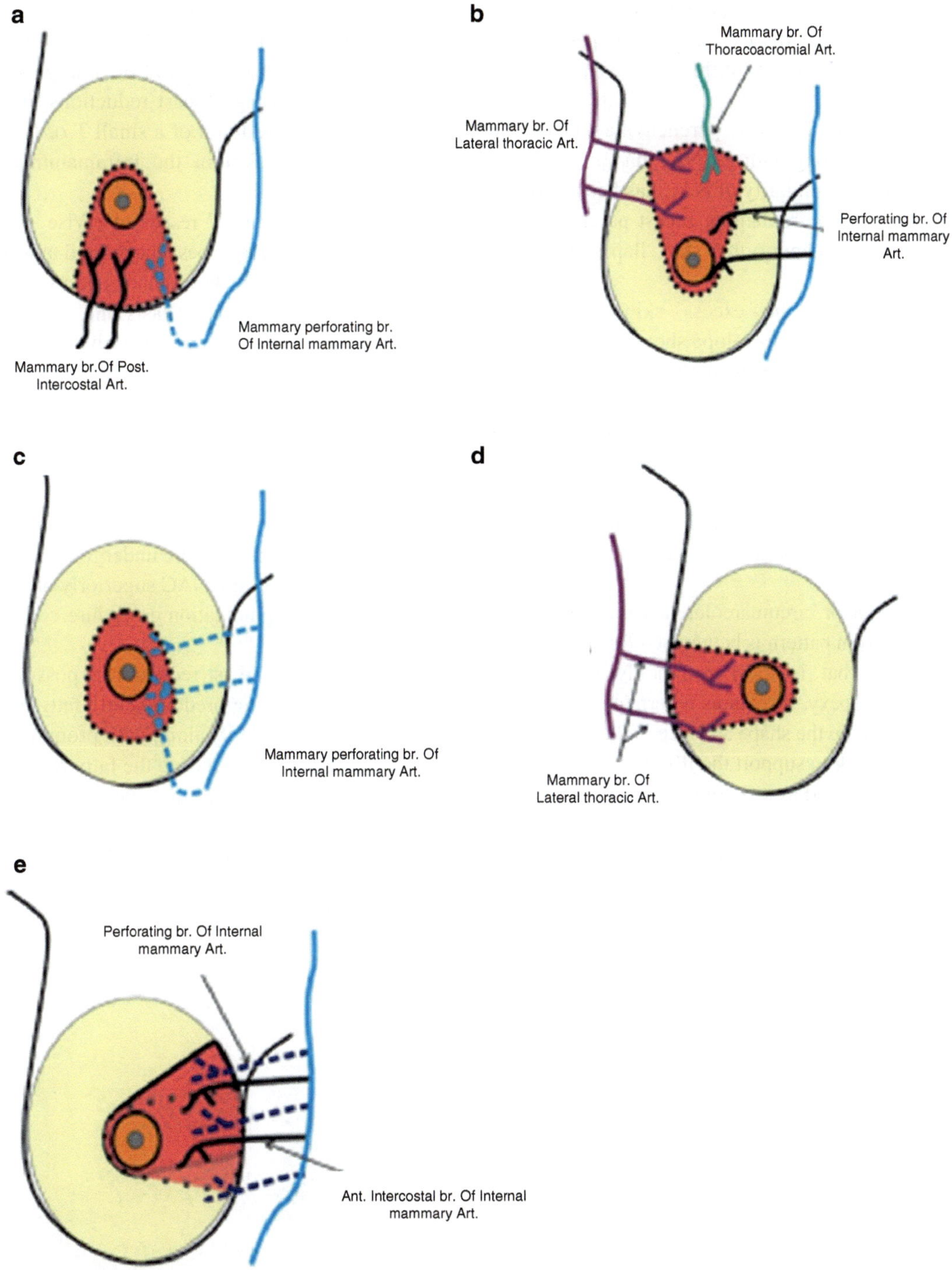

Fig. 20.4 Blood supply of each pedicle (**a**) Inferior Pedicle (**b**) Superior pedicle (**c**) Central pedicle (**d**) Lateral pedicle (**e**) Medial Pedicle

The choice of the pedicle with its various advantages and disadvantages is dependent on the other components of the operative plan.

2. Excision of Breast parenchyma—Once the pedicle along with the NAC has been created and the vascularity of NAC has been found to be safe, the remaining breast parenchyma is excised keeping the skin flap sufficiently thick.
3. Addressing the excess skin—The ptotic redundant skin envelope should be reduced in a way that the conical shape of the breast is created and at the same time the suture line is well hidden. This is where the recently described "short scar" technique has gained in popularity as it provides a more long-term conical shape to the breast over the traditional inverted T approach (Fig. 20.5).

(a) Donut or circumareolar incision—This skin incision pattern is better used for small resections but is most commonly used for a Mastopexy. It requires a permanent suture to maintain the shape and size of the areola. Use of mesh to support the lift is possible but not widespread as it is not easily available [16].
(b) Lollipop or Vertical resection—The vertical breast resection pattern creates a medial and a lateral pillar of the remaining breast parenchyma attached to the skin which will hold the breast up as well as provide a conical shape. The lower pole initially looks tighter and flat which over a period of time gets a natural shape. Large breast reductions may necessitate the addition of a small T or L to remove excess skin in the inframammary area [17].
(c) Anchor or Inverted T resection (Wise pattern)—This technique best shapes and tailors the excess skin in the breast and is best suited for massively enlarged breasts and patients having skin excess due to massive weight loss. It is best combined with the inferior pedicle technique of parenchymal reduction but can be combined with other pedicles too. This technique utilizes the skin brassiere to hold and shape the breast [18].
(d) Crescent incision—This incision is usually required in patients who are undergoing minimal movement of the NAC superiorly along with a breast augmentation procedure.

Liposuction only breast reduction is possible for ptotic breasts having predominantly fatty tissue (Grade I and Grade II) along with prominent breast extensions so as to reduce the fatty bulk of the breast along with giving it a narrower look. The presence of good skin tone and elasticity is required for adequate skin retraction. Liposuction only procedure tends to preserve nipple sensation and breastfeeding potential [19].

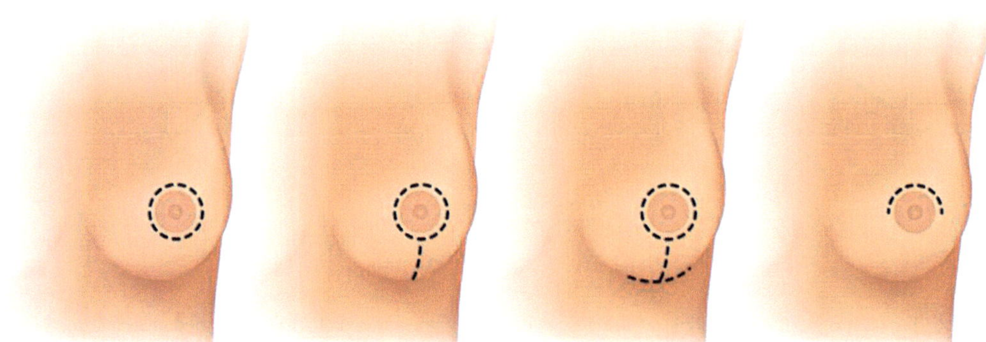

Fig. 20.5 Different skin incisions and scar lines being shown (**a**)–(**d**)

4. Management of shape—Once all resections have been completed and adequate breast volume has been achieved, it becomes important to hold the remaining tissue together so that the breast is not only smaller but also looks shapely. Various techniques to "shape" the breast have been described extending from breast flaps, use of internal suturing techniques, Gortex mesh, Alloderm and at times just relying on the skin envelope [20] (Figs. 20.6, 20.7, and 20.8).

20.1.5.1 Breast Ptosis

Ptosis of the breast in layman terms means "drooping of the breasts" which were at one time situated at a higher level. The treatment of breast ptosis involves reducing the discrepancy between the excess skin envelope and the deficient breast parenchyma which can be planned from the "inside-out" by increasing breast volume or the "outside-in" through tailoring and reducing the excess skin envelope.

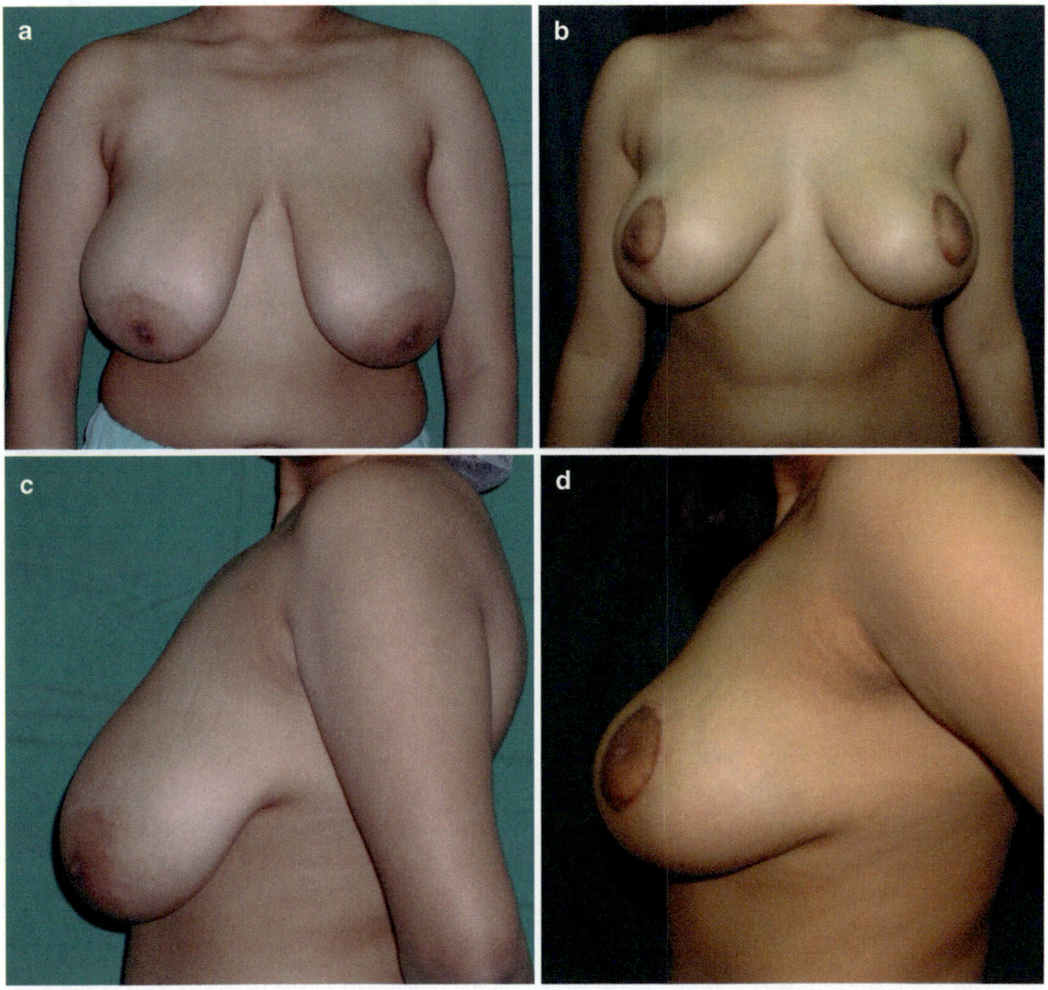

Fig. 20.6 (**a**) and (**c**)—Preoperative photographs of a 25-year-old unmarried female who wanted breast reduction as she had neck and back pain. She underwent resection of 550 gms of gland from the right and 625 gms from the left breast using a medial pedicle breast reduction with a vertical scar (Lollypop) skin excision. (**b**) and (**d**) are postoperative photographs, 2 months after surgery

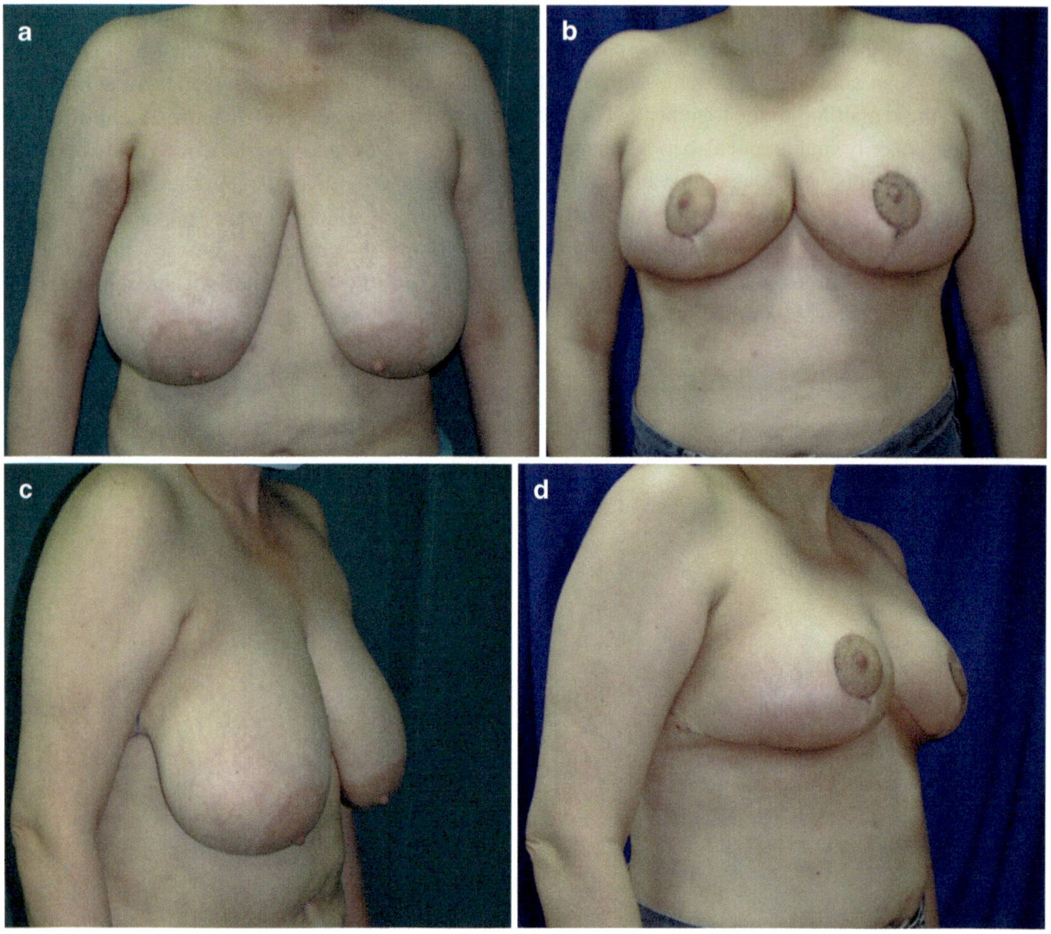

Fig. 20.7 (**a**), (**c**)—Pre-op photographs of post-menopausal lady who underwent breast reduction due to grade III ptosis with hyperplasia. 750 grams was reduced from the right and 700 grams was reduced from the left side. A supero-medial pedicle was used for NAC while the skin excision was planned as a 'Wise pattern'. (**b**) and (**d**)—postoperative photographs at 4 weeks

"Mastopexy" procedure aims to produce a perky, firm, and aesthetically beautiful breast by primarily reshaping it and differs from breast augmentation where the so-called lift is achieved by altering the breast volume. The two procedures may however be combined to treat different parts of the breast (skin vs parenchyma). There has been a sevenfold increase in the number of Mastopexiesas mentioned by the American Society for Aesthetic Plastic Surgery (ASAPS) in 2016 in comparison to 1997 [21].

In an adult woman, the "perfect" breast has to be an appropriate balance of fat, glandular tissue, connective tissue, and skin that hold everything in place. Any excess whether natural or pathologic of any one of these anatomical tissues may cause breast drooping (ptosis of breasts).

Aging and gravity cause changes in the connective tissue which causes stretching of the suspensory ligaments and skin stretching, especially in the lower part of the breast as well as the periareolar area where the drag of tissues is maximum.

The most commonly used classification to categorize ptosis of the breasts is by Regnault [22] who used a combination of the position of the nipple–areola complex (NAC), the IMF (infra-

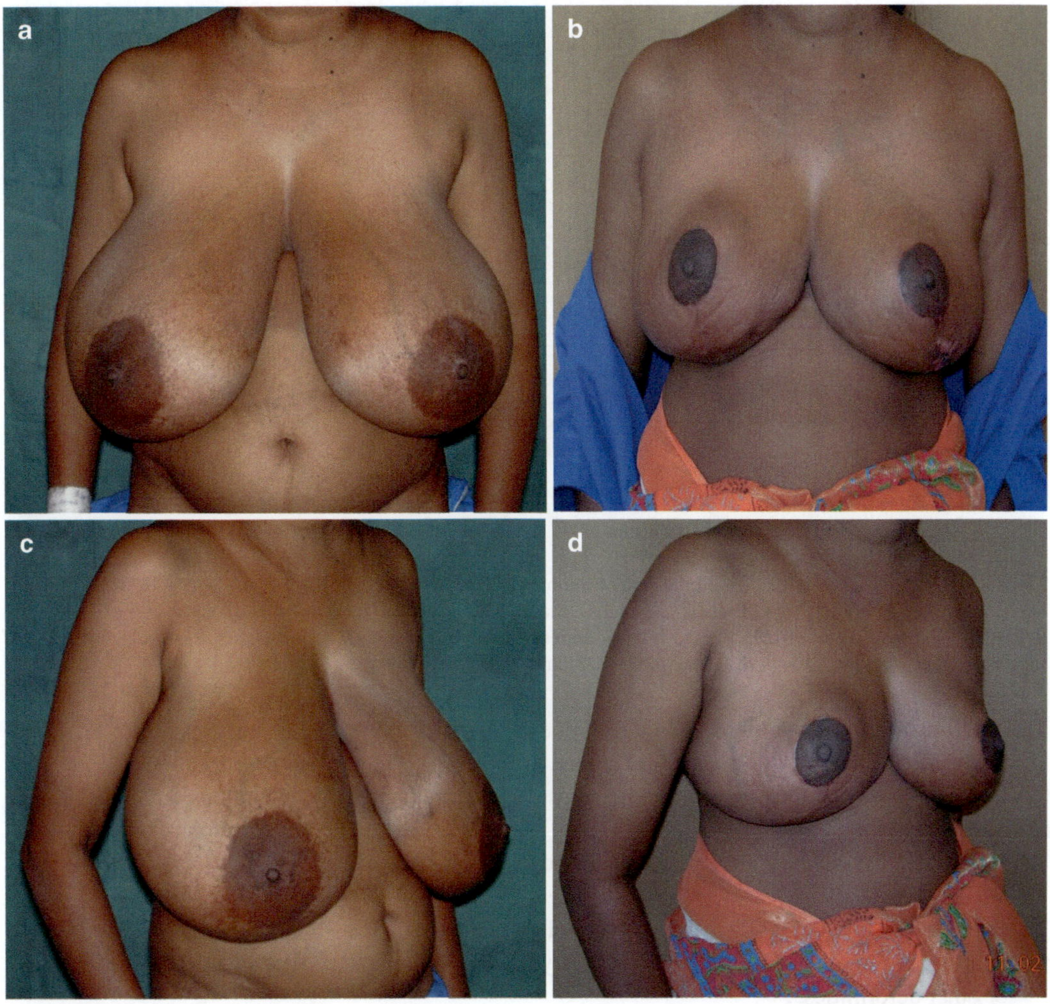

Fig. 20.8 (**a**), (**c**)—Preoperative photographs of 55-yearr-old lady with Gigantomastia after pregnancy. She underwent a 1500-gms reduction of breast tissue from each side. Inferior pedicle was used for NAC while the skin excision was in a "Wise pattern." (**b**) and (**d**) are post-surgery photographs at 6 months. She had a small area of superficial skin necrosis on the left side which healed secondarily

mammary fold), and the lowest point on the breast as landmarks. He avoided the use of numbers and centimeters and has individualized different types based on the glandular distribution such as the "normal breast," "pseudo ptosis," and "parenchymal maldistribution" and that is why this classification has been accepted worldwide (Fig. 20.9).

20.1.5.2 Principles of Mastopexy

Mastopexy is indicated in women who have less or adequate volume of breast tissue, excess skin envelope as well as droopy, and/ or a large areola.

The principles of Mastopexy include:

A. Discussion of the patient's requirements—She may want a larger and perkier breast in which case and an Implant may be required. Significant drooping and deflation of the breast may be due to massive weight loss in which case the nutritional status of the patient should be assessed or else the skin healing may not be adequate.

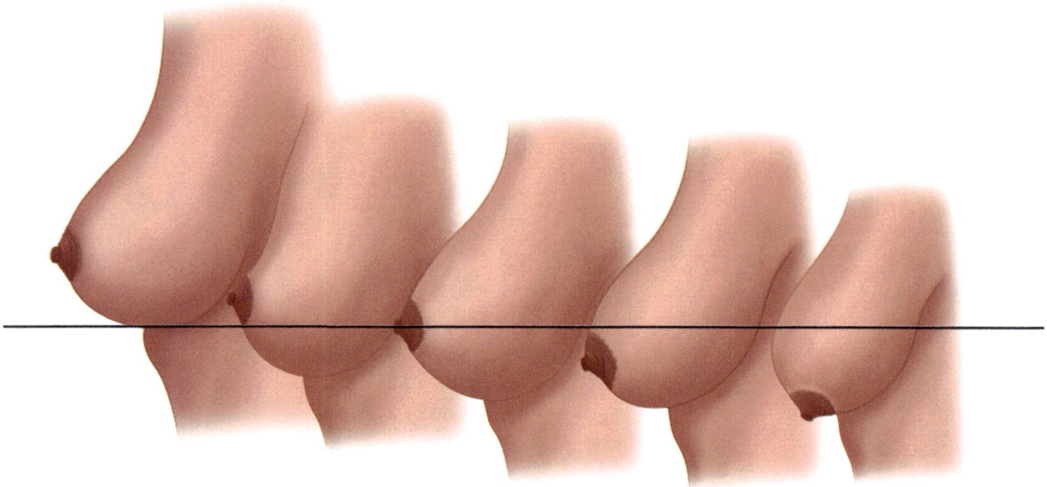

Fig. 20.9 Normal breast has the lower pole at the level of the inframammary fold (IMF), Pseudo ptosis has the lower pole below the level of IMF but the NAC is above the IMF, Grade I breast ptosis with areola at the IMF but above the contour of the gland, Grade II breast ptosis is NAC below the level of the IMF but cranial to the contour of the gland, Grade III ptosis is NAC is below the level of IMF and contour of the gland

B. Ask for history of smoking—It has to be stopped for at least 2 weeks prior to surgery.
C. Beware of any previous surgery on the breast and review the reports
D. A preoperative mammogram has to be carried out on all patients.
E. Preoperative photographs are a must—which is important to assess and record asymmetry or other deformities.
F. The following measurements have to be recorded preoperatively—a. Sternal notch and/or mid clavicle to nipples. b. Midline to the nipple c. Degree of ptosis: measurement from nipple to inframammary fold and inframammary fold to lowest point of the breast.
G. Assess the breast for herniation of the NAC and/ or contracted pole and rule out Tubular/ Tuberous breast.
H. All markings for skin incisions and excisions on the breast should be carried out in a sitting or standing position.
I. Choose the mastopexy technique according to the degree of ptosis. Mild ptosis can be treated using a crescent mastopexy or Benelli "Round Block" mastopexy while Grade III ptosis may need a short scar or a 'Wise pattern' skin tightening procedure.
J. Implant if used simultaneously should be placed before the closure of the glandular pillars.

20.1.6 Surgical Approach

A. Peri-areolar Mastopexy—This technique is usually used in ladies with Grade I or II breast ptosis and this technique can raise the nipple by 2–2.5 cm. The markings for this procedure include drawing an eccentric oval (long axis is craniocaudal) around the entire circumference of the areola as seen in Fig. 20.10. The upper pole of the oval should be the ideal upper end of the NAC from the mid-clavicular point on the breast meridian. This technique can be used to shift the NAC medially or laterally. A skin-only donut mastopexy differs from a Goes or Benelli mastopexy in that the parenchyma is plicated in Goes and Benelli procedure [23]. The advantage of this type of mastopexy is that the scar is hidden. Use of a

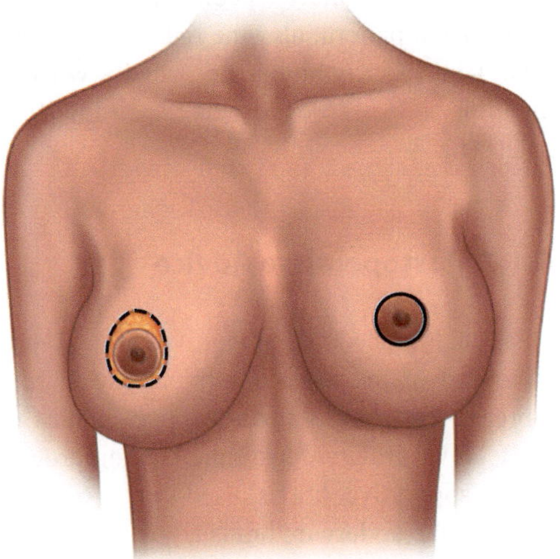

Fig. 20.10 Eccentric oval marking in the peri-areolar area with the closure seen on the right with scar merging with NAC

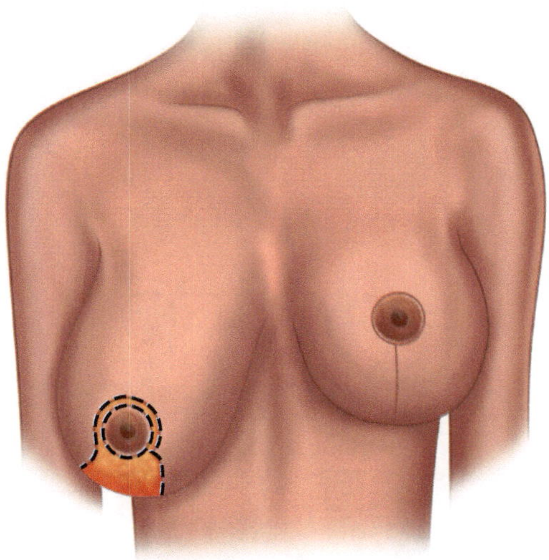

Fig. 20.11 Mosque pattern de-epithelization marked (shaded area on the left) on the lower pole of the ptotic breast area. Closure is seen on the right as a "Lollypop" appearance of the suture line

"cartwheel permanent suture" prevents widening of the NAC post healing but flattening of the breast has to be expected.

B. Crescent Mastopexy (Fig. 20.13)—This is a variant of the peri-areolar mastopexy where the incision is made superior to the areola in a crescent manner and is indicated in patients where the breast tissue is at the normal level but the NAC is lower.
C. Vertical mastopexy (Figs. 20.11 and 20.15)—This technique has evolved from the described techniques of Lassus, Peixoto, Arie, Pitanguy, Marchac, Lejour, Hammond (SPAIR), and the most recent Elizabeth Hall-Findlay technique for reduction mammoplasty [24].

These techniques use suspension sutures to pick the breast tissue superiorly and tightening of the remaining glandular tissue is achieved by the use of sutures inferiorly. Tailor-tacking of the skin is then done to mark the excision and achieve a vertical suture line inferiorly. Sometimes, the excess skin inferiorly may leave a dog ear which may require suturing in a "J" or "T" pattern. All possible types of pedicles can be used in The Hall-Findlay technique but usually a superomedial or medial pedicle for the nipple–areola complex (NAC) is used. Suturing of the medial and lateral pillars of the breast is required after an inferior wedge of tissue is removed. This glandular tightening provides support to the repositioned NAC and also narrows the breast. This is the most suitable technique for simultaneous breast augmentation using an implant below the muscle or in the subglandular plane.

The vertical breast lift technique initially causes increased fullness in the superior pole and a flattened lower pole which takes a few months to settle [25].

A. Inverted T or Wise Pattern Mastopexy (Figs. 20.12 and 20.16)—When there is a significant excess of skin relative to breast tissue such as in Grade III breast ptosis then this type of mastopexy is used.

De-epithelization of the marked area is undertaken followed by dissection of the medial and lateral skin flaps up to the edge of the breast. The NAC is then repositioned and the gland in plicated to tighten as well as lift it. The medial and lateral skin flaps are then brought together to get an inverted T scar. This technique has the greatest frequency of bottoming out which is because of

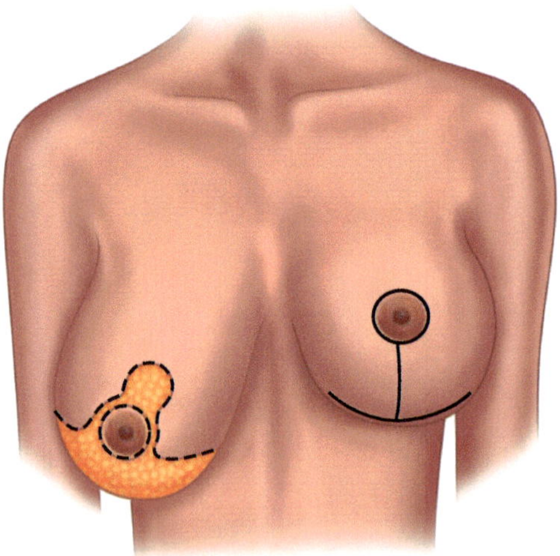

Fig. 20.12 Wise pattern de-epithelization marked (shaded area on the left) in the ptotic lower pole of the breast and the suture line is seen on the right as an "Anchor" appearance of the scar

the weight of the inferior pedicle rather than the skin excision pattern, also the risk of breakdown at the T-junction is highest, especially when an implant is added [26] (Figs. 20.13, 20.14, 20.15, and 20.16).

20.1.7 Postoperative Advice

1. Patient is placed in a supportive surgical bra at the end of the case to support the operated breast. This works like a support against the effects of gravity during the healing process. The supportive bra is used for 4–6 weeks before the underwire bra is advised.
2. Drains are not used routinely.
3. Broad-spectrum antibiotics may be advised at the Surgeon's discretion. Rarely advised beyond the hospital stay.

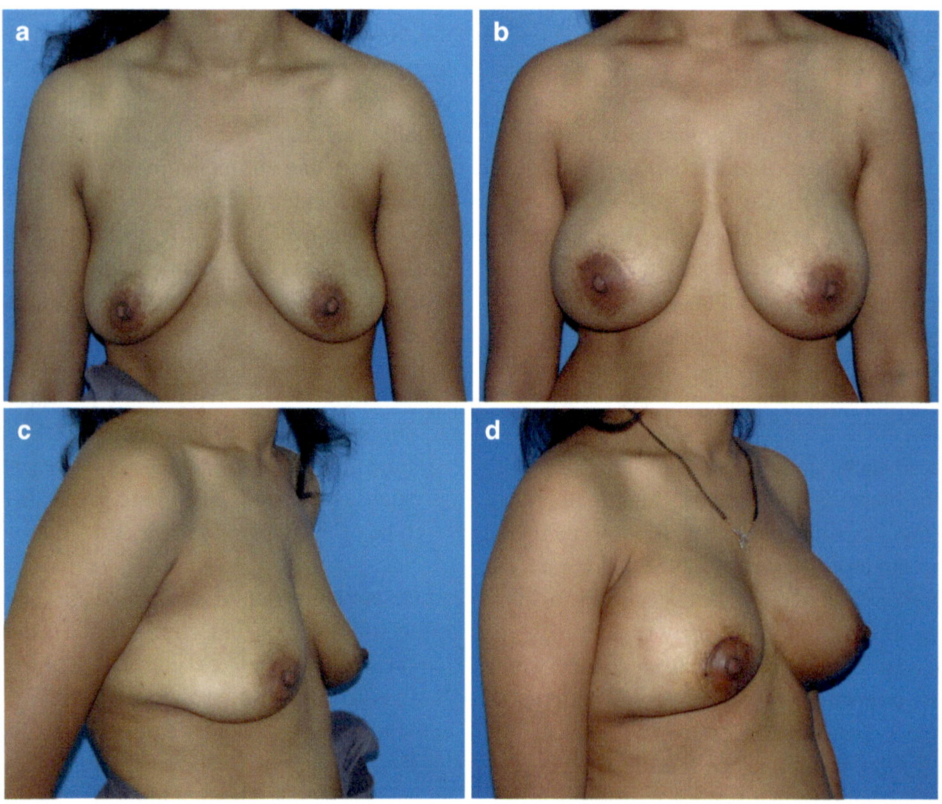

Fig 20.13 **a** and **c** are preoperative photographs of post-lactation deflated and ptotic breasts. She underwent augmentation with a crescent lift as seen in **b** and **d**

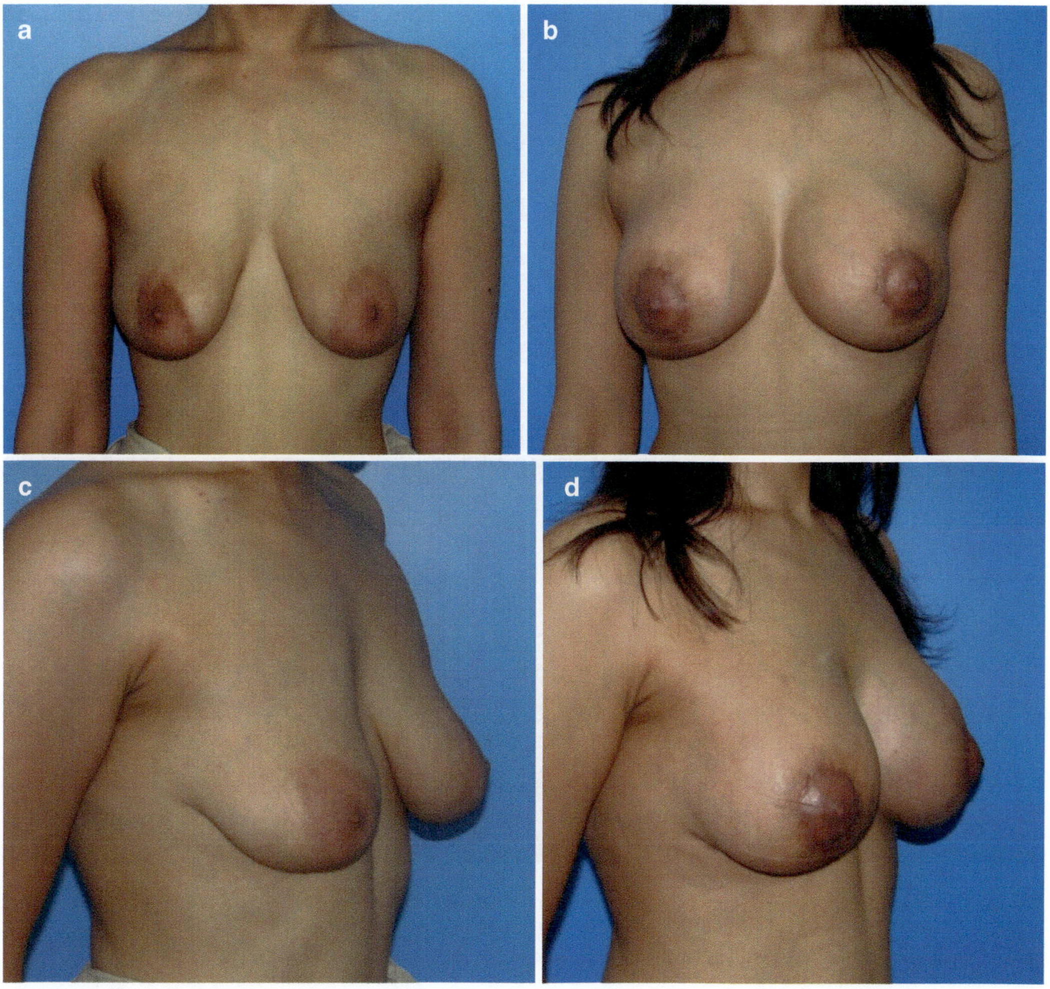

Fig. 20.14 **a** and **c** are preoperative photographs of a 24-year-old girl who lost 20 kgs with diet and exercise which improved her body shape but caused deflation and drooping of her breasts. She underwent a peri-areolar augmentation mastopexy as seen in **b** and **d**

4. If an implant has been used especially in the sub-pectoral plane then pain medication and/or muscle relaxants will be required.
5. Physical activity is advised only after 4–6 weeks depending on the type of mastopexy and reduction procedure.

20.1.8 Complications [27, 28]

20.1.8.1 Early

(a) Hematoma and/or Seroma: These are not very common and usually get absorbed spontaneously in a short period. Large hematomas may have to be evacuated.
(b) Flap Necrosis: is possible at the meeting of the three flaps with the inverted T technique, especially in smokers. Secondary procedures may be required.
(c) Infection: Usually seen due to underlying tissue necrosis.

20.1.8.2 Late

1. Asymmetry is possible. Preoperative asymmetry has to be assessed and resection planned accordingly.

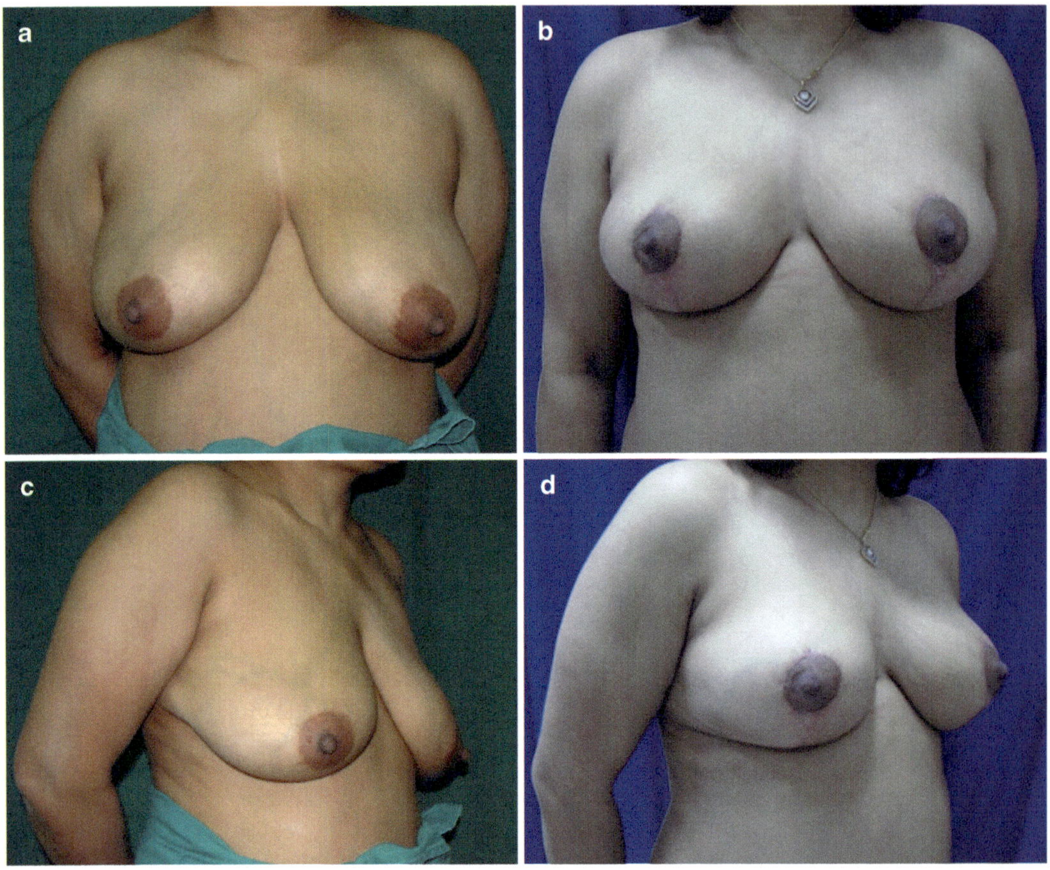

Fig. 20.15 Grade III drooping of the breasts along with deflation of the upper pole as seen in **a** and **c**. Patient wanted to maintain the breast volume and hence underwent a "Vertical mastopexy procedure" with a very good outcome as seen in **b** and **d**

2. Under resection: is usually seen in short scar vertical resection techniques. They may need repeat glandular resections or liposuction to achieve adequate reduction.
3. Over resection: Fortunately uncommon. Need to understand the patient's requirements before and counsel her accordingly.
4. Unsightly scars: Scars are usually prominently visible for 3–6 months and then reduce over a period of time but some are "scar formers" and they will need additional scar reduction procedures.

20.1.8.3 Summary

Over the years there has been a paradigm shift in breast shaping procedure which includes the rearrangement of breast tissue, use of mesh to support the breast and increasing publications and research showing the benefits of single stage mastopexy-augmentation even though it was once considered a highly litigated procedure. In addition to noninvasive devices such as plasma skin tightening, radiofrequency, and ultrasound energy technologies, patient's goals and expectations can now be met through minimal access procedures.

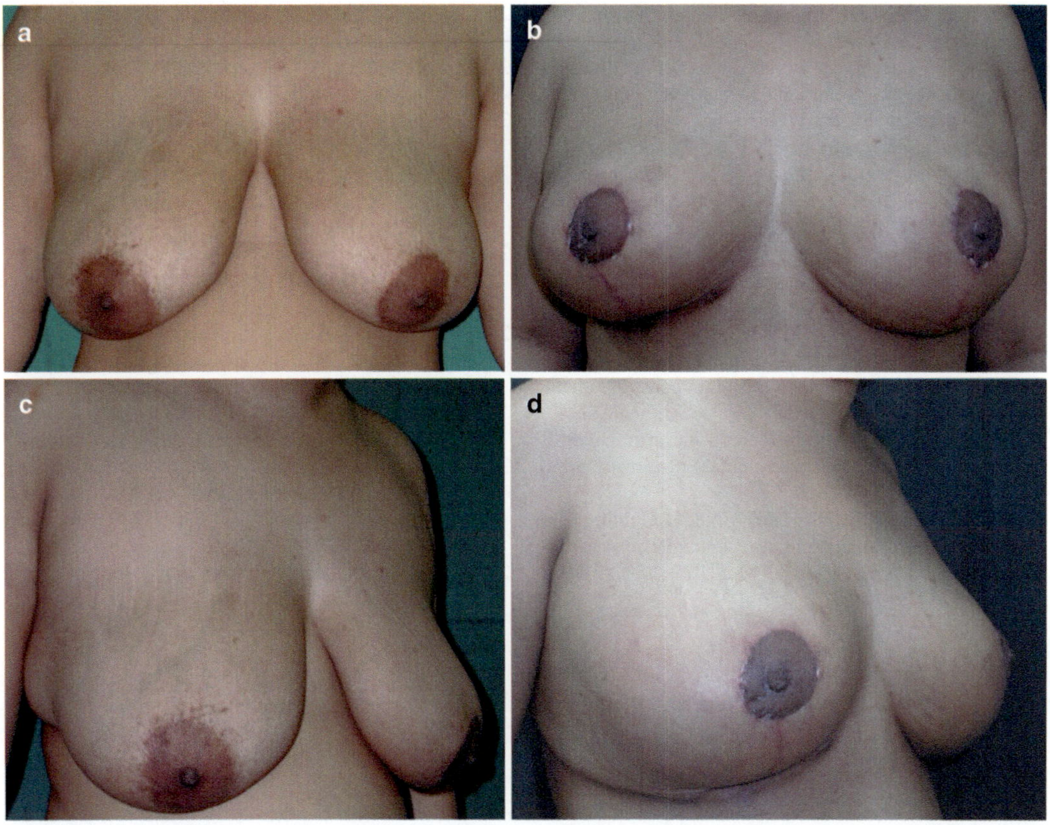

Fig. 20.16 52-year-old post-menopausal lady wanted a mastopexy procedure. She had grade III ptosis with significant excess skin envelope as seen in **a** and **c**. She underwent a "Wise pattern" mastopexy with outstanding results. Post-op pics **b** and **d** taken after 4 weeks

References

1. Bouraoui Kotti A. Simple classification and a simplified treatment's algorithm for ptotic breasts. Plast Aesthet Res. 2018;5:16. https://doi.org/10.20517/2347-9264.2018.11.
2. Brian Rinker, MD, FACS, Melissa Veneracion, BS, and Catherine P. Walsh, BS. Breast ptosis—causes and cure. Ann PlastSurg. 2010;64:579–84.
3. Arefanian, et al. A study on factors predisposing to breast ptosis. Arch Breast Cancer. 2018;5(2):63–7.
4. Morita A. Tobacco smoke causes premature skin aging. J Dermatol Sci. 2007;48:169–75.
5. Kopans D. Breast anatomy and basic histology, physiology, and pathology. In: Kopans D, editor. Breast imaging. 3rd ed. Philadelphia, PA: Lippincott Williams & Wilkins; 2007. p. 7–43.
6. Dufourmentel C, Mouly R. Chirurgieplastique. Paris: Flammarion; 1959. p. 327–70.
7. McCarty KS, Carpenter SA, Georgiade GS. The breast: embryology, anatomy, and physiology. In: Georgiade NG, Georgiade GS, Riefkohl R, editors. Aesthetic surgery of the breast. Philadelphia, PA: WB Saunders Co; 1990. p. 5.
8. Biesenberger H. Eineneue Methode der mammaplastik. ZentralblChir. 1928;55:2382.
9. Hall-Findlay EJ. Vertical breast reduction with a medially based pedicle. Oper Strategies Aesthetic Surg J. 2002;22:185–95.
10. Regnault P. Reduction mammaplasty by the "B" technique. Plast Reconstr Surg. 1974;53(1):19–24.
11. Robbins TH. A reduction mammaplasty with the areola-nipple based on an inferior dermal pedicle. Plast Reconstr Surg. 1977;59(1):64–7.
12. Bostwick J 3rd. Plastic and reconstructive breast. Surgery. 2000;1
13. Courtiss EH, Goldwyn RM. Reduction mammaplasty by the inferior pedicle technique. An alternative to free nipple and areola grafting for severe

macromastia or extreme ptosis. Plast Reconstr Surg. 1977;59(4):500–7.
14. Skoog T. A technique of breast reduction; transposition of the nipple on a cutaneous vascular pedicle. Acta Chir Scand. 1963;126:453–65.
15. Moustapha H, Dennis HC, Foad N. Vertical scar mammaplasty. Berlin Heidelberg: Springer-Verlag; 2005.
16. Weiner DL, Aiache AE, Silver L, Tittiranonda T. A single dermal pedicle for nipple transposition in subcutaneous mastectomy, reduction mammaplasty, or mastopexy. PlastReconstr Surg. 1973 Feb;51(2):115–20.
17. Hall-Findlay EJ. Vertical breast reduction with a medially based pedicle. Oper Strategies Aesthetic Surg J. 2002;22:185–95.
18. Wise RJ. A preliminary report on a method of planning the mammaplasty. Plast Reconstr Surg. 1946;17(5):367–75.
19. Hester TR Jr, Bostwick J 3rd, Miller L. Cunningham Breast reduction utilizing the maximally vascularized central breast pedicle. SJ Plast Reconstr Surg. 1985;76(6):890–900.
20. Purohit S. Reduction mammoplasty. Indian J Plast Surg. 2008;41(Suppl):S64–79.
21. Cosmetic Surgery National Data Bank Statistics. Aesthet Surg J. 2017;37(suppl 2):1–29.
22. Regnault P. Breast ptosis. Definition and treatment. Clin Plast Surg. 1976;3:193–203.
23. Benelli L. A new periareolarmammaplasty: the "round block" technique. Aesthet Plast Surg. 1990;14(2):93–100.
24. Góes JC. Periareolarmammaplasty: double skin technique with application of polyglactine or mixed mesh. Plast Reconstr Surg. 1996;97(5):959–68.
25. Hammond DC. Short scar periareolar inferior pedicle reduction (SPAIR) mammaplasty. Plast Reconstr Surg. 1999;103(3):890–901; discussion 902
26. Ali A, Qureshi MD, et al. Mastopexy and Mastopexy-Augmentation. Aesthetic Surg J. 2018;38(4):374–84.
27. Rohrich RJ, Gosman AA, Brown SA, Reisch J. Mastopexy preferences: a survey of board-certified plastic surgeons. Plast Reconstr Surg. 2006;118(7):1631–8.
28. Purohit S. Reduction mammoplasty. Indian J Plast Surg. 2008;41(Suppl):S64–79.

21 Butt Reshaping/Gluteal Recontouring Surgeries in Aesthetic and Regenerative Gynecology

A. Jayanthi Mala

21.1 Introduction

Buttock recontouring surgeries are major surgical procedures that carry with it risks, special surgical and instrument needs and postoperative complications. In this chapter, we will deal with the three most popular and acceptable methods used for buttock recontouring. Most common methods for buttock reshaping are with buttock implants or autologous fat grafting (AFG). Best method for acquiring ideal proportions is by combining buttock reshaping (implant or fat) along with liposuction.

In some patients, skin excess or buttock ptosis and lack of sufficient quantities of fat has given rise to other butt reshaping procedures. Traditional buttock lifting only deals with skin and soft tissue excess and does not tackle with loss of volume in these patients. In these patients, patients own tissues are used as various flaps (rotational, island or incremental/transposition flaps) to do the butt reshaping. Irrespective of the type of flaps these methods are broadly grouped under Autologous Gluteal Augmentation (AGA). All these methods require additional tissue dissection and precarious blood supply to flaps have made these procedures less popular. Autologous Gluteal Augmentation (AGA) can be performed with both circumferential body lift (CBL) and excisional buttock lifts (EBL).

We will be reviewing each of these procedures in this chapter, after evaluating gluteal anatomy to prevent complications and better patient satisfaction [1, 2].

21.2 Gluteal Aesthetics

Aesthetics of the buttock is not seen in isolation but in proportion to the waist. It is not about making the buttocks bigger, but about aesthetic redistribution of fat from unattractive zones to desirable location.

The ideal waist to hip ratio (WHR) of 0.7 is considered attractive, as per various studies done by Singh [3], Robert [4] and colleagues. This ratio is obtained by dividing the circumference of the waist by that of the hips. The waist circumference is measured at the level of 12th rib or elbows posteriorly and the hip circumference is obtained at the point of greatest projection in the gluteal frame.

As per the various studies done by Cuenca-Guerra and Quezada [5]. The characteristics of an aesthetically pleasing gluteal region are:

A. J. Mala (✉)
Department of Plastic Surgery, CMC, Ludhiana, India

P. Jindal et al. (eds.), *Aesthetic and Regenerative Gynecology*,
https://doi.org/10.1007/978-981-16-1743-0_21

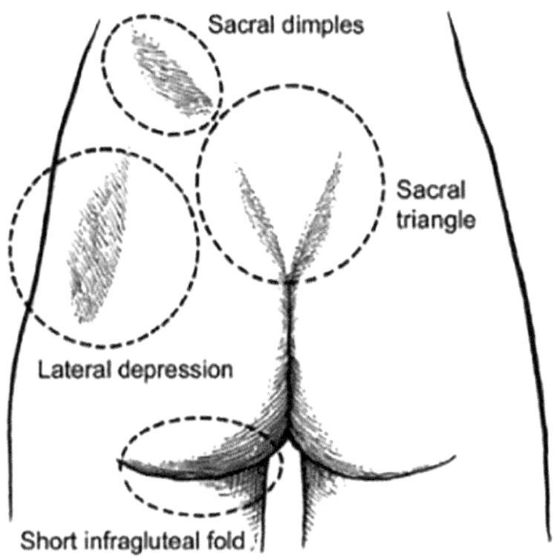

Fig. 21.1 Clinical Anatomy in Aesthetic gluteal contouring

1. A lumbosacral depression that helps to distinguish the back from the buttock
2. Two presacral dimples that correspond to the posterior superior iliac spines
3. Two mild lateral depressions that correspond to the greater trochanter of the femur
4. Short infragluteal creases that do not extend beyond the medial two thirds of the posterior thigh
5. An absence of excess fat in the lumbosacral, medial thigh and anterior thigh regions, and in areas commonly referred to as "love handles, saddlebags, and banana roll"
6. A point of maximum projection on the lateral view that corresponds to the level of mons pubis (Fig. 21.1)

21.2.1 Aesthetic Units/Zones

Posterior region is divided into 8 gluteal aesthetic units as shown in Fig. 21.2.

Surgical manipulation in and around these units will affect the final outcome of the gluteal contouring surgeries. These units should be considered during surgical planning and the junctions of the aesthetic units can act as a guide for incision placement during procedures (Fig. 21.2).

Gluteal Aesthetic Units

1. Sacrum V-Zone
2. Flank
3. Upper buttock
4. Lower back
5. Outer leg
6. Gluteus
7. Diamond Zone: Inner gluteal fold/Leg junction
8. Mid-lateral gluteal/hip junction
9. Inferior gluteal fold/posterior leg junction.
10. Upper back.

Understanding of aesthetic units and their relationship with underlying framework will guide and help us to determine the areas that will benefit from liposuction, fat transfer or gluteal augmentation.

21.2.2 Relevant Anatomy

Shape of the buttock is influenced by interplay of these anatomic variables:

1. Underlying bony framework
2. Gluteus maximus muscle
3. Subcutaneous fat topography
4. Skin
5. The interaction of these four variables gives the buttock an individualized and particular shape [6]. With muscle detached, what remains are the bony framework, fat and skin, which are collectively known as the frame. 1. The underlying bony framework influences the shape, but this structure cannot be surgically changed, so it does not play a major role in our classification system. Broadly the pelvic height can be referred to as tall, short or intermediate. 2. The skin plays a role in determining whether patient needs an upper buttock lift, inferior gluteal crease excision or inner gluteal fold excision. 3. Fat is the most important component and can be easily modified.

FRAME TYPES: Depending on the amount of fat present in three particular zones (Fig. 21.3).

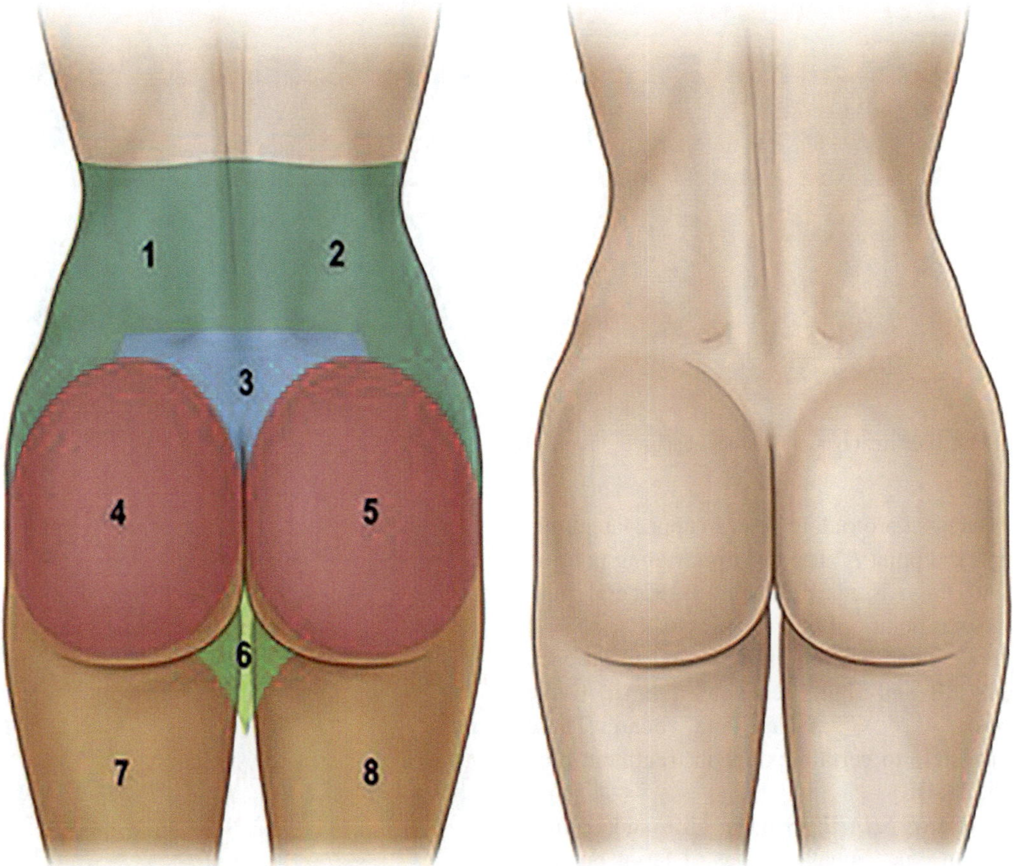

Fig. 21.2 The eight aesthetic units/zones

Fig. 21.3 Points A, B, and C, to be considered when evaluating frame type

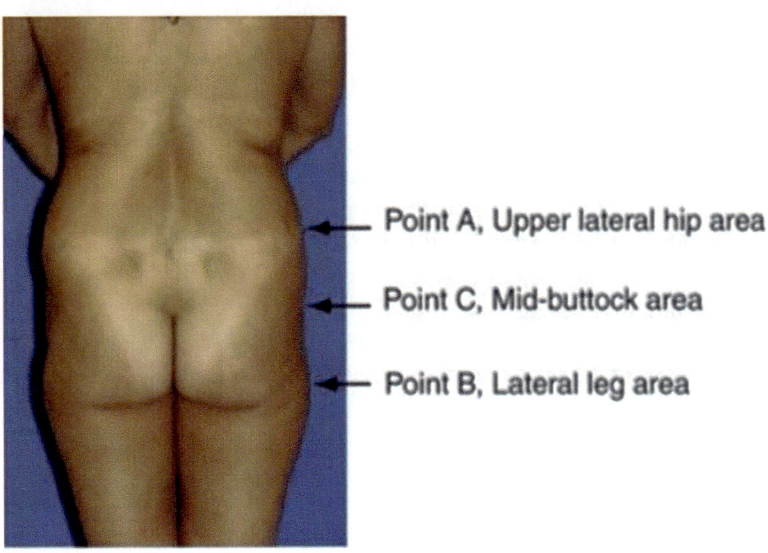

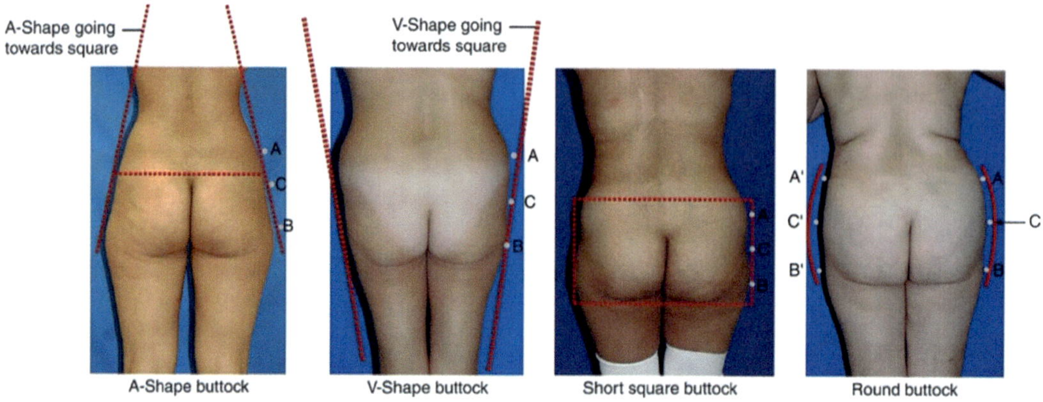

Fig. 21.4 Different types of buttock shapes

The most protruding point in upper lateral hip is marked point A. The most protruding point in lateral thigh is marked point B. Lateral mid-buttock is point C [7].

By connecting points A, B and C on each side (left and right) leads to 4 basic frame types. "A", "V", square and round. These points help to certain extent in treatment plan (Fig. 21.4).

SQUARE SHAPE: Seen in 40% and is most common type. Equal volumes of fat at point A and point B give this shape. This usually improves with liposuction at point A and B followed by with or without fat transfers.

ROUND SHAPE: Found in 15% of patients. In this by connection points A, B and C, gentle C shaped curve becomes apparent. These patients will do well with augmentation procedures.

PEAR SHAPE ('A' SHAPE): Seen in 30% of patents, there is more fat in the lateral upper thigh (point B) as compared to the lateral upper hip area (point A). These patients will improve with liposuction of point B and occasionally point A or lateral thigh.

APPLE SHAPE ('V' SHAPE): Seen in 15% of patients, most of the fat is located in point A and very little fat in point B. These patients have tendency towards central obesity. It is difficult to recontour, liposuction of point A and the flank along with implants or fat transfer will make an improvement.

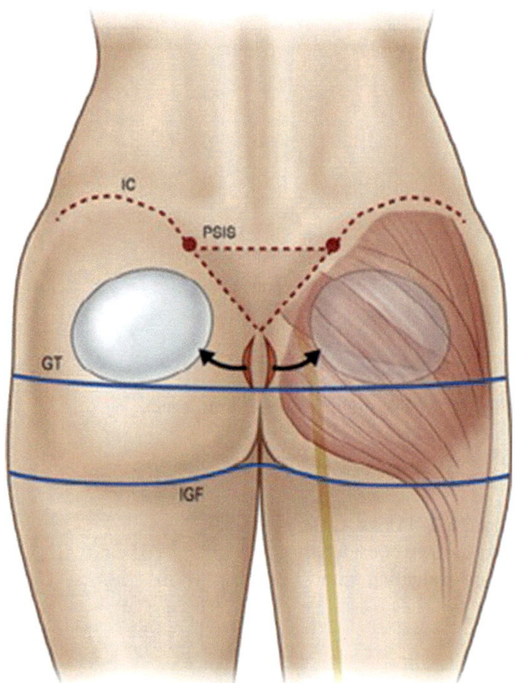

Fig. 21.5 Gluteal region: Bony Landmarks. IC, iliac crest; PSIS, posterior superior iliac spine; IGF, Inferior gluteal fold

MUSCLE ANATOMY: Most of Gluteal recontouring procedures and also implants used for augmentation are placed in relation to Gluteus maximus muscle, without disturbing gluteus medius and minimus. It is important to know its relevant anatomy (Fig. 21.5) [8].

ORIGIN: Fascia of gluteus medius, the external ilium, erector spinae fascia, dorsum of lower sacrum, lateral coccyx and sacrotuberous ligament.

INSERTION: Iliotibial tract and proximal femur.

After identifying superior, inferior, medial and lateral points of the gluteus muscle, the height and width of the muscle is apparent.

Ideal buttock falls in intermediate ratio. As most of the implants are placed in relation to gluteus maximus muscle, height-to-width ratio plays a critical role in choosing the type of implants [3, 8, 9].

21.3 Gluteal/Butt Reshaping Surgery [10]

Buttock recontouring surgeries [11] in three ways:

- Implants
- Autologous fat grafting (AFG)
- Autologous Gluteal flap Augmentation (AGA)

21.4 Implants

Indicated in, Patients with the following characteristics are best suited for augmentation using implants: 1. Lack of buttock volume (thin and lean patients); 2. Lack of projection; 3. Buttock asymmetry; 4. Contour deformity and 5. Limited fat for transfer. Three factors need to be considered for implant selection (Fig 21.6):

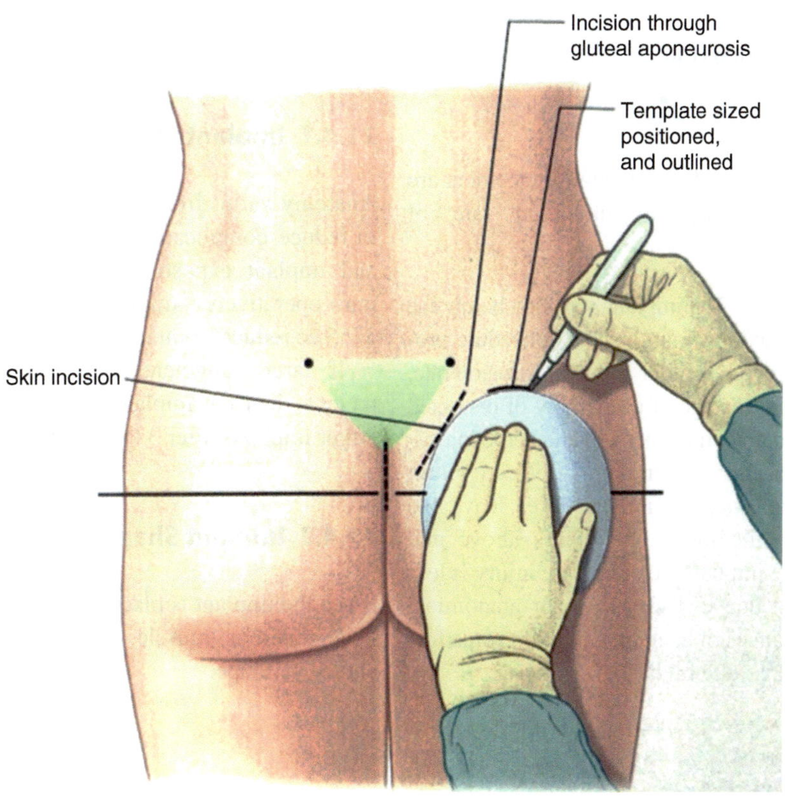

Fig. 21.6 Gluteal implant

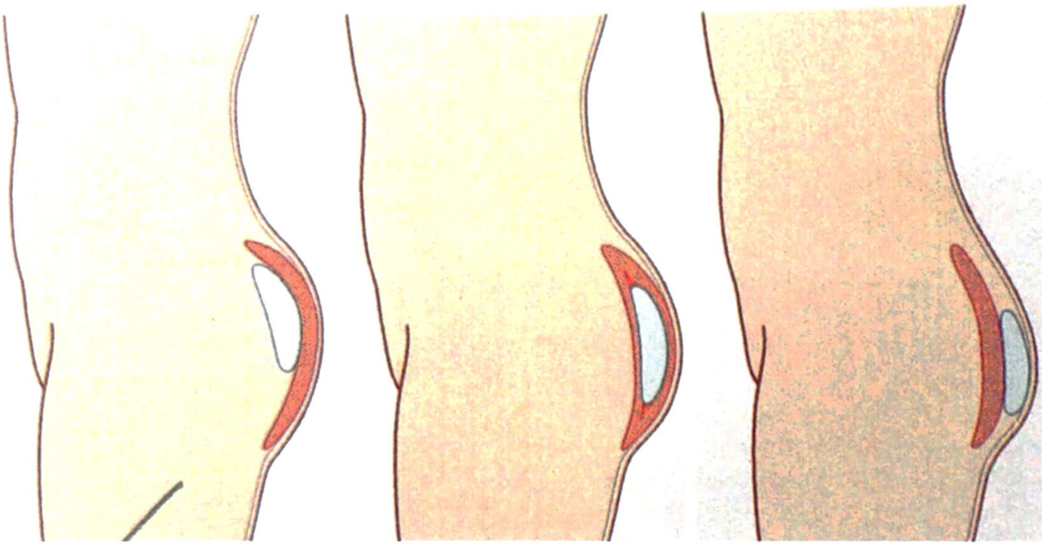

Fig. 21.7 Types of implant location

- Implant location/position
- Choice of incision
- Implant shape

Implant location (Fig. 21.7):

- Subfascial
- Intramuscular
- Submuscular and Subcutaneous placement are less popular due to increased rate of complications

(a) *Submuscular*: The implant is placed below the gluteus maximus muscle. This procedure provides a smooth contour and less extrusion rate. It has disadvantage of availability of restricted space and potential risk of sciatic nerve injury.
(b) *Intramuscular*: The implant is placed within the gluteus maximus muscle fibres, leave 2 to 3 cm of the muscle thickness above and below the implant. Sciatic nerve injury is less [12]. But due to lack of proper anatomical plane chances of haematoma and seroma formation are quite high.
(c) *Subfascial*: Gluteal fascia is strong and covers the entire surface of the muscle. Better understanding of anatomy, this technique is becoming more popular.

21.4.1 Implant Size

Anatomy varies from patient to patient, in order to reduce complications like wound dehiscence and implant exposure; various sizes are used intra-operatively. Implant size less than or 3755 cc reduces wound complications.

If larger augmentation is desired by the patient, then the implants can be replaced with larger implants after 3 to 6 months.

21.4.2 Implant Shape

Since the implant is placed in relation to the gluteus maximus muscle, two things decide the choice of shape.

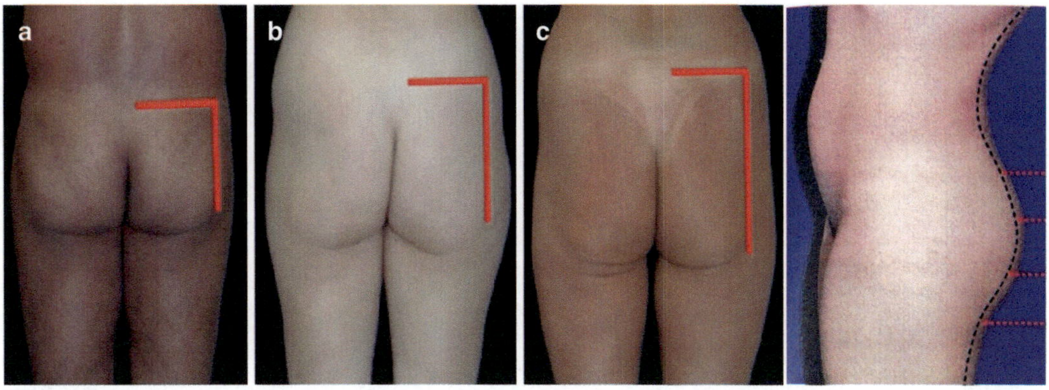

Fig. 21.8 (**a**) Short buttock, (**b**) Intermediate buttock, (**c**) Tall buttock

1. Gluteus maximus muscle height-to-width relation in the posteroanterior (PA) view. Superior, inferior, medial and lateral points of gluteus muscle are identified. These points fall into three ratios:

   - 1:1 (short muscle)
   - 2:1 (tall muscle)
   - 1.5: 1 (intermediate)

   Ideal buttock is intermediate [13]. Tall buttock (2:1) is augmented with anatomic implant, short buttock (1:1) is augmented with round implant. Intermediate (1.5:1) will require the lateral view to determine the final shape as shown in Fig. 21.8.

2. Volume distribution in lateral view. Evaluation is done regarding where most of the muscle bulk is located: Lower, upper and central (mid-buttock). Round implants are used in buttocks to increase fullness in the upper part of the buttock whereas anatomical implants is used to increase the projection inferiorly. In buttock with no significant projection, oval implants are used to even out the projection (Fig. 21.9).

Usually round or anatomic implants with varying textures is used. Soft-solid silicone elastomer implants are used in United States as cohesive gel implants is not approved by Food and Drug Administration (FDA).

Other ancillary contouring procedures that can be done along with gluteal implants are 1. Liposuction of back, flanks, hips and buttock; 2. Fat grafting to hips, buttock; 3. Infragluteal fold revision and 4. Buttock lift.

21.4.3 Post Operative Care

1. Patient to sleep prone for 3 weeks
2. Allowed to sit only while using bathroom
3. Normal activities can be resumed after 2 weeks
4. Wound dehiscence usually is seen between 12 and 16 days post-surgery. Restricted activities to be done to reduce wound pressure or friction
5. Return to work after 4 weeks
6. Driving is not permitted until 5 weeks
7. Physical exercise after 6 to 8 weeks.

If there is any dehiscence, then activities may be restricted for another 12 weeks.

Complications

1. Wound dehiscence
2. Implant exposure

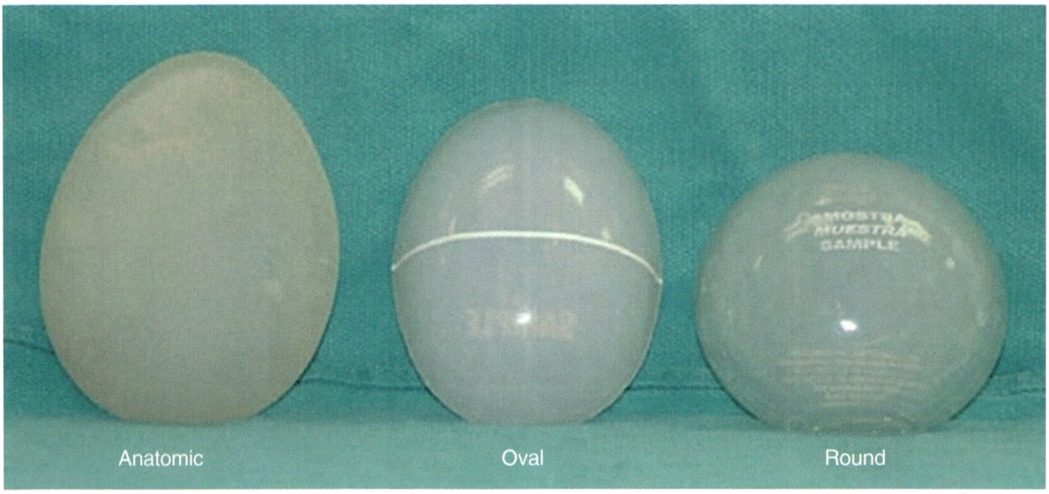

Fig. 21.9 Implant shape: anatomic, oval and round

3. Infection
4. Seroma
5. Capsular contracture
6. Neuropraxia
7. Implant rotation
8. Implant malposition
9. Hyperpigmentation/skin discoloration
10. Chronic pain

21.4.4 Buttock Augmentation with Fat Grafting (AFG)

Ideal candidates cannot be judged by only body mass index (BMI), some patients may have fat concentration in the viscera [5]. Simple method is pinch test and the surgeons experience will act as a rough guide to assess fat availability. Depending on the patient's body frame, our goal will be to graft fat between 500 to 1500 cc per buttock. Approximately 3000 to 4000 cc must be available from the patient's body because 1/3rd of the fat harvested gets damaged during the harvesting process.

Liposuction of flanks, lumbar area and lumbosacral area will increase the lower back lordosis and at the same time will also highlight the fullness of the buttocks.

Fat preparation is done by decantation and separating the aqueous component of the lipoaspirate. Commercial strainer is used to increase the fat content and also for removal of scar and fibrous tissue. Antibiotic solution is added to the fat globules before injected by syringes. Technical advancements have also led to closed systems with reinfusion pump, which prevents manipulation or exposure of fat cells to air.

Fat injection is performed in fanning fashion, rapidly in both antegrade and retrograde fashion (Fig. 21.10). Cannula size of 3.7 mm is preferred for injection and with each pass small aliquots of fat cells are deposited which are surrounded by healthy, well vascularized fat cells. It is important to avoid injection of large aliquots to prevent necrosis or liquefaction and formation of large cystic cavity.

During injections it is important that the cannula is parallel and not angled to avoid injury to gluteal vessels, which lie deep to the gluteus muscle and exit the piriformis. Planes of injection are superficial muscular/muscle fascial layer, deep subcutaneous and lastly superficial subcutaneous as shown in Fig. 21.11 [7]. Avoid injection into deeper muscular layers to avoid intravascular injection and injury to sciatic nerve.

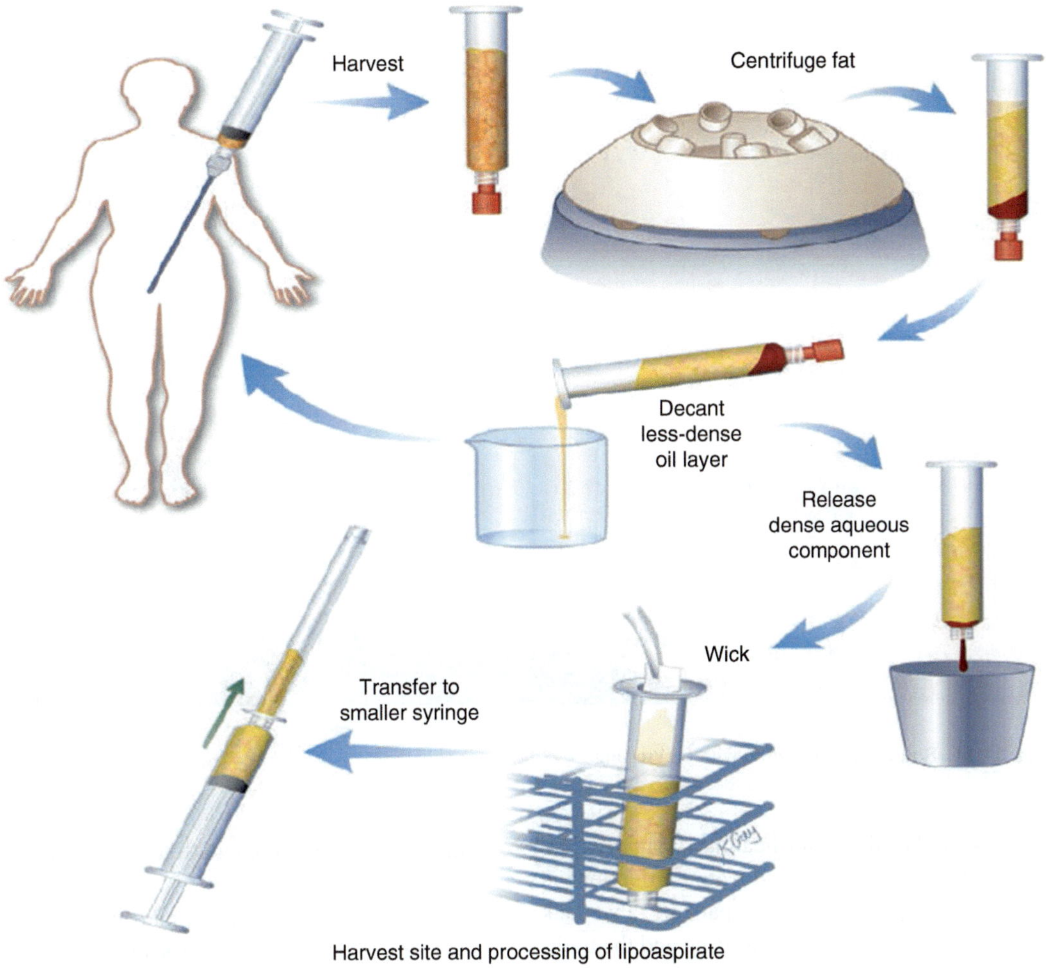

Fig. 21.10 Autologous fat processing

POSTOPERATIVE CARE:

1. Antibiotics cover given for 7 days
2. To avoid sitting or lying on the buttocks for 2 to 8 weeks, compression on the fat grafted area will cause fat death
3. Prone position while sleeping and sitting with pillows under thighs for 8 to 9 weeks
4. High caloric intake for the 1 to 3 months to ensure fat nutrition
5. Light exercise 4 weeks after the operation.

COMPLICATIONS:

1. Infection
2. Seroma
3. Neuropraxia
4. Haematologic and metabolic disturbances

21.5 Autologous Gluteal Flap (AGA) Surgeries

Increase in obesity and popularity and success of Massive Weight Loss (MWL) surgeries has in turn, led to increase in the need for various body lift and autologous gluteal flap (AGA) surgeries [6]. These patients have significant psychological distress due to body deformities that follow massive weight surgeries. Butt reshaping using implants has limited use due to various complica-

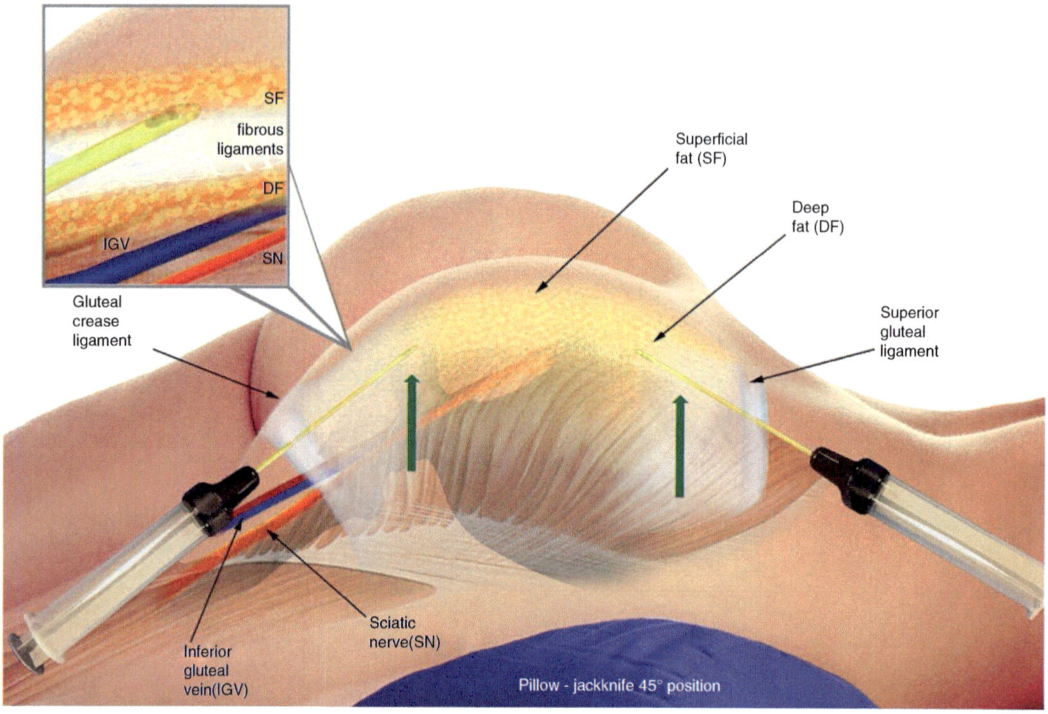

Fig. 21.11 Buttock augmentation with fat grafting

tion like delayed wound healing, wound dehiscence, implant extrusion and wound infections in these MWL patients. These patients also lack sufficient fat content for fat grafting and the traditional buttock lift and other circumferential body lift surgeries does not correct the lack of volume in these areas, making AGA an ideal procedure for these patients. And also, patients with normal ageing associated buttock atrophy and ptosis, also benefit from the AGA sugeries.

In AGA surgeries the redundant and excessive skin and soft tissues are used to create flaps to augment the buttock volume instead of discarding them. So, both buttock lift as well as augmentation is achieved.

Preoperative requirement:

1. Adequate cardiopulmonary status to tolerate prone position during surgery
2. Haemoglobin and coagulation status need to checked
3. Nutritional status evaluated and micro nutrients supplementation done for proper wound healing
4. Other co-morbid conditions like diabetes, vascular status need proper attention.

21.5.1 Operative Technique (Fig 21.12)

The amount of redundant skin and soft tissue is estimated by bimanual palpation. The redundant skin and soft tissue is deepithelialized instead of discarding to create the augmentation mound/island flap. The island flaps are dissected in a bevelled fashion with electrocautery down to the superficial fascial system (SFS), gluteal fascia and lumbosacral fascia. The dermis of the deepithelialized flap is anchored to the gluteal fascia,

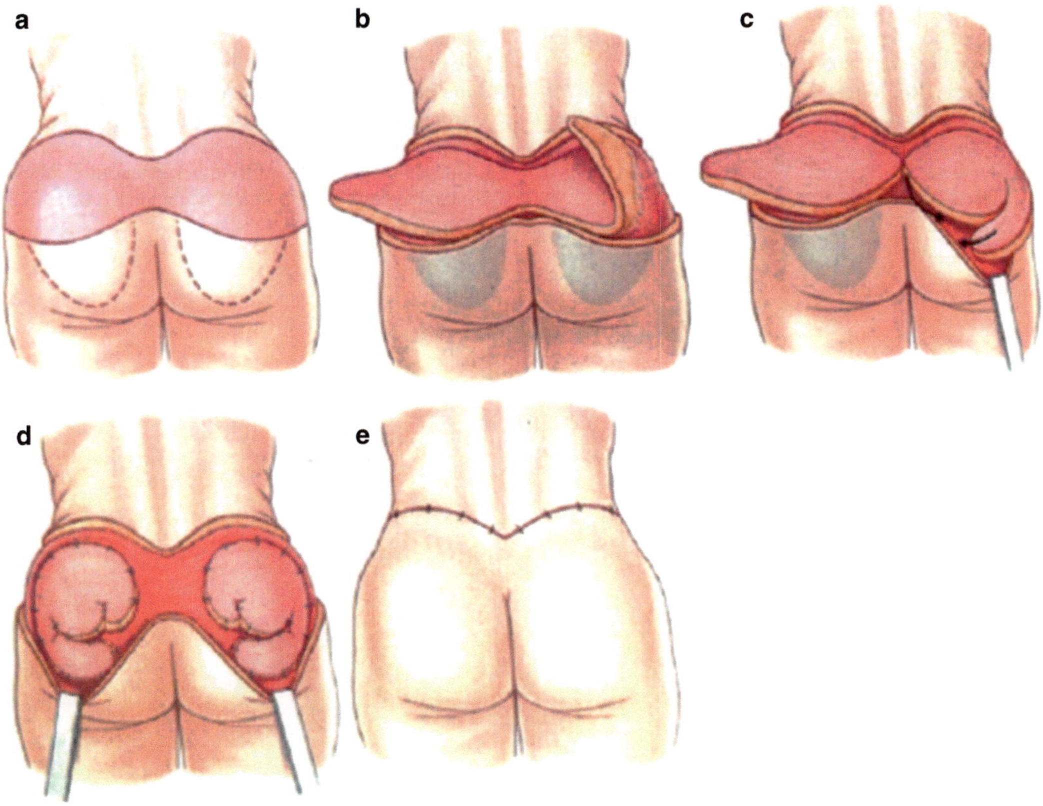

Fig. 21.12 Operative technique of flap surgeries

to give proper shape and to prevent displacement. Drains are placed and wound is closed. The final scar line is planned at or below the level of the gluteal apex from the greater trochanter to coccyx.

Early mobilization is encouraged. Drains are removed if less than 25 ml for 24 hrs. No particular position is advised. Patient can return to work in 10 to 15 days.

SUMMARY

Butt reshaping is constantly evolving and increasing in demand. Better understanding of the anatomy, proper clinical evaluation and careful surgical execution, keeping in mind patients expectations will help in better outcomes. Complications rates can also be reduce, some of which can be life threatening.

References

1. Kitzinger HB, Aayev S, Pittermann A. The prevalence of body contouring surgery after gastric bypass surgery. Obes Surg. 2012;22:8–12.
2. Azin A, Zhou C, Jackson T, et al. Body contouring surgery after bariatric surgery: a study of cost as a barrier and impact on psychological wellbeing. Plast Reconstr Surg. 2014;133:776e–82e.
3. Singh D. Universal allure of the hourglass figure, an evolutionary theory of female physical attractiveness. Clin Plast Surg. 2006;33:359–70.
4. Roberts TL, Weinfeld AB, Bruner TW, et al. "Universal" and ethnic ideals of beautiful buttocks are best obtained by autologous micro fat grafting and liposuction. Clin Plast Surg. 2006;33:371–94.
5. Cuenca-Guerra R, Quezada J. What makes buttocks beautiful? A review and classification of the determinants of gluteal beauty and the surgical techniques to achieve them. Aesth Plast Surg. 2004;28:340–7.
6. Centeno RF, Mendieta CG. Gluteal augmentation. Clinics in Plastic Surgery. 2018;45(2):145–294.

7. Centeno RF. Gluteal aesthetic unit classification: a tool to improve outcomes in body contouring. Aesthetic Surg J. 2006;26(2):200–8.
8. Centeno RF, Young VL. Clinical anatomy in aesthetic gluteal contouring surgery. Clin Plast Surg. 2006;33:347–58.
9. Mendieta CG. Gluteal reshaping. Aesthet Surg J. 2007;27(6):641–55.
10. Mendieta CG. Gluteoplasty. Aesthetic Surg J. 2003;23:441–55.
11. Centeno RF, Mendieta CG. Gluteal Augmentation. Clin Plast Surg. 2018;5(2):249–59.
12. Park TH, Whang KW. Buttock reshaping with intramuscular gluteal augmentation in an Asian Ethnic group: A six-year experience with 130 patients. Ann Plast Surg. 2016;77(3):272–9.
13. Cuenca-Guerra R, Lugo-Beltran I. Beautiful buttocks: characteristics and surgical techniques. Clin Plast Surg. 2006;33:321–32.

Hymenoplasty, Vaginoplasty, and Perineoplasty in Aesthetic and Regenerative Gynecology

22

Ragini Agrawal

22.1 Introduction

It has been scientifically shown that the appearance of a woman's genitalia affects her self-confidence and sexuality. Women now have been empowered with the choice to change the external appearance of their vulvovaginal region in case they are unhappy with its cosmetic appearance. Studies have shown that this is a trend driven by women themselves and not their sexual partners. Female cosmetic genital surgeries address the aesthetic, functional, and sexual concerns of women.

Addressing sexual dysfunction issues is important to improve the quality of life and self-esteem of women. Pelvic floor dysfunctions have a direct relationship with vaginal looseness and unsatisfying sex. Usually, these women complain of vaginal looseness, no sensation or friction during intercourse or low self-esteem due to the appearance of the vulva. Another reason for the desire for these surgeries, especially hymen repair, is its historical relation to virginity. Still in many cultures it has its value and is hence performed.

Another reason for wanting this Gyneplastic surgery, especially hymen repair, is its relation to virginity. Still in many cultures it has its value, and this is done for that reason.

In this chapter, two common female cosmetic gyne surgeries (FCGS) will be discussed:

1. Hymenoplasty
2. Vaginoplasty With Perineoplasty

22.1.1 Hymenoplasty

Many cosmetic vaginal and vulvar surgeries are well outside the traditional realm of medically indicated reconstructions. Hymenoplasty or revirginisation is one of them. The demand of this particular surgery is increasing day by day.

The term "**hymen**" comes from the Greek god of marriage and weddings, "Hymenaeus." Traditionally tearing of the hymen during intercourse is an indicator that the woman has never indulged in sexual activity and is a virgin. In many cultures around the world, an unmarried female who is found not to be a virgin is a subject of shame and ostracism. In such societies, many times, women are imprisoned or executed (honor killing) for premarital sex.

The integrity of the hymen is tested either by physical examination before marriage or confirmed by the occurrence of vaginal bleeding resulting from the tearing of hymen at the time of first sanctioned sexual contact. In some cultures, where displaying proof of a bride's virginity is

R. Agrawal (✉)
Pratiksha Hospital, Dr Ragini's Urogynae, Laser & Cosmetic Gynecology Centre, Gurgaon, India

P. Jindal et al. (eds.), *Aesthetic and Regenerative Gynecology*,
https://doi.org/10.1007/978-981-16-1743-0_22

customary, the nuptial blood-spotted bed sheet has to be exhibited as proof. Apart from sexual activity, other causes of rupture of the hymen are cycling, gymnastics, trauma, physical activity, tampoons, masturbation and many times hymen is so elastic that it may appear perforated.

22.1.1.1 Definition

Hymen repair or revirgination refers to a cosmetic surgery that restores the hymen.

Hymenoplasty is a simple procedure that repairs a torn hymen.This is one of the surgeries which have no medical benefit and is purely done for cosmetic reasons (Fig. 22.1).

22.1.1.2 Anatomical Consideration

Hymen: A thin membrane that surrounds the opening to the vagina. Hymens can come in different shapes. The most common hymen in young women is shaped like a half moon. This shape allows menstrual blood to flow out of the vagina (Fig. 22.2).

The hymen, despite its etymological derivative, is a Mullerian duct remnant that is variable in shape. Its appearance can and will change with hormonal changes and age; the hymen can become thin and, in some instances, near-transparent. It is a tissue that covers the external vaginal opening. The absence of the hymen is associated with agenesis of vagina, whereas the isolated total absence of the hymen is practically non-existent.

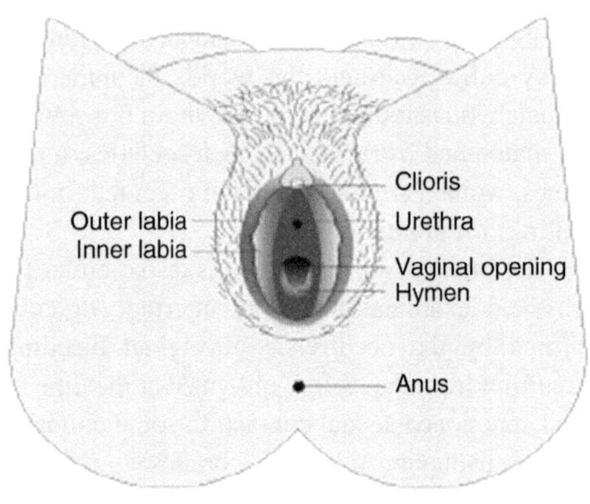

Fig. 22.1 Normal vulva

22.1.1.3 Why Females Request for Hymen Restoration?

There are several reasons why women request a hymenoplasty procedure.

- Cultural or religious reasons**:** In some cultures, a woman is physically inspected to make sure the hymen is intact (a sign she is a virgin).
- Injury to the hymen prior to marriage**:** Vigorous exercise, cycling, trauma to genitals or sexual intercourse/assault can cause a hymen to rupture.
- Revirginization/Revirgination**:** Women who want to give their partner a unique gift.

22.1.1.4 Patient Selection and Preoperative Considerations

Proper counseling and documentation are must. Women should be explained that in this procedure usually bleeding does not occur after penetration during the coital act as restored tags are fibrosed.

The ideal time of operation is 4–6 weeks before marriage.The patient's mental assessment is a prerequisite. Her real expectation should be clearly understood by repeated counseling.

Surgical steps and outcome should be discussed,financial counseling and consenting should be done.

As these women are sexually active, it is prudent to have viral markers to be done along with complete blood count, correlogram, viral markers, vaginal culture sensitivity, pap smear, HbA1c.

22.1.1.5 Contrindications

Any Active or untreated Pelvic Infections/Inflammatory Processes, malignancy, bleeding disorders, vulvodynia, or chronic pain in the area (relative contraindication), female sexual dysfunction, mental impairment, body-dysmorphic disorders.

22.1.1.6 Procedure

It is an OPD procedure done under local anesthetic cream applied for 30 min followed by 2%

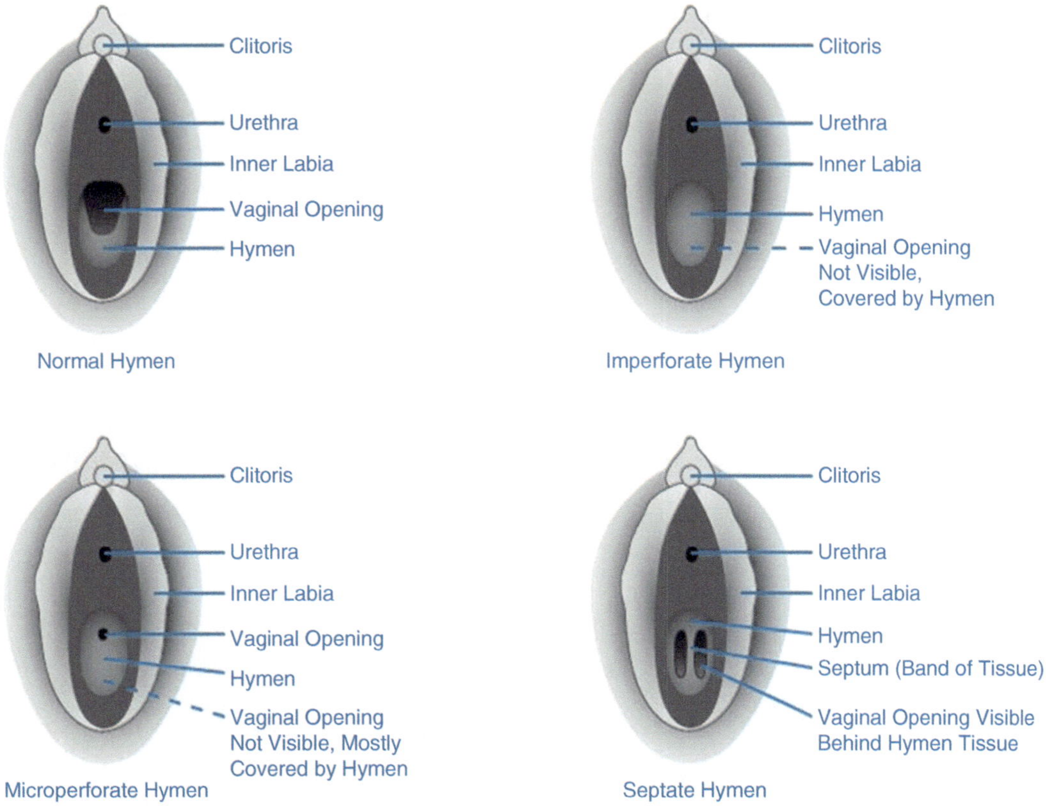

Fig. 22.2 Types of Hymen

lignocaine infiltration. The duration of the operation is approximately 40 min.

22.1.2 Technique

After hymen rupture, few torn remnants are left on the hymen ring.

The patient is placed in the lithotomy position. Full aseptic cleaning and draping, the Surgical area is examined, and bigger remnants are chosen. Their margins are freshened up, and remnants of the opposite side are stitched to each other in two layers with vicryl 3′0′, thus reducing the vaginal hole size.

22.1.2.1 Other Methods Discussed in Literature Are

1. **Flap Method**

The patient is placed in a lithotomy position. The procedure is carried out under spinal anesthesia. 2.5 cm long and 1 cm wide rectangular flaps are marked at 2, 5, 8, and 11 o'clock position on the anterior vaginal wall. Flaps at 2 and 5 o'clock positions are kept proximally based, and flaps at 8 and 11 o'clock positions are kept distally based or vice versa. Flaps are raised at the level of loose connective tissue below the mucosa. Donor areas are closed primarily. Opposing flaps are overlapped in a crisscross manner and sutured with 5/0 Polyglactin (Vicryl®) sutures, leaving no raw areas (Fig. 22.3).

2. **Cerclage**—Taiwan Method
3. **Luminal Reduction Hymenoplasty**—Canadian Method

Native hymen morphology is highly variable in configuration and size. The hymen remnants, also termed carunculae myrtiformes, should be examined to be of sufficient size for primary reapproximation. This is a prerequisite for the luminal reduction technique. Any

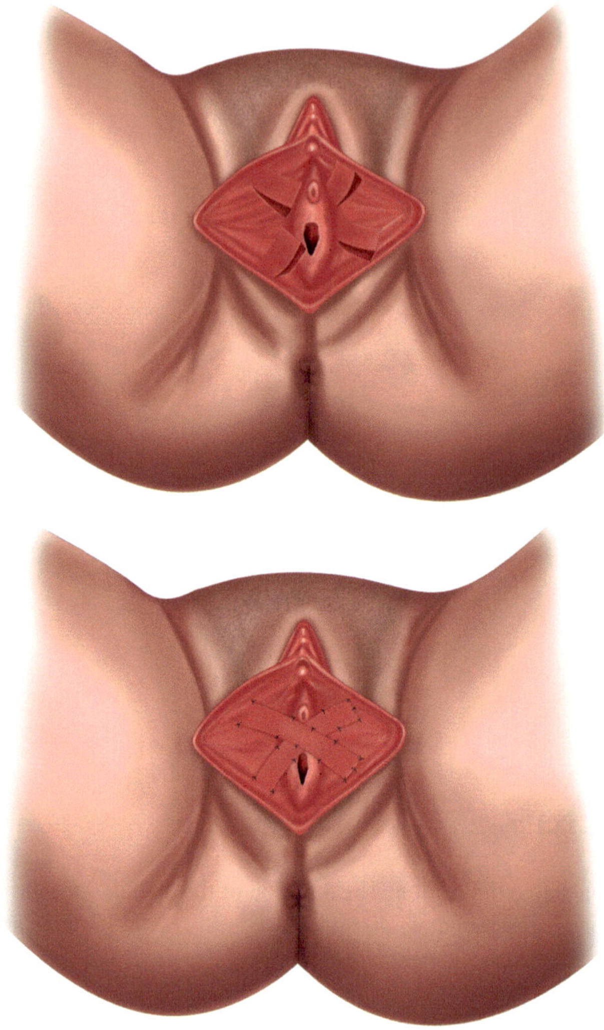

Fig. 22.3 Hymen Repair

masses or lesions to the surrounding soft tissues must be recognized and investigated accordingly.

The procedure is carried out under general anesthesia but can be performed under local anesthesia as well. After a dose of prophylactic antibiotics is given, the patient is placed in a lithotomy position with bilateral hips flexed and legs abducted in stirrups. The perineal area is prepped with 10% betadine solution.

Labia majora and minora are retracted bilaterally. The hymen remnants are identified at the outermost aspect of the vaginal introitus as an annular array of hymenal fragments separated by clefts. Remnants can appear flaccid and should be gently stretched to ascertain their actual length. Approximately 2 to 3 mL of 0.25% marcaine with 1:400,000 epinephrine is injected to the submucosal plane of the vagina deep to the hymen remnants so as not to distort and efface the fragments. The epithelialized free edges of each hymen fragment are gently excised along the clefts with scissors or a scalpel, leaving only the tip of the fragments intact. The free margin close to the tips of the remnants is sutured first using a simple interrupted 4-0 Vicrylrapide suture (Johnson & Johnson, Markham, Ontario, Canada), ensuring accurate approximation without step-off at the edge of the hymenal ring). This is followed by a simple interrupted 4-0 vicryl rapide suture on the internal surface and another suture on the external surface of the remnants as

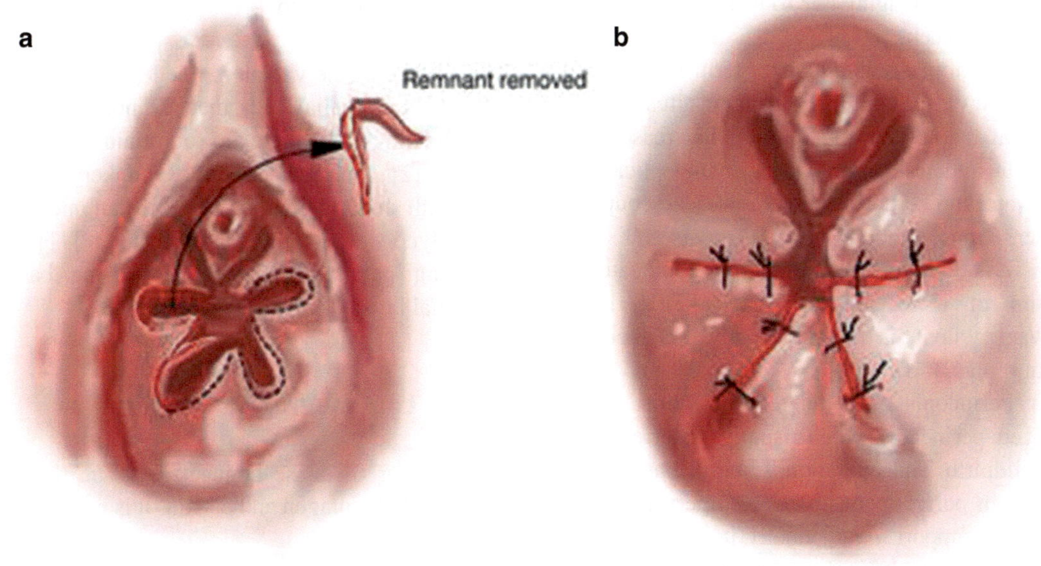

Fig. 22.4 Hymen repair illustrations

shown in Fig. 22.4. This is completed in a sequential fashion for clefts around the lateral and posterior aspects of the vaginal canal until the luminal reduction is accomplished by closure across all clefts. Sutures are not placed at the anterior aspect of the vaginal canal, where the hymen remnants can arise from adjacent to the urethra.

22.1.2.2 Postoperative Risks, Complications, and Follow-Up Instructions

Surgical risks of the procedure include wound dehiscence, infection, scarring, distortion of the external vaginal orifice and creating overly small introitus leading to obstruction of vaginal outflow and hematocolpos, dyspareunia and feelings of guilt.

There is no need to stay in hospital after the procedure and generally, you can walk out and resume normal activities the same day with the following **precautions**:

Abstain from sexual activity till marriage.

1. Avoid strenuous exercises and exertion and lifting of heavy weights for 3 weeks after the procedure.

22.2 Maintenance of Cleanliness and Local Hygiene

22.2.1 Medico Legal Consideration

Hymenoplasty should not be confused with female genital mutilation (FGM)—previously known as female circumcision. Unlike FGM, hymenoplasty is performed on women who have reached the age of consent and who request the surgery themselves, therefore the medical, ethical, and human rights arguments against FGM are not extendable to revirgination as it is another cosmetic surgery like mammoplasty.

In western society and in India, hymenoplasty is legal. This operation is done for the woman, and the principle of confidentiality is as old as medicine itself.

22.3 Vaginoplasty and Perineoplasty

22.3.1 Definitions

The concept of tightening the vaginal introitus, which has become lax due to childbirth and progress with aging due to loss of collagen is not a new concept.

Vaginoplasty is a blanket term for restructuring the vagina surgically from introitus to vaginal canal for aesthetic and functional reasons. The main aim is to reduce caliber and plication of levator ani so during intercourse, the couple can sense more friction.

It is done for ages in medically indicated reason of rectocele as posterior colporrhaphy. Vaginal laxity is the first step towards occurrence of prolapse.

Perineoplasty is the reconstruction of damaged perineum, which usually occurred during childbirth. This operation is being done traditionally in pelvic reconstructive surgery to increase pelvic support. Aesthetically it provides muscular support to vaginal tightening.

Vaginoplasty and perineoplasty are usually done together and a new name for it is surgical vaginal rejuvenation.

22.4 Anatomical Considerations

22.4.1 Pelvic Floor

22.4.1.1 The Vagina—Anatomy, Orientation, and Structural Support

The vagina is a hollow, fibromuscular tube with rugal folds that extends from the vestibule to the uterine cervix. In the standing woman, the upper two-thirds of the vagina is almost horizontal, whereas the lower one-third is nearly vertical—the upper two-thirds of the vagina rest on the levator muscles. The angle occurs as the vagina descends through the genital hiatus and is caused by the sling of levator muscles contracting at baseline [1].

The layers of the vagina include-

1. Epithelium
2. Lamina propria
3. Vaginal muscularis
4. Adventitia

22.4.2 Epithelium

The surface of the vagina is lined with stratified squamous epithelial cells that are nonkeratinizing. Histologically, there are four main epithelial layers: [2] most superficially, nonkeratinized cells with small, pyknotic nuclei; [3] an intermediate glycogen-containing cellular layer; [1] proliferating parabasal cells; and [4] the basal epithelial layer [4].

22.4.3 Lamina Propria

It is a layer of dense connective tissue that is composed of collagen and elastin. It is next to the epithelium.

22.4.4 Muscularis

It is largely composed of smooth muscle cells. The smooth muscle adopts a circular configuration for its inner layer and then a longitudinal orientation in its outer layer [5]. Blood vessels lie between the smooth muscle layers.

22.4.5 Adventitia

It is a layer of loose connective tissue and contains collagen, elastin, additional smooth muscle, and additional vasculature. The composition of the connective tissue here includes 84% collagen and 13% elastin [6].

The walls of the vagina are in contact except where its lumen is held open by the cervix. The vagina has an H-shaped lumen, with the principal dimension being transverse. In addition, the upper vagina is supported by connective tissue attachments to the sacrum, coccyx, and lateral pelvic sidewalls; these are identified at surgery as the cardinal and uterosacral ligament complex.

Anteriorly, the vagina lies adjacent to and supports the bladder base, from which it is separated by the vesicovaginal adventitia (endopelvic fascia). The urethra is fused with the anterior vagina, with no distinct adventitial layer separating them. The terminal portions of the ureters cross the lateral fornices of the vagina on their way to the bladder base. Posteriorly, the vagina is related to the cul-de-sac, to the rectal ampulla, and inferiorly to the perineal body.

22.5 Embryology and Innervation of the Vagina

Embryologically, the distal two-thirds of the vagina arise from sinovaginal bulbs. These are an outgrowth of the endoderm of the urogenital sinus. The proximal one-third of the vagina, as well as the cervix and the uterus, arise from the paramesonephric ducts, which fuse in the midline. The solid sinovaginal bulbs eventually fuse together, grow cephalad, and then fuse with the fused paramesonephric ducts. This whole structure then eventually canalizes to form the vagina. The original junction between the sinovaginal bulbs and the urogenital sinus is the site of the future hymen. Reflecting differential embryologic origins, innervation to the upper portion of the vagina is via the inferior hypogastric nerve plexus, and innervation to the lower vagina and skin is from the pudendal nerve. The spinal origin of all of these nerves is the sacral levels S2 to S4 [5]. The innervation differs between the proximal and distal portions of the vagina, and these differences begin early in embryogenesis (Fig. 22.5).

22.6 External Genital Anatomy

22.6.1 Labia Majora

Labia majora are two prominent cutaneous folds running from the mons pubis, merging at the perineum.

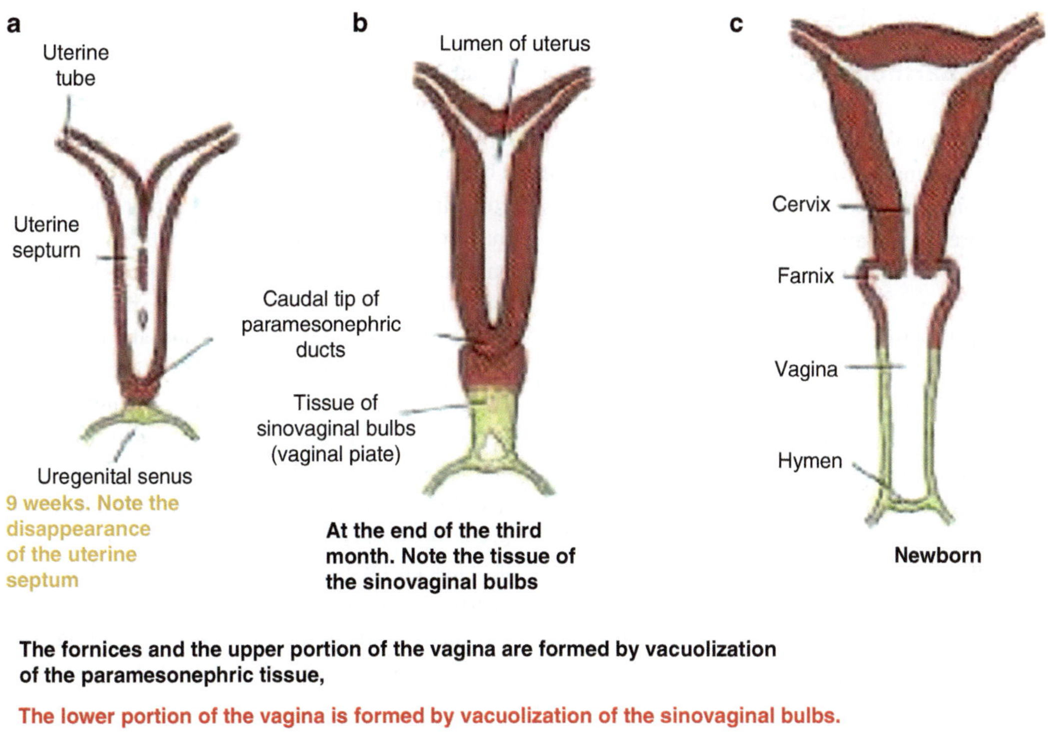

Fig. 22.5 Embryological development of the vaginal canal

22.6.2 Labia Minora

The labia minora are skin folds located medial to the labia majora and lateral to the vestibule.

22.6.3 Clitoris

The clitoris is composed of three erectile tissue components, the glans, body and crura. The body and the glans of the clitoris are covered by the prepuce or "hood" which is continuous with the labia minora. The prepuce acts as a protective cover over the clitoral glans.

22.6.4 The Vulvar Vestibule

The "vestibule" as its name implies, is the "entryway" from the vulva and perineum into the vagina. It is bordered by the posterior fourchette, hymenal ring and the labia minora.

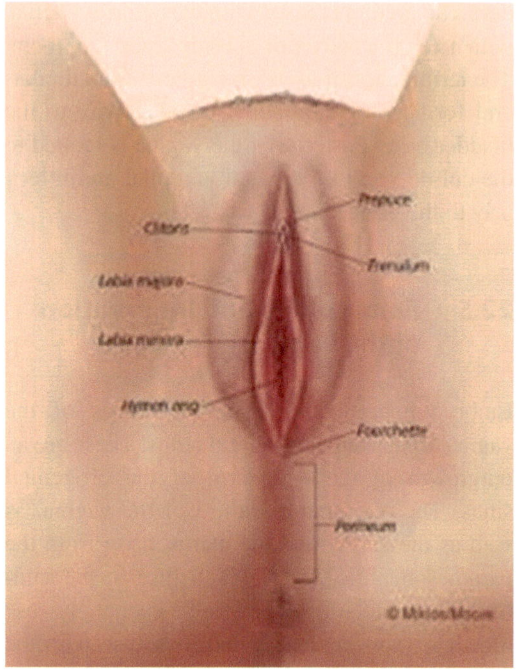

Fig. 22.6 Normal External Genitalia

22.6.5 Bartholin's Gland

The Bartholin's glands, or greater vestibular glands, originate from an embryonic urogenital sinus and correspond to the male's Cowper's (bulbourethral) gland.

22.6.6 Skene's Gland

These had a pair of periurethral glands lining the anterior vaginal wall beneath the epithelium proximal to the urethral meatus (Fig. 22.6).

B) Pelvic Floor

Pelvic floor muscles have two major functions; they provide

1. Support or act as a "floor" for the abdominal viscera, including the rectum
2. Constrictor or continence mechanism to the urethral, anal, and vaginal orifices (in females)

The pelvic diaphragm is a wide but thin muscular layer of tissue that forms the inferior border of the abdominopelvic cavity. Composed of a broad, funnel-shaped sling of fascia and muscle, it extends from the symphysis pubis to the coccyx and from one lateral sidewall to the other. The urogenital diaphragm, also called the triangular ligament, is a strong, muscular membrane that occupies the area between the symphysis pubis and ischial tuberosities and stretches across the triangular anterior portion of the pelvic outlet. The urogenital diaphragm is external and inferior to the pelvic diaphragm. The pelvic ligaments are not classic ligaments but are thickenings of retroperitoneal fascia and consist primarily of blood and lymphatic vessels, nerves, and fatty connective tissue. Anatomists call the retroperitoneal fascia subserous fascia, whereas surgeons refer to this fascial layer as endopelvic fascia. The connective tissue is denser immediately adjacent to the lateral walls of the cervix and the vagina. The broad ligaments are a thin, mesenteric-like double reflection of the peritoneum stretching from the lateral pelvic sidewalls to the uterus. The cardinal, or Mackenrodt's, ligaments extend from the lateral aspects of the upper part of the cervix and the vagina to the pelvic wall. The uterosacral ligaments extend from the upper portion of the cervix posteriorly to the third sacral vertebra.

The pelvic floor is comprised of the number of muscles, and they are organized into superficial and deep muscle layers. There is significant controversy with regards to the nomenclature, but generally speaking, the superficial muscle layer and the muscles relevant to the anal canal function are the external anal sphincter, perineal body and possibly the puboperineal (or transverse perinei) muscles (Fig. 22.7a and b). The deep pelvic floor muscles consist of pubococcygeus, iliococcygeus, coccygeus, and puborectalis muscles.

Muscles, in general, have relatively simple function; they shorten as they contract. Generally, the insertion point of the muscle moves towards the point of the origin (sphincter muscles being an exception). In the case of pubococcygeus, iliococcygeus and ischiococcygeus, such an action results in the movement of the coccyx anteriorly (ventrally) toward the pubic bone. In fact, during pelvic floor contraction the coccyx moves ventrally and cranially. The change in the shape of the pelvic floor during contraction, from a basin to dome, is due to the shortening of the pubococcygeus, iliococcygeus and ischiococcygeus muscles. At the same time, conversion from "basin" to the "dome" lifts the pelvic viscera (including the rectum) in the cranial direction and provides mechanical support or "floor" to the rectum and other pelvic floors viscera. Therefore, it is likely that the weakness of these muscles results in perineal descent.

22.6.7 Patient Selection

(a) **Indication**—Vaginal laxity and loss of friction during intercourse is the main indication for this procedure. If women complaints of urinary incontinence or something coming out of the vagina or a bulge in the vagina, then she requires specific operations for these corrections, and they can be done concomitantly.

22.6.8 Contraindications

1. Future desire of more childbirth

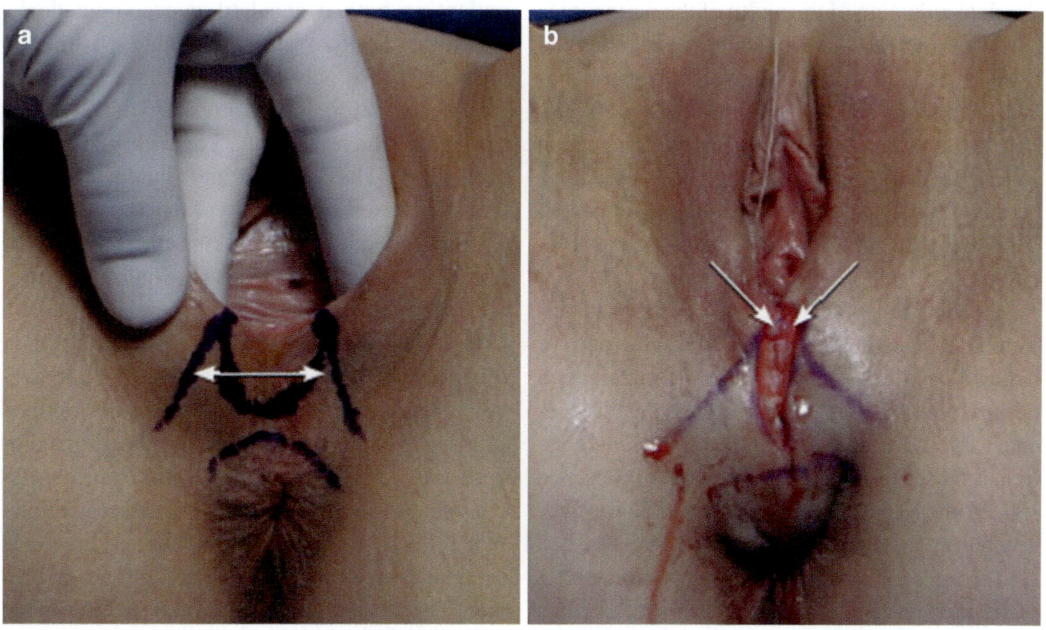

Fig. 22.7 (**a**) The medial edges of the bulbocavernosus and perineal muscles are widely separated before surgery (*arrow*). (**b**) Perineoplasty causes them to converge centrally (*arrows*) to tighten the introitus and improve muscle tone beneath the labia majora.Curtsey Mark pelosi

2. Active vulvovaginal infection
3. Undiagnosed vaginal Bleeding

22.6.9 Evaluation

Like any medical procedure, thorough medical history, physical and systemic examination, and basic investigation are required before opting for operation. A detailed counseling session and documentation of discussion are must before surgery. As it is a purely cosmetic procedure on demand of client psychoanalysis to rule out any-body dysmorphia and sexual history is also part of evaluation. As this procedure is not covered under any insurance, proper financial counseling, and documentation are must. Examination to rule out urinary incontinence,cystocele, rectocele, uterine descent is must as they require additional surgical correction, which can be done together. Psychosexual issues to be evaluated discussed and documented.

22.6.10 Operative Technique

Surgical reduction of the width of the perineum is called perineoplasty.Technically, it involves the excision of midline tissues spanning the perineum and lowers posterior vagina and the approximation of tissues immediately lateral to the excision zone. In therapeutic gynecologic operations, perineoplasty is performed to enhance support of the pelvic floor at the time of pelvic reconstructive surgery.

In cosmetic gynecology, perineoplasty is performed to create an external muscular cuff of reduced vaginal circumference to complement a deeper zone of caliber reduction, typically executed through conservative plication of the levator ani musculature and the resection of loosened skin from the posterior vaginal wall. A secondary, purely aesthetic effect of perineoplasty is a convergence of the labia majora posteriorly.

In most instances, perineoplasty is performed as a component of colpoperineoplasty—the tightening of the vaginal canal and levator ani musculature, previously described. The most common scenario for an isolated perineoplasty is the revision of a colpoperineoplasty for a complaint of persistent laxity where only the external portion of the vaginal canal displays residual laxity. The most common misapplication of the procedure is its use in the presence of vaginal canal laxity, and the surgeon fails to recognize the diagnosis. In therapeutic gynecology, perineoplasty is sometimes used in the process of wide local excision of localized neoplasms.

22.6.11 Patient Evaluation

As the patient is not suffering from any disease, it is necessary to have detailed counseling and documentation of discussion. Patient should have prior anesthesia fitness.

After the initial conversation regarding the cosmetic request and the requisite psychosocial analysis, a thorough gynecologic and sexual history is obtained. This information is useful to determine whether patients have symptoms of bladder, bowel, or pelvic floor dysfunction or sexual issues that may affect treatment.

A physical examination is conducted with the patient in both a standard gynecologic dorsal lithotomy position and a standing position to fully assess for pelvic organ prolapse. A speculum examination is performed to identify any infection, which may interfere with the surgical procedure. The clitoral and bulbocavernosus sacral reflexes are assessed at the perineum. Tapping the clitoris or stroking the labia majora should produce a reflex contraction of the external anal sphincter. The patient is asked to cough to assess for bladder hypermobility. The width of the levator hiatus and the quality of the puborectalis muscle tone is assessed digitally. A common practice is to measure the hiatus in fingerbreadths with the muscles at rest and with the muscles contracted. These data are converted to centimeters. If the levator muscles are either lax or widely separated, perineoplasty alone will be insufficient to satisfy the patient's request for vaginal tightening, and colpoperineoplasty is indicated. The thickness and dimensions of the perineal body are noted. An attenuated perineal body in need of repair frequently has a thin web of skin with little to no muscular tissue. A digital rectal examination is conducted to assess for a rectocele or a perineocele, which, if present, warrants repair at the time of surgery. Rectal fullness, pressure, and constipation are very common symptoms of these conditions.

Surgeons should explain the limits of the contemplated vaginal tightening by palpating the targeted structures and by displaying the anatomy with the use of a hand mirror. Markings are useful to define the boundaries of proposed treatments and to show untargeted structures at consultation and at the surgery. Dilators and fingers are used as aids to help patients visualize the degree of planned tightening. A discussion of the degree of tightness after surgery is essential.

A complete, documented medical evaluation should precede surgery in patients who have not had a recent examination. Any anatomic distortion that may increase the risk of injury should be assessed and managed by appropriate means preoperatively. Blood analysis includes testing for signs of infection, anemia, and coagulopathy. Pregnancy testing is performed or repeated on the day of surgery regardless of the history.

Medications, supplements, herbs, and other substances that could impair coagulation (for example, vitamin E, Ginkgo biloba, ibuprofen, and statins) should be discontinued 1 week prior to surgery. Substances that interact negatively with anesthetic agents, healing, and perioperative medications should also be withheld. If they cannot be discontinued or substituted, the surgical plan will need to be modified, delayed, or withheld. Cigarette smoking is not a contraindication to perineoplasty.

Expectations and motivations need to be explored in depth in cosmetic patients. Unrealistic expectations will never be fulfilled by surgery, even if procedures are performed to perfection by any medical or aesthetic standard. Cosmetic surgery "addicts," "perfectionists," and patients expecting cosmetic surgery to remedy interpersonal conflicts are examples of misguided personality types to be screened at the initial consultation.

The degree of planned tightening should be emphasized, the risks of overtightening should be explained thoroughly, and revision policies should be clearly explained.

22.6.12 Preoperative

If sedation or general anesthesia is planned, patients are advised not to take anything by mouth for 8 h before surgery. Informed consent is obtained in the patient's language.

22.6.12.1 Perioperative Care

Prophylactic broad-spectrum antibiotics are routinely given immediately before surgery. Patients are placed in a dorsal lithotomy position with the legs supported in boot-type stirrups and the knees mildly flexed. Intermittent pneumatic compression stockings are routinely employed. Indwelling bladder catheterization and vaginal packing are not typically used during perineoplasty. When a colpoperineoplasty is performed, a transurethral bladder catheter and vaginal packing are maintained postoperatively for 24 h.

22.6.12.2 Postoperative Care and Instructions

Postoperative medications and wound care instructions are reviewed, and Contact information is updated as necessary.

22.6.12.3 Take-Home Message

Cosmetic gynecological surgery is on the rise, and it should be done with proper training. Counseling and documentation play a very big role.

References

1. Goodman MP. Female cosmetic genital surgery. Obstet Gynecol. 2009;113:154–9.
2. Saraiya HA. Surgical revirgination: four vaginal mucosal flaps for reconstruction of a hymen. Indian J Plast Surg. 2015;48(2):192–5. https://doi.org/10.4103/0970-0358.163060.
3. Ahmadi A. Ethical issues in hymenoplasty: views from Tehran's physicians. J Med Ethics. 2014;40:429–30.

4. Heger A, Jean ES, David M. Evaluation of the sexually abused child: a medical textbook and photographic atlas. 2nd ed. New York: Oxford University Press; 2000. p. 116.
5. Kimberley N, Hutson JM, Southwell BR, Grover SR. Vaginal agenesis, the hymen, and associated anomalies. J PediatrAdolesc Gynecol. 2012;25:54–8.
6. Mor N, Merlob P. Congenital absence of the hymen only a rumor? Pediatrics. 1988;82:679–80.

23 Labiaplasty and Cliteroplasty

Shashi Joshi

23.1 Labiaplasty

23.1.1 Definition

Labiaplasty is a procedure that seeks to alter the labia minora and labia majora in specific ways to improve external aesthetics of female genitalia (Fig. 23.1). Practitioners utilise the technique to alter labia so that the genitalia appear proportionate, homogenous in colour and sexually desirable.

23.1.2 Introduction

Patient awareness has led to significant demand for genital aesthetic surgery. According to the American Society of Aesthetic Plastic Surgeons 2017 statistics, labiaplasty is up 217.2% from 2012 to 2017.

The labia minora and labia majora would be discussed separately as the kind of aesthetic procedures for they are different.

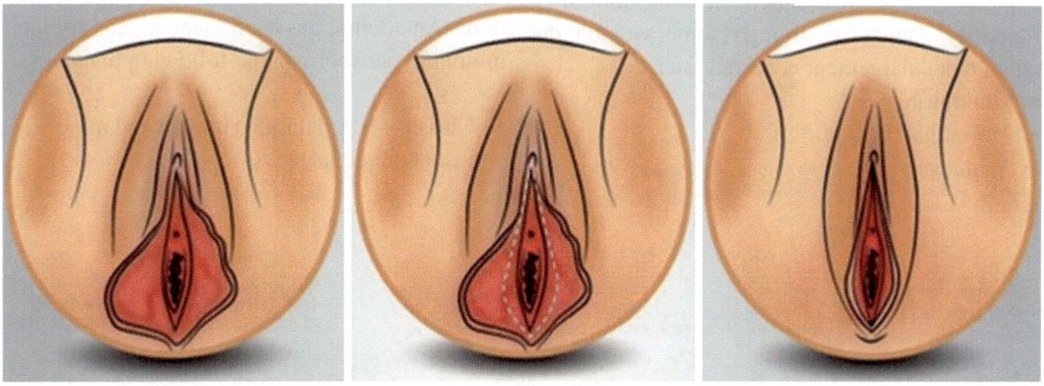

Fig. 23.1 Labiaplasty

S. Joshi (✉)
City Clinic, Mahboula, Kuwait

P. Jindal et al. (eds.), *Aesthetic and Regenerative Gynecology*,
https://doi.org/10.1007/978-981-16-1743-0_23

23.2 Anatomy of Labia Minora

The size, shape, colour, thickness of the labia minora vary greatly in normal women [1]. They may be hidden by labia majora or may be visible and may become larger with sexual arousal, sometimes two to three times their usual diameter [2].

The glans clitoris sits directly under the prepuce. The frenula are folds of skin extending from the glans clitoris and merging with the extension of clitoral hood to form labia minora as shown in Fig. 23.2. The blood supply of labia minora consists of branches from external pudendal artery, internal pudendal artery and internal circumflex artery. The labia minora is innervated by the posterior labial nerve, which is a continuation of the pudental nerve. The pudental nerve also branches off into the dorsal nerve of the clitoris.

23.2.1 Desire for Labiaplasty (L. Minora)

When the labia minora are enlarged, women seek labiaplasty. The enlargement of labia minora is primarily congenital, though some women do claim that enlargement occurred after:

- Childbirth.
- Hormonal therapy.
- With age.
- Genital piercing can increase labial size and asymmetry because of the weight of the ornament.
- Hypertrophy of the labia minora can cause discomfort and lead to dyspareunia.
- Chronic urinary tract infection.
- Local irritation.
- Difficulty in maintaining hygiene.
- Interference with tight clothes and sports such as horse riding.
- Aesthetically unpleasant although there is no standard.

All the above are a reason for a patient to opt. for labiaplasty. Cultural acceptance has further increased the incidence. Although there are no standard genital aesthetic ideals, basic guidelines for labia minoraplasty include symmetrical labia minora that do not protrude past the labia majora, especially when standing.

23.2.2 Labia Minor Hypertrophy Classification

Though there is no universally accepted classification of labia minora as there are wide variations. An attempt has been made to classify labia minora hypertrophy in the following manner:

- Class I—Normal Labia majora and minora are just about equal.

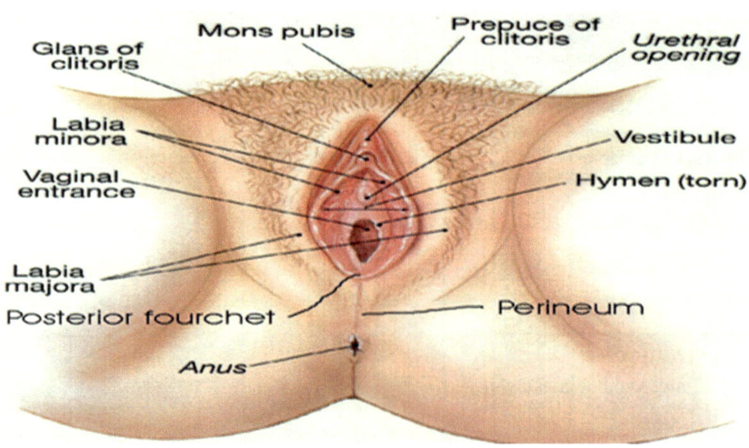

Fig. 23.2 Diagram of vulva

- Class II—Protrusion of labia minora beyond the labia majora.
- Class III—Includes an enlarged clitoral hood besides labia minora enlargement.
- Class IV—Labia minora extends to the perineum.

23.2.3 Contraindications

Even though the desire to have labiaplasty is an individual's choice, thorough medical and sexual history should be taken to rule out any contraindications such as:

1. Active genital infection.
2. Sexually transmitted disease.
3. Gynecological malignancies.
4. Unrealistic aesthetic goals.
5. Body dysmorphic syndrome.
6. Smokers who are unwilling to quit either temporarily or permanently (as this will interfere with optimal wound healing).
7. Pregnancy.
8. Severe coagulation disorder.

23.2.4 Techniques of Labioplasty for Labia Minora

1. Edge resection.
2. Wedge resection
3. De-epithelialization
4. Labiaplasty with ZIG-ZAG incision.
5. Labiaplasty is done with extended W plasty incision.
6. Laser labioplasty.
7. Labioplasty with clitoral hood resection.

23.2.4.1 Edge Resection

It is simple resection of tissue at the free edge of the labia minora. Clamp is placed across the area of labial tissue to the resected (clamp helps in haemostasis). The tissue to be removed is resected and the cut labium is sutured as shown in Fig. 23.3.

23.2.5 Advantages

- It is easy to perform.
- Remove the hyperpigmented, irregular edge of the labia minora.

23.2.6 Disadvantages

- There is a loss of natural rugosity of the free edge of labia minora.
- Possibility of everting the inner lining of the labia.
- Unable to excise the redundant tissue of clitoral hood, when present.
- Clitoral hood deformity can occur.

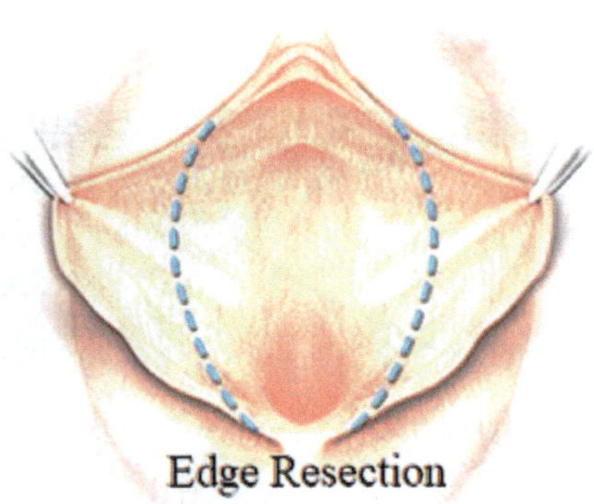

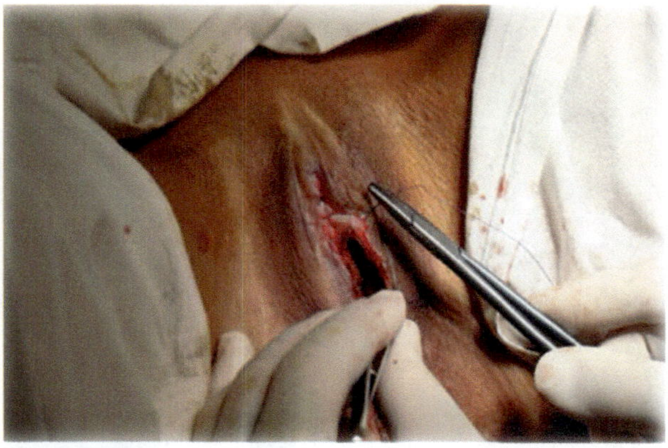

Fig. 23.3 Edge resection

23.3 Diagram

It is also important to note that the tips of the pedicles should be in symmetrical alignment with the transverse vulvar line mark that had been made preoperatively to ensure symmetry. The skin edges and medial edges of the flaps are then sewn with subcuticular stitches of the same suture.

23.3.1 Wedge Resection

It can be done either by Central wedge resection (Fig. 23.4) or Posterior wedge resection (Fig. 23.5).

It involves cutting and removing a partial thickness wedge of tissue from the thickest part of the labia minora [3]. Aesthetically it is superior to the edge resection because it preserves the natural rugosity of the labia. Removal of the skin and mucosa should be done, leaving the submucosa intact. This decreases the risk of damaging pertinent labial nerves that can result in painful neuromas or numbness.

F. Geraldo improved the central wedge resection technique with additional 90-degree Z plasty techniques. Produces a redefined surgical edge that is less tethered and there is less livelihood of scalloped edge scar [3, 4].

23.3.1.1 Advantages

- Central wedge can treat prominent clitoral without a separate incision [5].

23.3.1.2 Disadvantages

- Very demanding surgical procedure.
- Difficult to judge the correct amount of labial skin to resect. This can lead to either under or overcorrection.
- Increased probability of surgical wound dehiscence.

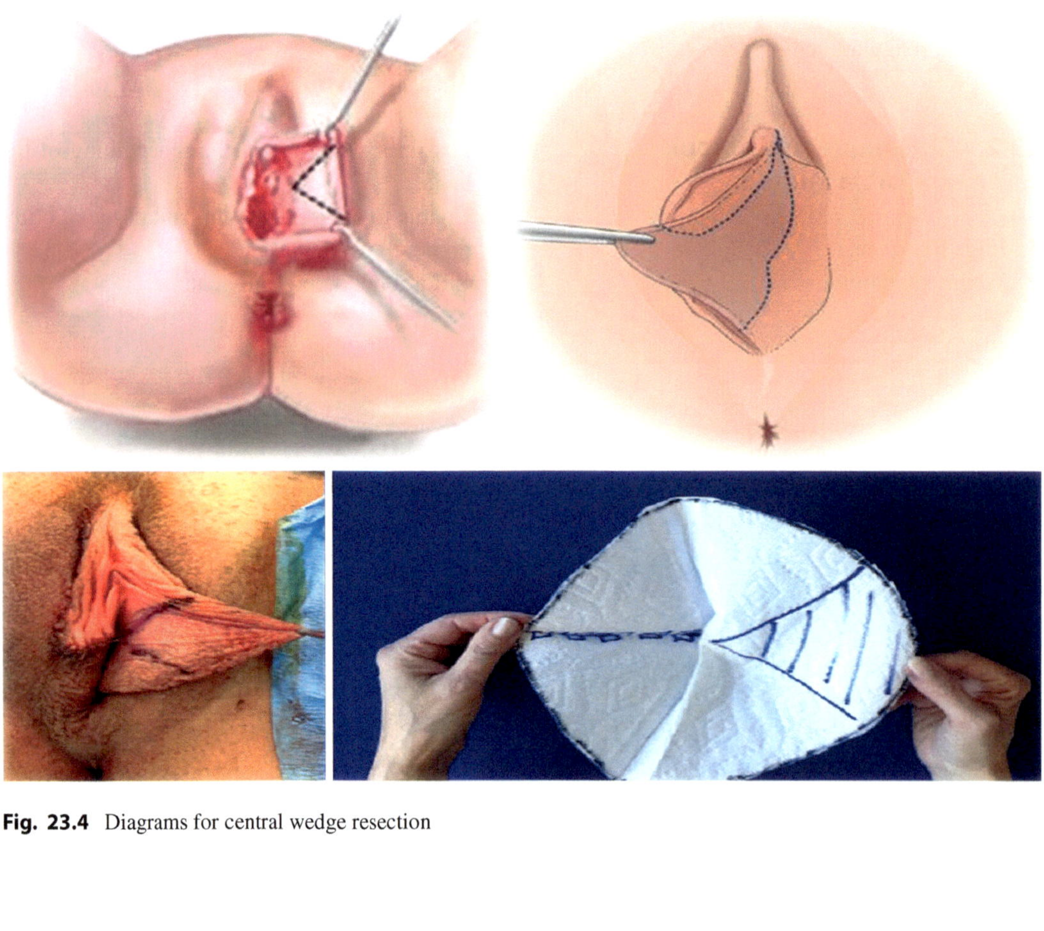

Fig. 23.4 Diagrams for central wedge resection

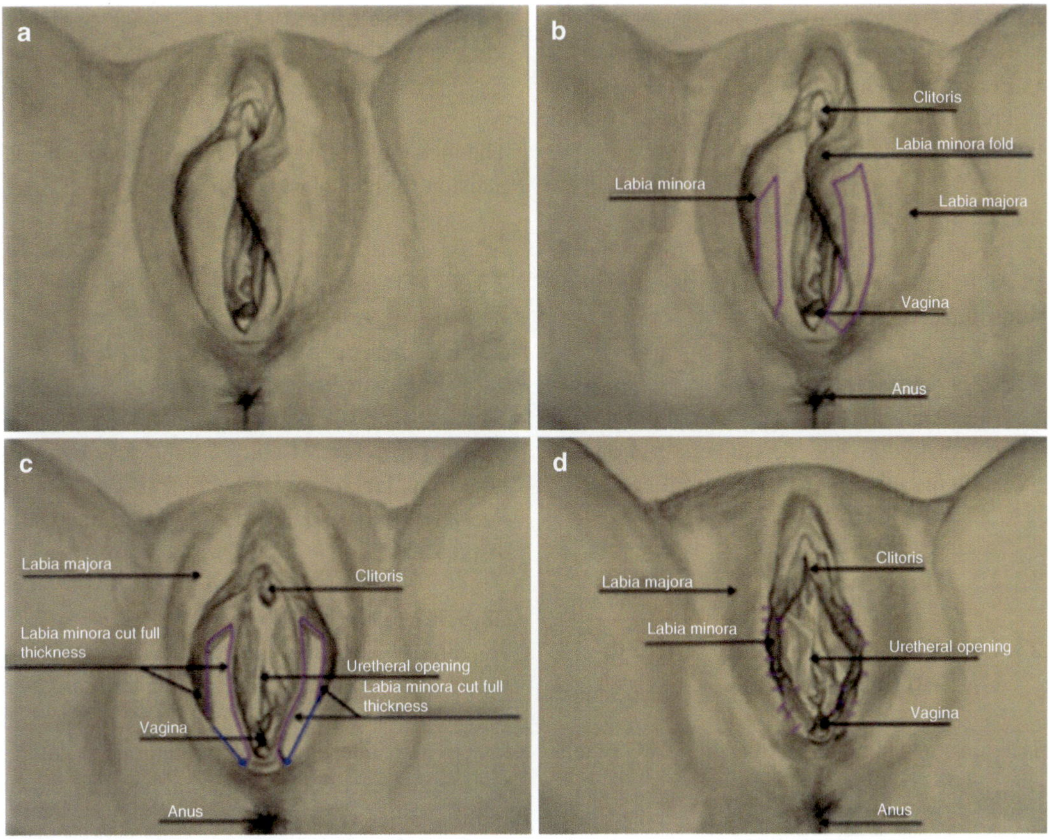

Fig. 23.5 Labiaplasty by posterior wedge resection

(A) The hypertrophic labia minora.
(B) Marks for posterior wedge resection.
(C) The blue arrows depict the superiorpedicle flap, which is created from the remaining labia minora after resection, being brought down toward the remaining labial base.
(D) The final appearance after placement of sutures.

23.3.2 De-Epithelialization Technique

This involves cutting the epithelium of central area of medial and lateral aspect of labia minora either with scalpel or with medical laser as shown in Fig. 23.6.

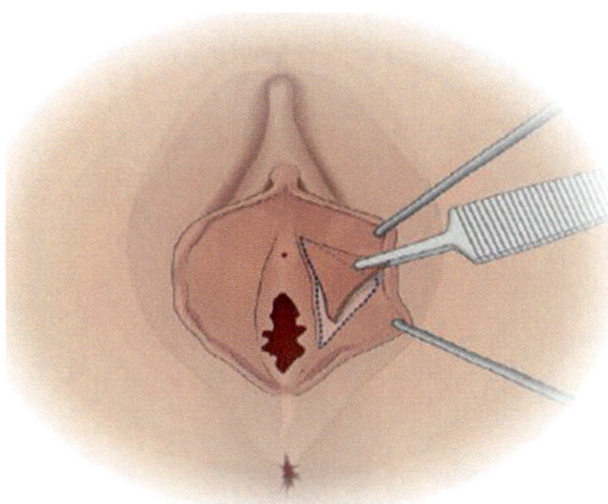

Fig. 23.6 De-epithelization

23.3.2.1 Advantages

- Reduces vertical excess tissue.
- Preserves the natural rugosity of labia minora.
- Preserves the sensory and erectile characteristics.

23.3.2.2 Disadvantages

- Large area of labial tissue must be de-epithelialized to achieve labial reduction, a technical disadvantage [6].

23.3.2.3 Special Notes

- When labial tissue is very large, a combination procedure of de-epithelialization and clamp resection may be more useful.

23.4 Diagram

23.4.1 Labia Minoraplasty with Zig-Zag Incision

The incision can be given in a ZIG-ZAG manner (Fig. 23.7).

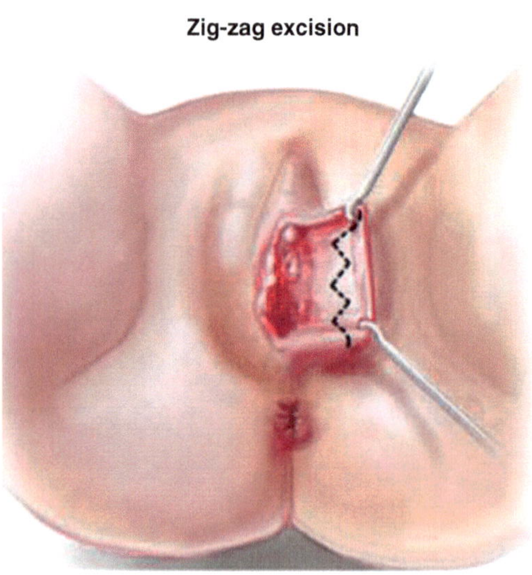

Fig. 23.7 Labia minoraplasty with zig-zag incision

23.4.2 Labiaplasty Done with Extended W Plasty Incision

The incision for this is given by extending the W incision for labiaplasty as shown in Fig. 23.8.

23.5 Diagram

23.5.1 Labioplasty with Clitoral Hood Resection

Surgical resection of clitoral hood may be combined with labiaplasty, if the thickness of the skin interferes with sexual response or is aesthetically displeasing [7, 8].

23.5.1.1 Method

A central wedge or V is removed from the most protuberant portion of each labia minora. The outer portion of V excision is normally curved lateral and anterior to excise redundant lateral labium and excess lateral clitoral hood as shown in Fig. 23.9.

Central wedge resection with clitoral hood reduction is safe, effective procedure with few complications and high patient satisfaction.

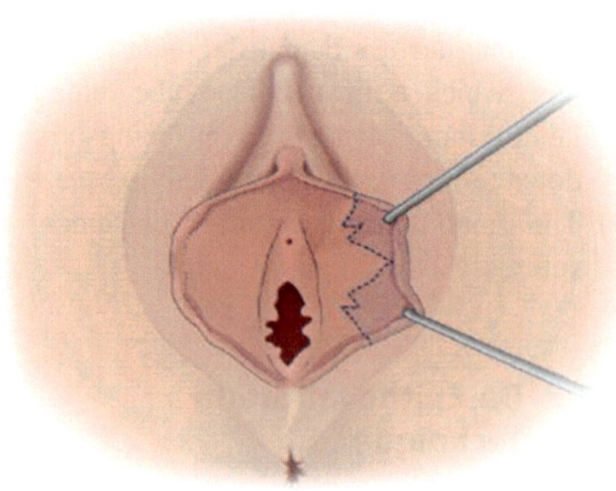

Fig. 23.8 Extended w plasty incision

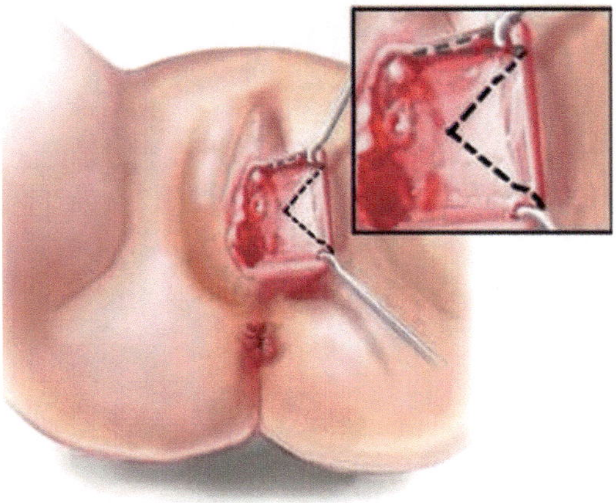

Fig. 23.9 Clitoral hood resection

23.6 Diagram

23.6.1 Laser Labiaplasty

Laser resection of the labia minora involves the de-epithelialization of labia. Laser is used instead of a surgical knife.

23.6.1.1 Advantage

- Bleeding is less.

23.6.1.2 Disadvantage

- Removal of excess labial epidermis risks causing epidermal inclusion cysts.

23.6.1.3 Post-Operative Care

- Patient is usually able to leave the hospital after a few hours.
- No vaginal pack is required.
- Wear a sanitary pad for comfort.
- Keep the area clean and dry.
- Local application of antibiotic ointment two to three times per day for several days.
- Follow-up visit to the surgeon after 1 week.
- Resume normal, physically unstrenous work 3–4days after surgery.
- Do not use tampons.
- Avoid tight clothes.
- Avoid sexual intercourse for four to 6 weeks.

23.6.1.4 Post-Operative Complications, Risks and Side Effects

Medical complications of labiaplasty procedure are uncommon. Rare complications that can occur include [9]:

- Complications or reactions of anaesthesia
- Bleeding
- Infection
- Excessive scarring
- Unfavourable cosmetic results
- Dyspareunia
- Reduced sensitivity
- Labial asymmetry
- Over or under correction
- Excess clitoral hood prominence after free edge resection labiaplasty

23.7 Labia Majora Plasty

23.7.1 Definition

It is the procedure to improve the aesthetics of labia majora, giving permanent results (Fig. 23.10).

23.7.2 Introduction

While many women are becoming more aware of labia minoraplasty surgery that means a reduction in size and reshaping the inner vaginal lips, less attention has been given to the labia majoraplasty or surgical rejuvenation of the labia majora this may also be because of non-surgical rejuvenation practices available for the labia majora. Most women prefer non-invasive rejuvenation as compared to surgical enhancement.

The labia majora can lose their original well-defined appearance for many reasons. These can be:

- Genetic causes for loosening or supportive tissue
- Ageing
- Child birth
- Weight loss

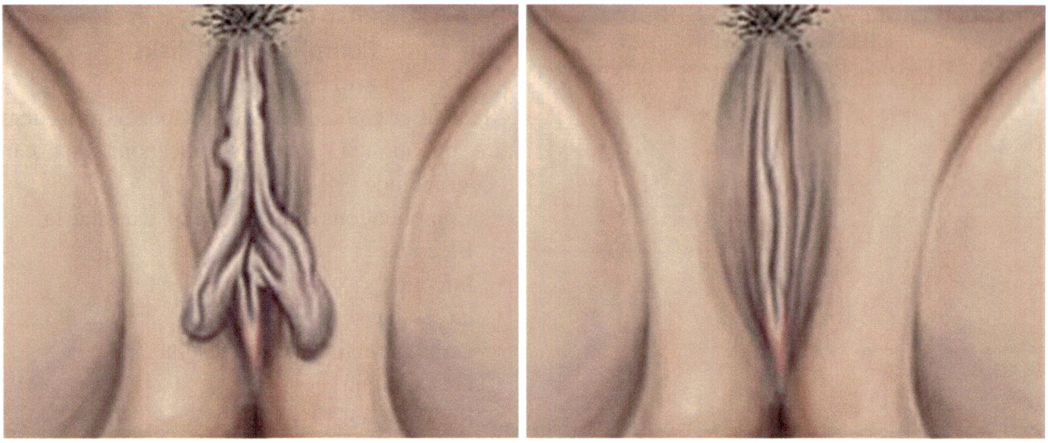

Fig. 23.10 Labia majora plasty

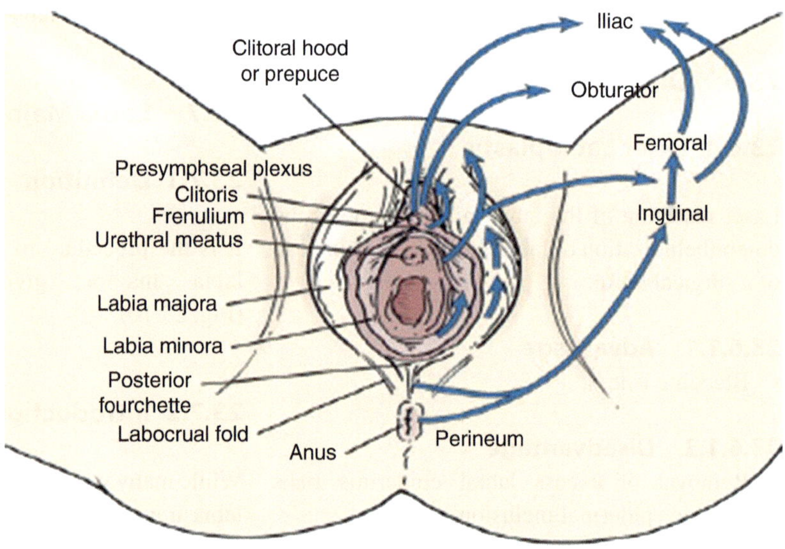

Fig. 23.11 Anatomy of labia majora

All the above lead to saggy, drooping labia, that are aesthetically unpleasing, sometimes sexually embarrassing and yet another reminder of ageing.

23.8 Diagram

23.8.1 Anatomy

The labia majora are commonly referred to as outer lips of the vulva. The labia majora are two longitudinal, adipose tissues filled cutaneous folds. They are composed of smooth muscle fibres, blood and lymphatic vessels, nerve ending and the dermis, where orifices of the numerous glands—sebaceous, sudoriferous and apocrine are located. Each labia majora has two surfaces, an outer pigmented and covered with long pubic hair and inner smooth beset with large sebaceous glands. They join anteriorly to form the anterior labial commissure and posteriorly to form the posterior labial commissure. Labia minora, urethral orifice, clitoris and vagina are located between the labia majora as shown in Fig. 23.11.

23.9 Diagram

23.9.1 Desire for Labiaplasty

- Decreased sensation during sexual intercourse.
- Discomfort during exercising or while wearing tight clothes.
- Asymmetry.
- Chronic urinary tract infection.
- Aesthetically unpleasant look and loss of volume and firmness.
- Sometimes sexually embarrassing.

All these factors motivate the patients for labia majora rejuvenation.

Patients often opt for a non-surgical procedure for labia majora rejuvenation, which could be done by injections and fillers such as:

- Fat fillers by fat harvested from another area
- Synthetic fillers
- Hyaluronic acid
- Platelet-rich plasma
- Stem cells
- Use of CO_2 laser
- Use of radio frequency laser
- Amniotic fluid injection

However, these non-invasive procedures have some limitations such as:

- They are expensive.
- Need to be repeated periodically.
- Irregular and slight bumps can occur if not uniformly injected.
- For this reason, some patients may opt for surgical correction.
- For performing surgical aesthetic procedure, knowledge of anatomy must be there.

CONTRAINDICATIONS: Same as labia minora.

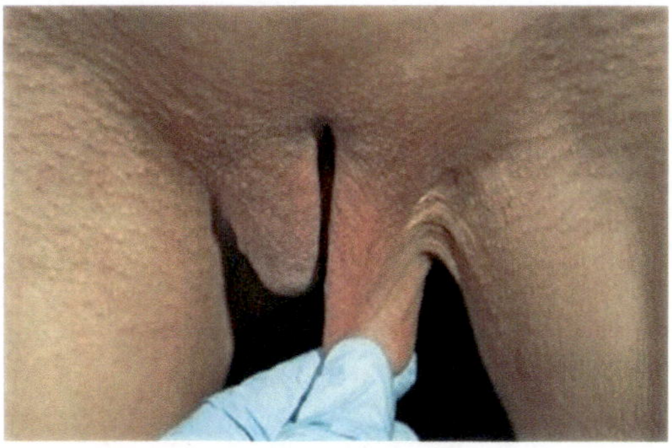

Fig. 23.12 Hanging lip of labia majora

23.9.2 Procedure

23.9.2.1 Hanging Lip of Labia Majora (Fig. 23.12)

- The patient is put in lithotomy position cleaned and draped. Unsightly hanging skin is removed by an incision at the lower end of labia majora. Remaining connective tissue is shrunken and firmed. This is followed by joining the refreshed skin edges with fine absorbable sutures.

23.10 Diagram

23.10.1 Labia Majora Plasty

- A vertical line is marked, beginning at the top level of the clitoral hood and extending in the labial crease, between the lateral labia majora and the medial clitoral hood and labia minora. The line is drawn down to the level of the introitus. A semi-elliptical mark is made from the top of the vertical line and extending laterally just past the top ridge of the labia majora, curving back down towards the lowest point of the vertical line. This is repeated on the opposite side for symmetry. Excising a moderate amount of labia majora tissue is recommended to prevent the appearance of a

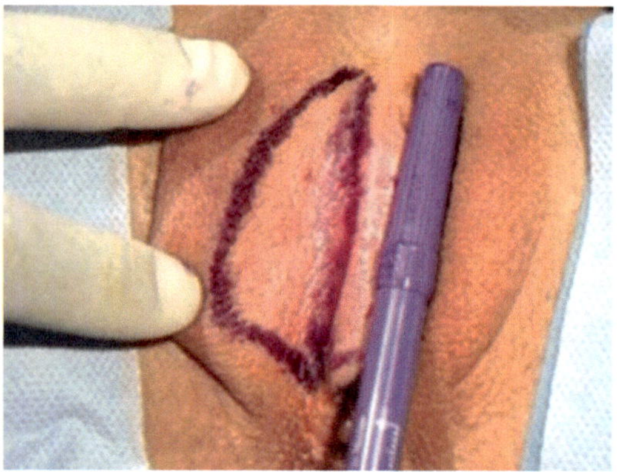

Fig. 23.13 Suturing labia majora after removal of excess tissue

gaping vaginal opening, which occurs if excess labia majora is removed as shown in Fig. 23.13.

23.11 Diagram

Incision for the labia majoraplasty

Labia majoraplasty is generally well tolerated and associated with favourable clinical outcomes.

- Removal of excess, loose skin does not appear to damage large vessels or nerves or affect sensitivity of the skin.
- Bleeding is an important potential complication when fat pad excision techniques are employed and larger deeper vessels are transected.
- Pain control is more challenging after fat pad removal.
- Preliminary findings suggest that high-frequency monopolar radio frequency (RF) and standard electrocautery are beneficial surgical tools in labia majora reduction surgery.
- Labia minoraplasty and labia majoraplasty may be performed as staged operations in the same patient. When performed by surgeons with advanced training, a combined and unified approach to simultaneous labia minoraplasty and labia majoraplasty is practical and safe in selected patients.

23.11.1 Key to Successful Healing

- Compliance with post-operative instructions as described under labia minoraplasty.
- Use of prescribed medications.
- Anti-ageing cream to prevent scar.

Labia majoraplasty can be combined with vaginal rejuvenation, labia minoraplasty and vaginoplasty depending on what the patient exact desire is.

23.11.2 Conclusion

The goal of labioplasty minora and majora is to eliminate functional problems, create aesthetic lips, reduce discomfort during some everyday activities (like wearing tight clothes, horse riding, physical exercise etc.) and sexual intercourse. It als gives psychological benefits for women who are anxious about the appearance of their genital organs.

- No criteria exist to define labial hypertrophy.
- Perception of normal labia defers amongst women, healthcare professionals and cultures.
- Many surgical approaches to labiaplasty have been described, meaning no gold standard technique exists.
- Complication rates are low and most complications are minor. But like with any other surgical procedure, severe complications with considerable consequences can occur.
- Most studies are small sized with lack of long-term data.
- Patient satisfaction evaluations are limited because of lack of long-term evaluation.

23.11.3 Rejuvenation of Labia Majora by Non-Surgical Procedures

23.11.3.1 Use of Hyaluronic Acid

Augmentation of Labia majora involves increasing their fulness as a result of injecting cross-linked Hyaluronic acid. The Hyaluronic acid fillers moisturise the skin and mucous membrane, stimulates fibroblasts to produce collagen. Hyaluronic acid fillers are recommended in lots of volume and firmness of Labia majora. Caused by a number of factors, but not limited to:

- Hypertrophy.
- Atrophy.
- Asymmetry of Labia majora.
- Mucocutaneous atrophy with associated symptoms like itching, burning and skin tightening.
- Women of low body weight or after weight loss.

These can develop due to vaginal birth, vulvitis, simply ageing and associated vulval and labial skin laxity [10–12]. Additionally, pain may be reported during coitus, sports, urination, discomfort with tight clothes, associated infection, inflammation, irritation and abrasion. Treatment can also be used for perineal repair after traumatic birth involving inferior part of Labia Majora. Filling Labia Majora with Hyaluronic acid seems to restore and repair contracture-related aesthetic defects [13].

The natural content of Hyaluronic acid in the body decreases with age, which results in skin being poorly moisturised and hydrated leading to disruption of collagen and elastic fibres [14]. Many women reporting dyspareunia suffer from tissue dehydration, mucosal dryness and cellular degeneration. Hyaluronic acid promotes fibroblasts to produce collagen, the structural protein of human skin leading to rejuvenation and improved elasticity.

23.11.3.2 Contraindications

1. Pregnancy
2. Breastfeeding
3. Active inflammation of injection site
4. Malignancy of vulva or perineum
5. Untreated mental disorders
6. Severe collagen disorders
7. History of allergy to Hyaluronic acid

23.11.4 Complications

Short-term complications include redness, haematoma, and local inflammation. Long-term complications fibroma, hypertrophy due to excess fillers and asymmetry (reversible with hyaluronidase).

23.11.4.1 Fat Augmentation

Labial augmentation can be achieved by dermal fillers. Fat grafting is an alternative with longer results. Before taking treatment, patient should stop taking blood thinners and birth control pills for at least 10 days. Stop smoking and avoid alcohol for 6 weeks before and after treatment.

During treatment, liposuction is done to harvest fat, which is purified for injection. Fat is injected at various depths in the Labia as shown in Fig. 23.14.

23.11.5 Side Effects

Immediate side effects include bruising and swelling, which usually subside within 5–10 days. Cool compressors can help to decrease swelling. Patient can resume work after 2 days and full activity by 1 week.

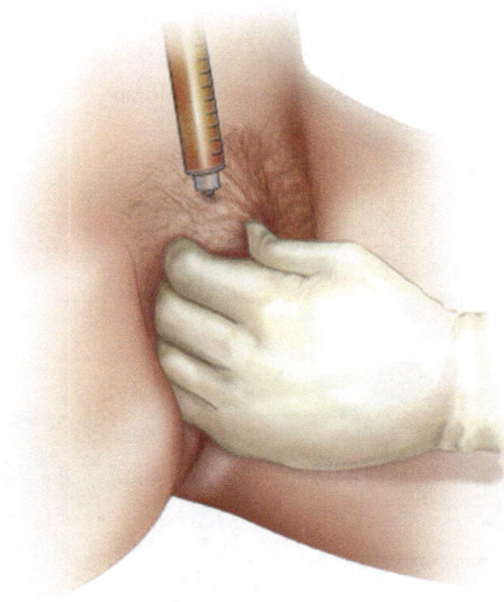

Fig. 23.14 Labial fat augmentation

Long-term side effects can include numbness, pain and asymmetry.

Some serious side effects can include excessive blood loss, DVT, fat necrosis, cardiac or pulmonary embolism. Revision of surgery may be demanded for unsatisfactory results.

23.11.5.1 CO_2 Laser

Carbon dioxide fractional laser has a dermal rejuvenation effect, stimulated by gentle heating and subsequent selective induction of collagen regeneration and deposition in deeper layers. This results in improvement of skin texture and wrinkles.

Carbon dioxide laser rejuvenates the skin tissue and promotes collagen production leading to tightening and shrinking of Labia Majora and Minora (Fig. 23.15).

23.11.5.2 Radio Frequency Laser Treatment of Labia Majora and Labia Minora

Radio frequency is temperature-controlled bipolar mechanism to heat tissues to target temperatures of 40–45 °C. Controlled energy delivery leads to inflammatory cascade leading to neocollagenesis, angiogenesis and elastogenesis over the next 3–4 months.

When performed correctly, the post-procedure recovery is good, quick and easy. Side effects are typically minimal and short lived, limited to mild redness, swelling and tingling. Radio frequency treatment can also be combined safely with injectables, microneedling or minimally invasive procedures (Fig. 23.16).

Rarely thermal burns may be caused, if it is not well controlled.

23.11.5.3 Platelet-Rich Plasma

Platelet-Rich Plasma (PRP) is a portion of the plasma fraction of autologous blood, which has a platelet fraction above the baseline (Fig. 23.17). PRP is an inexpensive and immunologically safe source of growth factors. PRP promotes tissue repair, influences angiogenesis and reduces inflammation [15].

AMNIOTIC FLUID AND STEM CELL INJECTION: These are the new methods for labial rejuvenation.

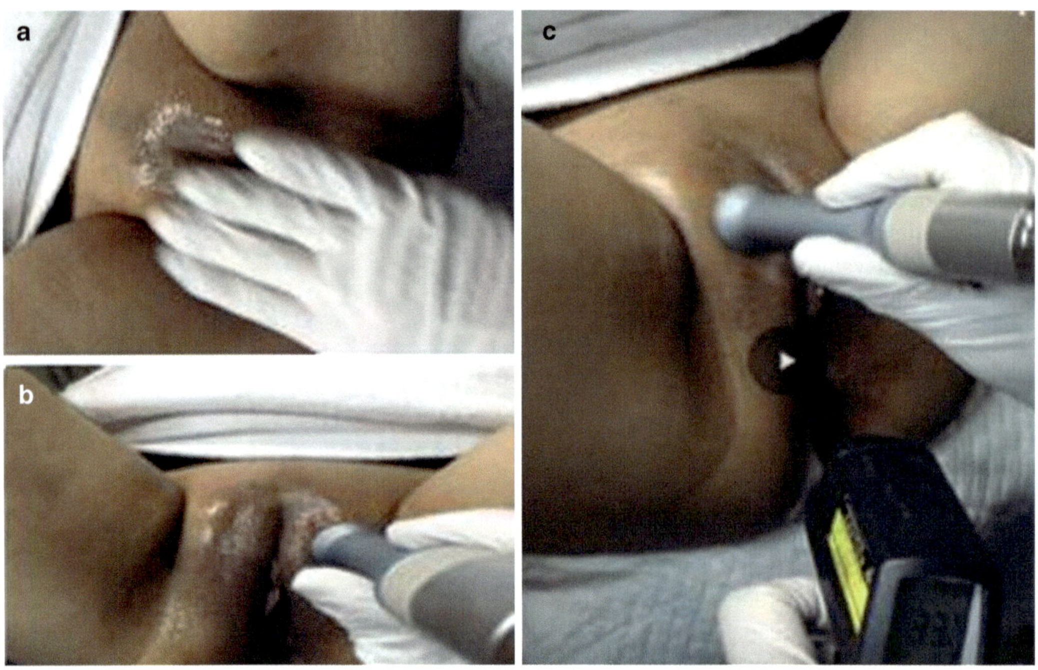

Fig. 23.15 CO_2 vulval laser demonstration

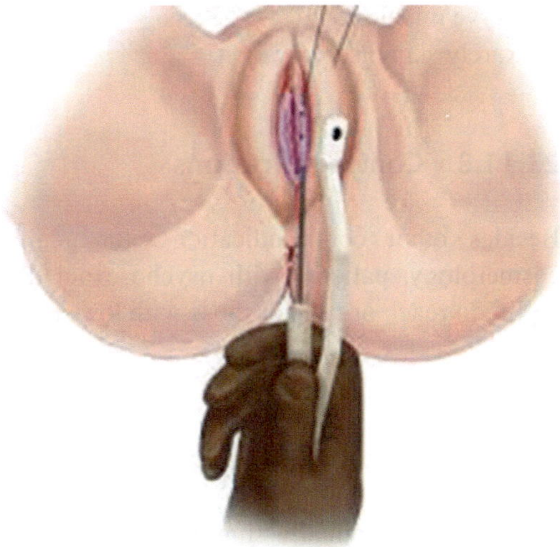

Fig. 23.16 Radio frequency laser

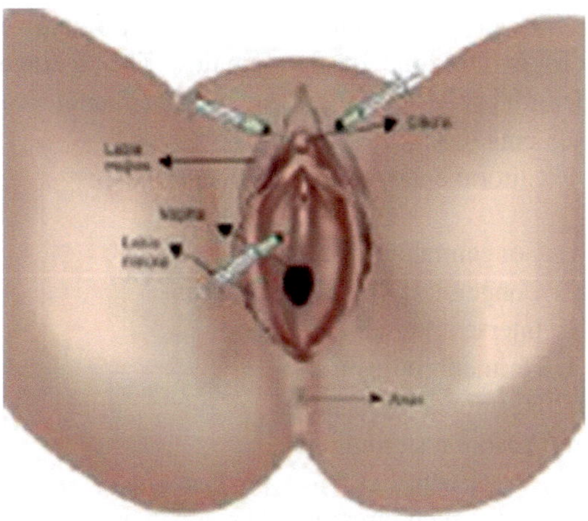

Fig. 23.17 PRP labial infiltration

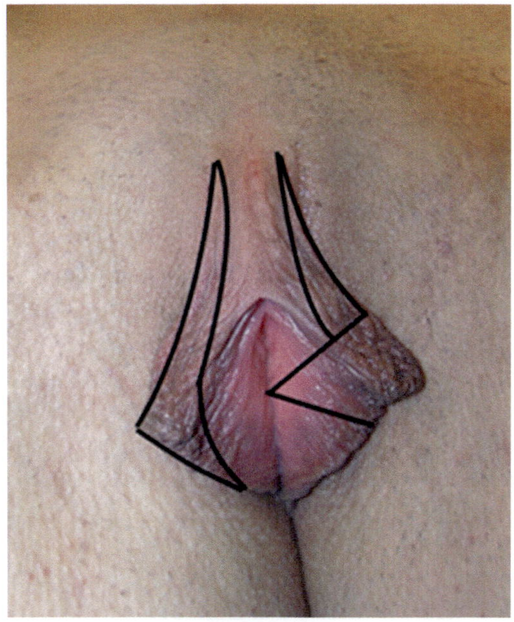

Fig. 23.18 Clitoral hood resection

23.11.6 Clitoral Hood Resection

Clitoral hood resection, also termed clitoral hoodectomy, clitoral unhooding, clirodotomy or partial hoodectomy is a procedure for reducing the size and area of clitoral hood (prepuce) in order to further expose the clitoral glans (Fig. 23.18). Therapeutic goals are to improve sexual gratification and aesthetic refinement of vulva.

The upper part of Labia Minora diverges into two branches. The superior branch of labia minora is termed as clitoral prepuce and the inferior branch as clitoral frenulum.

Clitoral hood reduction is rarely performed as an isolated procedure. It is most commonly performed in combination with Labia Minora reduction. The hood is prepuce of the glans clitoris. Clitoral hood redundancy, when present, maybe in the horizontal or vertical plane or in both.

Horizontal excess in the form of extra hood folds parallel and lateral to the portion of central portion of clitoral hood and is most commonly observed. Clitoral hood folds may be unilateral or bilateral and result in widened appearance. Vertical excess manifests as ptotic, elongated clitoral hood. If this is present, it should be dealt with during Labiaplasty. Not doing so may result in unnatural appearing genitalia [16, 17].

Vertical hood excess is addressed by transverse excision of a portion of the hood, usually by an inverted V wedge across its full width (Fig. 23.19). Excision is usually done cephalic to free margin.

In case of a very elongated hood, significantly overhanging the glans clitoris, the hood may be shortened conservatively by excision at the free margin itself.

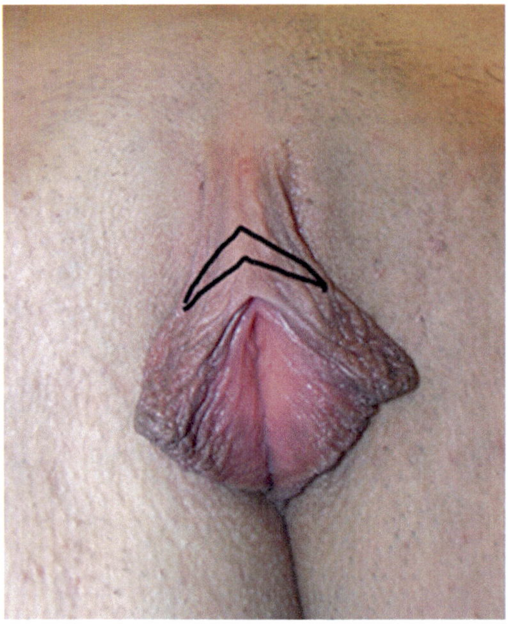

Fig. 23.19 Clitoral hood resection by inverted v wedge

23.11.7 Complications

1. Pain at or around the surgical wound. It is usually relieved by pain killers. Take rest and monitor symptoms. Active work may cause bruising and further pain.
2. Bleeding from surgical site can be caused by rubbing over the raw area, even by walking or other simple activities.
3. Abnormal bleeding and clot formation is rare occurring in less than 1%. Aspirin and other anticoagulants should be avoided 10–15 days before and after surgery.
4. Inflammation and infection because of close proximity of the surgical wound to vaginal opening, urethra and anus. To avoid it follow doctor's instructions carefully take care of coexisting diseases like diabetes and vaginal infection prior to surgery.
5. Irregularities at the surgical site due to wound deformation, suture scar or post-operative wound disruption.
6. Unsatisfactory results in the form of asymmetrical clitoral hood or non-proportional clitoral hood.
7. Nerve impairment.
8. Rarely skin discolouration.

23.11.8 Contraindications

Besides other contraindications, for genital cosmetology, patients with psychosexual disability to reach orgasm or those with low sexual drive.

23.11.9 Clitoral Reduction Surgery

Clitoris is anatomically similar to penis in the males. Glans, the head of the clitoris, is roughly the size of a pea, but can vary considerably in size. It is estimated to have about 8000 sensory nerve endings, making it the most sensitive erogenous zone of the females. The average length is 1–1.5 cms with an average diameter of 0.5 cms.

Clitoral enlargement can occur due to:

1. Congenital, idiopathic enlargement
2. Congenital adrenal hyperplasia
3. Intersex
4. Hermaphroditism
5. Hormonal imbalance
6. Anabolic steroid intake

Under such circumstances, a patient may opt for clitoral reduction surgery.

References

1. Masters Z, William H, Johnson VE, Kolodny RC. Human sexuality. Harper: Collins College Publishers; 1995. p. 47.
2. Sloane E. Biology of women. Cengage Learn. 2002:32.
3. Alter GJ. A new technique for aesthetic labia minora reduction. Ann Plast Surg. 1998;40(3):287–90. https://doi.org/10.1097/00000637-199803000-00016.
4. Giraldo F, González C, de Haro F. Central wedge nymphectomy with a 90-degree Z-plasty for aesthetic reduction of the labia minora. Plast Reconstr Surg. 2004;113(6):1820–5, discussion 1826–1827. https://doi.org/10.1097/01.PRS.0000117304.81182.96.

5. Alter GJ. Aesthetic labia minora and clitoral hood reduction using extended central wedge resection. Plast Reconstr Surg. 2008;122(6):1780–9. https://doi.org/10.1097/PRS.0b013e31818a9b25.
6. Choi HY, Kim KT. A new method for aesthetic reduction of labia minora (the deepithelialized reduction labioplasty). Plas Reconstruct Surg. 2000;105(1):419–22. discussion 423–424. https://doi.org/10.1097/00006534-200001000-00070.
7. Hamori CA. Postoperative clitoral hood deformity after labiaplasty. Aesthet Surg J. 2013;33(7):1030–6. https://doi.org/10.1177/1090820X13502202.
8. Hunter JG. Commentary on: postoperative clitoral hood deformity after labiaplasty. Aesthet Surg J. 2013;33(7):1037–8. https://doi.org/10.1177/1090820X13503476.
9. Davison S.P. et al. Labiaplasty and Labia Minora Reduction; 2008. eMedicine.com.
10. Origon M, Climmino C, et al. Post menopausal vulvovaginal atrophy is positively improved by topical Hyaluronic acid application. Eur Rev Med Pharmacol Sci. 2016;20(20):4190–5.
11. Nanette Santoro & Jame Komi "Prevalence and impact of vaginal symptoms among postmenopausal women. J Sex Med 2009,6(8), 2133 to 2142.
12. Rossel E. Nappi & Michele Lachowsky. Menopause and sexuality. Prevalence of symptoms and impact on quality of life. Maturitas 2009, 63(2), 138 to 141.
13. Bentazzi S, Londero AP, et al. Impact of episiotomy on pelvic floor disorders and their influence on women wellness, six months postpartum. A retrospective study. BMC Womens Health. 2011;
14. Fleishmajor R, et al. Aging of human dermis. J Dermatol. 1993;100:705–9.
15. Eppeley BL, Pietrzak WS, Blauton M. Platelet rich plasma –a review of biology and application in plastic surgery.
16. Hunter JG, Commentary on post operative clitoral hood deformity after labiaplasty. Aesthetic Surg J 2013; 30(7), 1037 to 1038.
17. Hamori C.A., Post- operative clitoral hood deformity after Labiaplasty 2013; 30 (7), 1030 to 1036.

Vulvodynia

24

Rashmi Bagga and Rimpi Singla

24.1 Background

Vulvar pain is a multifactorial complex disorder of varied etiologies. Specific disorders leading to vulvar pain include infections, inflammatory conditions, neoplastic lesions, neurological disorders, trauma or hormonal deficiency of menopause [1]. When vulvar pain cannot be attributed to a specific disorder, it is referred to as vulvodynia. The new consensus terminology defines Vulvodynia as chronic vulvar pain of a minimum of 3 months duration without a specific identifiable cause [1]. Vulvodynia is diagnosed when no other cause of vulvar pain is identified or if pain persists even after adequate management of potential cause(s). Despite having a significant negative effect on woman's health, self-esteem, relationships, and quality of life, vulvodynia is under-reported due to lack of awareness and is also under-recognized by medical practitioners. In one self-administered questionnaire-based study from Boston, 16% of women reported having vulvodynia [2]. Treatment is individualized based on the patient's presentation, functional impairment, and psychological impact. Its varied presentation and fluctuating symptoms make it difficult to compare various treatment modalities.

R. Bagga (✉) · R. Singla
Department of Obstetrics and Gynecology,
Post Graduate Institute of Medical Education and Research, Chandigarh, India

24.2 Classification of Vulvar Pain

Based on etiology, vulvar pain is classified according to ISSVD as [1]:

A. **Pain due to a specific disorder**
   - Infection (candidiasis, human papillomavirus related, herpes related)
   - Inflammatory (lichen sclerosus, lichen planus)
   - Neoplastic (Paget's disease, vulval intraepithelial neoplasia, and carcinoma)
   - Neurologic (post-herpetic neuralgia, nerve injury, neuromuscular disorders)
   - Trauma (female genital mutilation, fissures, obstetric trauma)
   - Iatrogenic (following surgery, chemotherapy, radiation)
   - Hormonal deficiency (vulvovaginal atrophy of menopause, lactational amenorrhea)

B. Vulvodynia (Vulvar pain without identifiable cause) may be described as:
   - Based on site of pain: Localized (e.g., vestibulodynia, clitorodynia) or generalized
   - Based on occurrence: Provoked (insertional, contact) or spontaneous
   - Based on onset: primary or secondary
   - Based on Temporal pattern: intermittent or constant, immediate or delayed

P. Jindal et al. (eds.), *Aesthetic and Regenerative Gynecology*,
https://doi.org/10.1007/978-981-16-1743-0_24

24.3 Pathophysiology

Its pathophysiology is not well characterized. It may result due to decreased peripheral threshold for pain sensation and heightened central response [3]. Individual's sensitivity to pain is determined by genetic, psychological, and environmental factors. Chronic vulvar pain is more prevalent among women with a history of sexual or physical abuse or co-existent pain disorders [4]. Localized provoked vestibulodynia, characterized by pain on touch or gentle pressure on the vulva, is the most common form of vulvodynia. It is attributed to neuroinflammatory changes in mucosa, increased concentration of pro-inflammatory peptides, and hyper-innervation with C-fibers which are responsible for prolonged burning sensation once stimulated [5]. Pain and particularly burning sensation continue even after intercourse for few hours but sometimes for several days [6]. Increased tone of pelvic floor muscles with consequent narrowing of introitus further contributes to pain. Typically, pain is provoked by sexual intercourse, the use of tampons, and tight clothing.

24.4 Evaluation

Description of symptoms: An adequate history of primary complaint, i.e., pain is essential including provoking factors, duration of symptoms, whether localized or generalized, and impact on the quality of life. The degree of pain can be quantified by visual analog pain scales and pain diaries.

Look for the cause: Vulvodynia is essentially a diagnosis of exclusion and it is imperative to exclude all treatable causes before attributing the symptoms to vulvodynia. In the absence of specific tests, diagnosis is made on the basis of clinical assessment. Hence, clinical assessment is of utmost importance in the evaluation of vulvar pain. During clinical examination, speculum examination should be avoided if possible as it is often very painful.

- History of recurrent vaginal discharge with redness, itching, and burning sensation is suggestive of infections. Genital herpes usually causes unilateral episodic pain. Initial examination involves inspection of the vulva and vagina. Simple tests like wet mount, fungal culture, and Gram stain may be done, if indicated [7].
- Chronic dryness with itching may be indicative of dermatoses (Eczema, lichen simplex chronicus, psoriasis, lichen sclerosus, lichen planus, etc.).
- Neoplastic lesions may be suspected on examination and a biopsy must be taken from the lesion and its healthy margin. White lesion of the vulva is usually biopsied.
- Hypoestrogenization in lactating women and in post-menopausal women may cause vulval dryness and dyspareunia. Examination would reveal atrophic vaginitis in post-menopausal women (pale and dry vagina with loss of rugosities).
- Localized vulvar pain during sexual intercourse may be due to tears of hymen or posterior fourchette and incompletely healed episiotomy wounds.
- Neuralgic pain referred from the pelvic girdle like pudendal neuralgia causes pain on sitting which is relieved by standing or lying down [8].
- Vulvar pain may sometimes just be a referred pain from other parts. Hence, musculoskeletal evaluation should be considered in the absence of any visible cause.

Evaluation of vulvodynia: Once other causes of pain are excluded; the diagnosis of vulvodynia is made. Vulvodynia may co-exist with other vulvar conditions and is diagnosed when pain persists even after the management of visible pathology.

- On inspection, the vulva is typically normal looking. Inner vestibule may show non-specific erythema.
- Cotton swab test [7, 9] helps to localize site and severity of pain. In this, gentle pressure is

applied with the help of a cotton wool tip bud on the thighs, followed by labia majora and interlabial sulci. Then vestibule is tested by applying pressure circumferentially at 2, 4, 6, 8, and 10-o'clock positions. Apart from localization this test helps to judge severity as the patient is asked to describe pain as mild, moderate, or severe. Diagrammatic documentation of the same would be helpful in follow-up over time.

- Women with vulvodynia should be assessed for pelvic floor dysfunction [10]. With one fingertip, palpate the pelvic floor muscles [7] provided it should not provoke the pain. The puborectalis, pubovaginalis muscles are often found to be contracted, taut, and tender.
- Biopsies from symptomatic areas are not necessary because of the lack of defined histological features. However, a biopsy is performed if other etiologies (especially neoplasia) are suspected. Vulvoscopy is rarely required as the acetic acid application may aggravate the pain and irritation. Rarely, magnetic resonance imaging may be required to diagnose sacral cysts that can cause referred pain [11].

Look for Co-existing pain conditions: Interstitial cystitis or irritable bladder, irritable bowel syndrome, endometriosis, fibromyalgia, and chronic low back pain.

Psychological assessment: Explore the patient's understanding and perceptions about pain, especially her fears. Women may be having fear of some serious underlying disorder like cancer that should be allayed through proper counseling. She should be asked how her day-to-day activities, professional life, and sexual relationship are affected by pain. Patients with dyspareunia should be asked about the presence of vaginismus and partner problems. Sexual dysfunction is common and frequently reported [12, 13] Many studies have demonstrated an increased prevalence of anxiety and depressive symptoms among women with vulvodynia [12, 13]. Identification and counseling for psychosexual disorders are important in addition to medical treatments.

24.5 Treatment

Clinical evaluation may diagnose some treatable cause. For example, chronic candidiasis is a frequent cause of the vulvar burning sensation. Look for the cause of recurrent vulvovaginal infections like uncontrolled diabetes and obesity. Vaginal estrogen preparation may relieve vulvovaginal irritation due to atrophic vaginitis. Women with specific dermatoses (e.g., lichen planus) are referred to a dermatologist. Chronic irritation must be treated as it is a risk factor for neoplasia. Neoplastic lesions are referred to a gyne-oncologist.

Due to limitations of existing literature like varied patient selection criteria, varied duration of follow-up, and paucity of randomized clinical trials (RCTs), treatment guidelines for vulvodynia are based primarily on clinical experience, expert opinions, and observational studies. There is no strong evidence of benefit for a particular intervention over the others [14, 15].

Vulvodynia is a multi-faceted disease that warrants a multidisciplinary approach [7, 10, 16, 17] including medical treatment, psychotherapy, physiotherapy, and dietary advice to tackle different facets of chronic vulval and sexual pain [16, 17].

It is important to individualize treatment plan and understand that it may need frequent revision, e.g., psychosexual approach would be beneficial for patients with sexual pain as a primary symptom [17]. For unprovoked pain, cognitive behavioral therapy and psychotherapy are needed [18].

A positive relationship between the patient and her clinician results in a better outcome and most women improve significantly over time. However, expectations must be realistic as the treatment is often prolonged. Explaining the rationale of the treatment plan to the patient ensures her active participation and improves the outcome.

A. ***General care***:

The basic principles (Box 24.1) are avoidance of potential irritants and improvement in skin moisture [7, 16].

Box 24.1 General Care Principles

General care principles:

- Use water with mild or no soap/ shampoo for cleansing vulva. Simple emollients can be used. Limited data suggest beneficial effect of coconut, olive, and vitamin E oil.
- Water-based non-scented and non-allergic lubricants that moisturise the skin are helpful in reducing pain during sexual contact but avoid alcohol-containing gels or warming agents.
- Applying cool gel packs before and after intercourse reduces pain soreness. Some women get symptomatic relief from burning sensation by regular application of ice packs at 6 hourly intervals.

Following should be avoided:

- Any activity such as bicycling that precipitates pain
- Constrictive garments like pantyhose, tight underpants, tight leggings and trousers
- Douching, tampons, fragrant wipes and talcum powder
- Fabric softener, preservatives
- Medicinal and herbal products (e.g., tea tree oil, benzocaine, antifungals)

B. ***Medical therapy***:

Medical therapy can be delivered by topical, oral, and intralesional routes. Sometimes nerve blocks and botulinum toxin are helpful. Patients with vulvodynia may be already taking one or more medications. It is preferable to stop all topical medications before starting a new course because of potential interactions between various medications and also because previous medication itself may be the cause of burning pain. Vehicles (base) used in creams and ointments for delivering the drug should be hypoallergenic.

1. **Topical therapy**:

Topical lidocaine: 2–5% ointment applied 15–20 min prior to sexual relation has benefitted women with provoked vestibulodynia. In a study by Zolnoun et al., patients with vulvodynia (mainly provoked pain) benefitted from overnight 5% lidocaine application. There was an improvement in the number of women reporting to be able to have intercourse [19]. However, others have not found topical lidocaine to be more effective than placebo [20, 21]). It is a low cost, safe option for episodic use that provides short-term symptom control [22].

Vestibular application of Topical estradiol: Topical application of estradiol as 0.01% cream alone or in with testosterone 0.05–0.1%, reduces symptoms in some women with atrophic changes in vagina [23]. Estriol (0.1%) cream commonly available as Evalon is also useful. These are used for a short duration (a few weeks) with gradually reducing frequency from daily to once or twice a week. Hormonal therapy has not proven to be uniformly effective in treating vulvodynia in menopausal women despite falling estrogen levels being the cause of symptoms. Due to conflicting data, the International Consensus of Sexual Medicine panel has suggested that further studies are needed before recommending topical hormones for treating vulvodynia [22].

Topical cromolyn: Cromolyn is a mast cell stabilizer, used as 5–10% in a petrolatum base, and is found to be helpful for women with itching. Data from a placebo-controlled trial regarding its efficacy in treating vulvodynia has not shown it to be more effective than placebo in one study [24].

Vaginal prasterone: Though not much studied, it is synthetic dehydroepiandrosterone (DHEA), used in the treatment of dyspareunia owing to vulvar or vaginal atrophy of menopause.

Capsaicin cream: Capsaicin is advisable when other medical treatments have failed [22]. Topical capsaicin has an agonist effect on vanilloid receptors VR1 in the peripheral terminals of the nociceptor, thereby producing desensitization to burning [25]. It is prescribed as compounded 0.025–0.05% capsaicin cream to be applied initially for 20 min daily for 12 weeks and then only as needed for symptom control. Compliance is enhanced with pre-treatment with 5% lidocaine ointment.

2. **Oral medications**:

Tricyclic Antidepressants (TCAs): These are used as first-line oral agents for pharmacological management of vulvodynia, in particular for unprovoked pain. TCAs affect levels of serotonin and norepinephrine and impact histamine receptors [26]. Pain relief is independent of antidepressant effects and, hence, may be achieved at much lower doses [26]. Amitriptyline, the best-studied tricyclic, is initially prescribed at a dose of 5–25 mg nightly which is titrated up as dictated by the patient's pain level by 10–25 mg every week to a maximum dose of 50–75 mg [10, 27]. Side effects like dry mouth, sedation, and constipation are common and often dose-limiting [27]. If effective, treatment with TCAs needs to be continued for 6–12 months. If there is no response even after 6 weeks of taking a maximum tolerated dose, it should be discontinued.

Anticonvulsants: Oral anticonvulsants including gabapentin, pregabalin, and carbamazepine can be used in patients who do not respond to TCAs or are intolerant to their side effects. Gabapentin is the most commonly used anticonvulsant for the treatment of vulvodynia [10]. The initial dose is 300 mg daily that can be increased over time to a maximum dose of 3600 mg daily (in three divided doses) [16]. Efficacy is reported to be quite high (50–80%) across small observational studies [28–30] but not in a placebo-controlled trial [31]. Contraception should be emphasized while prescribing such medicines to a woman in the reproductive age group while in elderly women, Gabapentin may exacerbate gait and balance problems.

Gabapentin and amitriptyline are also available as topical preparations for women who wish to avoid the side effects of oral medicines. These have been effective as well as well tolerated. In a prospective study, 56% of women with vulvodynia reported slight to excellent improvement with amitriptyline 2% cream [32].

Selective norepinephrine reuptake inhibitors (SNRI)—SNRIs (duloxetine, milnacipran, venlafaxine) are efficacious at treating neuropathic pain with fewer side effects than TCAs. Duloxetine is started at a dose of 20 mg daily and increased to 60 mg, if needed. Milnacipran is started with an oral dose of 50 mg twice a day and increased to 100 mg twice a day as needed [33].

3. **Injectable medications**:

Injectable steroids—Submucosal injection of corticosteroids with lidocaine or bupivacaine into the trigger point may have an anti-inflammatory effect (reduction of interleukin-beta) [34] and are effective in some patients with localized vulvodynia [35, 36].

Botulinum neurotoxin-A (BoNT-A)—Intrathecal delivery of BoNT-A blocks the release of glutamate and substance-P from nociceptive neurons [37] and is used for the treatment of chronic pain. Direct injection into primary site avoids the side effects of systemic medications [38]. Evidence for its use in the treatment of vulvodynia is not yet conclusive. Favorable results have been reported in small studies [39, 40]. but it has not proven to be more effective than placebo in randomized control trial [41] but an expert panel of the 2015 ICSM recognizes it as a reasonable treatment for vulvodynia when injected into hypertonic pelvic floor muscles. Benefits are mostly temporary. Adverse effects are usually self-limiting; however, rarely dysphagia and respiratory blockade may occur that can be life-threatening, hence requiring informed consent prior to injection.

4. **Biofeedback and Physical therapy**:

Pelvic floor physical therapy is an important intervention delivered alone or in combination with pharmacotherapy to treat vulvar pain [16] and is particularly helpful in vaginismus. Patients with vulvodynia and dyspareunia frequently have increased tone of pelvic floor muscles surrounding the vagina, urethra, and anus [42]. Basic mechanism of action of pelvic floor physical therapy is the restoration of proper muscle length that reduces tension. The main strategies are direct internal transvaginal or transrectal release of muscles [42] and stretching of muscles of hips, back, gluteus and abdomen. Other interventions are pelvic and core muscle mobilization, release of myofascial trigger point, connective tissue mobilization and electrical stimulation [42–44]. This is followed by strengthening techniques to restore muscle strength and stability. Strengthening exercises should be introduced gradually in order to avoid irritation of restricted neural and connective tissue and muscle spasm. In a study by Bergeron et al., 51% of the participants reported complete or great improvement and 20% reported a moderate improvement with physical therapy [45] Biofeedback is vital to physical therapy in order to establish mind–body control of the pelvic floor and down-train overactive muscles [46]. This aids in developing self-regulation to help overcome pelvic floor muscle dysfunction and hence, helpful in treating provoked vulvodynia [47].

For sustained impact, therapist-assisted physical therapy has to be complemented by home-based measures that include stretching exercises, relaxation, and breathing exercises. Patients should also be taught external and internal soft tissue self-massage, trigger point pressure, use of vaginal trainers, and biofeedback [43, 48]. Vaginal dilators may be helpful in normalizing the tone of muscles. Education about the rationale of these interventions is important to ensure compliance. The purpose of these therapies is to gradually desensitize the hypersensitive points in the vulva and vagina and restore sexual function. The incorporation of yoga and meditation further facilitates neural downregulation.

5. **Psychological therapy**:

Morbidity of vulvodynia is not restricted to physical pain but also affects self-esteem, intimacy, and marital relations and hence has toll on mental health. Sometimes intimate partner behavior or childhood trauma is the main or contributory factors. Thus, psychotherapy is an integral part of the treatment approach to vulvodynia. Psychological interventions include techniques to cope with symptoms. Women with vulvodynia have a higher level of stress, anxiety, and depression as compared to healthy women [12, 49]. So, stress management strategies are helpful in breaking the vicious cycle. Cognitive behavior therapy (CBT) is a validated non-invasive therapeutic approach for vulvar pain and specifically for sexual pain. The therapist helps to identify personal maladaptive coping strategies that increase pain, such as hypervigilance to pain, and excessive anxiety. Couple-based therapy is a component of cognitive behavior therapy whereby the couple is given psychological support along with education about impact of pain on sexual function and the couple's mutual relationship. This includes various strategies to enable better communication, reconnect through nonsexual physical and emotional intimacy, shift the focus away from sexual intercourse towards sharing mutual experiences of sexual intimacy. CBT has fared at least equally well as a treatment strategy when compared to other modalities like vulvar vestibulectomy, topical corticosteroids, and physical therapy across various comparative studies [18, 50, 51], and benefits were maintained in long term. Couple-based therapy has been shown to result in significant improvements in woman's pain and sexual outcomes [52, 53]. Maintenance of a daily pain diary helps the woman and her psychotherapist in identifying the factors that trigger or resolve the pain and also track her progress through treatment.

6. **Surgery**:

Pain localized to the vestibule and refractory to medical and physical therapies may require surgery in the form of vestibulectomy. It cannot be offered as initial treatment and is generally considered a treatment for patients with localized provoked vulvodynia who have failed to improve with conservative modalities [54]. It refers to the excision of painful portion of the vestibule from hymen to the lateral boundary of the vestibule, including all tender and painful parts extending to the anterior vestibule, if indicated [55]. The success rate for vestibulectomy has been reported to be 60 and 90%. However, there is no consensus on standardized definition of successful treatment and outcome measures are variable across different studies [7, 55, 56]. If there is associated vaginismus, it should be treated before a vestibulectomy is considered. It had been observed that women who respond to topical lidocaine prior to sex have a more successful outcome and hence, a failure to respond to lidocaine is considered as a relative contraindication to surgery [56]. Surgery can cause scarring and worsened pain and is reserved for women with localized symptoms who have not responded to conservative options. In a study by Bergeron et al., surgery achieved similar results in treating vestibulodynia when compared to behavioral therapy, sex education, partner therapy, and pelvic floor exercises but patients preferred behavioral approach [57].

Newer modalities (please refer to Chapter 26) Non**cosmetic uses of laser in aesthetic and** Regenerative Gynecology—vulvodynia, **Urinary incontinence,** infections**,** warts

Nowadays newer energy-based devices (EBDs) have shown significant results in treating vulvodynia patients. Editor has herself shown scientific evidence in her chapter to support the use of EBDs.

24.6 CO_2 Laser

Over recent years, few studies have been conducted evaluating the usefulness of LASER therapy in the treatment of vulvodynia. Laser-based treatments for managing cases of vulvodynia have been gaining popularity due to their minimal invasive nature and marked efficacy. In fact, laser therapy has reportedly been as effective as vestibulectomy, with significant response in >60% patients and symptomatic relief >90% of patients [58].

Laser therapy initiates a cascade of events that lead to disruption of vascular bed and thus promoting collagen remodeling without inducing any major anatomic alterations. Though its use for vulvodynia is not FDA approved yet.

24.7 Conclusion

Vulvodynia is a complex disorder of idiopathic vulvar pain with significant psychosexual attributes. Treatment is best provided through multidisciplinary approach tailored to individual needs.

Editor's remark: In an observational study done by Jindal et al. (2019), out of 29 patients who underwent CO_2 laser treatment for vulvodynia or vulval pain secondary to GSM, 79% showed significant improvement in pain scores and overall vestibular health index score with gradual improvement whereas 21% showed minimal or no response at all after completion of three laser sittings.

References

1. Bornstein J, Goldstein AT, Stockdale CK, et al. 2015 ISSVD, ISSWSH and IPPS consensus terminology and classification of persistent vulvar pain and vulvodynia. Obstet Gynecol. 2016;127(4):745–51.
2. Harlow BL, Stewart EG. A population-based assessment of chronic unexplained vulvar pain: have we underestimated the prevalence of vulvodynia? J Am Med Womens Assoc. 2003;58:82–8.
3. Siddall PJ, Cousins MJ. Persistent pain as a disease entity: implications for clinical management. Anesth Analg. 2004;99:510–20.
4. Reed BD, Harlow SD, Sen A, Edwards RM, Chen D, Haefner HK. Relationship between vulvodynia and chronic comorbid pain conditions. Obstet Gynecol. 2012;120:145–51.
5. Wesslemann U, Boham A, Foster D. Vulvodynia: current state of the biological science. Pain. 2014;155:1696–701.
6. Henzell H, Berzins K. Localised provoked vestibulodynia (vulvodynia). Assessment and Management. 2015;44(7):460–6.
7. American College of Obstetricians and Gynecologists' Committee on Gynecologic Practice; American Society for Colposcopy and Cervical Pathology (ASCCP). Committee opinion no 673: persistent vulvar pain. Obstet Gynecol. 2016;128(3):e78–84.
8. Labat JJ, Riant T, Robert R, et al. Diagnostic criteria for pudendal neuralgia by pudendal nerve entrapment (Nantes criteria). Neurourol Urodyn. 2008;27:306–10.
9. Haefner HK. Critique of new gynecologic surgical procedures: surgery for vulvar vestibulitis. Clin Obstet Gynecol. 2000;43(3):689–700.
10. Haefner HK, Collins ME, Davis GD, Edwards L, Foster DC, Hartmann ED, et al. The vulvodynia guideline. J Low Genit Tract Dis. 2005;9:40–51.

11. Lewis FM, Harrington CI. Use of magnetic resonance imaging in vulvodynia. J Reprod Med. 1997;42:169.
12. Schover LR, Youngs DD, Cannata RN. Psychosexual aspects of the evaluation and management of vulval vestibulitis. Am J Obstet Gynaecol. 1991;167:630–6.
13. Stewart D, Reicher A, Gerulath AH, Boydall K. Vulvodynia and psychological distress. Obstet Gynaecol. 1994;84:587–90.
14. Andrews JC. Vulvodynia: an evidence-based approach to medical management. JCOM. 2010;17:225–38.
15. Laudry T, Bergeron S, Dupuis MJ, Desrochers G. The treatment of provoked vestibulodynia: a critical review. Clin J Pain. 2008;24:155–71.
16. Mandal D, Nunns D, Byrne M, et al. Guidelines for the management of vulvodynia. Br J Dermatol. 2010;162:1180–5.
17. Munday P, Buchan A, Ravenhill G, et al. A qualitative study of women with vulvodynia: II. Response to a multidisciplinary approach to management. J Reprod Med. 2007;52:19–22.
18. Masheb RM, Kerns RD, Lozano C, et al. A randomized clinical trial for women with vulvodynia: cognitive-behavioral therapy vs. supportive psychotherapy. Pain. 2009;141:31–40.
19. Zolnoun DA, Hartmann KE, Steege JF. Overnight 5% lidocaine ointment for treatment of vulvar vestibulitis. Obstet Gynecol. 2003;102(1):84–7.
20. Foster DC, Kotok MB, Huang LS, et al. Oral desipramine and topical lidocaine for vulvodynia: a randomized controlled trial. Obstet Gynecol. 2010;116(3):583–93.
21. Bohm-Starke N, Brodda-Jansen G, Linder J, Danielsson I. The result of treatment on vestibular and general pain thresholds in women with provoked vestibulodynia. Clin J Pain. 2007;23:598.
22. Goldstein AT, Pukall CF, Brown C, et al. Vulvodynia: assessment and treatment. J Sex Med. 2016;13:572.
23. Burrows LJ, Goldstein AT. The treatment of vestibulodynia with topical estradiol and testosterone. Sex Med. 2013;1:30.
24. Nyirjesy P, Sobel JD, Weitz MV, Leaman DJ, Small MJ, Gelone SP. Cromolyn cream for recalcitrant idiopathic vulvar vestibulitis: results of a placebo controlled study. Sex Transm Infect. 2001;77(1):53–7.
25. Baron R. Capsaicin and nociception: from basic mechanisms to novel drugs. Lancet. 2000;356:785.
26. Sansone RA, Sansone LA. Pain, pain, go away: antidepressants and pain management. Psychiatry (Edgmont). 2008;5:16.
27. Reed BD, Caron AM, Gorenflo DW, Haefner HK. Treatment of vulvodynia with tricyclic antidepressants: efficacy and associated factors. J Low Genit Tract Dis. 2006;10:245–51.
28. Bates CM, Timmins DJ. Vulvodynia--new and more effective approaches to therapy. Int J STD AIDS. 2002;13:210.
29. Harris G, Horowitz B, Borgida A. Evaluation of gabapentin in the treatment of generalized vulvodynia, unprovoked. J Reprod Med. 2007;52:103.
30. Reed BD, Haefner HK, Cantor L. Vulvar dysesthesia (vulvodynia). A follow-up study. J Reprod Med. 2003;48:409.
31. Brown CS, Bachmann GA, Wan J, et al. Gabapentin for the treatment of vulvodynia: a randomized controlled trial. Obstet Gynecol. 2018;131:1000.
32. Pagano R, Wong S. Use of amitriptyline cream in the management of entry dyspareunia due to provoked vestibulodynia. J Low Genit Tract Dis. 2012;16:394.
33. Brown C, Bachmann G, Foster D, et al. Milnacipran in provoked vestibulodynia: efficacy and predictors of treatment success. J Low Genit Tract Dis. 2015;19:140.
34. Snyder DS, Unanue ER. Corticosteroids inhibit murine macrophage Ia expression and interleukin 1 production. J Immunol. 1982;129:1803.
35. Dede M, Yenen MC, Yilmaz A, Baser I. Successful treatment of persistent vulvodynia with submucous infiltration of betamethasone and lidocaine. Eur J Obstet Gynecol Reprod Biol. 2006;124:258–9.
36. Murina F, Tassan P, Roberti P, Bianco V. Treatment of vulvar vestibulitis with submucous infiltrations of methylprednisolone and lidocaine. An alternative approach J Reprod Med. 2001;46:713.
37. Cui M, Khanijou S, Rubino J, Aoki KR. Subcutaneous administration of botulinum toxin a reduces formalin-induced pain. Pain. 2004;107:125.
38. Beal BR, Wallace MS. An overview of pharmacologic management of chronic pain. Med Clin North Am. 2016;100:65–79.
39. Pelletier F, Parratte B, Penz S, et al. Efficacy of high doses of botulinum toxin a for treating provoked vestibulodynia. Br J Dermatol. 2011;164:617.
40. Dykstra DD, Presthus J. Botulinum toxin type a for the treatment of provoked vestibulodynia: an open-label, pilot study. J Reprod Med. 2006;51:467.
41. Petersen CD, Giraldi A, Lundvall L, Kristensen E. Botulinum toxin type A-a novel treatment for provoked vestibulodynia? Results from a randomized, placebo controlled, double blinded study. J Sex Med. 2009;6:2523.
42. Hartmann D. Chronic vulvar pain from a physical therapy perspective. Dermatol Ther. 2010;23:505–13.
43. Goldfinger C, Pukall CF, Gentilcore-Saulnier E, et al. A prospective study of pelvic floor physical therapy: pain and psychosexual outcomes in provoked vestibulodynia. J Sex Med. 2009;6:1955–68.
44. Murina F, Bianco V, Radici G, et al. Transcutaneous electrical nerve stimulation to treat vestibulodynia: a randomised controlled trial. Br J Obstet Gynaecol. 2008;115:1165–70.
45. Bergeron S, Brown C, Lord MJ, et al. Physical therapy for vulvar vestibulitis syndrome: a retrospective study. J Sex Marital Ther. 2002;28:183–92.
46. FitzGerald MP, Kotarinos R. Rehabilitation of the short pelvic floor. II: treatment of the patient with the short pelvic floor. Int Urogynecol J Pelvic Floor Dysfunct. 2003;14:269.
47. McKay E, Kaufman RH, Doctor U, et al. Treating vulvar vestibulitis with electromyographic biofeed-

back of pelvic floor musculature. J Reprod Med. 2001;46:337–42.
48. Murina F, Bernorio R, Palmiotto R. The use of amielle vaginal trainers as adjuvant in the treatment of vestibulodynia: an observational multicentric study. Medscape J Med. 2008;10:23.
49. Khandker M, Brady SS, Vitonis AF, et al. The influence of depression and anxiety on risk of adult onset vulvodynia. J Womens Health (Larchmt). 2011;20:1445.
50. Bergeron S, Khalifé S, Dupuis MJ, McDuff P. A randomized clinical trial comparing group cognitive-behavioral therapy and a topical steroid for women with dyspareunia. J Consult Clin Psychol. 2016;84:259.
51. Goldfinger C, Pukall CF, Thibault-Gagnon S, et al. Effectiveness of cognitive-Behavioral therapy and physical therapy for provoked Vestibulodynia: a randomized pilot study. J Sex Med. 2016;13:88.
52. Corsini-Munt S, Bergeron S, Rosen NO, et al. Feasibility and preliminary effectiveness of a novel cognitive-behavioral couple therapy for provoked vestibulodynia: a pilot study. J Sex Med. 2014;11:2515.
53. Brotto LA, Bergeron S, Zdaniuk B, Basson R. Mindfulness and cognitive behavior therapy for provoked vestibulodynia: mediators of treatment outcome and long-term effects. J Consult Clin Psychol. 2020;88:48.
54. Wu C, Goldstein A, Klebanoff JS, Moawad GN. Surgical management of neuroproliferative-associated vestibulodynia: a tutorial on vestibulectomy with vaginal advancement flap. Am J Obstet Gynecol. 2019;221:525.e1.
55. Tommola P, Unkila-Kallio L, Paavonen J. Surgical treatment of vulvar vestibulitis: a review. Acta Obstet Gynecol Scand. 2010;89:1385–95.
56. Kehoe S, Leusley D. An evaluation of modified vestibulectomy in the treatment of vulvar vestibulitis: preliminary results. Acta Obstet Gynecol Scand. 1996;75:676–7.
57. Bergeron S, Binik YM, Khalife' S, et al. A randomized comparison of group cognitive behavioral therapy, surface electromyographic biofeedback, and vestibulectomy in the treatment of dyspareunia resulting from vulvar vestibulitis. Pain. 2001;91:297–306.
58. Leclair CM, Goetsch MF, Lee KK, et al. KTP-nd: Yag laser therapy for the treatment of vestibulodynia, a follow up study. J Repro Med. 2007;52:53–8.

Non Energy Based Modalities in Cosmetic Gynaecology

25

Madhuri Agarwal and Sejal K. Shah

25.1 Introduction

The beautification of the female form is no longer restricted to the exposed areas like face, neck, arms, legs, but has gone beyond these to encompass the covered areas like the genitalia. Over the last decade or so, the demand for aesthetic and medical enhancement of the female genitalia has grown substantially, aided by the perceptions created by the social media.

This has shown a rise in demand due to women's increased inclination towards improvement of their genital appearance, namely 'Barbie doll look', in which the labia minora are narrow and not visible and the vagina opening appears very tight [1, 2].

25.2 Physiological Changes in the Female Genitals

Various extrinsic and intrinsic factors are responsible for ageing of the human body. Factors like ultraviolet radiation, pollution, smoking, poor nutrition, poor physical activity, stress, mainly contribute to fast ageing of the skin. In women, hormonal changes in the body, primarily menopause also contributes to the same [3].

As ovarian function diminishes with age, there is a decline in levels of collagen, elastin, hyaluronic acid, capillaries and subcutaneous adipose tissue leading to dried, rougher, firm, sagged skin and fine wrinkles [4]. Atrophic changes also occur in the genital area due to reduction in levels of oestrogen and androgen, leading to loss of function and also the aesthetic look, of both external and internal genitalia.

At advanced age, there is a decline in levels of oestrogen and lack of testosterone leading to loss of libido, vaginal dryness, vaginal infections and incontinence symptoms. In addition to this, there occurs loss of hyaluronic acid which makes labia slack and atrophied.

25.2.1 Indications

Due to the reasons mentioned above, women tend to seek medical advice for both cosmetic and medical reasons. Usually, the first-line approach to improve vaginal atrophy and dryness are non-invasive therapies like moisturizers, lubricants and hormone replacement medications or pelvic strengthening exercises.

A different approach in untreatable cases requires invasive, surgical procedures such as labiaplasty to alter the labia minora and majora or vaginoplasty, which involves surgery to the pelvic floor. Recently, however, newer non-invasive modalities have been introduced for

M. Agarwal (✉) · S. K. Shah
Mumbai, India
e-mail: drmadhuri@yavana.in

P. Jindal et al. (eds.), *Aesthetic and Regenerative Gynecology*,
https://doi.org/10.1007/978-981-16-1743-0_25

women wary of surgery due to the risk, expense and downtime involved. These treatments have more advantages, i.e.; less time consuming, minimal side effects, and patient compliance.

Some of the indications, where these non-surgical treatments could be considered are [5–7]:

1. Genitourinary syndrome of menopause (GSM)
2. Vulvovaginal laxity
3. Vulvovaginal atrophy
4. Sagging or discolouration of the labia majora
5. Decreased sensation during coitus
6. Early stress urinary incontinence

25.2.2 Evaluation for Procedures: [7]

- A thorough history taking, local examination and evaluation of other factors can help identify suitable candidates.
- History of urodynamic testing and voiding diary should be obtained.
- HRT, oral contraceptive use, active infection, sexually transmitted diseases and pregnancy history must be assessed.
- Detailed sexual history of both partners helps in understanding the cause root.
- Any absolute and relative contraindication should be excluded.
- A detailed examination should be performed by a trained medical professional to rule out pelvic organ prolapse.
- An informed consent should be taken from the patient, which explains all about procedure, safety profile and its side effects.

Patient-Reported Questionnaires: In addition to examination, diagnosis of GSM or Vaginal laxity largely depends on patient-reported symptomatology. Several self-reported questionnaires exist that assess individual perceptions of VL and vulvovaginal symptoms (Table 25.1) [7].

Table 25.1 Questionnaires used in NVR evaluation and studies

Questionnaire	Validated	Aim of questionnaire
Female Sexual Distress Scale Revised (FSDS-R)	Yes	Distress with sexual dysfunction
Female Sexual Function Index (FSFI)	Yes	Multiple domains of sexual function
Incontinence Impact Questionnaire (IIQ-7)	Yes	Impact of urinary leakage on quality of life
Sexual Satisfaction Questionnaire (SSQ)	No	Sexual satisfaction
Urogenital Distress Inventory (UDI-6)	Yes	Assesses frequency of urination, urgency and incontinence
Vaginal Laxity Questionnaire (VLQ)	No	Degree of patient reported vaginal laxity
Vulvovaginal Symptom Questionnaire (VSQ)	Yes	Quality of life impact from vulvovaginal symptoms, emotional and sexual concerns

Table 25.2 Contraindications for procedure

Absolute contraindications	Relative contraindications
Exclude vulvar, vaginal, cervical lesions	Trans vaginal mesh
Current pelvic malignancy	Mid-urethral sling mesh
Recent pelvic surgery	Prior pelvic irradiation
Active infection	

25.2.3 Contraindications

The American Urogynecologic Society (AUGS) (Table 25.2) compiled a clinical consensus statement based on rigorous criteria of the most important agreed upon expert opinion statements supported by the available literature as it pertains to the use of vulvovaginal EBDs [8].

Table 25.3 Different modalities for vulvovaginal rejuvenation

Energy-based devices	Aesthetic medicine	Regenerative medicine
CO_2 Laser	Fillers	PRP
Erb Yag Laser	Neurotoxin	Fat transfer
Diode Laser	Chemical peels	Nano Fat
Radio frequency	PDO threads	Stem cells
HIFU	Carboxy therapy	Bone marrow aspirate
	Ozone therapy	Bioidentical hormonal replacement therapy
	HBOT	Growth factors and cytokine-based applications

25.2.4 Modalities for Vulvovaginal Rejuvenation

The various non invasive modalities for vulvovaginal rejuvenation are enumerated in Table 25.3. We shall discuss only the non energy based modalities in this chapter.

25.2.5 Aesthetic Medicine

25.2.5.1 Fillers

Fillers are used in aesthetic gynecology for various indications:

1. Labia majora augmentation (labial puff)
2. Vulvovaginal atrophy
3. Genitourinary syndrome of menopause (GSM)

Different materials are used as fillers:

- Dermal fillers like Hyaluronic acid
- Fat—microfat, nanofat

Hyaluronic acid is a safe molecule to be used in the vulvovaginal area. It is naturally present in the body and acts as space filler and has absorbant and hygroscopic properties [9]. The injection of hyaluronic acid acts as space filler and by binding to water, it increases moisture and decreases dryness.

25.2.5.2 Labia Majora Augmentation:

De Lorenzi et al. described the use of hyaluronic acid injections to correct the physical appearance of the mons pubis, labia minora or labia majora [10, 11]. Labial fillers are used to increase the volume and outlook of labia majora. It provides more youthful appearance, while augmentation of the mons pubis could improve discomfort during intercourse reducing trauma along the pubic bone [11, 12].

Treatment protocol: 1–2 ml cross-linked, non-animal stabilized HA 19 mg/21 mg HA injected per side of labia with an 18G/20G cannula. Injection lignocaine 2% can be infiltrated at entry point. The filler is injected in the area between the lip dartos and the fibrous tunic layer [11]. Filler can be topped up after 6 months if needed.

Treatment outcome: 'Labial puff' treatment helps to volumize the labia majora, thereby concealing the labia minora in order to give a more young looking appearance. It aims to restore tone and elasticity, strengthen the intra-vaginal muscles, and improve sensitivity, while also reducing mucosal dryness [11, 12].

Adverse effects: Pain, tenderness, infection, haematoma and uneven labial symmetry. Other side effects include nerve paralysis, bleeding and loss of sensation during sex, due to the abundance of nerves and blood vessels around the clitoris, labia and urethra.

HA fillers for Genitourinary syndrome: Hyaluronic acid fillers have also been tried in Genitourinary syndrome in post and pre menopausal women [13].

Treatment protocol: Cross-linked hyaluronic acid filler, 1–2 ml is injected deep intradermal all around the vulvar vestibule after anaesthetizing with topical lidocaine-prilocaine cream and injection lignocaine 2%. The area is gently massaged to prevent lumps.

Treatment outcome: Studies have shown that there is some improvement in the symptoms and the quality of life post-treatment with HA fillers in GSM [13].

25.2.6 G-Shot

The "G-Shot" is a trademarked term that refers to the injection of filler into the G-Spot, repeated every 3–5 months, to improve female sexual gratification during intercourse by increasing friction. The Grafenberg spot, or G-spot, is a term coined by Addiego et al. in 1981 to credit Dr. Ernest Grafenberg who is believed to have first described an erogenous zone on the anterior vaginal wall, located 1–2 cm from the urethra in 1950. There is significant debate in the medical literate as to whether the G-spot actually exists or plays a critical role in female sexual satisfaction [14].

25.3 Fat Transfer

Labia majora augmentation by autologous fat transfer enhances the volume, shape, symmetry, firmness and contour of atrophied labia majora. Microfat is used to enhance trophicity as well as for filling. A study by Kim et al. reported that injecting PRP and lipofilling was effective in a patient with vulvovaginal atrophy and lichen sclerosus. Microfat and nanofat grafting seems to offer significant and long-lasting improvement with one treatment and is an autologous procedure. Moreover, they naturally integrate into host tissues, and there are no major side effects [15].

Vulvovaginal lipo-filling of the posterior vaginal wall injects hyaluronic acid and PRP subcutaneously in the perineum [16]. Microfat injected into the posterior and lateral vaginal walls can also help in vaginal tightening [17]. It also helps reduce vaginal dryness and dyspareunia.

25.4 Botulinum Toxin

Botulinum Toxin A (BoNT-A) injections are used to treat various gynecological conditions such as chronic pelvic pain, vaginism, dyspareunia, overactive bladder and other pelvic neuropathies. The dosage of BoNT-A used usually ranges from 40 U to 400 U in a single administration. In refractory cases, repeated injection cycles may be necessary, depending on the symptomatology [18–21].

25.4.1 Regenerative Medicine

25.4.1.1 Platelet-Rich Plasma Therapy

Platelet-Rich Plasma (PRP) application has already gained popularity in the treatment of various medical and aesthetic concerns. Autologous PRP is derived from an individual's whole blood, and then centrifuged to separate plasma and red blood cells. Extracted plasma has a higher growth factor concentration 5–10 fold higher than whole blood that promotes tissue [22, 23]. PRP has been utilized by aesthetic gynecologists in treatments such as vaginal rejuvenation and O-shot therapy. Based on this mechanism of action, PRP has been used in various gynecological and dermatological autoimmune disorders, one of which being Lichen Sclerosus.

PRP has also been used in labial augmentation, breast reconstruction [24, 25].

25.4.1.2 O-Shot Therapy

The O-shot, or 'orgasm shot', is used for female sexual dysfunction and vaginal rejuvenation procedure in which platelet-rich plasma (PRP) is injected into the clitoris and the upper vaginal wall. It improves both urinary incontinence and sexual dysfunction by using your own growth factors [15, 26, 27].

25.4.1.3 PRP in Vaginal Rejuvenation

PRP has been used for vaginal regeneration in postmenopausal women to treat vaginal dryness, improves lubrication and help in improving vagina senstivity.The ligaments and muscles supporting the urethra become stronger, alleviating urinary incontinence. Kim et al. reported the use of PRP in a case of vaginal rejuvenation. They concluded that the application of autologous lipofilling mixed with PRP in a patient with vaginal atrophy produced relief of symptoms and contour restoration. The rejuvenated appearance of the external genitalia provided a pleasing cosmetic outcome to the patient [28].

25.4.2 Stem Cells

The potential use of stem cell-based therapies for the repair and regeneration of various tissues and organs is an upcoming modality in plastic and reconstructive surgery. Among the adult mesenchymal stem cell population, adipose-derived stem cells (ADSC) have the potential to differentiate the mesenchymal, ectodermal and endodermal lines and are easy to harvest. Additionally, adipose tissue yields a high number of ADSC per volume of tissue 29, [30].

It helps in the treatment of physiological and pathological vulvovaginal dystrophies, vaginal atrophy and SUI. There is a reduction of the typical pain and burning sensation associated with vulvovaginal atrophy, thanks to the immunomodulatory properties of the ADSCs [31, 32].

The fat is harvested under tumescent anaesthesia from the lower abdomen or thigh or buttock area. This is then centrifuged, washed and processed to obtain ADSCs/stromal vascular fraction, which is then injected in the concerned areas. ADSCs treatment leads to the reacquisition of vaginal and periurethral tonicity, thus solving problems related to the hypotrophic involutive processes affecting the genitourinary tract. Combination treatments like stem cell-enriched lipofilling along with hyaluronic acid-based filler which provides a base for the cells to grow is also a good option.

25.4.3 Chemical Peels:

Alpha hydroxyl acid peels, TCA peels, Retinol-based peels, peels with lightening agents like kojic acid and glutathione are routinely used for the face and other body parts. They exfoliate the superficial dead skin layer and help regenerate fresh cells. These have also been tried on the vulva-mons pubis, labia majora, inguinal area to lighten these areas and give a smoother, less wrinkled appearance.

25.4.4 Other Therapies

Various other treatments are being used for vulvovaginal rejuvenation like PDO threads, carboxytherapy and ozone therapy. In regenerative medicine, upcoming treatments like Bone marrow aspirate concentrate infiltration, growth factors and cytokine-based applications, Bio-identical hormone replacement therapy etc. are being studied for their use in vulvovaginal atrophy.

25.5 Conclusion

With increasing awareness, the demand for vulvovaginal rejuvenation is gradually increasing. Currently, there is no approved device or modality available for the aesthetic gynecology indications, but there are many in the pipeline. Larger, double-blinded, randomized control trials are required to establish protocols and standardize these devices as preventive, first-line treatments.

References

1. Schick VR, Rina BN, Calabrese SK. Vaginal rejuvenation: current perspectives. Int J Women's Health. 2017;9:9.
2. Evaluation: the portrayal of women's external genitalia and physique across time and the current Barbie doll ideals. J SexRes. 2010; 48(1):74-81.
3. Bajurna B, Marcinkowski JT. The role of cosmetic gynecology treatments in women in perimenopausal period Anna Galęba. Open J Nurs. 2015;5:153–7. Published Online February 2015 in SciRes. http://www.scirp.org/journal/ojn https://doi.org/10.4236/ojn.2015.52018.
4. Galęba, A.. Evaluation of quality life of patients before and after selected treatments from aesthetic medicine. Doctoral Dissertation, Poznan University of Medical Sciences, Poznan; 2011.
5. Yona Tadir MD. Light and energy based therapeutics for genitourinary syndrome of menopause: consensus and controversies. Lasers Surg Med. 2017;49(2):137–59.
6. Karcher C, Sadick N. Vaginal rejuvenation using energy-based devices. Int J Womens Dermatol. 2016; 2(3): 85–8. Published online 2016 Jun 21. https://doi.org/10.1016/j.ijwd.2016.05.003.
7. Qureshi AA, Tenenbaum MM, Myckatyn TM. Nonsurgical vulvovaginal rejuvenation with

radiofrequency and laser devices: a literature review and comprehensive update for aesthetic surgeons. Aesthet Surg J. 2018;38(3):302–11. https://doi.org/10.1093/asj/sjx138.
8. Alshiek J. Augs clinical consensus statement. Vaginal energy-based devices. Female Pelvic Med Reconstr Surg. 2020;26:287–98.
9. Papakonstantinou E, Roth M, Karakiulakis G. Hyaluronic acid: a key molecule in skin aging. Dermatoendocrinolv. 2012;4(3)
10. De Lorenzi F, Mascolo E, Albani F, Sideri M. Female Genital Surgery. In: Alam M, Pongprutthipan M, eds. Body rejuvenation. Springer; 2010:204-205.
11. Fasola E, Gazzola R. Aesthet Surg J. 2016;36(10):1155–63.
12. Barnes T. Using dermal fillers for vaginal skin rejuvenation. Journal of Aesthetic Nursing. 2018;7(9)
13. Garavaglia E, Sala C, Busato M, Bellia G, Tamburlin N, Massirone A. First use of thermal stabilized hyaluronic acid injection in one-year follow-up patients with genitourinary syndrome. Med Devices (Auckl). 2020;13:399–410.
14. Vanaman. Emerging trends in nonsurgical female genital rejuvenation. Dermatol Surg. 42(9):1019–29.
15. Menkes S, SidAhmed-Mezi M, Meningaud JP. Microfat and Nanofat grafting in genital rejuvenation. Aesthet Surg J. 2020:sjaa118.
16. Aguilar P, Hersant B, SidAhmed-Mezi M, et al. Novel technique of vulvo-vaginal rejuvenation by lipofilling and injection of combined platelet-rich-plasma and hyaluronic acid: a case-report. Springerplus. 2016;5:1184.
17. Triana L. Non-surgical vaginal tightening procedures. Aesthetic Vaginal Plastic Surgery. 2020;
18. Moga MA, Dimienescu OG, Bălan A, Scârneciu I, Barabaş B, Pleş L. Therapeutic approaches of botulinum toxin in Gynecology. Toxins (Basel). 2018;10(4):169.
19. Anbazhagan A. Nonurological uses of botulinum toxin in gynaecology. MRCOGRalph Roberts MD FRCP MRCOG.
20. Mahajan ST, Brubaker L. Botulinum toxin: From life-threatening disease to novel medical therapy. Botulinum toxin in the treatment of refractory vaginismus. Obstetr Gynecol. 2004;104(5 Pt 1):922–5. Non -Surgical Aesthetic Procedures In Gynaecology.
21. Dawood AS, Salem HA. Current clinical applications of platelet-rich plasma in various gynecological disorders: an appraisal of theory and practice. Clin Exp Reprod Med. 2018;45(2):67–74.
22. Everts PA, Hoogbergen MM, Weber TA, Devilee RJ, van Monftort G, de Hingh IH. Is the use of autologous platelet-rich plasma gels in gynecologic, cardiac, and general, reconstructive surgery beneficial? Curr Pharm Biotechnol. 2012;13:1163–72.
23. Behnia-Willison F, Pour NR, Mohamadi B, Willison N, Rock M, Holten IW, et al. Use of platelet-rich plasma for vulvovaginal autoimmune conditions like lichen sclerosus. Plast Reconstr Surg Glob Open. 2016;4:e1124.
24. Salgarello M, Visconti G, Rusciani A. Breast fat grafting with platelet-rich plasma: a comparative clinical study and current state of the art. Plast Reconstr Surg. 2011;127:2176–85.
25. Runels C, Melnick H, Debourbon E, Roy L. A pilot study of the effect of localized injections of autologous platelet rich plasma (PRP) for the treatment of female sexual dysfunction. J Womens Health Care. 2014;3:169.
26. Jain A, Bedi RK, Mittal K. Platelet-rich plasma therapy: a novel application in regenerative medicine. Asian J Transfus Sci. 2015;9:113–4.
27. Kim SH, Park ES, Kim TH. Rejuvenation using platelet-rich plasma and lipofilling for vaginal atrophy and lichen sclerosus. J Menopausal Med. 2017;23:63–8.
28. Naderi N, Combellack EJ, Griffin M, Sedaghati T, Javed M, Findlay MW, Wallace CG, Mosahebi A, Butler PE, Seifalian AM, Whitaker IS. The regenerative role of adipose-derived stem cells (ADSC) in plastic and reconstructive surgery. Int Wound J. 2017;14(1):112–24.
29. Mehrabani D, Mehrabani G, Zare S, Manafi A. Adipose-derived stem cells (ADSC) and aesthetic surgery: a Mini review. World J Plast Surg. 2013;2(2):65–70.
30. Giuseppina Onesti M, Carella S, Ceccarelli S, Marchese C, Scuderi N. The use of human adipose-derived stem cells in the treatment of physiological and pathological vulvar dystrophies. Stem Cells Int. 2016;2016:2561461.
31. Stark L, Razzaque M, Yoon J, Aref-Adib M, Banwell M, Beski S. Safety and feasibility of autologous Micro-fragmented adipose tissue injections for the treatment of vaginal atrophy, vulvovaginal dystrophy, and stress urinary incontinence: an observational case series. Urology. 2020;
32. Desai SA, Kroumpouzos G, Sadick N. Vaginal rejuvenation: From scalpel to wands. Int J Womens Dermatol. 2019;5(2):79-84. Published 2019 Mar 7.

Genitourinary Syndrome of Menopause

26

Preeti Jindal, Isha Kundal, Ridhi Kalra, and Sumitu Sahdev

26.1 Introduction

Modern medicine has increased the longevity. The average life expectancy for a woman in developed countries is expected to be around 80 years and in developing countries 70 years. If we assume that the average for menopause is 50 years, then most women can expect about 30 years of postmenopausal life. With increasing years added to life, there are more chances for women to suffer from somatic and psychological symptoms due to aging and estrogen deficiency. It is very important to offer these women a good quality of life and not only quantity.

The **genitourinary syndrome of** menopause (GSM) is a new term coined in 2013 by—The International Society for Study of women's Sexual Health and the North American Menopause Society; that describes various menopausal symptoms and signs. It includes genital somatic symptoms (vaginal dryness, burning, and irritation), sexual symptoms (lack of lubrication, discomfort, or pain during intercourse), and urinary symptoms (urgency, dysuria, and recurrent urinary tract infections) that negatively affect the quality of life of postmenopausal women [1].

It is estimated that more than half of postmenopausal women will experience GSM, in one or another way. Of these, one-third of women avoid seeking medical advice as they feel too embarrassed to talk about their symptoms. Further more vaginal atrophy is often not recognized by many clinicians and these symptoms are brushed under the carpet adding to the distress of patients suffering with them.

GSM results from the effect of estrogen deficiency during postmenopausal lifespan. This includes a spectrum of symptoms like vaginal dryness, burning, and irritation; urinary symptoms like dysuria, urgency, stress urinary incontinence (SUI), recurrent urinary tract infections (UTIs), and vaginal infections; vaginal dryness and vulvodynia. Physical changes and signs are varied [2].

Vulvovaginal atrophy (VVA) is also a component of GSM which only describes about the changes in the genital tissue but not the associated symptoms.

GSM is a hypoestrogenic state associated with the onset of natural menopause. Other hypoestrogenic states occur in primary ovarian insufficiency (POI), surgical menopause (bilateral oophorectomy with or without hysterectomy), hypothalamic amenorrhea, the postpartum state and breastfeeding, use of gonadotropin-releasing hormone (GnRH) agonists or aromatase inhibitors (AIs), and cancer treatments such as surgery, pelvic radiation therapy, or chemotherapy that render ovaries inactive, either temporarily or permanently.

P. Jindal (✉) · I. Kundal · R. Kalra · S. Sahdev
The Touch, Advanced Obstetrics, IVF & Cosmetic Gynae Centre, Mohali, India

P. Jindal et al. (eds.), *Aesthetic and Regenerative Gynecology*,
https://doi.org/10.1007/978-981-16-1743-0_26

26.1.1 Anatomy and Physiology

Due to common embryological origin, female genitalia and lower urinary tract share common distribution of estrogen receptors. Hence, hypoestrogenism has both vulvovaginal and urological effects [3].

After menopause, the number of estrogen receptors continues to decrease but never fully disappear. Therefore, in the presence of exogenous administration of estrogen, one can replenish lost estrogen receptors [4]. Estrogen deficiency also reduces the collagen, elastin, and hyaluronic acid in the vaginal layers leading to thinning of epithelial layers, lack of elastin leads to altered functioning of smooth muscle cells hence dyspareunia.

The vaginal and urethral epithelium is lined with nonkeratinized stratified squamous epithelium which inhabits a variety of organisms, namely Lactobacilli. Lactobacillus helps in the conversion of glycogen to lactic acid thus making vaginal pH acidic, i.e., 3.5–4.5. The acidity of the vagina provides a natural defence mechanism against urinary tract infections (UTI) and vaginitis, discouraging the growth of pathogenic bacteria and infection [5].

In a postmenopausal state, hypoestrogenism leads to a decrease in Lactobacilli dominance as atrophy of superficial epithelial layers (Fig. 26.1) of the vagina causes thinning of vaginal epithelium. The decreased cells produce less glycogen for lactobacilli to convert to lactic acid. This also decreases the number of Lactobacilli and also increases vaginal pH. The higher pH disturbs the natural defence mechanism which impairs the viability of healthy vaginal flora and promotes the overgrowth of organisms causing recurrent vaginal infections, UTI, and inflammation [6]. It starts the vicious cycle of vaginal alkalinity, infection, vaginal atrophy more thinning, less Lactobacilli, and more rise in vaginal pH. A direct relationship between vaginal estrogenization and candidiasis has been reported in postmenopausal women [7].

Vaginal thinning also reduces vaginal elasticity which is accompanied by decreased secretions from Bartholin glands, leading to dyspareunia and dryness. The bladder and urethra also become atrophic, causing urinary incontinence

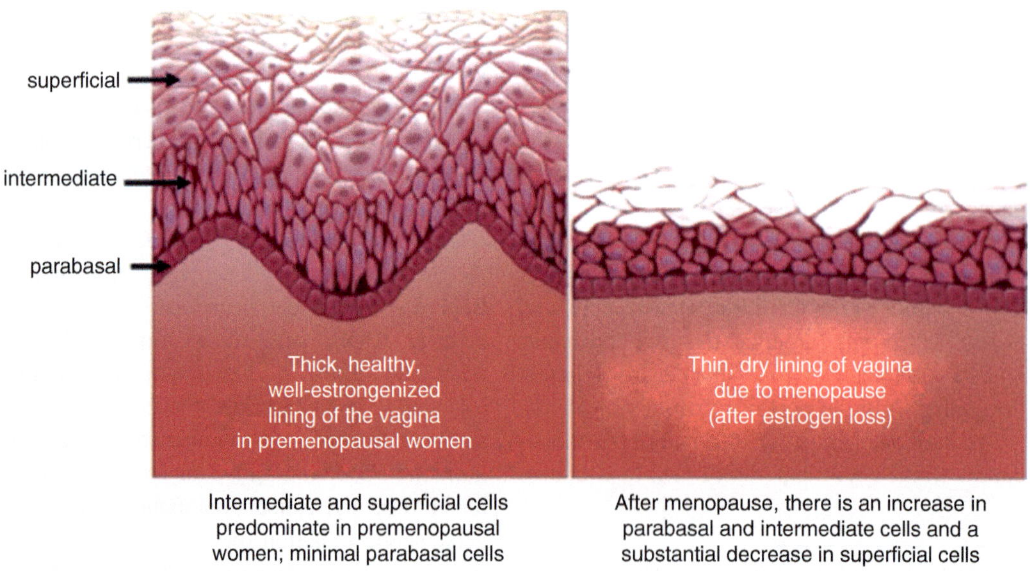

Fig. 26.1 Vaginal atrophy pathophysiology: Cellular changes

and frequency. One study reported that 20% of postmenopausal women experienced urge incontinence while roughly 50% experienced stress urinary incontinence [8].

26.2 Clinical Manifestations

Most women avoid reporting GSM symptoms to clinicians, thus making this problem very challenging to treat. Thorough history taking and examination are required to diagnose this problem. The clinical manifestations are divided into external genital, urological and sexual symptoms of GSM (Table 26.1). Common signs and symptoms include vaginal dryness, dyspareunia, vaginal itching, abnormal discharge, urinary incontinence, recurrent vaginal and urinary tract infections.

26.2.1 Assessment

A detailed history taking and local examination should be performed to rule out any other pelvic pathology. Pelvic USG for uterus, tubes, ovaries, Pap smear, and mammogram above the age of 40 years will help in ruling out major pathologies, infections, cancer, etc. Acknowledging that GSM symptoms are common in 33% of that specific age group, direct questions can be asked which helps in diagnosing and treating the problem. Identify and document the onset of symptoms, duration, relieving and aggravating factors, and their impact on the quality of life of women. Doctors should also emphasize on patient's psychological and sexual history as well.

26.2.2 Treatment

Our aim of treating GSM is to provide symptomatic relief and improving the quality of life of women. Various treatment options can be offered to these women to alleviate their suffering.

Lifestyle modifications like personal hygiene practices, diet, exercise counseling, smoking cessation, reducing alcohol consumption. Women should be advised to refrain from habits of vaginal douching, washing vulva frequently as wetness

Table 26.1 External genital, urological, and sexual manifestation of genitourinary syndrome of menopause

TABLE 1

External genital, urological, and sexual manifestations of genitourinary syndrome of menopause

External genital		Urological		Sexual
Signs and symptoms	**Complications**	**Signs and symptoms**	**Complications**	**Signs and symptoms**
Vaginal/pelvic pain and pressure	Labial atrophy	Frequency	Ischemia of vesical trigone	Loss of libido
Dryness	Vulvar atrophy and lesions	Urgency	Meatal stenosis	Loss of arousal
Irritation/burning	Atrophy of Bartholin glands	Postvoid dribbling	Cystocele and rectocele	Lack of lubrication
Tenderness	Intravaginal retraction of urethra	Nocturia	Urethral prolapse	Dyspareunia
Pruritus vulvae	Alkaline pH (5–7)	Stress/urgency incontinence	Urethral atrophy	Dysorgasmia
Decreased turgor and elasticity	Reduced vaginal and cervical secretions	Dysuria	Retraction of urethral meatus inside vagina associated with vaginal voiding	Pelvic pain
Suprapubic pain	Pelvic organ prolapse	Hematuria	Uterine prolapse	Bleeding or spotting during intercourse
Leukorrhea	Vaginal vault prolapse	Recurrent urinary tract infection	Urethral polyp or caruncle	
Ecchymosis	Vaginal stenosis and shortening			
Erythema	Introital stenosis			
Thinning/graying pubic hair				
Thinning/pallor of vaginal epithelium				
Pale vaginal mucous membrane				
Fusion of labia minora				
Labial shrinking				
Leukoplakic patches on vaginal mucosa				
Presence of petechiae				
Fewer vaginal rugae				
Increased vaginal friability				

Gandhi. Genitourinary syndrome of menopause. Am J Obstet Gynecol 2016.

predisposes to infections, avoiding chemicals in vulval areas as it alters vaginal pH, thus increasing the risk of recurrent vaginal infections.

Cognitive behavior therapy: Sexual counseling and stress reduction therapy help in immediately improving the quality of life.

Nonpharmacologic therapies: Lubricants and moisturizers are first-line therapies for less severe symptoms. It helps in alleviating symptoms of vaginal dryness and painful intercourse. Both lubricants and moisturizers also contain a wide range of additional ingredients such as humectants, emollients, and preservatives which regulate viscosity, pH of vagina and thus prevent bacterial and fungal infections.

Pharmacologic therapies include low-dose vaginal estrogens in form of gel, suppositories or transdermal estrogen, vaginal DHEA inserts, and oral ospemifene. Low-dose vaginal estrogens are preferred over systemic estrogen therapy due to their safety and effectiveness.

Hormonal replacement therapy (HRT): For women with moderate to severe dyspareunia, transdermal and oral Hormonal replacement therapy are effective options until and unless contraindicated. In addition to the above indications, HRT can also be used for additional menopause symptoms like hot flashes and risk of low bone density.

Hyaluronic acid: Hyaluronic acid (HA) is a commonly found material in the body that is added to many skincare and wound-healing products because of its hygroscopic properties. As we age, the hyaluronic acid our body naturally produces starts to deplete. So, synthetic HA can act as a lubricant and moisturizer. In gynecology, HA has been widely used as an important ingredient for conditions such as dyspareunia. It may be used in labial majora augmentation as thin majora may lead to dyspareunia due to the absent cushioning effect.

Energy-based devices (EBD): Newer technologies such as laser, electromagnetic waves (HIFEM), Radiofrequency (RF) and high intensity-focused ultrasounds (HIFU) are noninvasive treatment modalities for GSM that focus energy in the vaginal wall, thereby heating the targeted tissue at various depths without causing any pain and with no downtime [9].

The basic mechanism of energy-based devices aims to heat the connective tissue of the vaginal wall, thus raising local temperature up to 40 °C to 42 °C, leading to neocollagenesis and growth factor infiltration that ultimately revitalizes and restores the elasticity and moisture of the vaginal mucosa [10].

The different energy-based devices have different mechanisms of raising the temperature. In laser-based devices, the energy of accelerated electrons is used. In HIFEM, electromagnetic vibrations are used. In HIFU, focused ultrasound waves are used to raise the temperature. In radiofrequency, the frequency of vibrating particles is used.

Vulvovaginal energy-based devices including lasers (fractional CO_2, Erbium: YAG, diode, etc.) and radio-frequency devices are under investigation as treatments for GSM, but none have FDA approval for this indication (till date of writing of this article). Ironically, fractional CO_2 laser has demonstrated safety and efficacy in tissues of the skin, face, and neck [11–14]. It is believed that it is a matter of time when these devices will be approved for GSM also.

In our private setup, more than 150 CO_2 laser procedures in last 2 years for women affected with GSM symptoms were performed; it has been seen that there was a significant decrease in the occurrence of GSM symptoms and improvement in female sexual function index (FSFI) and vaginal health index scores (VHIS).

Lets revise them one by one:

(a) **Laser** (Ablative **and Non-Ablative Laser):**

It is suggested that microablative fractional CO_2 laser helps in restoring vaginal flora with treatment involving three to five sessions. Several small studies and round table discussions have reported that laser intervention with intravaginal use is a safe and potentially effective nonpharmacologic intervention for GSM [15, 16].

26.2.3 Principle

The primary role of using the nonablative laser technology is to achieve new collagen remodeling and elastic fibers recreation in the tissues under the mid-urethra. Their penetration depth of 0.03 mm is very safe as vaginal mucosa is 4 to 6 mm thick. With the application of laser, thermal response is generated within the tissues of the mid-urethra (Fig. 26.2). The cellular reaction is determined by the production of heat shock proteins (HSP) which play a major role in transforming the TGF-beta growth factor. The recruitment of fibroblasts initiates neocollagenesis. As a result of the temperature increase, intermolecular cross-links that stabilize collagen triple-helix structure are broken, which leads to the shrinkage of collagen fibrils and improvement in tissue firmness [17] (Fig. 26.3).

(b) **Radiofrequency (RF):**
In gynecology, monopolar radiofrequency is commonly used with two electrodes. The aim of treatment is to generate an electrical field resulting in the creation of heat delivered to tissue to approximately 40–45 °C for a defined treatment time, with each zone being treated for 3–5 minutes, for a total time of 25–30 minutes per treatment session. These sessions may be repeated at 4–6 week intervals [18].

At a local temperature of 40–45 °C, the inflammation cascade is initiated that results in the re-tensioning of collagen fibers and recruitment of fibroblasts to create new collagen and elastin fibers. The generation of heat in the local tissues further promotes blood vessels proliferation which improves the nutrition and oxygenation of skin, as well as its metabolism, thus making the skin and mucosa firmer.

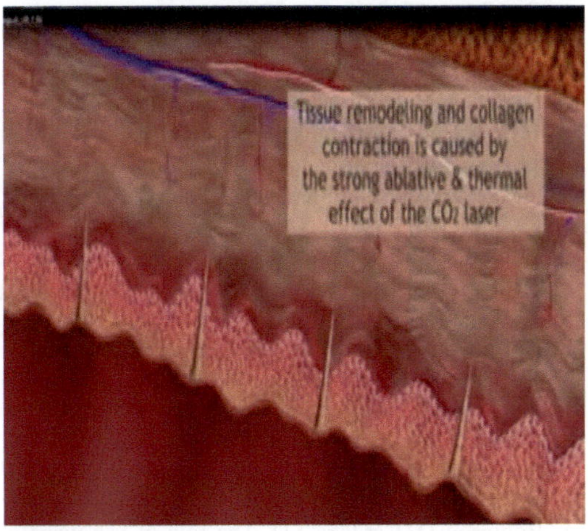

Fig. 26.2 Thermal and ablative effect on vaginal mucosa © Femopase et al.

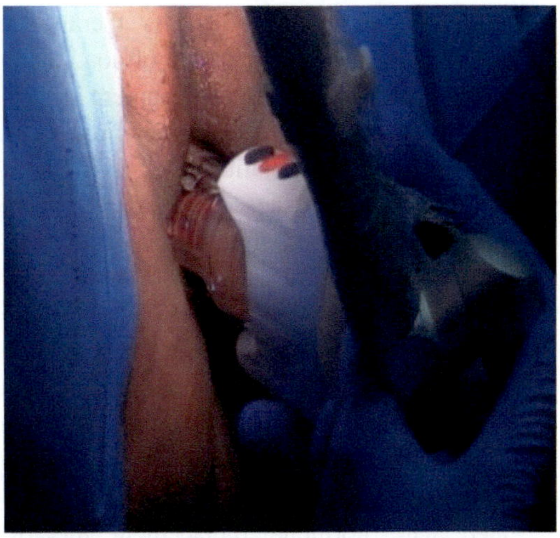

Fig. 26.3 Application of CO_2 laser in the vagina © Jindal et al.

(c) **High—Intensity Focused Ultrasound (HIFU):**
In pursuit of developing a more effective approach to treat GSM & SUI, a transvaginal and transurethral ultrasound energy source that can target tissue regions adjacent to the mid-urethra was required. HIFU applicators aim to precisely target heating the endopelvic fascia and surrounding tissues along the mid-urethral zone, with the potential to generate tissue stiffening and remodeling. Thermal treatment that tightens this tissue may produce a biologic "hammock" by remodeling collagen and connective tissue and increasing hydrostatic pressure. This leads to vaginal tightening and rejuvenation and improvement in SUI symptoms. The treatment protocol includes two sessions at an interval of 4–6 weeks. No anesthesia is required. To be repeated after a year. Positive outcome has been seen in GSM, mild SUI, and improved sexual satisfaction. No significant side effects reported.

(d) **Carboxytherapy:** Carboxytherapy, also known as carbon dioxide therapy, is an FDA-approved popular treatment for many diseases. Its usage in medicine has a long tradition since the 1950s. Its therapeutic effect uses principle of oxygen off loading by hemoglobin in CO_2 rich environment is associated with improved microcirculation, angiogenesis, vasodilation, and as a consequence better tissue oxygenation. There occurs improvement of pelvic floor microvasculature due to stimulation of neovascularization and neocollagenases. Stimulation of the pelvic floor blood flow contributes to the vaginal microflora normalization and improvement in vaginal pH in patients with vaginal dryness. Carboxytherapy slows down physiological aging without the risks associated with hormone replacement therapy, thereby improving the quality of life. Its safety profile allows it to be used as an alternative prevention and treatment option in patients who cannot use other treatment modalities [19].

(e) **Platelet-rich plasma (PRP)**: Autologous PRP is derived from whole blood that is subjected to centrifugation which separates plasma from red blood cells. The obtained plasma consists of a higher concentration of growth factors (PDGF, TGF, IL, PDAF, VEGF, EGF, IGF), five- to tenfolds greater than whole blood which modulate cell proliferation, differentiation, angiogenesis, and chemotaxis. The principle behind this treatment modality was derived from the natural healing process wherein injury to any tissue attracts platelets and stem cells to the site of injury leading to remodeling and healing of tissue.

Currently, PRP has become popular as a nonoperative treatment option for a broad spectrum of gynecological disorders ranging from facial to ovarian to uterine rejuvenation. Based on the experience of using PRP to improve smoothness and to decrease wrinkles in the facial area, aesthetic practitioners have also used PRP for the regeneration of the vaginal mucosa and skin in patients with vaginal atrophy and symptoms of dyspareunia. Following PRP injection, vaginal vascularity seems to increase with a subsequent improvement of sexual function. In addition, the mucosa becomes thicker and firmer, making the vagina looks much more youthful [20, 21].

26.3 Conclusion

GSM is a chronic complex syndrome of multiple changes in the genitourinary tissues in response to estrogen deficiency developing with the onset of menopause. Previously treatments included simple measures that provided less relief. Nowadays, technology has given safe, noninvasive and effective options. Newer noninvasive modalities can help in curing the patients with GSM symptoms in very patient-friendly ways, thus improving the quality of life. These modalities are painless, non- invasive, user friendly procedures with quick results. Hence these can be offered to patients suffering with GSM.

References

1. Kim H-K, et al. The recent review of the genitourinary syndrome of menopause. J Menopausal Med. 2015;21(2):65–71. https://doi.org/10.6118/jmm.2015.21.2.65.
2. Faubion SS, et al. The Genitourinary syndrome of menopause position statement of the North American Menopause Society. Menopause J NAMS. 2020;27(9):976–92. https://doi.org/10.1097/GME.0000000000001609.
3. Goldstein I. Recognizing and treating urogenital atrophy in postmenopausal women. J Womens Health (Larchmt). 2010;19:425–32.
4. Palacios S. Managing urogenital atrophy. Maturitas. 2009;63:315–8.
5. North American Menopause Society. The role of local vaginal estrogen for treatment of vaginal atrophy in postmenopausal women: 2007 position statement of the North American Menopause Society. Menopause. 2007;14:355–69. quiz 370-1
6. Willhite LA, O'Connell MB. Urogenital atrophy: prevention and treatment. Pharmacotherapy. 2001;21:464–80.
7. Hoffmann JN, You HM, Hedberg EC, Jordan JA, McClintock MK. Prevalence of bacterial vaginosis and Candida among postmenopausal women in the United States. J Gerontol B Psychol Sci Soc Sci. 2014;69(Suppl. 2):S205–14. https://doi.org/10.1093/geronb/gbu105.
8. Robinson D, Cardozo LD. The role of estrogens in female lower urinary tract dysfunction. Urology. 2003;62(Suppl):45–51.

9. Tadir Y, Gaspar A, Lev-Sagie A, Alexiades M, Alinsod R, Bader A, et al. Light and energy based therapeutics for genitourinary syndrome of menopause: consensus and controversies. Lasers Surg Med. 2017;49:137–59.
10. Karcher C, Sadick N. Vaginal rejuvenation using energy-based devices. Int J Women's Dermatol. 2016 May;2:85–8.
11. Ong MW, Bashir SJ. Fractional laser resurfacing for acne scars: a review. Br J Dermatol. 2012;166:1160–9.
12. Tierney EP, Hanke CW. Fractionated carbon dioxide laser treatment of photoaging: prospective study in 45 patients and review of the literature. Dermatol Surg. 2011;37:1279–90.
13. Tierney EP, Hanke CW. Ablative fractionated CO2, laser resurfacing for the neck: prospective study and review of the literature. J Drugs Dermatol. 2009;8:723–31.
14. Peterson JD, Goldman MP. Rejuvenation of the aging chest: a review and our experience. Dermatol Surg. 2011;37:555–71.
15. Angelou K, Grigoriadis T, Diakosavvas M, Zacharakis D, Athanasiou S. The genitourinary syndrome of menopause: an overview of the recent data. Cureus. 2020;12 4:e7586. Published 2020 Apr 8. https://doi.org/10.7759/cureus.7586.
16. Menachem A, Alexander B, Martinec KA, et al. The effect of vaginal CO_2 laser treatment on stress urinary incontinence symptoms. Alma Surgical. 2016;
17. Eric S, Rovner MD, Alan J, Wein MD. Treatment Options for Stress Urinary Incontinence. Rev Urol. 2004;6(Suppl 3):S29–47.
18. Gold M, Andriessen A, Bader A, et al. Review and clinical experience exploring evidence, clinical efficacy, and safety regarding nonsurgical treatment of feminine rejuvenation. J Cosmet Dermatol. 2018;17:289–97.
19. Saidova AS, Sencha AN, Apolikhina IA. Carboxytherapy of genitourinary syndrome of menopause. Akusherstvo i Ginekologiya/ Obstet Gynecol. 2020;5:113–21.
20. Styles R. Would you plump up your vagina with fillers? Top cosmetic doctor warns of risky new trend after being inundated with clients looking for 'genital enhancements' [internet]. London: Daily Mail; 2015. [cited 2018 Jun 17]
21. Kim SH, Park ES, Kim TH. Rejuvenation using platelet-rich plasma and lipofilling for vaginal atrophy and lichen sclerosus. J Menopausal Med. 2017;23:63–8.

Management of Urinary Incontinence in Aesthetic & Regenerative Gynecology

27

Ronen Gold

27.1 Introduction

Worldwide, urinary incontinence (UI) is a major health burden that affects 17–45% of adult women with high costs in terms of well-being and financial expenditure for both patients and society. The most prevalent type is stress incontinence, contributing 48% of all cases (1–3).

Stress urinary incontinence (SUI) is defined by the International Continence Society (ICS) as "the complaint of any involuntary loss of urine on effort or physical exertion (e.g., sporting activities) or on sneezing or coughing" [1].

SUI is the unintended, accidental leakage of urine that happens during strenuous activity (such as coughing, sneezing, laughing, changing position, running, heavy lifting), conditions that increase the abdominal pressure (stress) on the bladder, anatomic variations (urethral hypermobility, intrinsic sphincter deficiency ISD) or, most commonly, both. There is no relation to psychological stress.

The definition of incontinence was not modified in 2002 when the ICS standardization subcommittee reviewed the definitions of LUTS [2].

R. Gold (✉)
Department of Uroygnecology and Pelvic Floor Rehabilitation, Aya Medical Advanced Gynecology, Tel Aviv-Yafo, Israel

Service for Obstetric Anal Sphincter Injury at Birth, Lis Women's Hospital, Tel Aviv Medical Center, Tel Aviv Medical School, Tel Aviv-Yafo, Israel

The term implies an element of psychological distress, which is not a cause of the condition, whereas the term "effort incontinence" does not include some of the common involuntary triggering factors for stress incontinence such as coughing or sneezing. Since the mid-1940s, the term "stress incontinence" had been however in use in English literature [3].

When urodynamic testing is performed on women with UI, there is often a discordance between subjective and objective findings [4–6]. By utilizing symptoms alone, a misdiagnosis would be made in 13% of women reporting only stress incontinence and 59% of women reporting only urge incontinence [7]. This is explained by the artificial nature of office urodynamics, the patient's poor understanding of the different types of UI, and incomplete history taking. Stress incontinence was found to be more prevalent in younger women aged 30–49 years (78%) versus those aged 50–89 years (57%). Urge incontinence occupies the older population (67%) versus women under the age of 50 years (56%) [8]. With advancing age, the rate of urge incontinence tends to increase, whereas due to lower physical activity, the rate of stress incontinence tends to decreases in the older age groups [9] (Fig. 27.1). A meta-analysis of studies of UI epidemiology reports that stress incontinence is the most common type of incontinence found in women [10]. Urge incontinence constitutes 29% of the population; however, the analysis showed

P. Jindal et al. (eds.), *Aesthetic and Regenerative Gynecology*,
https://doi.org/10.1007/978-981-16-1743-0_27

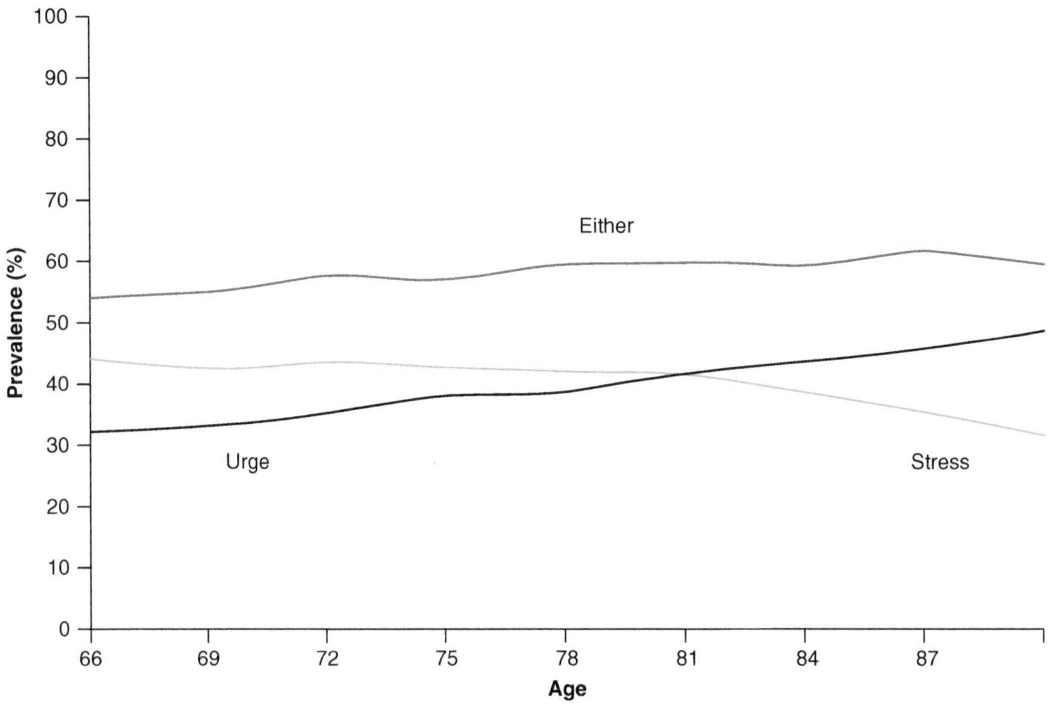

Fig. 27.1 Prevalence of incontinence by age groups. "Urge" and "stress" refer to women who answered affirmatively to the urge and stress incontinence questions, respectively. "Either" refers to women who reported any incontinence (either urge or stress). (McGrother CW et al., Br J Urol Int, 93, 763, 2004)

that urge incontinence was less common as compared to others.

The direct and indirect financial impact of UI on the healthcare system is substantial.

27.1.1 Pathophysiology of SUI

Pregnancy, childbirth, pelvic floor trauma at delivery, age, body mass index (BMI), chronic cough, and genetic factors have an effect on developing SUI [11]. During pregnancy, the endopelvic fascial and distal sphincter are weakened due to hormonal influences [12]. Progesterone reduces urethral closure pressures and produces connective tissue changes [13, 14] causing a high incidence of antenatal incontinence. In most cases, incontinence improves after delivery [15]. In women with damage to the endopelvic fascial attachments and sphincter, recovery might not happen. Antenatal SUI, and especially, incontinence before a first pregnancy is a risk factor for the development of incontinence in later years.

The correlation between SUI and parity is well established (Fig. 27.2). In women reviewed 6 years after childbirth, the rate of new-onset incontinence is about 30% in women who had been continent at 3 months postpartum. However, in 27% who were incontinent at 3 months, there was spontaneous remission at 6 years. In women who were incontinent prior to pregnancy, there was a markedly increased risk for UI at 6 years. These findings imply that there are women at risk of incontinence, while in others there is spontaneous remission. In the first 3–12 months postpartum, the prevalence of postpartum incontinence was found to be 33% [15].

There is an association between connective tissue disorders like Marfan and Ehlers–Danlos syndromes and SUI [16].

Urethra is the primary factor responsible for stress incontinence. Data analysis of women

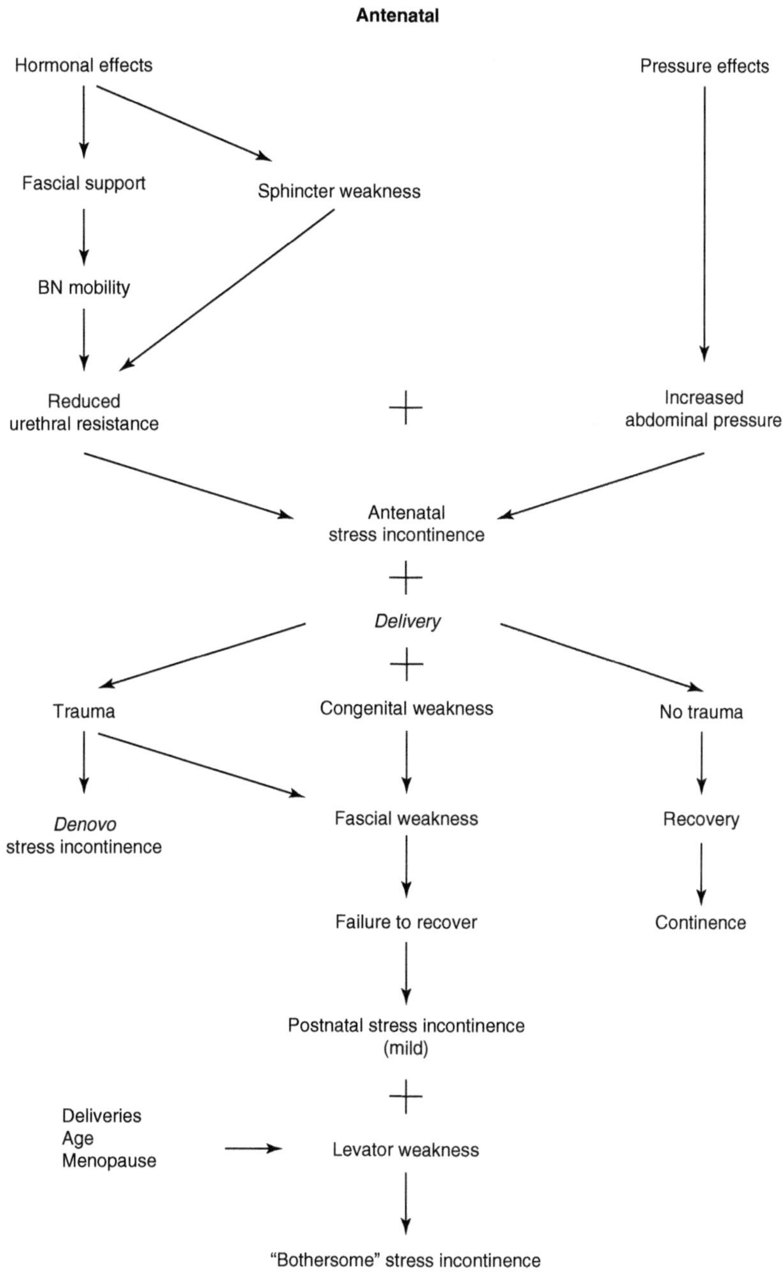

Fig. 27.2 Hypothesis for the genesis and natural history of stress urinary incontinence. (Maclean AB and Cardozo L eds., Incontinence in Women, RCOG Press, London, U.K., 2002)

with stress incontinence and asymptomatic factors revealed that more than half of stress incontinence is due to elevated urethral closure pressure [14]. The volume of striated muscle and innervating nerves declines significantly with age [17, 18]. During the first six months of vaginal delivery, there occur changes in the pattern of striated sphincter musculature with reduced motor function [19].

Urethral function (maximal urethral closure pressure) is the key factor of whether or not a woman has stress incontinence [20]. As urethral function declines with age, the role that maximum urethral closure pressure begins to dominate. After first birth, injury to the levator ani muscle is twice in women with incontinence as compared to individuals who deliver and are continent. There is a decrease in levels of collagen in premenopausal women with stress incontinence, suggesting loss of vaginal support.

Types of Urinary Incontinence according to pathophysiology.

I. Outlet dysfunction: bladder outlet and pelvic floor A. Underactive outlet (decreased urethral resistance)
   Symptomatic: SUI.
   1. Anatomical support defects (SUI-A) (types I and II SUI): pathophysiology—anatomical motion creates inequities in transmission pressures to bladder and outlet, overcoming urethral resistance, and/or and conformational changes caused by vaginal wall motion disrupt outlet integrity:
      (a) Anatomical defects of fascia, muscles, ligaments, and bony pelvis
      (b) Functioning support: muscular contraction—denervation or the loss of identification, strength, or coordination of levator musculature
      (c) A degree of ISD (see 2) that in combination with SUI-A I or II contributes to critical decreases in resistance during effort to permit urinary leakage
   2. ISD (SUI-ISD) (type III) (LUCP) Pathophysiology: deficiency of the urethral closure mechanism secondary to decreased innervation, vascularization or trauma to mucosa, submucosa, or smooth, nonstriated skeletal or skeletal musculature of the urethra—intrinsic deficiency of the closure mechanism.
   (a) Proximal urethral sphincter: bladder neck (smooth muscle) and proximal urethra (SUI-ISD-p)
   (b) External sphincter: denervation or loss of resting tone, voluntary or reflex contraction (SUI-ISD-d)
   (c) Combined or total proximal and external sphincter deficiency (SUI-ISD-t)
   3. Combined SUI (SUI-A-ISD) Pathophysiology: a degree of both anatomical motion and sphincter dysfunction
   (a) Leakage from ISD may exist without hypermobility
   (b) Hypermobility alone without a degree of ISD is not sufficient to result in urinary loss.
   4. Failure to inhibit the detrusor: decreased pelvic floor inhibitory activity of bladder (etiology for OAB—see C). Pathophysiology: failure to contract pelvic floor releases detrusor reflex and decreases the ability to inhibit active contraction.
   (a) Neurological: (infrasacral) denervation (areflexia of pelvic floor).
   (b) Behavioral: failure to contract pelvic floor (lack of identification/strength/coordination).
   (c) Mechanical: damage to pelvic floor structures with intact innervation.

27.1.2 History

History taking must use the woman's own words. Women often don't understand terms such as stress incontinence thinking that this relates to mental stress and leaking [21]. This can be simplified into an easily understandable list of graded symptoms using a standard questionnaire. A diagnosis based on history corresponds to urodynamic diagnosis in only up to 55% of women [22]. The use of a self-completed symptom questionnaire produces a better relationship between urinary symptoms and urodynamic diagnosis [23]. Twenty-four-hours bladder training diary is helpful in the initial assessment and evaluation of women with urinary incontinence. It helps as a therapeutic measure also by estimating the effectiveness of treatment.

Even though symptoms are not of diagnostic value, they are valuable as a guide in assessing treatment. Symptoms should be taken into consideration during the urodynamic test as the provocative maneuvers should mimic conditions encountered by the woman in her normal daily activities causing her urinary symptoms.

27.2 Medical Diagnosis

The International Consultation on Incontinence (ICI) distinguishes between initial management and, in case of failure, specialized management of UI [21, 22]. For initial management, the ICI recommends a simple clinical assessment leading to a presumed medical diagnosis. First-line healthcare providers should use simple diagnostic tools like structured history taking; micturition or voiding diaries and physical examination [23–25]. To differentiate between stress, urgency, and mixed incontinence, a short questionnaire (two-item Stress/Urge Incontinence Questionnaire) can be helpful (S/UIQ) [26].

27.2.1 Incontinence Symptom Classification System with S/UIQ

- SUI ≥ 4; UUI = 0 pure SUI.
- SUI > 0; SUI > UUI; UUI > 0 stress predominant mixed urinary incontinence.
- SUI > 0; SUI = UUI balanced mixed urinary incontinence.
- SUI > 0; UUI > 0; UUI > SUI urge predominant mixed urinary incontinence.
- UUI ≥ 4; SUI = 0 pure urge urinary incontinence.

For the GP, who, in many cases, is the first physician the patient is referred to for consultation, the use of these simple tools is associated with difficulty to find the exact cause of the urinary incontinence. Specialists, like the urologist or the gynecologist, mostly rely on specific diagnostic tests such as urodynamic evaluation. Whereas, GPs mostly rely on their history taking, physical examination, and questionnaires like the S/UIQ. Since urodynamics has invasive character and doubts about its usefulness, reliability, and validity, the need for urodynamic testing is currently being questioned. Evidence showed that if surgery has been planned for women with uncomplicated, demonstrable SUI, preoperative evaluation alone was not inferior to evaluation with urodynamic testing for surgical outcomes at 1 year [27], suggesting that urodynamic tests are only needed to determine the type of incontinence and choice of treatment. The history taking and physical examination sensitivity and specificity respectively incontinences are 78% and 84%, with a positive value of 87% [24]. Using simple diagnostic tools such as the S/UIQ, these percentages even can be enhanced [22]. A meta-analysis of primary care diagnostic methods of urinary incontinence (initial management) showed a sensitivity of 0.92 (95% confidence interval [CI] 0.91–0.93) and specificity of 0.56 (0.53–0.60) to correctly identify women with urodynamic stress incontinence [28].

27.3 Examination

Before examining, it is important to reassure the possibility of urinary leakage and explain to the patient that she should not be embarrassed as a result of this. Abdominal examination heads vaginal assessment to evaluate: evidence of previous abdominal surgery, bladder status (fully or partially distended) as well as other abdominal or pelvic masses. Neurological factors should also be evaluated before confirming the diagnosis The anal tone should be assessed. In women with neurological disease, a positive response of the bulbocavernosus reflex indicates that sensory and motor pathways related to the urogenital tract are intact. By stroking the skin lateral to the anus, which elicit contraction of the external anal sphincter indicates that sacral reflex is intact.

The objective most useful test in clinical practice is the pad test (Fig. 27.3). A short version has been standardized by the ICS, making its use more uniform for research and data collection. The 1-hour pad test is less likely found to have good reproducibility, improved with standard bladder volume. The short-term pad test was found to be valid in differentiating normal from abnormal continence mechanisms; however, its validity is limited because of a significant false-negative rate. Finally, the ability of the short-term pad test (≤1-hour) to categorize the severity of incontinence is poor. The long-term pad test (≥24 hours) is valid in detecting incontinence, with a good sensitivity and a lower false-negative results. Hence, a 24-hour home pad test represents a good tool in detecting and quantifying incontinence. Neither test can distinguish between different types of urodynamic diagnosis.

In high-grade Pelvic organ prolapse, it has been seen that the descent of the anterior vaginal wall may cause urethral kinking and compression [29]. Documenting the presence of stress urinary incontinence (SUI) type in patients with the reduction of the prolapse helps preoperative assessment of Occult SUI (SUI occurring after prolapse correction). The prolapse may be reduced with vaginal packing, rectal swabs, a speculum blade, or pessary. In women with high-stage anterior compartment prolapse, the 2012 AUA Guidelines on Adult Urodynamics recommends that all patients undergoing repair surgery should undergo stress testing with reduction of the prolapse [30].

The abdominal leak point pressure (ALPP) helps in determining urethral function. The lowest intravesical pressure (Pves) at which urine leakage occurs is because of increased abdominal pressure in the absence of a detrusor contraction [15]. The ALPP represents measuring the vesical pressure at which leakage is associated with increased abdominal pressure, it may happen by Valsalva (VLPP) as shown in Fig. 27.4 or by cough (CLPP). The CLPP is usually measured in the absence of a VLPP, because it is harder to quantify and is usually higher. Thus, VLPP and CLPP are two types of ALPP.

ALPP measures the sphincter strength. In a normally positioned and closed urethral sphincter, increased abdominal pressure does not open the sphincter—so there is no normal ALPP. Urine leakage could also be caused by an increase in abdominal pressure when the urethra is abnormal.

0 minute
- Apply preweighed pad
- Drink 500 cc sodium-free liquid
- Sit and rest

30 minutes
- Walk and stairs climbing

45 minutes
- Activities: Sit/stand × 10
- Cough × 10
- Run in place × 1 minute
- Pick up objects from floor
- Wash hands under running water × 1 minute

60 minutes
- Collect and weigh pad
- Patient voids, volume measured

Fig. 27.3 Standardized 1-Hour Pad Test depicts at

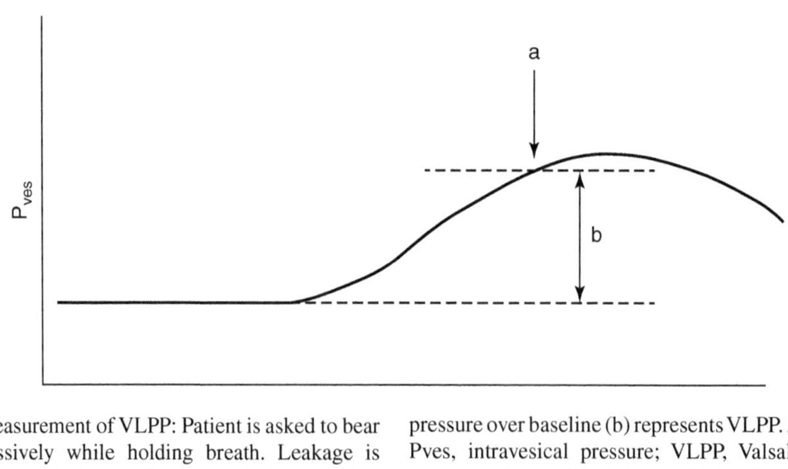

Fig. 27.4 Measurement of VLPP: Patient is asked to bear down progressively while holding breath. Leakage is recorded (arrow) at the precise moment that fluid is observed at the urethral meatus. The rise in intravesical pressure over baseline (b) represents VLPP. Abbreviations: Pves, intravesical pressure; VLPP, Valsalva leak point pressure. (Bump RC et al., Am J Obstet Gynecol, 173(2), 551, August 1995)

Low values for ALPP imply less urethral resistance to protect against leakage during increased intra-abdominal pressure; therefore, lower ALPP values imply a weak sphincter. The detrusor pressure at the time of ALPP testing must be low. High detrusor pressure (due to detrusor overactivity, or poor bladder compliance) at the time of ALPP testing may result in the false diagnosis of urethral dysfunction [31].

Patient position is important, because if urine leaks while standing, then the test may require to be done in the standing position. Patients with suspected SUI but do not leak during UDS testing with the catheter in place, the AUA Adult Urodynamics Guidelines recommend that stress testing be repeated with the catheter removed, because the urethral catheter may mask the presence of SUI in certain patients [30].

In most continent women, the functional urethral length is approximately 3 cm and the maximum urethral closure pressure (MUCP) is 40–60 cm water, but there is considerable variability [32]. It has been suggested that UPP can be helpful in women undergoing stress incontinence surgery. McGuire and Sand found that patients with urethral closure pressures less than 20 cm water had bigger failure rates for their incontinence repairs [33, 34], other studies also showed that a low MUCP is a predictor of poor outcomes for retropubic suspension surgery [35, 36] making sling procedures the anti-incontinence procedure of choice for patients with low urethral closure pressure [37, 38]. Utilizing sling procedures (particularly mid-urethral slings) for all cases of stress incontinence makes the UPP increasingly irrelevant to the choice of surgical procedure [38].

Imaging of the pelvic floor in patients with SUI is limited by a lack of complete understanding of the physiopathology of these conditions. Measuring postvoid residual (PVR) in patients with UI and POP is part of the assessment before and after surgery. Elevated PVR is associated with an increased risk of urinary tract infection. It may develop following surgery for SUI and POP. US imaging is regarded as the gold standard compared to catheterization [32, 39–41]. Various formulae have been proposed, in clinical practice, the residual volume is probably best approximated by height × width × depth in centimeters × 0.5. Automated US systems are widely used and they have been shown to be accurate and usable by nursing staff [42]. Data from asymptomatic peri- and postmenopausal women suggest a median PVR of 19 mL with 95% of subjects having a PVR < 100 mL [8], another study, showed an average PVR < 30 mL in 81% of women with LUT symptoms; larger volumes were associated with increasing age, advanced grades of POP, and increased prevalence of recurrent urinary tract infections [43].

Both US and MRI proved to be comparable to physical examination and sometimes superior but this rarely translates into a real clinical benefit.

27.3.1 Endoscopy

Is utilized to evaluate erosion of Mesh to the urethra or the bladder following sling procedures or sutures in the bladder after Burch operations for stress incontinence.

It is also the route of injecting bulking agents to the bladder neck.

27.3.2 Treatment

27.3.2.1 Behavioral Therapies

Behavioral therapies help to improve bladder control by teaching skills for preventing urine leakage or changing their daily voiding habits. Behavioral interventions are comprised of multiple components, tailored to the individual needs of the patient, characteristics of her symptoms, and her life circumstances. The main aim of behavioral treatment is to improve bladder function by changing voiding habits, bladder training and pelvic floor muscle training and strengthening exercise. Among the techniques included in the behavioral treatment are keeping a record of bladder diary, pelvic floor muscle training techniques (including biofeedback), pelvic floor muscle exercise, urge avoidance strategies, scheduled voiding (including bladder training), delayed voiding, teaching normal voiding techniques, fluid management, dietary changes involving restriction of bladder irritants (including caffeine), weight loss, and other lifestyle modification. These behavioral measures are safe and without the risks or side effects.

27.3.3 Weight Loss

Obesity is a predisposing risk factor for urinary incontinence. Elevated body mass index (BMI) by five-unit increases the risk of daily urinary incontinence by 60% [44, 45]. Moderately obese women report significant improvement in symptoms of incontinence with weight loss of 45–50 kg following bariatric surgery [46–48]. Significant improvements in incontinence have been demonstrated with as little as 5% weight reduction [49]. The most definitive randomized trial of weight loss for incontinence uses the Program to Reduce Incontinence with Diet and Exercise (PRIDE) study. PRIDE compared an intensive 6-month group-administered, weight loss diet program, exercise, and behavioral modification to a structured education control program in obese women with incontinence [50]. The weight loss program resulting in a mean weight loss of 8%, showed noticeable reductions in the number of incontinence symptoms compared to the control group, which had a mean weight loss of 1.6% (47% vs. 28%). Most women prefer to achieve the goal of moderate weight loss. Thus, weight loss is recommended as a first-line treatment to treat incontinence in overweight and obese women.

27.4 Who Is a Candidate for Behavioral Therapy?

Most patients who went with behavioral treatment for incontinence experience some level of improvement, yet there is striking variation in outcomes. We cannot predict which patients will respond best. Studies showed that outcomes are not related to the type of incontinence. Evidence on the effect of severity of UI or age are inconsistent [51–58]. Outcomes are not associated with patient race, parity, body mass index, cystocele, uterine prolapse, hysterectomy, hormone therapy, use of diuretics, or urodynamic parameters. Thus, given that behavioral therapies are without risk and most adherent patients experience symptom improvement, offering behavioral treatment as first-line therapy is appropriate for any woman with urinary incontinence.

27.4.1 Physiotherapy

Physiotherapy treatment modalities to treat SUI are: PFMT with or without biofeedback, electrical stimulation, magnetic stimulation, and/or vaginal cones. The rationale for using PFMT is that a pelvic floor muscle contraction will clamp the urethra leading to increased urethral pressure, thus preventing leakage during a sharp increase in intra-abdominal pressure during any strenuous activity [59]. Timing plays a crucial role; in healthy continent women, activation of the PFM before or during physical exertion is an automatic anatomic response, [60, 61] that precede bladder pressure rise by 200–240 milliseconds [62, 63], that have been lost in women with SUI. Bø suggested that a well-timed, fast, and strong pelvic floor muscle contraction may prevent urethral descent during the rise in an intra-abdominal pressure [64, 65]. Pelvic floor muscle training is focused on adequate timing, strength improvement, and coordination of the periurethral and pelvic floor muscles. Assessment of pelvic floor muscle contraction and relaxation is important, as the effect of PFMT is fully dependent on whether the contractions and relaxations are performed correctly [65]. For PFMT to become effective, repeated correct contractions of the pelvic floor, strengthening the pelvic floor muscles in a regular, intensive, and long-lasting training program, are essential [65–67]. Pelvic floor muscle training should include short- and long-duration exercises. Daily recommended regimes of increasing repetitions to the point of fatigue are: 8–12 maximal pelvic floor muscles contractions, 1–3 seconds to 6–8 seconds hold/relax, three extra quick peak contractions superimposed on the maximal contraction, three times a day for at least 6 months [67]. An individually tailored home exercise program during daily life activity is necessary [65]. Because the urethral closing mechanism also depends on a competent intrinsic urethral sphincter, there is no guarantee of improvement in timing, absolute strength, and endurance of the pelvic floor, with respect to the extrinsic part of the total urethral closing mechanism, will fully restore continence. Evidences have showed that PFMT is effective in the reduction of involuntary loss of urine in patients with stress incontinence in those who are sustaining it for the long term [68–71]. Intensive training results are better than a low-intensity program [72, 73]. One-thirds of females have become dry after 5 years, while on follow-up two-thirds of them indicate that they are very satisfied with their present state and that they wish no further intervention. Biofeedback refers to audiovisual techniques supplying information regarding "concealed" physiological processes to the patient, for permitting self-regulation of these events. To record the vaginal or rectal pressure, a sensor or electrode is introduced into the vagina or rectum. The patient receives visual or acoustic information about the pressures measured and/or the EMG signals, enabling them to see what force has been generated by the pelvic floor muscles and whether it has reached its maximum.

In patients with SUI without dysfunction of the pelvic floor- ISD—pelvic floor training is less effective if at all.

27.5 Pessaries and Devices for Stress Urinary Incontinence

Supportive pessaries are also used for treating SUI. Pessaries used to treat incontinence frequently have a knob that is placed under the urethra (Fig. 27.5). Some of the most commonly used incontinence pessaries include the incontinence dish (with or without support) and the incontinence ring with knob (with or without support) (Figure 10), but other types are also

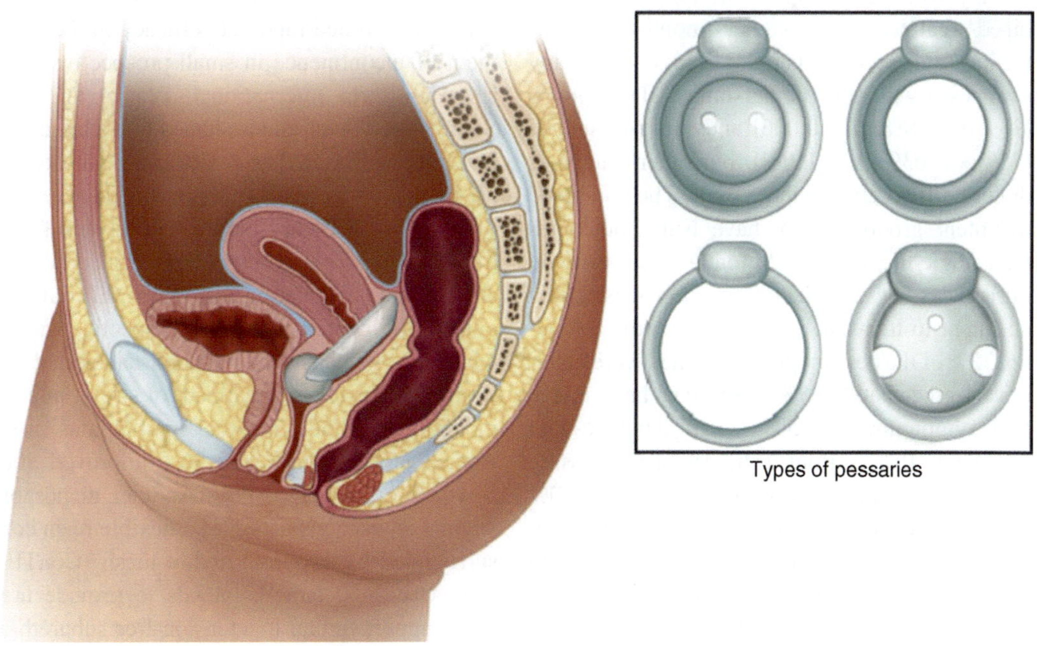

Fig. 27.5 Pessary placement for the treatment of stress urinary incontinence and (inset) commonly used incontinence pessaries (clockwise from top left: an incontinence dish with support, an incontinence dish without support, an incontinence ring with support and knob, and an incontinence ring without support). (Rogers RG, N Eng J Med, 358, 1029, 2008; Copyright © 2008 Massachusetts Medical Society)

available. Clinical protocols for fitting and following patients with incontinence pessaries are similar to those used in patients who use pessaries for POP. Other vaginal or urethral devices may be used to treat SUI, including tampons, diaphragms, disposable vaginal devices, and urethral inserts.

Nygaard et al. [74] conducted a randomized, crossover, clinic-based study on 18 women who completed three standardized aerobics sessions, wearing either a Hodge with support pessary, a large menstrual tampon, or no device. Both devices significantly diminished urine loss (measured with a pad test) during exercise when compared to the control session. Better outcomes were seen with both devices in women who had milder urine loss [74]. The National Institutes of Health Pelvic Floor Disorders Network conducted a large, multicenter, randomized trial (the Ambulatory Treatments for Leakage Associated with Stress Incontinence [ATLAS] trial) to test mechanical devices for SUI [75]. In total, 450 women with stress urinary incontinence were randomized to incontinence pessary, behavioral therapy with pelvic muscle training, or a combined treatment arm. After 3 months of treatment, some outcomes favored behavioral therapy over pessary therapy (bothersome SUI symptoms [33% vs. 49%, $p = 0.006$] and satisfaction rate [75% vs. 63%, $p = 0.03$]), but other outcomes did not differ between the groups. The combined treatment group did not have better outcomes than single therapy. At 12 months, there were no differences between the treatment groups. Using an intent-to-treat analysis at 12 months follow-up, 32% of the pessary group felt they were better or very much better, 35% reported no bothersome SUI symptoms, and 50% were satisfied with treatment, 46% continued the use of the pessary at 12 months, and among this group, about 60% reported they were better or very much better, and 90% were satisfied with treatment [75]. In two retrospective studies involving 100 and 190 women respectively, about 60% chose to undergo a pessary fitting for stress or mixed urinary incontinence, and 85–90% were successfully fit [76, 77]. Of those successfully treated with a pessary, 55–60% continued using the pessary for the next few months or years (median duration of follow-up 11–13 months). In contrast, a small prospective study was conducted, where only 16% of 38 women fit with an incontinence ring with support pessary continue to use up to 1 year. In the few that continued the use, the pessary resulted in fewer leaking episodes and 9 (24%) were subjectively "dry."

Fewer and smaller studies have tested other types of vaginal and urethral devices for the treatment of SUI. Placement of a menstrual tampon shows moderate levels of success (57% continent during use) in treating exercise-induced incontinence [75]. In a small study where 32 women were enrolled, a novel bell-shaped self-positioning incontinence pessary (Uresta, EastMed, Inc., Dartmouth, Nova Scotia, Canada) has significantly improved incontinence symptoms as well as daily leaking episodes and urinary pad weights [5/34]. Similar to the larger retrospective studies of traditional incontinence pessaries described earlier, 50% of women beginning the study chose to continue the use of the pessary for SUI treatment after 12 months. Other internal vaginal devices have been developed and demonstrated moderate efficacy in the treatment of incontinence in small prospective studies. Contiform International, Blacktown, New South Wales, Australia is shaped like a large hollow tampon [5/35]. The device can be reused for 30–60 days. In a small study of this device, 54% of women who completed the course were found to be dry. Single-use, disposable vaginal devices are available for treating SUI in several countries outside the United States. A prospective trial in which 94 women were included had tested the Controller device for treatment of stress-induced incontinence finds 48% of participants were subjectively continent and 36% improved after 5 weeks of treatment [78]. Another disposable vaginal device, consisting of a flexible resin core surrounded by a porous nylon mesh (ConTIPI Ltd., Caesarea, Israel), intends to provide urethral support similar to a tension-free suburethral sling. Used by 60 women with severe SUI, 85% had a $\geq 70\%$ reduction of urine loss on daily pad weights. Half of the women who has completed the 4-week trial, out of them 92% were subjectively continent / [79]. Urethral inserts and external occlusive devices function as mechanical barriers to prevent loss of urine. These devices tend to need highly motivated and manually dexterous patients as the devices must be removed to urinate and then has to be replaced after each void. Studies suggest they have lower overall success rates than other vaginal devices, partly because of higher dropout rates [80]. Urethral inserts are sterile, single-use devices placed into the urethra by the patient and held in place by an inflated balloon at the bladder neck. Such inserts are advisable for women with no history of recurrent urinary tract infections and no contraindications to bacteriuria (e.g., artificial heart valves). Multicenter studies showed an increased incidence of continence with urethral inserts in place (80–95%) and higher satisfaction rates in women with continuous use, but overall results are limited by high withdrawal rates and frequent adverse events [81–83]. While inserts remain an appropriate treatment option for many women with SUI, the need to frequently insert the device into the urethra makes this option noncompliant for many patients. External urethral occlusive devices fit over the external urethral meatus and are held in place by adhesive or suction. These devices have fewer reported side effects than the urethral inserts, but reported continence rates are lower (40–50%) [83, 84]. Patient acceptance of this type of device, similar to urethral inserts, are limited. In past, several types have been marketed in the United States, but none are currently available in the market. In summary, published studies suggest most women who are interested can be successfully fit with an incontinence pessary and about half of those women find it an acceptable and satisfactory long-term therapy for SUI. A Cochrane review of vaginal and urethral incontinence devices found there is insufficient evidence to recommend any specific pessary or device or to show that mechanical devices are better than others [85]. Urethral inserts have higher rates of adverse events than vaginal devices. The most common complications reported by patients include urinary tract infections, hematuria, and urethral and/or bladder irritation and discomfort [81, 82, 85]. These complications appear to occur most often in the first month of device use. In a long-term study of a urethral insert, symptomatic cystitis occurred in 10.5% of participants during the first month of use, but in only 2–3% in subsequent months [83, 85].

27.5.1 Increasing Outlet Resistance (Table 27.1)

Alpha-Adrenergic Agonists (Ephedrine, Midodrine, Norfenefrine, Phenylpropanolamine, Pseudoephedrine).

The bladder neck and proximal urethra consist mainly of α-adrenergic receptor, which, when stimulated, produce contraction of smooth mus-

Table 27.1 Drugs used in the treatment of stress urinary incontinence

Drug	Level of evidence	Grade of recommendation
Clenbuterol	3	C
Duloxetine	1	B
Ephedrine	3	D
Estrogen	2	D
Imipramine	3	D
Methoxamine	2	D
Midodrine	2	C
Phenylpropanolamine	3	D

cle and an increase in maximal urethral pressure (MUP) and maximal urethral closure pressure (MUCP) [86]. *Stimulation of α-Adrenergic generally increases outlet resistance to a variable degree but is most often limited by their potential side effects including rise in blood pressure, anxiety, insomnia, headache, tremor, weakness, palpitations, cardiac arrhythmia, and respiratory difficulties, thus makes it anecdotal.*

A Cochrane review evaluated randomized and quasi-RCTs were conducted in 1099 adult women with their SUI treated by at least one arm of the trial with an adrenergic agonist drug [87] *in 22 trials. The conclusion that was drawn from that is: "There was weak evidence to suggest that use of an adrenergic agonist was better than placebo treatment." Similar adverse events profile was reported for adrenergic agonists and placebo.*

27.5.1.1 Antidepressants

Duloxetine

FDA approved, Duloxetine is a combined SNRI drug for depression, diabetic neuropathic pain, and generalized anxiety disorder and is also licensed to use in the European Union for the treatment of SUI. During the voiding phase, there is an increase in sphincter muscle activity during the filling/storage phase of micturition with no effect on sphincter function [88]. *Duloxetine exerts its effects on serotonin and norepinephrine nerve terminals by prevention of reuptake. These nerve terminals are dense in spinal areas associated with LUT function (pudendal nerve and Onuf's nucleus) and activation leads to rhabdosphincter contraction. A Cochrane review of the effects of duloxetine on SUI summarizes data from nine RCTs that included 3060 women* [89]. *The cure rate in the duloxetine 40 mg twice daily group was higher as compared to the placebo group (10.8%* vs. *7.7%, $p = 0.04$). Significant reduction in episodes of incontinence and improvement in HRQL and patient global impression of improvement were seen. However, no data were available on the sustainability of treatment. The approximate size of the effect showed that for every 100 patients treated, three patients were cured. Only one trial reported objective cure data and showed no clear difference between drug and placebo. Nausea (23–25%) was the most common complaint and thus the reason for discontinuation. Other side effects reported with the drug were vomiting, constipation, dry mouth, fatigue, dizziness, and insomnia. Across these six trials, 17% in the drug group withdrew* versus *4% in the placebo arm. The authors conclude that more data and research would be required to determine whether duloxetine is clinically and cost-effective compared to the other minimally or noninvasive treatment options available.*

There is a "black box" warning of "increased risk of suicidal thinking and behavior in children, adolescents, and young adults taking antidepressants for major depressive disorder and other psychiatric disorders" [90]. *Other warnings include risk for hepatotoxicity in patients with substantial alcohol use or chronic liver disease, orthostatic hypotension, serotonin syndrome, withdrawal symptoms with abrupt discontinuation, and increase the concentration of drug with inhibitors of CYP1A2 and CYP2D6. The most common adverse effects compared to placebo are nausea (24%* vs. *8%), dry mouth (13%* vs. *5%), fatigue (10%* vs. *5%), somnolence (10%* vs. *3%), insomnia (10%* vs. *6%), constipation (10%* vs. *4%), and dizziness (10%* vs. *5%)* [89] *were the primary adverse effects.*

Imipramine

An increase in urethral resistance by enhanced α-adrenergic effect resulted from an inhibition of noradrenaline reuptake. However, imipramine also causes α-adrenergic blocking effects. In patients treated with imipramine primarily for DO but who had also some component of sphincteric incontinence, imipramine showed efficacy. In a study of 30 women with SUI who were treated with imipramine 75 mg daily for 4 weeks, 21 women reported continence; the mean MUCP for the group increased from 34.06 to 48.23 cmH2O [90]. *There are no RCTs on the use of imipramine for SUI.*

Estrogens

Although estrogens have been recommended for the treatment of incontinence for many years, there remains considerable controversy for this purpose. Numerous clinical studies exist; however, because of little consistency in the methodology or delivered drug makes analysis difficult. After menopause, urethral pressure parameters decrease, and although this change is believed to be related somehow to decreased estrogen levels, it is still a matter of speculation whether the actual changes occur in the smooth muscle, blood circulation, supporting tissues, or mucosal seal mechanism [91]. The Hormones and Urogenital Therapy Committee reviewed 166 articles containing 6 controlled and 17 uncontrolled trials [92]. They were able to show subjective improvement in urinary continence, but when objective urine loss was measured, there was no significant change. Only one study showed an increase in MUCP. A second meta-analysis concluded that "recent trials do not provide enough evidence to prove estrogen replacement as efficacious therapy for SUI" [93]. HERS study that randomly assigned hormone therapy revealed worsening of incontinence symptoms in the treatment group (39% vs. 27%, $p < 0.001$). WHI study on 23,296 postmenopausal women revealed that 1 year of use hormone therapy increased all types of urinary incontinence, with SUI being the most common [94]. Estrogen alone, or in a combined form with progesterone, was likely seen in increasing the risk of incontinence among continent women and worsen incontinence in mildly symptomatic women. To conclude—there is no convincing evidence that estrogen improves SUI [95].

27.5.2 Injectable Bulking Agents

Bulking agents offer a simple minimally invasive in-office injection technique with immediate, but not permanent results for patients who are not candidates for other procedures due to fragility or medical comorbidities. Less common, patients may have a poorly coaptable, fixed urethra "(pipestem urethra") [96, 97], therein reducing the efficacy of procedures dependent upon the augmentation of closure mechanisms by reducing urethral hypermobility with stress events.

It is recommended to determine postvoid residual volume prior to the use and reinjection of bulking agents [98], both diagnostically and to reduce adverse events, as use of bulking agents in raised postvoid residual volume may lead to obstruction of outlet and overflow incontinence [99].

Bulking agents are most likely used as primary treatment and do not hinder upcoming surgical procedures [100]. There are many review articles and few primary RCTs. In addition to use as a primary treatment, bulking agents are effective following failed surgical procedures [6–8/32–34], following radical pelvic surgery [101, 102], in patients with injury of the spinal cord [103, 104], or for continent stomas [105].

Bulking agents act by raising resistance to intra-abdominal forces via the soft-tissue filler properties of the injectable agent [106]. The capacity to bulk soft tissue is intrinsically limited. Thus, improvement in symptoms of continence is not satisfactory in case the degree of incontinence is severe. Injected autologous cells of skeletal muscle may provide an additional benefit to soft-tissue bulking including a dynamic element of muscular contraction and potentially enhanced resting tone. Bulking agents' effectiveness decline over time from 48% at 12–23 months to 32% at 24–47 months [107]. None of the current bulking agents have proven durable. Thus, bulking agents have their own disadvantages as they are less effective and less durable than other procedures but have lower complication rates due to their less invasive nature [108].

27.6 Which Bulking Agent to Choose?

Each of the available bulking agents is similar in benefits/risks/burdens [109]. In a trial of multiple agents, a 12-month analysis of over 800 cases of transurethral applied glutaraldehyde-cross-linked collagen (n = 312), nonanimal stabilized hyaluronic acid/dextranomer (NASHA/Dx) (*n* = 54), ethylene vinyl alcohol (EVOH) (*n* = 104), and polyacrylamide hydrogel (*n* = 44) demonstrated that the objective and subjective measures were similar for each group, with negative pad tests in over 70% of these women [110]. Reduction in the durability of the bulking agents is a bigger challenge than current safety issues.

Commercially available agents include the following:

- ***Polydimethylsiloxane macroparticles suspended within a bioexcretable carrier hydrogel of polyvinylpyrrolidone (MacroplastiqueTM, Cogentix Medical, Minnetonka, MN).***

At 12 months, the cure/dry rate was statistically better at 36.9% for those treated with Macroplastique versus 24.8% in the control group. In a study following a study group during a time period of 24 months, 33 out of 38 patients achieving dry/continence at 12 months remained dry at 24 months. An additional 12 of 29 patients, who were improved at 12 months, remain dry at 24 months [111]. The Macroplastique Implantation Device, a specialized pressured syringe and applicator, allows for outpatient transurethral cystoscopic injection under direct vision.

- **Calciumhydroxyapatite (CaHA) (CoaptiteTM, Boston Scientific Corp., Natick, MA)**.

A multicenter RCT of 296 women, also using the poorly discriminatory Stamey Urinary Incontinence Scale to assess outcomes, found that 63.4% of patients injected with CaHA achieved an improvement compared with 57% of patients injected with ContigenTM [112].

- **Zirconium-coated Carbon beads (Durasphere® EXP, Coloplast Corp, Minneapolis, MN).**

In a noninferiority trial with ContigenTM, Durasphere® achieved a nonstatistically significant improvement in 48 of the 61 women treated, compared with 47 out of 68 women treated with ContigenTM [113].

- **Polyacrylamide *hydrogel (PAHG) (BulkamidTM, Contura, International A/S, Soeborg, Denmark)*.**

Periurethral injections of BulkamidTM appear safe and effective in two single-arm 12-month studies of 135 and 82 women demonstrating objective improvement on pad tests and in subjective measures [114, 115]. A multicenter RCT of BulkamidTM vs. ContigenTM demonstrated noninferiority of Bulkamid TM to ContigenTM with 47.2% and 50% of patients treated with BulkamidTM versus ContigenTM achieving zero incontinence episodes, respectively, and 77.1% versus 70% of patients treated with BulkamidTM versus ContigenTM, respectively, cured/improved.

- **Porcine *dermal implant (PermacolTM, Covidien plc, Dublin, Ireland)*.**

In a 6-month RCT of 50 women, PermacolTM formulated for injection was compared with MacroplastiqueTM again using the weakly discriminating Stamey Urinary Incontinence Scale coupled with a validated questionnaire. They reported that 64% and 60% of the Permacol-treated patients achieved improvements in these respective outcome measures versus 46% and 42% improvement in the MacroplastiqueTM group [116].

Bulking agents have been injected periurethrally (peristomal), transurethrally (transluminally), or combined in the mid-urethra and at the bladder neck, both in retrograde and antegrade fashion, mostly with cystoscopic guidance. Most clinicians prefer that agents should be injected on the luminal side and at the high-pressure region of the sphincter [106]. A decrease in success rates has been reported with an increased number of injection sites because of extravasation [117]. Material should be injected slowly in order to maximize fill and reduce extravasation in fragile soft tissue. The volume injected needs to be balanced: too little can be ineffective, and too much will burst the envelope filled with the bulking agent. Transurethral injections administered close to the bladder neck may be associated with less urinary retention compared with periurethral injections with higher efficacy [118].

Contraindications include an active urinary tract infection or hypersensitivity to the agent or any of its components. Bulking agents have been successfully and safely injected in fully anticoagulated patients.

27.7 Complications

Bulking agents have an acceptably low rate of acute and chronic complications. The most common complication is temporary urinary retention with major adverse events being rare [119–121]. Acute urinary retention is seen in up to 24% of patients and it is transient, lasting 7 days or less in most series [24, 26]. Side effects associated with bulking agents include dysuria, hematuria, uncomplicated urinary tract infection, and de novo urinary urgency, which typically resolves with conservative management [119, 120, 122, 123]. Rare long-term complications include urethral prolapse, potentially from disruption of support between the urethral mucosal and muscle wall with urethral bulking [124–126]. Periurethral masses specifically include periurethral pseudocyst [127] or sterile pseudo abscess [127–135], may be large and symptomatic and are associated with outlet obstruction [136] and/or pain [129]. Aspiration is performed for these lesions [137] which is further associated with symptomatic pseudo abscess reoccurrence and thereby may require definitive transurethral, transvaginal, or retropubic incision and drainage [130]. An autologous fat has been seen to result in pulmonary embolism [129] and death [138]. Bulking agents can be inadvertently injected into the periurethral venous plexus. All bulking agents could be erosive, results either from a property of the injected agent [139] or of the surrounding tissue [140] and in rare instances may result in the formation of a fistula.

27.7.1 Surgery for Stress Incontinence

The primary principle of any surgical procedure for SUI is to provide urinary continence without disturbing micturition and with an acceptable rate of complications. Current and past surgical procedures have failed in achieving universal success in urinary continence outcomes. Approximately 30% of the surgical procedures are done for recurrent SUI patients [141]. Failure rates vary depending on the way continence is measured or defined. The failure rates are reflective, in part, of the shortcomings in the understanding of the pathophysiology of SUI [142, 143].

In 2003 Waetjen [144] reviewed the 1998 NHDS and 1998 National Census databases to estimate prevalence, morbidity, and mortality rates for SUI inpatient surgeries in the United States. Adjusting for population growth, the study found a 45% increase in the frequency of such surgeries between 1988 and 1998. Retropubic suspension was the most common procedure performed (46%), followed by anterior repair (27%) and other SUI procedures (14%). Mortality from stress incontinence surgery was low at 1 per 10,000 surgeries. However, 18.3% had 1 or more complications of which nearly half were due to infection (44%), surgical injury (24%), bleeding (16%).

27.7.2 Retropubic Suspension

The common step among all described procedures is the lifting and fixation of the urethra and bladder neck to the pubic bone or ligaments. The most commonly performed procedures are Marshall–Marchetti–Krantz and the Burch colposuspension. It was first described in 1949 in a man with urinary incontinence after a perineal prostate resection [145]. The original description in females included the placement of vaginal and periurethral sutures. Reports have shown an 82% success rate. The Burch procedure was first described in 1961 as a modification of the MMK, and it involved the approximation of the periurethral fascia to the Cooper's (iliopectineal) ligament by using three pairs of sutures [146]. The procedure underwent a modification by Tanagho that involved an additional approximation of the anterior vaginal wall to the lateral pelvic walls because overcorrection at the bladder neck did not improve success rates and carried a higher risk of voiding dysfunction [147]. The initial success rate was 100%. The 9-year results showed a 93% success rate but an 8% rate of enterocele. These procedures can be performed laparoscopically, but long-term studies have failed to demonstrate complete success [148–151].

Different perspective case series and cohort studies which assess the efficacy of colposuspension with some studies provide long-term follow-up data up to 20 years. A meta-analysis of 2196 women was reported by the American Urological Association in 1997. Mean objective cure rates were found to be 84% at 48 months (CI: 79%–88%) with a follow-up to at least 4 years [146]. A meta-analysis of 39 randomized controlled trials published by the Cochrane group on 2403 women with a mean follow-up of 1 year. Objective cure rates were found to be 85%–90%, and there was a slow decline in cure rates to 70% over 5 years [152].

Traditional sling is associated with a higher incidence of voiding difficulties [153]. Trials comparing colposuspension with the TVT revealed similar efficacy for both procedures, operation time, hospital stay, and time to resume normal activities were significantly shorter with the TVT [11–13/76,79,81]. However, the number of bladder injuries was found to be higher in those women having a TVT when compared to colposuspension [142, 154]. Conversely, significantly, more patients were found to have voiding difficulties following colposuspension [142] and subsequently were taught self-catheterization. In addition, significantly more women required surgery for urogenital prolapse following colposuspension [155]. The 4th ICI has suggested, based on the available evidence, that TVT is more effective than colposuspension and equally effective as the more traditional sling procedures [153]. These findings are also supported by the more recently published 5th report [156]. To date, there has been only one randomized controlled trial comparing colposuspension to a TOT procedure [157] in 100 women followed up at 12 and 24 months. Success rates at 12 months were 80% and 86% in the colposuspension and TOT, respectively, and were no different at 24 months (84% and 88%, respectively).

27.7.2.1 Colposuspension Versus Laparoscopic Colposuspension

Several studies were reported by comparing open and laparoscopic colposuspension. In addition, the Cochrane group has also reported a meta-analysis of 22 studies including laparoscopic colposuspension [158].

There are only two small trials regarding long-term data at 5 years comparing laparoscopic and open colposuspension, one favoring the open procedure [159] and the other the laparoscopic [160]. In light of the available evidence, the 5th ICI concluded that laparoscopic colposuspension has comparable subjective and objective outcomes when compared with open colposuspension in the short to medium term, although longer-term outcomes are unknown. Laparoscopic colposuspension has a similar subjective outcome to TVT, although the objective outcome is less good [156].

27.8 Complications

27.8.1 Perioperative Complications

Common perioperative complications include hemorrhage and lower urinary tract injury. While major hemorrhage is rare [161], one series of 151 patients reported blood loss of over 1000 mL in 2% of cases [162], and transfusion rates were seen to be approximately 5% (CI: 3%–8%) [156]. Bladder and urethral trauma may occur during dissection in the retropubic space, the reported incidence of bladder injury is 1.2% [163] to 2.0% [161]. In addition, the ureters may be injured or kinked at the level of the most cephalad suture with one series of 60 cases reporting an incidence of 6.7% [164].

27.8.2 Postoperative Complications

27.8.2.1 Overactive Bladder

"De novo" urgency and urge incontinence may be reported. While the cause may be multifactorial and partially explained by increased outflow resistance, postoperatively, there is some evidence to suggest that damage to the autonomic innervation may occur following dissection and mobilization of the bladder neck [164]. Overall, the incidence of "de novo" symptoms suggestive of overactive bladder (OAB) has been reported as 11% (CI: 8–16) [156]. These findings support the findings of earlier studies that reported an incidence of detrusor overactivity between 12% and 18.5% [165, 166]. Unfortunately, there are no reliable predictors of outcome with regard to OAB, although there is some evidence to suggest that increased bladder wall thickness may be associated with a higher risk of "de novo" detrusor overactivity postoperatively [167].

27.8.2.2 Voiding Difficulties

In the early or late postoperative period, Voiding difficulty may occur. While the former may be caused by overelevation of the bladder neck at the time of the procedure [168], the latter may be related to gradual detrusor muscle compensation secondary to prolonged outflow obstruction. In addition, many women also notice that their urinary stream is slower and may need to change position to empty their bladder completely. Overall, the incidence of difficulty in voiding lasting over 1 month is reported as 5% (CI: 3–5) [156], although other series have reported rates as high as 21% [142].

27.9 Conclusions

Retropubic urethropexy remains a durable and efficacious surgical treatment for stress urinary incontinence in women. The description of Burch colposuspension in 1961 revolutionized the surgical approach to stress incontinence and rapidly replaced the Marshall–Marchetti–Krantz procedure. Almost 50 years later, the available evidence demonstrates that open retropubic suspension is considered an effective treatment for the treatment of stress incontinence in both long- and short-term trials. Comparative studies have demonstrated that colposuspension is superior to anterior colporrhaphy and needle suspension procedures and is comparable to traditional sling procedures and laparoscopic colposuspension. The evidence from individual trials suggests that colposuspension is as effective as both retropubic and transobturator mid-urethral tape procedures, although a meta-analysis has shown that overall cure rates (OR 0.61; 95% CI: 0.46–0.82; $p = 0.00009$) and objective cure rates (OR 0.38; 95% CI: 0.25–0.57; $p < 0.0001$) are lower for colposuspension when compared to mid-urethral tapes [169]. These findings are also supported by the Cochrane analysis, which has also shown a higher cure rate with mid-urethral slings when compared to colposuspension (RR 0.75; 95% CI: 0.62–0.90) [170]. Colposuspension has also been shown to be a durable and long-lasting procedure. A recent long-term study has reported outcomes in 155,458 women over 10 years following surgery for stress urinary incontinence. The 9-year cumulative incidence of repeat surgery after all procedures was 14.5% (95% CI: 13.4–15.5), after sling procedures 13.0% (95% CI: 11.7–14.3), and colposuspension 10.8% (95% CI: 9.3–12.3). Overall, the rate of repeat surgery was 28% higher for sling procedures when compared to colposuspension (adjusted HR 1.28; 95% CI: 1.19–1.37) [171]. In the 5th report of the ICI, colposuspension continues to be recommended as an effective treatment for primary stress incontinence, and, while superseded by the less invasive mid-urethral tapes, it should still be considered for those women who are having a concomitant abdominal procedure [156]. Colposuspension still has a role in women having concomitant surgery such as abdominal hysterectomy, oophorectomy, and open abdominal sacrocolpopexy. In addition, colposuspension may offer an alternative to a mid-urethral tape procedure following urethral diverticulectomy or repair of a urethra–vaginal fistula, where it may be preferable to avoid the interposition of a synthetic mesh.

27.9.1 Transvaginal Suspension

The search for less invasive procedures led to the development of transvaginal procedures. In 1959, Pereyra described the transvaginal needle suspension. The aim of the procedure was to achieve continence using the same supportive periurethral tissue as the MMK or the Burch colposuspension [172]. Suture fixation stabilized pubocervial fascia to the anterior rectus muscle. A modification by Stamey included the use of an endoscope to verify the correct position of the sutures and synthetic bolters in the vaginal suture to prevent pull-through [173]. Raz in 1981 described his modification to the Pereyra operation [174]. The goal of the procedure was to include the endopelvic fascia for urethral support. The Gittes procedure described in 1987 applied the same technique as Raz with no vaginal dissection [175].

The thought was to mimic the mechanism of the retropubic suspensions. Long-term results of transvaginal suspensions are poor, and the evidence does not support the use of these techniques [176].

27.9.2 Pubovaginal Slings

Over 100 years ago, the principle of placing a strip of autologous tissue beneath the urethra to provide an obstructive effect was established. The person who popularized the concept of placing RF around the urethra was Aldridge in 1942 [177]. Modifications of this technique were practiced until MUS procedures became the surgery of choice since the late 1990s. Nevertheless, since the FDA warning, PVS returned to the main stage [178].

The mechanism of action of sling procedures has been a subject of debate and uncertainty, simply provide additional passive resistance, or passive occlusion during straining or a dynamic component as well in which abdominal straining with rectus muscle contraction, which will pull upward on the attached sling, consequently elevating the bladder neck for the duration of the contraction. All slings thus have the potential to obstruct the urethra. Slings, whatever their maid of, have the potential to shrink/contract by up to 30% of their initial length. Shrinkage of this degree will further elevate the bladder neck or urethra and may cause voiding difficulty. The tendency may be more marked with some tissues than others. The material used matters in terms of its capacity for shrinkage, infection, erosion, and subsequent stretching or failure, the amount of tension applied to the sling and its positioning adversely affect filling and voiding function. Aldridge described a transverse suprapubic incision, two strips of RF were mobilized in the line of the external oblique fibers, detached at their lateral extremities, and left attached to the RF close to the midline creating strips of the fascia of at least 10 cm in length on each side A midline suburethral vaginal incision was then made to expose the pubocervical fascia and dissection continued cranially toward the retropubic space. A midline colporrhaphy was performed to provide bulking of tissue under the urethra, and the ends of the rectus strips were then passed downward through the retropubic space on each side to emerge into the vaginal incision on either side of the urethra sewn together under the urethra to provide elevation and support of the urethra. The procedure has been modified; dissection of the medial component of the rectus strip was angled downward to create additional length, reducing the need for lateral dissection. Lateral dissection in the original procedure was complicated by the thinness of the external oblique fascia beyond the lateral border of the rectus sheath and the risk of lateral angle pain postoperatively. Leaving the sling attached at one end obviated any need for fixation except under the urethra, but this created difficulties in establishing the correct amount of tension; in 1978, McGuire reported the use of a free graft of RF, which facilitated adjustment of tension Blaivas also reported the use of a free graft [179], the reasons for this were the growing awareness of poor long-term outcomes from needle suspensions and that a fascial sling would be effective in women with either hypermobility or intrinsic sphincter weakness or both. Slings with less tension remained effective but with a diminished likelihood of postoperative voiding dysfunction, which remained a problem for retropubic suspensions. Modifications took place with the realization that tension should be minimal and the mid-urethral rather than bladder neck support was the objective.

A disadvantage with RF is its varying thickness and length, to overcome this problem, the use of host fascia lata was described. A length of up to 20 cm of fascia can be mobilized and separated from its ends. The resulting defect does not widen (as is the case with the RF defect) and so does not need to be closed. However, pain and hematoma can still occur so a small suction drain should be placed. The way in which autologous sling surgery is performed has changed gradually over the years.

The single large-scale RCT [142] shows that the efficacy of RF is superior to Burch colposuspension but the risk of adverse events is higher. Guerrero et al. [180] showed that the efficacy and risk of shorter suspended slings equate to full-length slings, and this evidence reflects the changing use of rectus fascial slings in which the use of full lengths has become unusual and shorter lengths the norm. Rates of urinary retention range from 0% to 18%. Postoperative pain ranges from 0% to 25% and de novo urgency from 2% to 30%. A Cochrane Review in 2005 of traditional sling surgery [181] failed to draw any conclusions about the efficacy or risk of various sling procedures. Complications of the autologous sling have included hemorrhage, wound infection, and bladder perforation. Long-term complications include de novo urgency, voiding difficulty, lower abdominal pain, incisional hernia, and erosion. Additionally, according to the AUA guidelines, the estimated dry rates for an autologous sling (rectus fascia or fascia lata) ranged between 90% to 82% at 12–23 months and 48 months, respectively [182]. This is supported by the Stress Incontinence Surgical Treatment Efficacy Trial, which was a multicenter RCT of 655 women randomized to autologous sling versus Burch. The study showed a higher success rate in the treatment of stress incontinence with the autologous fascial sling (47% vs. 38%, $p = 0.01$) but also a higher complication profile [183, 184].

27.9.3 Synthetics

The long-term durability of these procedures with graft materials has raised many questions, due to reports of complications [182]. As such, the mid-urethral synthetic sling was developed, replacing the pubovaginal sling as the gold standard for stress incontinence. In 2008, the FDA issued a statement cautioning the use of mesh for both pelvic organ prolapse and urinary incontinence. In 2011, a statement was given by the FDA warning about mesh-related complications which recommend against the use of mesh for SUI [185]. Patients should be informed that synthetic slings are considered a first-line treatment option for stress incontinence. In fact, the American Urogynecologic Society and Society of Urodynamics, Female Pelvic Medicine, and Urogenital Reconstruction issued an opinion in 2014 that strongly support the use of polypropylene mesh for the treatment of SUI, stating that it is safe and effective and remains the standard of care for the treatment of SUI [186].

27.9.3.1 Tension-Free Vaginal Tapes

These procedures have gained the position as the first line of treatment for SUI as they are minimally invasive that can be done also under local analgesia as day care procedure with good long-term results. It was first described in 1996 by Ulmsten et al. [187]. The initial cure rate was 84% in the first 2 years after surgery. An incision at the anterior vaginal wall with a submucosal paraurethral dissection was made immediately behind the symphysis pubis. Then, a mesh was introduced on each side of the urethra using two pairs of specially designed needles and the mesh is then left in place as such without suturing and incisions are closed [175].

27.9.3.2 Author's Conclusion

The safety and effectiveness of mini-slings for female SUI have not been adequately demonstrated. Presently, it is unclear how mini-slings compare to multi-incision slings with respect to safety and effectiveness for treating SUI. Additional studies may help the agency to better understand the safety and effectiveness of these devices.

The long-term complications of surgical mesh sling repair for SUI that are reported in the literature are consistent with the adverse events reported to the FDA.

The complications associated with the use of surgical mesh slings currently on the market for SUI repair are not linked to a single brand of mesh.

The most common complications reported through MDRs for surgical mesh slings for SUI repair, in descending order of frequency, include: pain, mesh erosion through the vagina (also called exposure, extrusion, or protrusion), infection, urinary problems, recurrent incontinence, pain during sexual intercourse (dyspareunia), bleeding, organ perforation, neuro-muscular problems, and vaginal scarring. Many of these complications require additional medical intervention and sometimes require surgical treatment and/or hospitalization. With the exception of mesh erosion, the above complications can occur following a non-mesh surgical repair for SUI.

MUS operations have been the most extensively researched surgical treatment for stress urinary incontinence (SUI) in women and have a good safety profile.

References

1. Haylen BT, et al. An International Urogynecological Association (IUGA)/International Continence Society (ICS) Joint Report on the Terminology for Female Pelvic Floor Dysfunction. Neurourol Urodyn. 2010;29:4–20.
2. Abrams P, Cardozo L, Fall M, Griffiths D, Rosier P, Ulmsten U, et al. The standardisation of terminology in lower urinary tract function: report from the standardisation sub-committee of the International Continence Society. Urology. 2003 Jan;61(1):37–49.
3. Reich WJ, Wilkey JL, Silverman HE. Urinary stress incontinence in the female; a combined gyno-urological approach to its correction. Am J Surg. 1945 Dec;70:341–53.
4. Nager CW, Albo ME, Fitzgerald MP, et al. Reference urodynamic values for stress incontinent women. Neurourol Urodynam. 2007;26:333–40.
5. Summitt RL Jr, Stovall TG, Bent AE, Ostergard DR. Urinary incontinence: correlation of history and brief office evaluation with multichannel urodynamic testing. Am J Obstet Gynecol. 1992;166:1835–40.
6. Walters MD, Shields LE. The diagnostic value of history, physical examination, and the Q-tip cotton swab test in women with urinary incontinence. Am J Obstet Gynecol. 1988;159:145–9.
7. Cundiff GW, Harris RL, Coates KW, Bump RC. Clinical predictors of urinary incontinence in women. Am J Obstet Gynecol. 1997;177:262–6.
8. Luber KM, Boero S, Choe JY, et al. The demographics of pelvic floor disorders: current observations and future projections. Am J Obstet Gynecol. 2001;184:1496–501. discussion 1501–1503
9. Nygaard IE, Lemke JH. Urinary incontinence in rural older women; prevalence, incidence, and remission. J Am Geriatr Soc. 1996;44:1049–54.
10. Hampel C, Wienhold D, Benken N, et al. Definition of overactive bladder and epidemiology of urinary incontinence. Urology. 1997;50(6A Suppl):4–14.
11. Foldspang A, Mommsen S, Djurhuus JC. Prevalent urinary incontinence as a correlate of pregnancy, vaginal childbirth, and obstetric techniques. Am J Public Health. 1999;89(2):209–12.
12. Swift SE, Ostergard DR. Effects of progesterone on the urinary tract. Int Urogynecol J. 1993;4(4):232–6.

13. Van Geelen JM, Lemmens WA, Eskes TK, Martin CB Jr. The urethral pressure profile in pregnancy and after delivery in healthy nulliparous women. Am J Obstet Gynecol. 1982;144(6):636–49.
14. Landon CR, Crofts CE, Smith ARB, Trowbridge EA. Mechanical properties of fascia during pregnancy: a possible factor in the development of stress incontinence of urine. Contemp Rev Obstet Gynaecol. 1990;2:40–6.
15. Thom DH, Rortveit G. Prevalence of postpartum urinary incontinence: a systematic review. Acta Obstet Gynecol Scand. December 2010;89(12):1511–22.
16. Carley ME, Schaffer J. Urinary incontinence and pelvic organ prolapse in women with Marfan or Ehlers-Danlos syndrome. Am J Obstet Gynecol. 2000;182(5):1021–3.
17. Dolan LM, Hosker GL, Mallett VT, Allen RE, Smith ARB. Stress incontinence and pelvic floor neurophysiology 15 years after the first delivery. Br J Obstet Gynaecol. 2003;110(12):1107–14.
18. Viktrup L, Lose G. The risk of stress incontinence 5 years after first delivery. Am J Obstet Gynecol. 2001;185(1):82–7.
19. Glazener C, Elders A, MacArthur C, et al. Childbirth and prolapse: long-term associations with the symptoms and objective measurement of pelvic organ prolapse. Br J Obstet Gynaecol. 2013;120(2):161–8.
20. DeLancey JO, Trowbridge ER, Miller JM. Stress urinary incontinence relative importance to urethra support and urethra closure pressure. J Urol. 2008;179:2286–90.
21. Viktrup L, Summers KH, Dennett SL. Clinical practice guidelines for the initial management of urinary incontinence in women: a European-focused review. Br J Urol Int. 2004;94(Suppl 1):14–22.
22. Abrams P, Cardozo L, Khoury S, Wein A, eds. Incontinence, 5th International Consultation on Incontinence, Paris, France, February 2012. ICUD-EAU 2013.
23. Abrams P. Assessment and treatment of urinary incontinence. Lancet. 2000;355:2153–8.
24. Lagro-Janssen ALM, Debruyne FMJ, Smits AJA, van Weel C. The effects of treatment of urinary incontinence in general practice. Fam Pract. 1992;9(3):284–9.
25. Berghmans LCM, Hendriks HJM, De Bie RA, Van Waalwijk van Doorn ESC, Bø K, Van Kerrebroeck PEV. Conservative treatment of urge urinary incontinence in woman: a systematic review of randomized clinical trials. Br J Urol Int. 2000;85:254–63.
26. Bent AE, Gousse AE, Hendrix SL, et al. Validation of a two-item quantitative questionnaire for the triage of women with urinary incontinence. Obstet Gynecol. 2005;106:767–73.
27. Nager CW, Brubaker L, Litman HJ, et al. A randomized trial of urodynamic testing before stress-incontinence surgery. N Engl J Med. 2012;366(21):1987–97.
28. Martin JL, Williams KS, Sutton AJ, Abrams KR, Assassa RP. Systematic review and meta-analysis of methods of diagnostic assessment for urinary incontinence. Neurourol Urodyn. 2006;25:674–83.
29. Gallentine ML, Cespedes RD. Occult stress urinary incontinence and the effect of vaginal vault prolapse on abdominal leak point pressures. Urology. January 2001;57(1):40–4.
30. Winters JC, Dmochowski RR, Goldman HB, Herndon CDA, Kobashi KC, Kraus SR, Lemack GE, Nitti VW, Rovner ES, Wein AJ. Adult urodynamics: AUA/SUFU guidelines. Linthicum, MD: American Urological Association Education and Research Inc.; April 2012.
31. McGuire EJ, Fitzpatrick CC, Wan J, et al. Clinical assessment of urethral sphincter function. J Urol. November 1993;150(5 Pt 1):1452–4.
32. Choe JH, Lee JY, Lee KS. Accuracy and precision of a new portable ultrasound scanner, the BME-150A, in residual urine volume measurement: a comparison with the BladderScan BVI 3000. Int Urogynecol J Pelvic Floor Dysfunct. 2007;18:641–4.
33. McGuire EJ. Urodynamic findings in patients after failure of stress incontinence operations. Prog Clin Biol Res. 1981;78:351–60.
34. Sand PK, Bowen LW, Panganiban R, Ostergard DR. The low pressure urethra as a factor in failed retropubic urethropexy. Obstet Gynecol. March 1987;69(3 Pt 1):399–402.
35. Koonings PP, Bergman A, Ballard CA. Low urethral pressure and stress urinary incontinence in women: risk factor for failed retropubic surgical procedure. Urology. September 1990;36(3):245–8.
36. Bowen LW, Sand PK, Ostergard DR, Franti CE. Unsuccessful Burch retropubic urethropexy: a case-controlled urodynamic study. Am J Obstet Gynecol. February 1989;160(2):452–8.
37. Rezapour M, Falconer C, Ulmsten U. Tension-Free vaginal tape (TVT) in stress incontinent women with intrinsic sphincter deficiency (ISD)—A long-term follow-up. Int Urogynecol J Pelvic Floor Dysfunct. 2001;12(Suppl. 2):S12–4.
38. Appell RA. Primary slings for everyone with genuine stress incontinence? The argument for. Int Urogynecol J Pelvic Floor Dysfunct. 1998;9(5):249–51.
39. Nwosu CR, Khan KS, Chien PF, Honest MR. Is real-time ultrasonic bladder volume estimation reliable and valid? A systematic overview. Scand J Urol Nephrol. 1998;32:325–30.
40. Gehrich A, Stany MP, Fischer JR, Buller J, Zahn CM. Establishing a mean postvoid residual volume in asymptomatic perimenopausal and postmenopausal women. Obstet Gynecol. 2007;110:827–32.

41. Teng CH, Huang YH, Kuo BJ, Bih LI. Application of portable ultrasound scanners in the measurement of post-void residual urine. J Nurs Res. 2005;13:216–24.
42. Resnick B. A bladder scan trial in geriatric rehabilitation. Rehabil Nurs. 1995;20:19.
43. Haylen BT, Lee J, Logan V, et al. Immediate postvoid residual volumes in women with symptoms of pelvic floor dysfunction. Obstet Gynecol. 2008;111:1305–12.
44. Brown J, Seeley D, Feng J, et al. Urinary incontinence in older women: who is at risk? Study of Osteoparotic Fractures Research Group. Obstet Gynecol. 1996;87:715.
45. Brown J, Grady D, Ouslander J, Herzog A, Varner RE, Posner S. Prevalence of urinary incontinence and associated risk factors in postmenopausal women. Heart & Estrogen/Progestin Replacement Study (HERS) Research Group. Obstet Gynecol. 1999;94:66.
46. Bump R, Sugerman H, Fantl J, et al. Obesity and lower urinary tract function in women: effect of surgically induced weight loss. Am J Obstet Gynecol. 1992;166:392.
47. Deitel M, Stone E, Kassam HA, et al. Gynecologic-obstetric changes after loss of massive excess weight following bariatric surgery. J Am Coll Nutr. 1988;7:147.
48. Burgio KL, Richter HE, Clements RH, Redden DT, Goode PS. Changes in urinary and fecal incontinence symptoms with weight loss surgery in morbidly obese women. Obstet Gynecol. 2007;110:1034–40.
49. Subak LL, Johnson CEW, Boban D, et al. Does weight loss improve incontinence in moderately obese women? Int Urogynecol J Pelvic Floor Dysfunct. 2002;13:40.
50. Subak LL, Wing R, West DS, et al. for the Program to Reduce Incontinence by Diet and Exercise (PRIDE). Weight loss to treat urinary incontinence in overweight and obese women. N Engl J Med. 2009;360:481–90.
51. Nygaard IE, Kreder KJ, Lepic MM, Fountain KA, Rhomberg AT. Efficacy of pelvic floor muscle exercises in women with stress, urge, and mixed urinary incontinence. Am J Obstet Gynecol. 1996;174:120–5.
52. Baigis-Smith Smith J, DAJ S, Rose M, Newman DK. Managing urinary incontinence in community-residing elderly persons. Gerontologist. 1989;229:33.
53. Fantl JA, Wyman JF, McClish DK, et al. Efficacy of bladder training in older women with urinary incontinence. J Am Med Assoc. 1991;265:609–13.
54. Wyman JF, Fantl JA, McClish DK, et al. Comparative efficacy of behavioral interventions in the management of female urinary incontinence. Am J Obstet Gynecol. 1998;179:999–1007.
55. Weinberger MW, Goodman BM, Carnes M. Long-term efficacy of nonsurgical urinary incontinence treatment in elderly women. J Gerontol. 1999;54:M117–21.
56. Subak LL, Quesenberry CP, Posner SF, Cattolica E, Soghikian K. The effect of behavioral therapy on urinary incontinence: a randomized controlled trial. Obstet Gynecol. 2002;100:72–8.
57. Burgio KL, Goode PS, Locher JL, et al. Predictors of outcome in the behavioral treatment of urinary incontinence in women. Obstet Gynecol. 2003; 102:940.
58. Theofrastous JP, Wyman JF, Bump RC, McClish DK, Elser DM, Bland DR, Fantl JA. Effects of pelvic floor muscle training on strength and predictors of response in the treatment of urinary incontinence. Neurourol Urodynam. 2002;21:486–90.
59. Howard D, Miller J, DeLancey J, et al. Differential effects of cough, valsalva, and continence status on vesical neck movement. Obstetr Gybecol. 2000;95:535–40.
60. Deindl FM, Vodusek DB, Hesse U, Schussler B. Activity patterns of pubococcygeal muscles in nulliparous continent women. Br J Urol. July 1993;72(1):46–51.
61. Bø K, Stien R, Kulseng-Hanssen S, Kristofferson M. Clinical and urodynamic assessment of nulliparous young women with and without stress incontinence symptoms: a case–control study. Obstet Gynecol. December 1994;84(6):1028–32.
62. Constantinou CE, Govan DE. Spatial distribution and timing of transmitted and reflexly generated urethral pressures in healthy women. J Urol. May 1982;127(5):964–9.
63. Thind P, Lose G, Jorgensen L, Colstrup H. Variations in urethral and bladder pressure during stress episodes in healthy women. Br J Urol. October 1990;66(4):389–92.
64. Bø K. Pelvic floor muscle exercise for the treatment of stress urinary incontinence: an exercise physiology perspective. Int Urogynecol J Pelvic Floor Dysfunct. 1995;6:282–91.
65. Bø K. Physiotherapy to treat genuine stress incontinence. Int Cont Surv. 1996;6:2–8.
66. DiNubile NA. Strength training. Clin Sports Med. 1991;10:33–62.
67. Bø K, Hage RH, Kvarstein B, Jorgensen J, Larsen S. Pelvic floor muscle exercise for the treatment of female Stress urinary incontinence. III Effects of two different degrees of pelvic floor muscle exercises. Neurourol Urodyn. 1990;9:489–502.
68. Berghmans LCM, Hendriks HJM, Bø K, Hay-Smith EJ, De Bie RA, Van Waalwijk van Doorn ESC. Conservative treatment of stress urinary incontinence in women: a systematic review of randomized clinical trials. Br J Urol. 1998;82:181–91.

69. Hay-Smith EJC, Bø K, Berghmans LCM, Hendriks HJM, de Bie RA, van Waalwijk van Doorn ESC. Pelvic floor muscle training for urinary incontinence in women (Cochrane review). In: The Cochrane Library, Issue 4. Chichester, UK: John Wiley & Sons, Ltd.; 2003. CD001407.
70. Bø K, Kvarstein B, Nygaard I. Lower urinary tract symptoms and pelvic floor muscle exercise adherence after 15 years. Obstetr Gynecol. 2005;105(5 Pt 1):999–1005.
71. Cammu H, Van Nijlen M, Amy J. A ten-year follow-up after Kegel pelvic floor muscle exercises for genuine stress incontinence. Br J Urol. 2000;85:655–8.
72. Hay-Smith Smith J, Herderschee R, Dumoulin C, Herbison P. Comparisons of approaches to pelvic floor muscle training for urinary incontinence in women: an abridged Cochrane systematic review. Eur J Phys Rehabil Med. December 2012;48(4):689–705.
73. Dumoulin C, Hay-Smith J. Pelvic floor muscle training versus no treatment, or inactive control treatments, for urinary incontinence in women. https://doi.org/10.1002/14651858.CD005654.pub2.
74. Nygaard I. Prevention of exercise incontinence with mechanical devices. J Reprod Med. 1995;40(2):89–94.
75. Richter HE, Brubaker L, Nygaard IE, et al. Continence pessary compared with behavioral therapy or combined therapy for stress incontinence: a randomized controlled trial. Obstet Gynecol. 2010;115(3):609–17.
76. Donnelly MJ, Powell-Morgan S, Olsen AL, Nygaard IE. Vaginal pessaries for the management of stress and mixed urinary incontinence. Int Urogynecol J Pelvic Floor Dysfunct. 2004;15(5):302–7.
77. Farrell SA, Singh B, Aldakhil L. Continence pessaries in the management of urinary incontinence in women. J Obstet Gynaecol Can. 2004;26(2):113–7.
78. Allen W, Leek H, Izurieta A, Moore K. Update: the "Contiform" intravaginal device in four sizes for the treatment of stress incontinence. Int Urogynecol J. 2008;19(6):757.
79. Thyssen H, Bidmead J, Lose G, Møller Bek K, Dwyer P, Cardozo L. A new intravaginal device for stress incontinence in women. Br J Urol Int. 2001;88(9):889–92.
80. Ziv E, Stanton SL, Abarbanel J. Efficacy and safety of a novel disposable intravaginal device for treating stress urinary incontinence. Am J Obstet Gynecol. 2008;198(5):594.e1.
81. Vierhout ME, Lose G. Preventive vaginal and intra-urethral devices in the treatment of female urinary stress incontinence. Curr Opin Obstet Gynecol. 1997;9(5):325–8.
82. Staskin D, Bavendam T, Miller J, et al. Effectiveness of a urinary control insert in the management of stress urinary incontinence: early results of a multicenter study. Urology. 1996;47(5):629–36.
83. Sirls LT, Foote JE, Kaufman JM, et al. Long-term results of the FemSoft urethral insert for the management of female stress urinary incontinence. Int Urogynecol J Pelvic Floor Dysfunct. 2002;13(2):88–95. discussion 95
84. Brubaker L, Harris T, Gleason D, Newman D, North B. The external urethral barrier for stress incontinence: a multicenter trial of safety and efficacy. Obstet Gynecol. 1999;93(6):932.
85. Lipp A, Shaw C, Glavind K. Mechanical devices for urinary incontinence in women. Cochrane Database Syst Rev. 2011;7:CD001756.
86. Wein AJ, Rovner ES. Pathophysiology and Classification of Voiding Dysfunction. In: Appell RA, editor. Voiding Dysfunction. Totowa, NJ: Current Clinical Urology. Humana Press; 2000. https://doi.org/10.1007/978-1-59259-198-5_1.
87. Alhasso A, Glazener CMA, Pickard R, et al. Adrenergic drugs for urinary incontinence in adults. Cochrane Database for Syst Rev. 2005;3:CD001842.
88. Thor KB, Katofiasc MA. Effects of duloxetine, a combined serotonin and norepinephrine reuptake inhibitor, on central neural control of lower urinary tract function in the chloralose-anesthetized female cat. J Pharmacol Exp Ther. 1995;274(2): 1014–24.
89. Mariappan P, Alhasso A, Ballantyne Z, et al. Duloxetine, a serotonin and noradrenaline reuptake inhibitor (SNRI) for the treatment of stress urinary incontinence: a systematic review. Eur Urol. 2007;51:67–74.
90. Gilja I, Radej M, Kovacic M, et al. Conservative treatment of female stress incontinence with imipramine. J Urol. 1984;132:909–11.
91. Rud T. Urethral pressure profile in continent women from childhood to old age. Acta Obstet Gynecol Scand. 1980;59:331–5.
92. Fantl JA, Cardozo L, McClish DK. Estrogen therapy in the management of urinary incontinence in postmenopausal women: a meta-analysis. First report of the Hormones and Urogenital Therapy Committee. Obstet Gynecol. 1994;83:12–8.
93. Sultana CJ, Walters MD. Estrogen and urinary incontinence in women. Maturitas. 1994;20: 129–38.
94. Hendrix SL, Cochrane BB, Nygaard IE, et al. Effects of estrogen with and without progestin on urinary incontinence. J Am Med Assoc. 2005;293:935–48.
95. Robinson D, Cardozo LD. The role of estrogens in female lower urinary tract dysfunction. Urology. 2003;62(4 Suppl 1):45–51.
96. Zinner NR, Sterling AM, Ritter RC. Role of inner urethral softness in urinary continence. Urology. 1980;16:115–7.

97. Raz S, Maggio AJ Jr, Kaufman JJ. Why Marshall–Marchetti operation works ... or does not. Urology. 1979;14:154–9.
98. Winters JC, Dmochowski RR, Goldman HB, et al. Urodynamic studies in adults: AUA/SUFU guideline. J Urol. 2012;188:2464–72.
99. Chrouser KL, Fick F, Goel A, Itano NB, Sweat SD, Lightner DJ. Carbon coated zirconium beads in beta-glucan gel and bovine glutaraldehyde cross-linked collagen injections for intrinsic sphincter deficiency: continence and satisfaction after extended follow up. J Urol. 2004;171:1152–5.
100. Koski ME, Enemchukwu EA, Padmanabhan P, Kaufman MR, Scarpero HM, Dmochowski RR. Safety and efficacy of sling for persistent stress urinary incontinence after bulking injection. Urology 2011; 77:1076–1Lee HN, Lee Y-S, Han J-Y, Jeong JY, Choo M-S, Lee K-S. Transurethral injection of bulking agent for treatment of failed mid-urethral sling procedures. Int Urogynecol J Pelvic Floor Dysfunct. 2010;21:1479–83.
101. Plotti F, Zullo MA, Sansone M, et al. Post radical hysterectomy urinary incontinence: a prospective study of transurethral bulking agents injection. Gynecol Oncol. 2009;112:90–4.
102. Westney OL, Bevan-Thomas R, Palmer JL, Cespedes RD, McGuire EJ. Transurethral collagen injections for male intrinsic sphincter deficiency: the University of Texas-Houston experience. J Urol. 2005;174:994–7.
103. Bennett JK, Green BG, Foote JE, Gray M. Collagen injections for intrinsic sphincter deficiency in the neuropathic urethra. Paraplegia. 1995;33:697–700.
104. De Vocht TF, Chrzan R, Dik P, Klijn AJ, De Jong TPVM. Long-term results of bulking agent injection for persistent incontinence in cases of neurogenic bladder dysfunction. J Urol. 2010;183:719–23.
105. Welk BK, Afshar K, Rapoport D, MacNeily AE. Complications of the catheterizable channel following continent urinary diversion: their nature and timing. J Urol. 2008;180:1856–60.
106. Klarskov N, Lose G. Urethral injection therapy: What is the mechanism of action? Neurourol Urodyn. 2008;27:789–92.
107. Dmochowski RR, Appell RA. Injectable agents in the treatment of stress urinary incontinence in women: Where are we now? Urology. 2000;56:32–40.
108. Maher CF, O'Reilly BA, Dwyer PL, Carey MP, Cornish A, Schluter P. Pubovaginal sling versus transurethral Macroplastique for stress urinary incontinence and intrinsic sphincter deficiency: a prospective randomised controlled trial. Br J Obstet Gynaecol. 2005;112:797–801.
109. Lucas MG, Bosch RJ, Burkhard FC, et al. EAU guidelines on surgical treatment of urinary incontinence. Eur Urol. 2012;62:1118–29.
110. Mohr S, Siegenthaler M, Mueller MD, Kuhn A. Bulking agents: an analysis of 500 cases and review of the literature. Int Urogynecol J Pelvic Floor Dysfunct. 2013;24:241–7.
111. Ghoniem G, Corcos J, Comiter C, Westney OL, Herschorn S. Durability of urethral bulking agent injection for female stress urinary incontinence: 2-year multicenter study results. J Urol. 2010;183:1444–9.
112. Mayer RD, Dmochowski RR, Appell RA, et al. Multicenter prospective randomized 52-week trial of calcium hydroxylapatite versus bovine dermal collagen for treatment of stress urinary incontinence. Urology. 2007;69:876–80.
113. Lightner D, Calvosa C, Andersen R, et al. A new injectable bulking agent for treatment of stress urinary incontinence: results of a multicenter, randomized, controlled, double-blind study of Durasphere. Urology. 2001;58:12–5.
114. Hobson T-H, Al-Singary W, Fynes M, Tegerstedt G, Lose G. Two-year follow-up of an open-label multicenter study of polyacrylamide hydrogel (Bulkamid) for female stress and stress-predominant mixed incontinence. Int Urogynecol J Pelvic Floor Dysfunct. 2012;23:1373–8.
115. Leone Roberti Maggiore U, Alessandri F, Medica M, Gabelli M, Venturini PL, Ferrero S. Outpatient periurethral injections of polyacrylamide hydrogel for the treatment of female stress urinary incontinence: effectiveness and safety. Arch Gynecol Obstet. 2013;288:131–7.
116. Bano F, Barrington JW, Dyer R. Comparison between porcine dermal implant (Permacol) and silicone injection (Macroplastique) for urodynamic stress incontinence. Int Urogynecol J Pelvic Floor Dysfunct. 2005;16:147–50. discussion 50
117. Barranger E, Fritel X, Kadoch O, Liou Y, Pigne A. Results of transurethral injection of silicone micro-implants for females with intrinsic sphincter deficiency. J Urol. 2000;164:1619–22.
118. Schulz JA, Nager CW, Stanton SL, Baessler K. Bulking agents for stress urinary incontinence: short-term results and complications in a randomized comparison of periurethral and transurethral injections. Int Urogynecol J Pelvic Floor Dysfunct. 2004;15:261–5.
119. TA S. Urinary Incontinence in the female: stress incontinence. In: Harrison JHGR, Perlmutter AD, editors. Campbell's Urology. 4th ed. Philadelphia, PA: WB Saunders; 1979. p. 2272–93.
120. D'Agostino R, Nam BH. Evaluation of the performance of survival analysis models: discrimination and calibration measures. In: Balakrishnan N, Rao CR, editors. Handbook of Statistics: advance in Survival Analysis. Amsterdam, the Netherlands: Elsevier B.V; 2004. p. 1–25.

121. Wainstein MA, Klutke CG. Antegrade techniques of collagen injection for post-prostatectomy stress urinary incontinence: the Washington University experience. World J Urol. 1997;15:310–5.
122. Frauscher F. Stem cells for urinary incontinence and coauthor responsibility. Lancet. 2008;372:1732.
123. Salles AG, Lotierzo PH, Gimenez R, Camargo CP, Ferreira MC. Evaluation of the poly-L-lactic acid implant for treatment of the nasolabial fold: 3-year follow-up evaluation. Aesthetic Plast Surg. 2008;32:753–6.
124. Harris RL, Cundiff GW, Coates KW, Addison WA, Bump RC. Urethral prolapse after collagen injection. Am J Obstet Gynecol. 1998;178:614–5.
125. Palma PC, Riccetto CL, Martins MH, et al. Massive prolapse of the urethral mucosa following periurethral injection of calcium hydroxylapatite for stress urinary incontinence. Int Urogynecol J Pelvic Floor Dysfunct. 2006;17:670–1.
126. Ko EY, Williams BF, Petrou SP. Bulking agent induced early urethral prolapse after distal urethrectomy. Int Urogynecol J Pelvic Floor Dysfunct. 2007;18:1511–3.
127. Wainstein MA, Klutke CG. Periurethral pseudocyst following cystoscopic collagen injection. Urology. 1998;51:835–6.
128. Graf W, Mellgren A, Matzel KE, et al. Efficacy of dextranomer in stabilised hyaluronic acid for treatment of faecal incontinence: a randomised, sham-controlled trial. Lancet. 2011;377:997–1003.
129. Sweat SD, Lightner DJ. Complications of sterile abscess formation and pulmonary embolism following periurethral bulking agents. J Urol. 1999;161:93–6.
130. Lightner DJ, Fox J, Klingele C. Cystoscopic injections of dextranomer hyaluronic acid into proximal urethra for urethral incompetence: efficacy and adverse outcomes. Urology. 2010;75:1310–4.
131. Berger MB, Morgan DM. Delayed presentation of pseudoabscess secondary to injection of pyrolytic carbon-coated beads bulking agent. Female Pelvic Med. 2012;18:303–5.
132. Gopinath D, Smith AR, Reid FM. Periurethral abscess following polyacrylamide hydrogel (Bulkamid) for stress urinary incontinence. Int Urogynecol J Pelvic Floor Dysfunct. 2012;23:1645–8.
133. Hanke CW, Higley HR, Jolivette DM, Swanson NA, Stegman SJ. Abscess formation and local necrosis after treatment with Zyderm or Zyplast collagen implant. J Am Acad Dermatol. 1991;25:319–26.
134. McLennan MT, Bent AE. Suburethral abscess: a complication of periurethral collagen injection therapy. Obstet Gynecol. 1998;92:650–2.
135. Madjar S, Sharma AK, Waltzer WC, Frischer Z, Secrest CL. Periurethral mass formations following bulking agent injection for the treatment of urinary incontinence. J Urol. 2006;175:1408–10.
136. Bernier PA, Zimmern PE, Saboorian MH, Chassagne S. Female outlet obstruction after repeated collagen injections. Urology. 1997;50:618–21.
137. Petrou SP, Pak RW, Lightner DJ. Simple aspiration technique to address voiding dysfunction associated with transurethral injection of dextranomer/hyaluronic acid copolymer. Urology. 2006;68:186–8.
138. Currie I, Drutz HP, Deck J, Oxorn D. Adipose tissue and lipid droplet embolism following periurethral injection of autologous fat: case report and review of the literature. Int Urogynecol J Pelvic Floor Dysfunct. 1997;8:377–80.
139. Carlin BI, Klutke CG. Development of urethrovaginal fistula following periurethral collagen injection. J Urol. 2000;164:124.
140. Pruthi RS, Petrus CD, Bundrick Ws JR. New onset vesicovaginal fistula after transurethral collagen injection in women who underwent cystectomy and orthotopic neobladder creation: presentation and definitive treatment. J Urol. 2000;164:1638–9.
141. Olsen AL, Smith VJ, Bergstrom JO, Colling JC, Clark AL. Epidemiology of surgically managed pelvic organ prolapse and urinary incontinence. Obstet Gynecol. 1997;89:501–6.
142. Albo ME, Richter HE, Brubaker L, et al. Burch colposuspension versus fascial sling to reduce urinary stress incontinence. N Engl J Med. 2007;356:2143–55.
143. DeLancey JO. Structural aspects of the extrinsic continence mechanism. Obstet Gynecol. 1988;72:296–301.
144. Waetjen LE, Subak LL, Shen H, Lin F, Wang TH, Vittinghoff E, Brown JS. Stress urinary incontinence surgery in the United States. Obstet Gynecol 2003 Apr; 101(4):671–676. https://doi.org/10.1016/s0029-7844(02)03124-1. PMID: 12681869.
145. Marshall VF, Marchetti AA, Krantz KE. The correction of stress incontinence by simple vesicourethral suspension. Surg Gynecol Obstet. 1949;88:509–18.
146. Leach GE, Dmochowski RR, Appell RA, et al. Female Stress Urinary Incontinence Clinical Guidelines Panel summary report on surgical management of female stress urinary incontinence. The American Urological Association. J Urol. 1997;158:875–80.
147. Tanagho EA. Colpocystourethropexy: the way we do it. J Urol. 1976;116:751–3.
148. Prezioso D, Iacono F, Di Lauro G, et al. Stress urinary incontinence: long term results of laparoscopic Burch colposuspension. BMC Surg. 2013;13(S2):S38.
149. Vancaillie TG, Schuessler W. Laparoscopic bladder neck suspension. J Laparoendosc Surg. 1991;1:169–73.
150. Burton G. A randomised comparison of laparoscopic and open colposuspension. Neurourol Urodyn. 1994;13:497–8.

151. Su T, Wang K, Hsu C, Wei H, Hong B. Prospective comparison of laparoscopic and traditional colposuspension in the treatment of genuine stress incontinence. Acta Obstet Gynecol Scand. 1997;76:576–82.
152. Lapitan MC, Cody DJ, Grant AM. Open retropubic colposuspension for urinary incontinence in women. Cochrane Database Syst Rev. 2005;3:CD002912.
153. Smith ARB, Dmochowski R, Hilton P, Rovner E, Nilsson CG, Reid FM, Chang D. Surgery for urinary incontinence in women. In: Abrams P, Cardozo L, Khoury S, Wein A, editors. WHO-ICUD International Consultation on Incontinence. 4th ed. Paris: France; 2009. p. 1193–272.
154. Ward KL, Hilton P. On behalf of the UK and Ireland TVT Trial Group. Tension free vaginal tape versus colposuspension for primary urodynamic stress incontinence: 5 year follow up. Br J Obstet Gynaecol. 2008;115:226–33.
155. Wang AC, Chen MC. Comparison of tension free vaginal taping versus modified Burch colposuspension on urethral obstruction: a randomised controlled trial. Neurourol Urodyn. 2003;22:185–90.
156. Dmochowski R, Athanasiou S, Reid FM, Kraus S, Nitti V, Gomelsky A, Robinson D, Smith ARB. Surgery for urinary incontinence in women. In: Abrams P, Cardozo L, Khoury S, Wein A, editors. WHO-ICUD International Consultation on Incontinence. 5th ed. Paris: France; 2013. p. 1307–75.
157. Sivaslioglu AA, Caliskan E, Dolen I, Haberal A. A randomised comparison of transobturator tape and Burch colposuspension in the treatment of female stress urinary incontinence. Int Urogynaecol J Pelvic Floor Dysfunct. 2007;18:1015–9.
158. Dean NM, Ellis G, Wilson PD, Herbison GP. Laparoscopic colposuspension for urinary incontinence in women. Cochrane Database Syst Rev. 2002;3:CD002239.
159. Burton G. A five year prospective randomised urodynamic study comparing open and laparoscopic colposuspension. Neurourol Urodyn. 1999;18:295–6.
160. Morris A, Reilly E, Hassan A, Ramsay I, Hawthorn R. 5–7 year follow-up of a randomised trial comparing laparoscopic colposuspension and open colposuspension in the treatment of genuine stress incontinence. Int Urogynaecol J Pelvic Floor Dysfunct. 2001;12:s6.
161. Cardozo L, Hextall A, Bailey J, Boos K. Colposuspension after previous failed continence surgery: a prospective observational study. Br J Obstet Gynaecol. 1999;106:340–4.
162. Kenton K, Oldham L, Brubaker L. Open Burch Urethropexy has a low rate of perioperative complications. Am J Obstet Gynaecol. 2002;187:107–10.
163. Cosson M, Boukerrou M, Narducci F, et al. Long term results of the Burch procedure combined with abdominal sacrocolpopexy for treatment of vault prolapse. Int Urogynecol J Pelvic Floor Dysfunct. June 2003;14(2):104–7.
164. Harris RL, Cundiff GW, Theofrastous JP. The value of intraoperative cystoscopy in urogynaecologic and reconstructive pelvic surgery. Am J Obstet Gynaecol. 1997;177:1367–71.
165. Feyereisl J, Dreher E, Haenggi W, et al. Long term results after Burch colposuspension. Am J Obstet Gynaecol. 1994;171:647–52.
166. Usten Y, Engin-Ustin Y, Gungor M, Tezcan S. Randomised comparison of Burch urethropexy procedures concomitant with gynaecologic operations. Gynecol Obstet Invest. 2005;59:19–23.
167. Robinson D, Khullar V, Cardozo L. Assessing surgical outcome: Urethral function revisited. Neurourol Urodyn. 2005;24:512–4.
168. Dundas D, Hilton P, Williams JE, Stanton SL. Aetiology of voiding difficulty following colposuspension. Proceedings of the 12th Meeting of the International Continence Society, Leiden, the Netherlands, 1982. p. 132.
169. Novara G, Artibani W, Barber MD, Chapple CR, Constantini E, Ficarra V, Hilton P, Nilsson CG, Waltregny D. Updated systematic review and meta-analysis of the comparative data on colposuspensions, pubovaginal slings, and mid-urethral tapes in the surgical treatment of female stress urinary incontinence. Eur Urol. 2010;58:218–38.
170. Rehman H, Bezerra CC, Bruschini H, Cody JD. Traditional sling operations for urinary incontinence in women. Cochrane Database Syst Rev. January 19, 2011;3:CD001754.
171. Jonsson Funk M, Siddiqui NY, Kawasaki A, Wu JM. Long-term outcomes after stress urinary incontinence surgery. Obstet Gynaecol. 2012;120:83–90.
172. Pereyra AJ. A simplified surgical procedure for the correction of stress incontinence in women. West J Surg Obstet Gynecol. 1959;67:223–36.
173. Stamey TA. Endoscopic suspension of the vesical neck for urinary incontinence. Surg Gynecol Obstet. 1973;136:547–54.
174. Raz S. Modified bladder neck suspension for female stress incontinence. Urology. 1981;17:82–5.
175. Gittes RF, Loughlin KR. No-incision pubovaginal suspension for stress incontinence. J Urol. 1987;138:568–70.
176. Leach GE, Dmochowski RR, Appell RA, et al. Female stress urinary incontinence clinical guidelines panel summary report on surgical management of female stress urinary incontinence. J Urol. 1997;158:875–80.
177. Aldridge A. Transplantation of fascia for relief of urinary incontinence. Am J Obstet Gynecol. 1942;44:398–411.
178. Goodman C. Personal communication.
179. Blaivas JG, Jacobs BZ. Pubovaginal fascial sling for the treatment of complicated stress urinary incontinence. J Urol. 1991;145:1214–8.
180. Guerrero K, Watkins A, Emery S, et al. A randomised controlled trial comparing two autologous fascial sling techniques for the treatment of stress

urinary incontinence in women: short, medium and long-term follow-up. Int Urogynecol J Pelvic Floor Dysfunct. 2007;18:1263–70.
181. Bezerra CA, Bruschini H, Cody DJ. Traditional suburethral sling operations for urinary incontinence in women. Cochrane Database Syst Rev. 2005;3:CD001754.
182. Dmochowski R, Blaivas J, Gormley E. Female Stress Urinary Incontinence Update Panel of the American Urological Association Education and Research, Inc, Whetter LE. J Urol. May 2010;183(5):1906–14.
183. Rutner AB, Levine SR, Schmaelzle JF. Processed porcine small intestine submucosa as a graft material for pubovaginal slings: durability and results. Urology. 2003;62:805–9.
184. Jones JS, Rackley RR, Berglund R, et al. Porcine small intestinal submucosa as a percutaneous mid-urethral sling: 2-year results. Br J Urol Int. 2005;96:103–6.
185. Togami JM, Brown ET, Winters JC. Vaginal mesh-the controversy. F1000 Med Rep. 2012;4:21.
186. Wood L, Anger J. Urinary incontinence in women. Br Med J. 2014;349:g4531.
187. Bergman A, Koonings PP, Ballard CA. Primary stress urinary incontinence and pelvic relaxation: prospective randomized comparison of three different operations. Am J Obstet Gynecol. 1989;161:97–101.

28 Non-cosmetic Uses of Laser in Aesthetic and Regenerative Gynecology: Vulvodynia, Urinary Incontinence, Infections, Warts

Preeti Jindal, Isha Kundal, and Sonam Goyal

28.1 Introduction

"Light Amplification by Stimulated Emission of Radiation" (LASER) has been worldwide applied in medical science in past decades.

Laser is a very powerful tool. It is basically harnessed energy that can be channelized for cutting, coagulating, warming, heating, and vaporizing. The mechanism of action depends on the energy levels, dimensions of the area on which it is concentrated, duration of use, and type of laser.

Different types of laser are currently in common use for cosmetic reasons to enhance the beauty and health of intimate areas. But over recent years, laser is being used in treatment of various non-cosmetic gynecological problems. Laser is being used as cutting and coagulating device in endoscopic procedures during laparoscopy and hysteroscopy, but that is core surgical use as an adjuvant to other modalities. If we look at non-cosmetic uses of laser in gynecology to cure diseases then these five major conditions come to mind.

Vaginal atrophy, vulvodynia, non-neurological urinary incontinence, recurrent vaginal infections, and genital warts. Let us look at them one by one.

P. Jindal (✉) · I. Kundal · S. Goyal
The Touch, Advanced Obstetrics, IVF & Cosmetic Gynae Centre, Mohali, India

28.1.1 Vaginal Atrophy

Atrophy of vaginal tissue is more commonly seen in the postmenopausal phase due to predisposing factors such as hypoestrogenism and aging. Other causes of vaginal atrophy also include menopause, hormonal changes due to pills, breastfeeding, breast cancer treatments, pelvic radiation, and chemotherapy. As the life expectancy is increasing world over, the postmenopausal population is also increasing, hence we are seeing increasing number of women suffering from vaginal dryness, urinary incontinence, recurrent vaginal infections, and dyspareunia. It is often recognized as a leading cause of morbidity in the late 60s and above. Patients specifically on anti-estrogen treatment such as aromatase inhibitors, often suffer from vaginal dryness, itching, irritation, dyspareunia, and dysuria, collectively known as genitourinary syndrome of menopause (GSM) [1].

It is high time to take this problem seriously and start treating our women by asking them direct questions and offering them different non-invasive modalities of treatment like Laser and other energy-based devices. In 2014 these symptoms were integrated into the broader terminol-

P. Jindal et al. (eds.), *Aesthetic and Regenerative Gynecology*,
https://doi.org/10.1007/978-981-16-1743-0_28

ogy of "genitourinary syndrome of menopause" (GSM) [2]. GSM comprises signs and symptoms as a syndrome, previously known as vulvovaginal atrophy.

In recent years, treatment of vaginal atrophy with energy-based devices (EBDs) has gained popularity. In cancer patients also these modalities are safer alternatives as compared to hormones.

28.2 Pathophysiology

During menopause, with decreased estrogen levels, thinning of vaginal epithelium occurs. Fewer epithelial cells result in less exfoliation of cells into the vagina. Exfoliated cells produce lactic acid with action of lactobacilli which causes disturbance in vaginal pH, resulting in a loss of lactobacilli and an overgrowth of other bacteria, including group B streptococcus, staphylococci, coliforms, and diphtheroid [3].

Hypoestrogenism along with advanced aging causes changes in physiological and structural modifications within the genital structures and vaginal mucosa. Common drugs causing hypoestrogenism are Tamoxifen and Anastrozole (aromatase inhibitors).

28.3 Treatment

Therapies range include pharmacological (vaginal moisturizers, estrogen creams, and tablets) and non-pharmacological methods. Nowadays, Laser has been marketed as a non-invasive treatment modality for vaginal atrophy and other symptoms related to menopause and sexual function along with other pharmacological measures. Especially in breast cancer survivors, laser can be used to treat GSM symptoms as vaginal estrogen is not recommended for use.

28.4 Principle of Laser

In laser-based devices, the laser energy of different particles is used. The basic mechanism of action is lasing the tissue stimulates the neocollagenesis by thermomodulation. The primary role is to achieve neocollagenesis and fibroblast recruitment in the superficial layer, i.e., lamina propria. The penetration depth of CO_2 laser is 0.1–0.5 mm and wavelength of 10,699 nm which is very safe as vaginal mucosa is 4 mm in thickness. Photothermal waves reach up to 0.5 mm below the vaginal mucosa surface.

28.4.1 Procedure

It is a coffee break procedure. Done in outpatient setting. Requires 10 minutes. No anesthesia is required, at the most if patient is very sensitive local anesthetic cream can be applied in lower one-third of the vagina half-hour before procedure. Laser probe settings are done as per company guidelines and 3 passes of it are done. One pass means probe going from intoitus to posterior fornix, second pass is probe coming out, third pass is again going in. As the probe moves in, it lases in circular motion and lases the vaginal wall 1 cm apart circumferentially as shown in Fig. 28.1. Once 360-degree rotation is complete, it is advanced 1 cm and again rotated 360-degree lasing vaginal tissue 1 cm apart. While withdrawing same process is repeated. There are markings on probe 1 cm apart circularly and vertically to guide the operator. The vaginal probe is covered with either disposable sheath or sterilizable sheath for hygienic purposes. It is recommended that 3 cycles of treatment to be given one month apart. Point of caution is that patients should be informed that these treatments though widely popular and very safe; are not FDA approved and written consent should be taken before procedures. Patient may experience minor side effects like transient dysuria, slight blood-stained discharge or mild pain which will disappear in 2–3 days. They are advised not to have intercourse for 4 days and not to wear tampoons for 4 days.

28.5 Vulvodynia

In this chapter, we will focus on non-cosmetic use of only laser in vulvodynia, as a separate chapter on vulvodynia is already been written

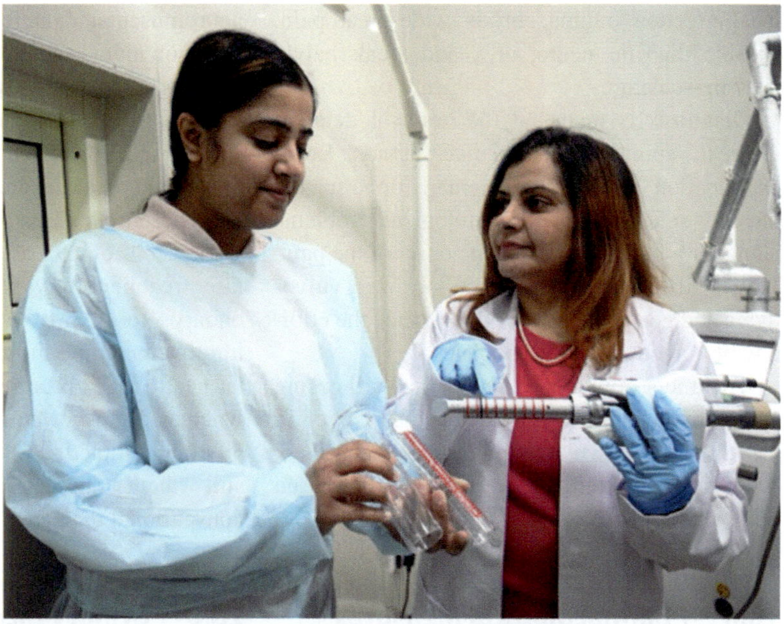

Fig. 28.1 CO_2 Laser probe showing disposable covers with markings © *Jindal P.* et al.*; 2019*

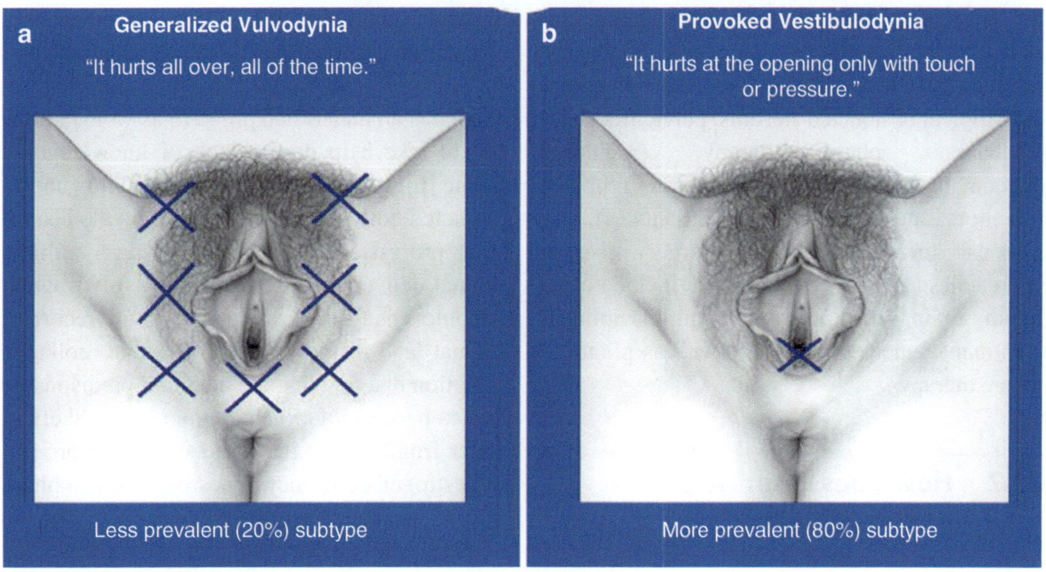

Fig. 28.2 Subtypes of vulvodynia: (**a**) Generalized and (**b**) Provoked

(Chap. 25). Vulvodynia refers to chronic vulval discomfort, burning, stinging, irritation, or rawness for more than 3 months without other dermatological or gynecological causes. According to the International Society for the Study of Vulvovaginal Disease, the classification of vulvodynia is based on the site of the pain, whether it is localized or generalized; whether the pain is provoked, unprovoked, or mixed and the onset of the pain symptoms whether it is primary or secondary [4] Fig. 28.2.

This condition has a notable impact on the quality of life in women as incidence rates of vulvodynia have been highly underreported due to failure of many women to seek medical help [5].

The contributing factors that may be responsible for vulvodynia are multifactorial such as nerve injury, over sensitivity of pain nerve fibers, local infection, trauma, chronic widespread pain, chronic diabetic neuropathy, and weak pelvic floor musculature.

Diagnosis is made on the basis of detailed medical, surgical, and sexual history. Local examination of vulva and vagina is done to rule out any infections or dermatological problems along with that determining the onset of the pain. It is difficult to discern if it is localized vulvodynia and generalized vulvodynia since the etiology of vulvodynia is still unclear.

Treatment options are directed toward easing symptoms and provide moderate pain relief in case the cause of vulvodynia has been identified. Women show different spectrums of relief based on type and site of origin.

28.6 Pharmacological Measures

Treatment of vulvodynia includes sexual counselling of both partners, acupuncture, cognitive behavioral therapy (CBT), sacral neuromodulation, topical or injected steroids, pelvic floor biofeedback, pharmacotherapy including anticonvulsants, selective serotonin reuptake inhibitors, and tricyclic antidepressants (TCAs), physical therapy (PT) and surgery. Newer approaches to a vulvodynia patient include botulinum toxin, hypnotherapy, multidisciplinary pain/management program, as well as photodynamic therapy.

28.7 How Does Laser Help in Vulvodynia?

Laser-based treatment has gained popularity as it is minimally or non-invasive and has better patient compliance. This has shown significant results in enhancing sexual health life and thus improving quality of life.

The possible explanations for the clinical benefit of laser in vulvodynia include anti-inflammatory effects with a reduction in levels of inflammatory markers (prostaglandin E2, IL 1β, and TNF α), oxidative stress and skeletal muscle fatigue, and the inhibition of transmission at the neuromuscular junction, thus having a direct effect on myofascial pain and trigger points.

Approach to vulvodynia treatment is multifactorial as its pathogenesis is not clearly understood, and, therefore, treatment is generally predicated on a trial and error basis.

In an observational study done by Jindal et al. (2019); out of 29 patients who underwent CO_2 laser treatment for vulvodynia or vulval pain secondary to GSM, 79% showed significant improvement in pain scores and overall vestibular health index score with gradual improvement whereas 21% showed minimal or no response at all after completion of three laser sittings. It was still a better success rate as compared to other conventional therapies.

28.8 Urinary Incontinence

In this chapter, we will focus less on urinary incontinence as there is a whole separate chapter already on it (Chap. 25). Urinary incontinence in women is an underrated problem as women hesitate to take help due to lack of knowledge or shame [6]. It is a hidden problem and it is underestimated both by the sufferer and by the healthcare professionals. Women having multiple vaginal deliveries, postmenopausal vulvovaginal atrophic changes, weak pelvic floor, recurrent vaginal and urinary infections, and collagen depletion due to aging are the main predisposing factors because of which they are more likely to suffer from urinary incontinence. Aging process and estrogen deficiency in postmenopausal phase lead to decrease collagen production leading to weakened pelvic floor support. Other Risk factors for SUI in women are obesity; vaginal deliveries (traumatic and instrumental), multiparous and late pregnancy; chronic asthma; neurological dysfunction, smoking; diabetes; collagen disorders; and history of hysterectomy [7].

Due to these advancing issues, there will be more adult diapers sold per year than baby diapers within the nearby future.

In past, surgical approach was the only option for treating women with such problems. These surgeries included sling procedures like retropubic mid-urethral sling, trans-obturator mid-urethral sling and colpo-suspension. However, a recent meta-analysis has reported that long-term data on patients undergoing such procedures regarding effectiveness and adverse events was lacking. Moreover, all surgeries have inherent risks of infection, injury to other pelvic organs including urinary bladder caused by the mesh tapes for the slings and issues related to anesthesia [8].

28.9 Treatment of Urinary Incontinence

The non-pharmacological methods focus on lifestyle modification, bladder training, pelvic floor muscle strengthening exercises, and electronic stimulation of pelvic floor muscles. Nowadays, non-invasive approach has been shown promising results in the treatment of urinary incontinence and thus improving the quality of life.

Different types of lasers are used for different causes depending on specific wavelengths. CO_2 laser falls in the infrared spectrum with a wavelength of 10,600 nm and it is highly absorbed by water. These two characteristics determine its superficial action [9]. Micropixeleted CO_2 laser is a novel, non-invasive, and with minimal side effects approach. It works by lasing the tissue with CO_2 in blocks of 9 × 9 columns leaving health tissues between columns as shown in Fig. 28.3. That limits its harmful ablative effects leaving a column of healthy tissues in between lased tissue to help in regeneration that stimulates the collagenesis. Laser helps in producing new collagen remodeling and elastic fibers recreation in the tissues under the mid urethra with the penetration depth of 0.03 mm. With the application of laser, thermal response is generated within the tissues of the mid urethra (Fig. 28.4). The cellular reaction is determined by the production of heat shock proteins (HSP), which plays a major role in transforming TGF-beta growth factor. The recruitment of fibroblasts initiates neocollagenesis. As local temperature rises, intermolecular cross-links that stabilize collagen triple-helix structure are broken, which leads to the shrinkage of collagen fibrils and improvement in tissue firmness thus improving incontinence symptoms [8].

The main goal is to ultimately increase the thickness of the vaginal wall in the area by

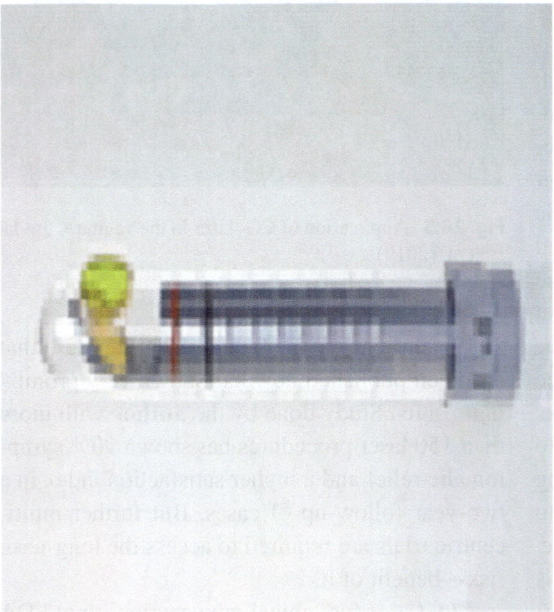

Fig. 28.3 CO_2 laser energy is delivered to the vaginal tissue using a pixel pattern, producing microscopic lesions thus promoting collagen remodeling

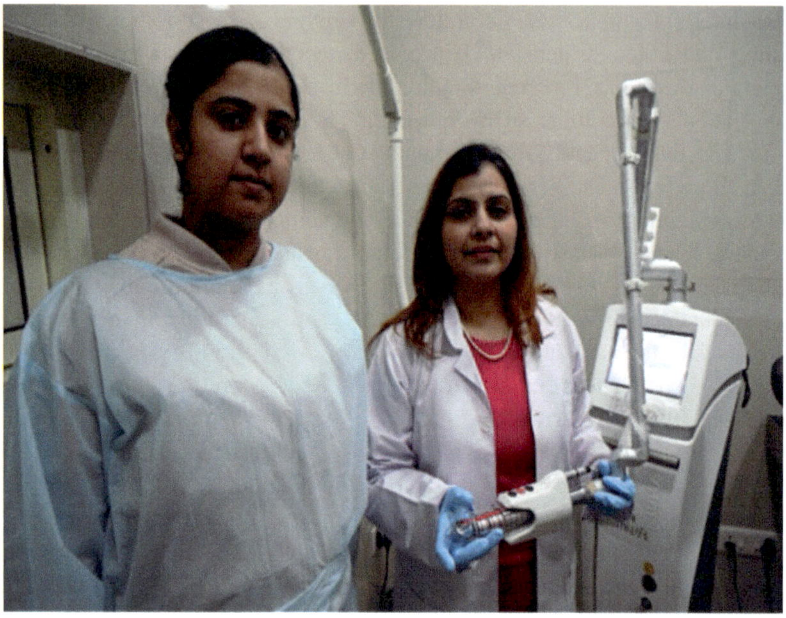

Fig. 28.4 CO_2 laser demonstration © Jindal P. et al. (2020)

increasing the collagen and elastin formation in order to increase the support of the mid-urethra underside and bladder neck. It also increases the elasticity and hammock effect preventing urinary leakage.

28.10 Procedure Protocol

After taking a detailed history, informed consent and pelvic examination to rule out any anatomical variations are completed. During the procedure patient is given a lithotomy position. Under aseptic conditions, speculum is inserted to rule out any local infection or bleeding. A 90° vaginal laser-emitting probe is then inserted up to the level of the posterior fornix, rotated lasing at 1 cm distance and withdrawn (Fig. 28.5). Three to five passes were given. Patient receives three treatments at an interval of four to six weeks apart. After the procedure, patient is advised to avoid intercourse for five days, avoid swimming for at least 4 days, avoid application of bleach to vulval area for 7 days, spotting can be there, which is completely normal. Hydrating oil is applied to minimize post-laser dryness and burning. Memory single sitting bis advised annually for continuing the benefit.

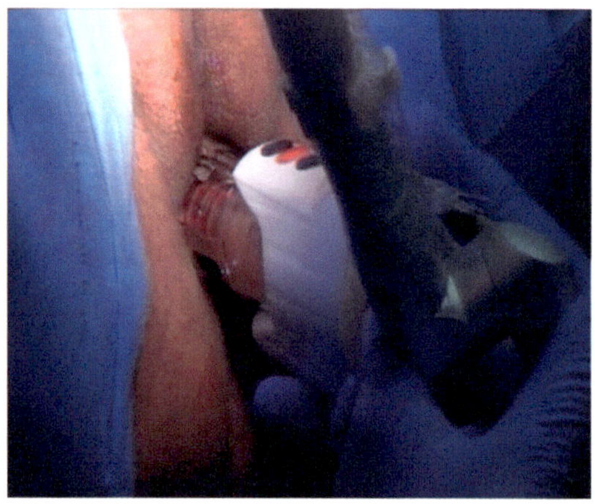

Fig. 28.5 Application of CO_2 laser in the vagina © Jindal P. et al. (2019)

It is an innovative strategy. The literature that has been published now has shown very promising results. Study done by the **author** with more than 150 laser procedures has shown 90% symptomatic relief and a higher satisfaction index in a two-year follow-up of cases. But further multicentric trials are required to access the long-term risk—benefit of it.

CO_2 Laser for vaginal rejuvenation is not FDA approved yet, but then CO_2 laser is FDA approved for face, surgical treatments, and other procedures. So it appears paradoxical that why it should not be approved for this particular reason.

28.11 Recurrent Vaginal Infections

Postmenopausal estrogen and testosterone deficiency in vaginal tissue leading to decrease vaginal wall thickness and reduced vascularity is the main underlying pathology owing to atrophic changes of the vaginal wall which contributes to the vaginal dryness and imbalance in vaginal bacterial flora leading to recurrent vaginal infections caused by different pathogens (Table 28.1).

Vulvovaginal candidiasis (VVC) is considered the second most common genital infection in women, after bacterial vaginosis. However, these agents may lead to adverse reactions and their chronic use might lead to resistance to antifungal agents. Due to recurrent infections, microabrasions take place in the vaginal wall leading to predisposition to infection as soon as pH alters after intercourse, menstruation, pregnancy, menopause, chronic antibiotic use for systemic illnesses, etc.

Current therapeutic approaches include non-hormone and hormonal therapies. Non-hormonal therapies include vaginal moisturizers and lubricants for mild symptoms, while hormonal therapy includes estrogen therapy for moderate to severe symptoms. These medications provide temporary relief for a shorter span.

CO_2-Laser therapy is nowadays promising treatment for improving the vaginal health of postmenopausal women by helping repopulate the vagina with normally existing *Lactobacillus* species and reconstituting the normal flora to pre-menopausal status [10].

CO_2 laser helps in restoring the pH of the vaginal mucosa by increasing the glycogen levels that rebalances lactobacilli species, which inhibit the growth and virulence of pathogenic bacteria [11].

Post-laser treatment, histologically vaginal epithelium becomes thick, rich in glycogen, newer blood vessels in sub-epithelia and collagen and elastin fibers formation in lamina propria. The mucosal layer heals reinforcing the effective barrier against infections.

In patients with recurrent infections, it is mandatory to treat any active infection with conventional therapy first. Once the vagina is free of any visible infection and the patient is symptomatically fine then giving one sitting of vaginal lasing has been found to be very effective in preventing recurrent infections. Lasing should not be done if an active infection is present.

Table 28.1 Different types of vaginal infections

Type of vaginal infections	Bacterial Vaginosis	Trichomoniasis Vaginosis	Candidias
Causative organism	Over growth of Gardnerella Vaginalis	Trichomonus Vaginalis	Yeast and Candida Albicans
Clinical Picture	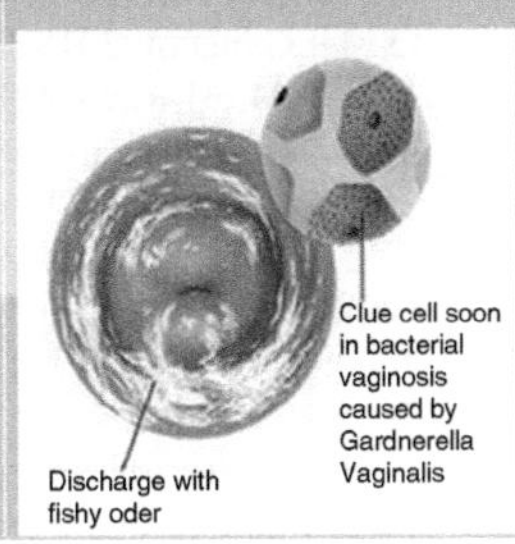	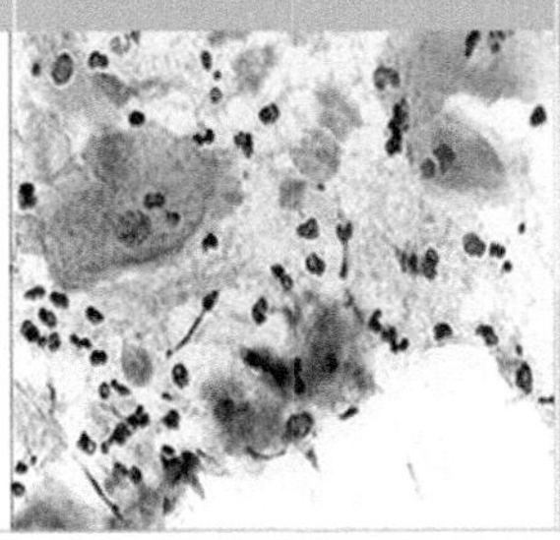	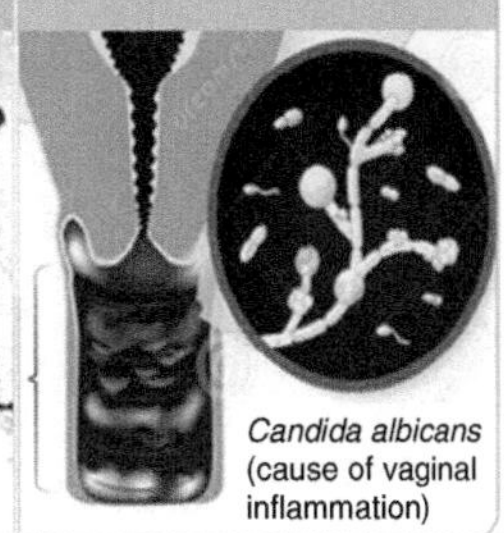

28.12 Warts

Genital warts are one of the most common types of sexually transmitted diseases and are also referred to as venereal warts or condylomata acuminata. Anogenital warts are caused by human papillomaviruses (HPVs) which are transmitted through oral, anal, and genital sexual contact with rare cases of autoinoculation and vertical transmission have been reported.

Risk factors are early sexual intercourse, multiple sexual partners, unprotected intercourse, infection by other STD at the same time, immunosuppressive drugs, or conditions like HIV.

Pharmacological therapy includes topical application of podophyllin, podofilox, 5-fluorouracil, and trichloroacetic acid (TCA) which are tedious processes.

Carbon dioxide laser vaporization is typically used for treatment of genital warts. Laser therapy is an efficient therapeutic modality because of its precision and rapid healing without scarring. It can be done in outpatient clinic with local anesthesia.

28.12.1 Procedure

Patient is given lithotomy position. Sterilization of the lesion is done with povidone-iodine solution. Application of topical lidocaine cream 10% and in case of wide lesion and big size wart local infiltration of 2% xylocaine is performed (Fig. 28.6).

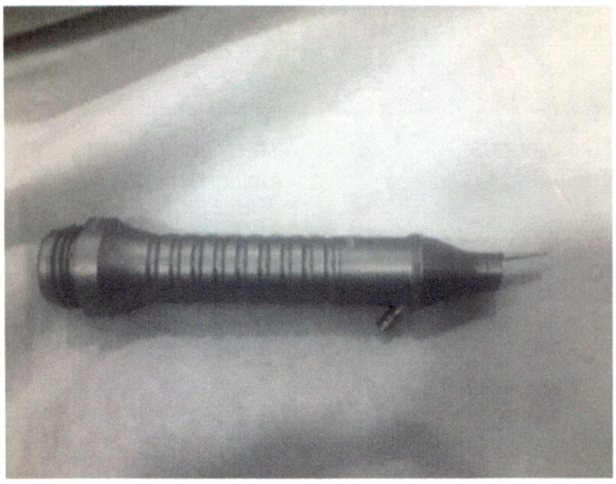

Fig. 28.6 Laser surgical handpiece © *Jindal P.* et al.; *2019*

Treatment is done by CO_2 laser vaporization of warts and residual debris wiped away with a piece of gauze. Laser vaporization of the whole wart to the level of normal skin is done. Settings of the machine are kept as per company guidelines. It is important to wear protective eye cover both by operator, patient, and people in room. Treatment is done in one session except for patients with large numbers and multiple locations who are treated in multiple sessions.

CO_2 laser vaporization has a low complication and side effect profile including postoperative oozing, pain, and scarring, also it is an effective and safe method for treatment of genital warts [12]. It is a quick, effective, and painless modality of treatment for this distressing problem.

References

1. Sussman TA, Kruse ML, Thacker HL, Abraham J. Managing genitourinary syndrome of menopause in breast cancer survivors receiving endocrine therapy. J Oncol Pract. 2019;15(7):363–70.
2. Preti M, et al. The clinical role of LASER for vulvar and vaginal treatments in Gynecology and female urology: an ICS/ISSVD best practice consensus document. Journal of lower genital tract disease vol. 2019;23(2):151–60. https://doi.org/10.1097/LGT.0000000000000462.
3. Roy S, Caillouette JC, Roy T, Faden JS. Vaginal pH is similar to follicle-stimulating hormone for menopause diagnosis. Am J Obstet Gynecol. 2004;190:1272–7.
4. Sadownik LA. Etiology, diagnosis, and clinical management of vulvodynia. Int J Women's Health. 2014;6:437–449. Published 2014 May 2. https://doi.org/10.2147/IJWH.S37660.
5. Harlow B, Stewart E. A population based assessment of chronic unexplained vulvar pain: have we underestimated the prevalence of vulvodynia? Journal of American Medical Women's Association. 2003;58(2):82–8.
6. Guin G, et al. Prevalence of stress urinary incontinence and its associated risk factors amongst females attending tertiary referral Centre. Int J Reprod Contracept Obstet Gynecol. 2018 Jun;7(6):2115–9.
7. Jasim SM, Al-Rawi RYK. Efficacy and safety of vaginal CO_2 laser treatment in female stress urinary incontinence. Med J Babyln. 2018 Sept;15(3):251–7.
8. Eric S, Rovner MD, Alan J, Wein MD. Treatment Options for Stress Urinary Incontinence Rev Urol. 2004;6(Suppl 3):S29–47.
9. Stefano S, Stavros A, Massimo C. The use of pulsed CO2 lasers for the treatment of vulvovaginal atrophy. Curr Opin Obstet Gynecol. 2016;3(6):S504–8.

10. Athanasiou S, Pitsouni E, Antonopoulou S, et al. The effect of microablative fractional CO_2 laser on vaginal flora of postmenopausal women. Climacteric. 2016 Oct;19(5):512–8.
11. Pillai R, Rahim S. Fractional CO2 laser treatment for vaginal rejuvenation in post-menopausal Indian women. Trichol Cosmetol Open J. 2019;1(2):7–10. https://doi.org/10.17140/TCOJ-1-111.
12. Mehdi MH, Mahmood AS. Treatment of genital warts by using CO2 laser. Med J Babylon. 2019;16:94–8.

29 Gender Reassignment Surgery in Aesthetic and Regenerative Gynecology

Rakesh Kain

29.1 Introduction

The description of hermaphrodite and Gender dysphoria is available in mythology. The "Hermaphrodite" was considered the son of "Hermes" and the Goddess "Aphrodite" consists of human and nymph Salmacis in one body. However, similar meaning terms were in existence like "Inter sex" since ancient time. From the time of evolution and since the time of tribes, its long been considered as naturally occurring variants of uncommon sexual behaviour in many cultures and herds.

"Gender Identity" in general describes a person's sense of being "male", "Female", "Nighther" or a combination of both.

By definition, **gender dysphoria syndrome (GDS)** is a heterogeneous group of disorder comprising dissatisfaction with their anatomic gender, and desire to have opposite-sex characters. This group constitutes of all sexual ambiguity, physiologically, environmental and chromosomal anomalies, etc. However other common terms in use are Transsexual, Transgender, Trans, Gender nonbinary, gender Incongruent and gender queer are used as adjectives. Cisgender is the term for a person who is not transgender; sex recorded at birth aligns with their sex identity.

The transgender men have male gender identity and were recorded as female at birth; the transgender women have female gender identity and are recorded as male at birth. The **Gender-Non confirming People** refer to the extent to which a person's gender identity, role, or expression differ from the cultural norms prescribed for people of a particular sex.

Eunuchs or Hijra constitutes the remarkable population of Gender dysphoria syndrome (GDS). By being castrated just early period of teenage, many avoid the development of male secondary sex characteristics and their bodies can remain permanently like a girl.

Even today, very large numbers of desperate young transsexuals in India from other countries run away from home to join the "Hijra" caste. To become Hijra, these teens voluntarily undergo fully primitive emasculating surgeries even under compromised facilities, just as they would have in ancient times. With the increased awareness and invention of more precisely performed surgical procedures now, they have started coming for the surgery in their teens shortly after the onset of puberty. Transsexual individuals have long had the choice to cross-dress and live in their opposite gender.

R. Kain (✉)
AIIMS, Rishikesh, India

P. Jindal et al. (eds.), *Aesthetic and Regenerative Gynecology*,
https://doi.org/10.1007/978-981-16-1743-0_29

29.2 Magnitude of Problem

The magnitude of the problem was first recognised by the American Psychiatrist **Harry Benzamin** [1, 2] in the 1950s. He formulated detailed guideline for diagnosis and treatment. Since then and now in 2001 the details of guidelines published by the World Professional Association for Transgender Health (WPATH) laid down standards of care for transgender. The exact incidence of the problem is not known in India. But a large number of patients reported in clinics and sporadically for surgical correction. The reported incidence is 1:60000 with broad geographical variation. The cultural difference from one country to another would alter the behavioural pattern in gender identities in a population. Most of the studies conducted in European countries (Zucker and Lawrence 2009) found 1:11,900 to 1:1,45000 for male to female and 1:30,400 to 1:200000 for female to male [3]. In general, the male-to-female transsexual is four times higher than female to male.

29.3 Etiopathogenesis

The physiology and etiopathogenesis of GDS is not fully understood. Most of the theories of causes are based mainly on psychological perspectives, without much supportive evidences. The common factors that contribute to the problem are as below.

1. **Environmental factors**
2. **Neurological factors**: The red nucleus of stria terminalis in the brain responsible for sexual behaviour is larger in males. In 6–8% of transsexuals, this area was found small in male-to-female transsexuals. Researchers have shown a relationship of Hypothalamus area for sexual behaviour abnormality in transsexuals.
3. **Psychological factors**: are considered superior to all other factors which determined the sex, e.g., chromosomal or phenotype.

29.4 Embryology

The urogenital system in human beings develops from Cloaca which forms a primitive urogenital sinus. This primitive urogenital sinus differentiates into urogenital sinus and vesicourethral canal, further joined by pronephros and mesonephric duct to complete the system (Table 29.1).

Urogenital sinus ↗ Ventral Part (Pelvic)-Bladder, Part of Male urethra
↘ Urethral Part-receive the mesonephric duct & fused to mullarian duct, forms part of male urethra, lower $1/5^{th}$ of Vagina

Table 29.1 Male and female homologous structure and embryological origin

Embryonic tissue	Male	Female
Mesonephric duct	Renal pelvis trigone, Ureter, Epididymis, Vas deference, Ejaculatory duct, appendix, testis	Renal pelvis trigone, Ureter, duct of epiphooron, Gartner's duct, Vesicular appendages
Mullerian duct	Appendix testis, prostatic utricles	Uterine tube, Uterus, upper 4/5th vagina
Junction of UGS and Mullerian duct	Part of prostatic utricles	Lower 1/5th vagina
Junction of sinovaginal bulb to UGS	Disappears (Post-urethral valve)	Hymen
Ventral part of UGS	Urinary bladder (except trigone), prostatic urethra, membranous urethra	Urinary bladder, whole urethra, vaginal vestibule
Gubernaculum	Gubernaculum testis	Ligament of ovary, round lig. of uterus
Genital tubercle	Penis	Clitoris
Urethral fold	Penile urethra	Labia major and minor

Vesicourethral canal: forms urinary bladder and primitive urethra, and whole female urethra (except lower 1/5th) and prostatic urethra.

Diagnosis: the patient presents with:

1. Different sexual orientation
2. Different patterns of erotic responses
3. Dysmorphic behaviour

According to Harry Benjamin, "as a rule sex assignment should be made before the 18 month of age." What he stressed on is "sex of rearing." The sex of rearing profoundly influences the gender role, which may be quite independent of the chromosomal sex, the gonadal sex, or genital appearance.

WAPTH Guideline [4] (World Professional Association for Transgender Health): Provides for healthcare professionals to assist transgender, transsexual, and gender–nonconforming group with a safe and effective pathway to achieving lasting effects. It emphasises personnel comfort, in order to maximise their overall health, psychological well-being and self-fulfilment. This assistance may include primary care, gynecologic care, urologic care, reproductive options, voice and communication options, mental health services (assessment, counselling, psychotherapy, etc.) and hormonal and surgical treatment. [5]

29.5 Surgical Procedure

The management of gender dysphoria consists of a combination of psychotherapy, hormonal therapy, and surgery [6]. However, gender reassignment surgery is the series of surgical procedures (non-genital and genital) performed for the management of GDS have two clinical entities: [7]

1. **Gender reassignment surgery (GRS)**: includes all surgical procedures for conversion of the gender of patients including non-genital and genital surgeries.
2. **Sex reassignment surgery**: is the part of GRS that refers to genital surgery procedures only.

Very rarely do patients come for only penectomy and orchidectomy without reconstruction of female genital organs and for mastectomy or salpingo-oophorectomy in the opposite group.

Indications & Contraindications: GDS with a psychological orientation towards the other sex is the greatest indication for SRS. Psychiatric evaluation is essential before gender reassignment surgery undertaken. According to Harry Benjamin's guideline, the candidate should undergo at least 1 year "real life experience" in cross living, full-day and seven days in a week. Cross life is not to test endurance but an opportunity to feel what life is like in the opposite gender. Failure to show satisfaction in cross life situations is considered not a suitable candidate; however, a further trial of prolongation of cross life period is advocated. The candidate is evaluated by a psychiatrist and one psychologist for suitability. Simultaneously, the candidate should start hormonal therapy under the supervision of an endocrinologist.

Hormonal therapy: Feminisation through hormonal therapy is achieved by suppression of androgen by suppressing Gonadotropin-releasing hormone (GnRH) or suppression of GnRH antagonist which suppress the luteinising hormone (e.g., progesterone substance, cyproterone acetate). The substance which interferes with testosterone metabolism to dihydrotestosterone (Spironolactone, Finasteride) is also used by endocrinologists. The hormones used for male-to-female conversion are given in Table 29.2.

Table 29.2 Hormonal therapies in male-to-female conversion (before and after)

	Before orchidectomy	After orchidectomy
17B estradiol (Aestrogen Analogue)	0.1 mg/dose, twice a week to gradually increase after 1 month; 0.2 mg/dose twice a week, transdermally or ORAL 1–2 mg QID gradually increase up to 4 mg QID.	0.400 to 0.1 mg twice daily
Spironolactone (Androgen Antagonist)	Oral therapy 50–100 mg QID, daily, gradually increase to 200–400 mg in 3 months periods	25–50 mg QID
Finasteride (Androgen Antagonist)	2.5–5 mg on alternate days	2.5 mg QID

A combination of one estradiol and one antiandrogen is given. While on estradiol therapy, one should be aware of symptoms and signs of DVT and other side effects. Serum Potassium and renal function should be closely monitored if the patient is on spironolactone therapy.

29.6 Male to Female

1. **Genital surgeries**: includes bilateral orchidectomy, penectomy, labioplasty, clitoroplasty and vaginoplasty.
2. **Non-genital surgeries**: Breast augmentation, female feminisation surgeries—frontal Bossing reduction, reduction of brow ridge, reduction of the mandibular angle, liposuction and fat injections for body sculpturing, hair transplant, Adam's apple reduction, dermal fillers, nose reduction, change of frontonasal angle, chin reduction or contouring, Malar reduction, augmentation or reshaping. Apart from these, the vocal cord and throat surgery [2] to raise the pitch of voice by approximating thyroid and cricoids which causes stretching and pull of vocal cord in a way to increase the pitch.

   Vaginoplasty is used by creating a pocket in between the bladder neck, prostate and rectum. The pocket is lined by penile skin (superior method) or scrotal skin or a skin graft. Hairy skin should be avoided to line the vaginal cavity. Other methods are colon (right colon, sigmoid colon, Ileum) [8]. Transposition is not a preferred technique due to excessive intestinal secretion. Inverted penoscrotal vaginoplasty is another popular technique [9].

   Clitoroplasty [10], the part of glans penis dissected off the tunica and corpora keeping intact the dorsal neurovascular bundle intact deep to bucks fascia, up to the base of penis and folded on it and glans tissue stitched in a way to make mound just anterior to the urethra. It provides the erogenous sensations [11].

   Labioplasty: Labia majus is satisfactorily reconstructed by scrotal skin without much effort. The meticulous use of penile skin and scrotal tissue to line the vaginal cavity and reconstruct the labia minora. Some surgeons preserve some part of prepuce attached with glans penis while using for the clitoris; this penile skin is used to make folds of labia minora [6].

   Urethra reconstruction: shorten the male urethra, make an opening in the anterior wall of vagina and stitch to the mucosa of urethra.

29.7 Female to Male

The genital correction includes urethroplasty to urinate in standing, metoidioplasty, vaginal obliteration, phalloplasty and scrotoplasty.

The nongenital corrections include a variety of procedures like mastectomy, hysterectomy, oophorectomy (prevent to develop endometrial carcinoma). The facial profile correction is required after assessment of hormonal therapy and its effect, like male pattern skin texture, baldness, voice, musculature, etc. Chest masculinisation is also required. [12]

1. **Hormonal therapy**: Testis of adult males produce 12–15 mg of testosterone daily, which is converted into Dihydrotestosterone by 5α-reductase. A smaller amount of testosterone is secreted by the adrenal cortex. Androsterone is the metabolite of testosterone excreted in the urine. The plasma level of testosterone in adult males ranges from 0.3 to 1 μg/dl. Plasma half-life of testosterone is 10–20 min. Methyltestosterone and Fluoxytestosterone (Synthetic) metabolise slowly and have a longer duration of action. Inj testosterone depot 100 mg (testosterone propionate 25 mg + testosterone enanthate 100 mg) in 1 ml I M/weekly are used to achieve secondary sexual characters.

   Caution must be taken to use in kidney and liver disease. Side effects are oedema, cholestatic jaundice, lowering of HDL and rising of LDL and hepatocellular carcinoma.

   Different synthetic testosterones are available and they are also used with their different dose protocol.
2. **Metoidioplasty**: This is the procedure to use hypertrophied clitoris to reconstruct the penis

with the elongation of urethra at the tip by using vaginal epithelium and labia minor flap. It preserves the erectile tissue and sensations, but small size is unsatisfactory for most patients.

3. **Phalloplasty**: The common methods to reconstruct the penis are radial artery forearm flap (RAFF), abdominal tube flap, groin flap, fibula osteocutaneous flap and anterolateral thigh flap. The free flap required the shaping of the phallus, totally detached from the donor site (ex. Radial artery flap) [11] and anastomose radial artery and its veins at the recipient sites to restore circulation in flap. Now a day's anterolateral thigh flap (ALT) [13] with intact blood supply by Lateral circumflex femoral artery and its branches can satisfactorily transfer at the pubic area after giving shape and reconstructing urethra [14]. Microsurgical flap failure and urethral fistula [14, 15] formations are common complications. The erectile function of penis can be restored by inflatable implants, bone graft, silicon rods and vascular bone graft. The failure rates for implants are very high.
4. **Mastectomy**: Subcutaneous mastectomy with preservation of nipple areola sensation is the main aim of surgery. The nipple areola can be reduced in size or can be used as a free skin graft. The circumareolar incision is the preferred method.
5. **Scrotoplasty**: Scrotum reconstruction is done either by labia minora or medial thigh flap skin. Silicon testicular implants can be inserted inside. Normally, this step is done separately.

29.8 Practical Tips from Author

Construction of realistic labia minora is a challenge; secondary surgical procedures to readjust the tissue are helpful. To get a forward movement of the urinary stream, the suturing of the urethral mucosa should be done in a way to open more downward. Otherwise, it comes far interiorly to fall beyond the toilet seat. The stripping of excessive erectile tissue from urethra leads to profuse bleeding and leaving behind the excessive tissue leads to pain during sexual activity. To spatulate the urethral opening reduces the chances of meatal stenosis.

29.9 Discussion

The surgical alteration of genitalia to relieve intense cross-gender feelings was also carried out in past. In some cultures, even ancient ones, many transsexuals have voluntarily undergone surgeries to modify their bodies. Performing such surgeries on normal post-pubertal males does not change their gender feelings or gender identity, although it lessens their sexual drives somewhat and sharply reduces their ability to develop male musculature [16].

During the late 1950s and into the 1960s, several hundred transsexuals in the United States came under the care of Harry Benjamin, M.D, a compassionate physician and endocrinologist who had offices in New York, N.Y. and San Francisco, CA.

Dr. Benjamin [2] was the first physician/researcher to sort out the distinction between cross-gender identity and homosexuality. Instead of viewing transsexuals as mentally ill deviants, as did most psychiatrists of the day, he began to visualise transsexuals as truly suffering from a genuine mis-gendering condition of unknown origins [17, 18].

Then, in the late 1950s, a French plastic surgeon named Georges Burou, M.D. invented the modern form of penile inversion in MtF sex reassignment surgery for MtF transsexuals. Variations of Dr. Burou's technique have been used ever since.

With the rapid advances in knowledge of sex hormones and plastic surgery following World War II, it finally became possible to contemplate complete medical and surgical solutions for transsexualism.

Christine Jorgensen, a U.S. citizen, was among the first small group of transsexuals to undergo such a surgical "change of sex". She was "outed" in 1952 by U.S. print media shortly after her initial surgery, and her story became a national sensation.

The surgeons constructed the patient's vagina by using skin grafts taken from her thighs or buttocks. Through her story, many transsexuals for the first time learned of the existence of the new hormonal and surgical treatments

29.10 Conclusion

29.10.1 Myths versus Reality, and the Decision to Undergo SRS

Sex is not a single biological entity, but the summation of morphogenetic, functional and psychological well-being. The psychological sex and sex of rearing are paramount. Once the sex rearing is assigned, efforts should be made to reinforce it by surgical methods. And if any ambiguity of genitalia is diagnosed, it should be assigned before 2 yr of age; he/she should be reared accordingly. The sexual identity depends upon gender pattern, for example, when a child starts comparison in peers, clothes, genitals, games and activities. Forcing a child in a different gender makes him/her in an uncomfortable environment, which is difficult to adjust to.

Many people simply assume that the loss of the external male genitalia will result in a complete loss of sexuality. This very naive myth unnecessarily frightens many pre-op women, and it also furthers prejudice against post-op transsexual women, who are often thought of by the general public as having "desexed themselves".

The surgery can be liberating and can enable fuller expression of sensuality and libidinous feelings. Just as in the case of modern post-operative transsexual women, many Hijra can have strong feelings of sexual arousal in the inner remnants of their genitalia (even though they lack the external nerve tissue left by modern SRS, they retain the internal portions of the erectile corpora cavernosa and of course the prostate, with its spasmodic orgasmic capabilities) [1, 19].

Gender reassignment surgery needs lots of support from friends, family, sexual partners and co-workers as well [20].

Illustration:

Male to Female

1. Male Genitalia in MtF
2. Marking for scrotal flap for vagina (Post wall)
3. Dissection of penile skin, urethra and glans tissue based on dorsal neurovascular Bundle.
4. Reconstruction of Labia, Clitoris, vagina (Gauge Packed)

Illustration:

Female to Male

5. Vaginal mucosa striping
6. Phallus reconstructed from radial Artery Forearm flap
7. Phallus anastomosed to Vein and Artery

Bibliography

1. Birnbaum R. First physiologic study of orgasm in postoperative male-to-female transsexuals. Dissertation, The Institute for Advanced Study of Human Sexuality, San Francisco. 18 October 2000.
2. Meyer, W. 3rd., et al. The Harry Benjamin International Gender Dysphoria Association's standards of care for gender identity disorders. Sixth version. World Professional Association for Transgender Health [online]. 2001. http://www.wpath.org/documents2/socv6.pdf.
3. Bakker A, Van Kesteren P, Gooren L, Bezemer P. The prevalence of transsexualism in the Netherlands. Acta Psychiatr Scand. 1993;87:237–8.
4. World Professional Association for Transgender Health [online]. 2010. http://www.wpath.org/.
5. Colman E, et al. Standards of care: WPATH. Int J Transgender. 13(4):165–232. https://doi.org/10.1080/15532739.2011.700873.
6. Selvaggi G, Bellringer J. Gender reassignment surgery: an overview. J Nat Rev Urol. 2011;8:274–81.
7. Edgerton MT. The surgical treatment of male transsexuals. Clin Plast Surg. 1974;1:285.
8. LiguoriG TC, Bucci S, et al. Laparoscopic mobilization of neovagina to assist secondary ileal vaginoplasty in male-to-female transsexuals. Urology. 2005;662:293–8.
9. Selvaggi G, Ceulemnas P, De Cuypere G, et al. Gender identity disorder: general overview and surgical treatment for vaginoplasty in male-to-female transsexuals. Plast Reconstr Surg. 2005;116:135e–45e.

10. Wroblewski P, et al. Sex reassignment surgery for transsexuals. J Co-Endocrineol. 2013;20(6):570–4.
11. Selvaggi G, Monstrey S, Hoebeke P, et al. The donor site morbidity of the radial forearm free flap after 125 phalloplasties in gender identity disorder. Plast Reconstr Surg. 2006;118:1171–7.
12. Agarwal CA, Scheefer MF, Wright LN, et al. Quality of life improvement after chest wall masculinization in female-to-male transgender patients: a prospective study using the BREAST-Q and Body Uneasiness Test. J Plast Reconstr Aesthet Surg. 2018;71(5):651–7.
13. Wong C-h, Wei F-c. Anterolateral thigh flap. Head Neck. 2010 Apr;32(4):529–40. https://doi.org/10.1002/hed.21204.
14. Young J, Purohit R. Retained vaginal remnants and Urethrocutaneous fistula in transgender man after phalloplasty. Clin Urol. 2019;00:1–2.
15. Hoebeke P, et al. Impact of sex reassignment surgery on lower urinary tract function. Eur Urol. 2005;47:398–402.
16. de Vries AL, Doreleijers TA, Steensma TD, et al. Psychiatric comorbidity in gender dysphoric adolescents. J Child Pscyhol Psychiatry. 2011;52:1195–202.
17. Wewrnick JA, Samantha B, et al. A systematic review of the psychological benefits of the gender-affirming surgery. Urol Clin N Am. 2019;46:475–486. Elsevier Inc.
18. Millet N, Longworth J, Arcelus J. Prevalence of anxiety symptoms and disorders in the transgender population: a systematic review of the literature. Int J Transgend. 2016;18:27–38.
19. Massie JP, Morrison SD, Wilson SC, Crane CN, Chen ML. Phalloplasty with urethral lengthening: addition of a vascularized bulbospongiosus flap from vaginectomy reduces postoperative urethral complications. Plast Reconstr Surg. 2017:140;551e–558e.
20. Josua SD, et al. Care of transgender person. N Engl J Med. 2019;381:2451–60. https://doi.org/10.1056/NEJMcp1903650.

Anaesthesia in Procedures of Aesthetic and Regenerative Gynecology

30

Amit Kumar, Neerja Bharti, and Adarsh Chandra Swami

30.1 Introduction

Aesthetic and Regenerative gynecological procedures are mostly done in office-based or day-care centres by gynecologists, plastic surgeons, dermatologists or a multidisciplinary team. These are mostly painless and minimally invasive, requiring only local anaesthesia. Recently, some of the more invasive cosmetic gynecological procedures are being developed, requiring anaesthesia.

The administration of anaesthesia is becoming more safer, because of advances in regional and general anaesthesia techniques, monitors and drugs. It is imperative to know the basic neuroanatomy of external female genitalia and perineum to understand the anaesthetic requirements and prevent any complications.

A. Kumar (✉) · A. C. Swami
Department of Anaesthesia and Critical Care, Fortis Hospital, Mohali, India

N. Bharti
Department of Anaesthesia, PGIMER, Chandigarh, India

30.1.1 Nerve Supply of External Female Genitalia and Perineum (Fig. 30.1)

- The main nerve supply of the external genitalia and perineum is the Pudendal nerve, which is made up of the second, third, and fourth sacral (S 2,3,4) spinal roots.

  The Pudendal nerve has three main branches:

1. Dorsal nerve for the clitoris—supplies the sensory innervation to clitoris.
2. Perineal nerve for the external genitalia—provides sensory innervation to the external genitalia through posterior labial nerves and motor innervation to the external urethral sphincter and muscles of the perineum.
3. Inferior rectal nerve—supplies sensory innervation to the perianal skin and motor innervation to the external anal sphincter.

   In addition:

- Cavernous nerves from the uterovaginal plexus—supply parasympathetic innervation to the clitoris.
- Anterior labial nerves, which are branches of the Ilioinguinal nerve (L1)—supply labia majora.
- Genitofemoral nerve (L1)—provides innervation to mons pubis.

P. Jindal et al. (eds.), *Aesthetic and Regenerative Gynecology*,
https://doi.org/10.1007/978-981-16-1743-0_30

Fig. 30.1 Nerve supply of external female genitalia and perineum

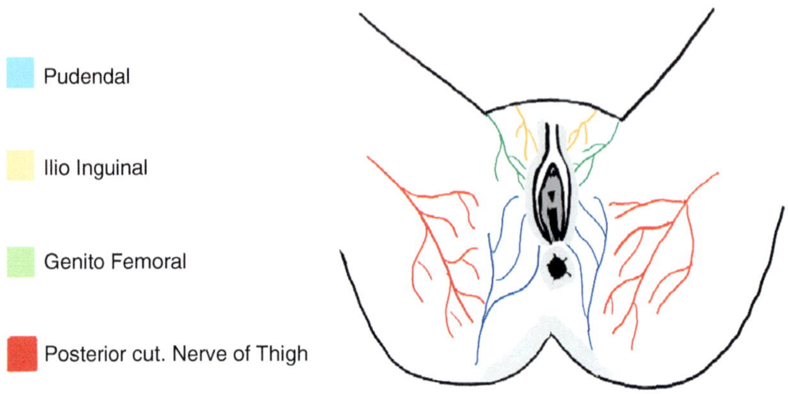

- Posterior cutaneous nerve of thigh (S1, 2, 3)—also contributes to nerve supply of posterior vulva and perineal region.

So, the nerve supply to external female genitalia and perineum is basically derived from branches of sacral plexus (pudendal and posterior cutaneous nerve of thigh) and lumbar plexus (ilioinguinal and genitofemoral nerve).

30.2 Anaesthesia for Aesthetic Gynecology

Anaesthesia for aesthetic gynecology can be divided into three broad subsectors. These are:

A. Anaesthesia for invasive/major cosmetic gynecological procedures.
B. Anaesthesia for minimally invasive cosmetic gynecological procedures.
C. Anaesthesia for non-invasive cosmetic gynecological procedures.

30.2.1 Anaesthesia for Invasive/Major Cosmetic Gynecological Procedures

1. Gender reassignment surgeries (GRS)
2. Abdominoplasty, Butt or Thigh reshaping
3. Breast lift/reduction
4. Liposuction

These surgeries should only be undertaken in a fully equipped hospital. Pre-anaesthetic check-up (PAC) is done to ascertain the health status of the individual. History of any comorbidities, any allergic reaction in the past, or any previous surgeries is recorded. General physical and systemic examination are done to determine the fitness for the surgery.

Minimum investigations are complete blood count, blood sugar, coagulation profile, ECG, and others (renal/liver function tests, lipid profile, chest X-ray, Echocardiography) depending on the patient's cardiorespiratory, metabolic and endocrine status.

The patient is given anxiolytic (alprazolam) and antiemetic (ondansetron) as premedication before surgery. Six to eight hours of Nil per Oral (NPO) instruction is given.

During the surgery, ECG, Pulse oximeter, End-tidal CO_2, NIBP (Non-invasive blood pressure) and temperature monitoring are done.

The majority of cases are done in general anaesthesia (GA). A balanced anaesthesia based on halogenated inhaled anaesthetics like desflurane, sevoflurane or isoflurane is used, as these have favourable recovery profiles. Intravenous anaesthetics like propofol, etomidate and fentanyl are also used with safe outcomes. Muscle relaxants like rocuronium, vecuronium and atracurium are sparingly used, that is, only for prolonged surgeries.

Post-operative analgesia is given with adequate doses of opioids and NSAIDS. PCA (Patient-controlled analgesia) is frequently employed for post-operative analgesia, in which either opioids (morphine/fentanyl) or non-opioid analgesics are selected. When PCA is used, the bolus demand dose, background infusion rate and lockout interval are set, to ensure safe delivery of analgesics.

Some of the patients may be good candidates for regional anaesthesia (RA), depending on the extent of surgery. Techniques like spinal, epidural or combined spinal-epidural anaesthesia can provide better perioperative analgesia. Continuous epidural infusion can be given along with GA, especially for surgeries like abdominoplasty and gluteal/thigh surgical reshaping.

Plexus blocks (lumbar and/or sacral plexus blocks) or the Nerve block of their branches like ilioinguinal, genitofemoral and pudendal nerve are also given as sole or adjunct anaesthesia for surgeries on the vulva and perineal area.

Ultrasound, nerve stimulator and landmark guidance are used to provide accurate anaesthesia of these nerves or plexuses.

A. Special anaesthetic considerations in GRS

GRS is a multidisciplinary venture, requiring specialized training and expertise of surgeon as well as anaesthesiologist to perform genital/non-genital reconstruction. Detailed protocols are used to ensure that the patient is a suitable candidate, so that appropriate surgical and anaesthesia plans are selected and executed with perfection. Patient optimization is done by preoperative risk reduction and informed consent.

Psychiatric and endocrine evaluation is specially done for gender reassignment surgeries. Details of prior hormonal therapy and its effect on liver and other organs are evaluated. Oestrogen therapy makes the patients prone to venous thromboembolism, migraine, dyslipidaemia and post-operative nausea and vomiting. Likewise, testosterone therapy increases the risk of cerebrovascular accidents/stroke (due to high haematocrit values) and liver dysfunction (high LFT values). Discontinuation of hormonal therapy may be undertaken after the endocrinologist's opinion.

GRS includes TOP and BOTTOM surgery, as well as other cosmetic surgeries, which may be done in a single sitting or in multiple stages, a plan of which has to be discussed with anaesthesiologist. These are:

Feminizing Surgery (Male to Female)

- Genital reconstruction (Vaginoplasty, Orchiectomy, Penectomy)
- Breast augmentation
- Facial feminizing surgery (Rhinoplasty, Jaw reduction, Cheek augmentation, Forehead surgery)

Masculinizing Surgery (Female to Male)

- Mastectomy /Breast reduction
- Genital reconstruction (Vaginectomy, Phalloplasty, Scrotoplasty, Testicular implants, Erectile prosthesis)
- Hysterectomy and Oophorectomy

Depending upon the patient's physical status, it may be safe to perform two or more surgeries in the same operative sitting, but it will increase the surgical time and may require a coordinated effort from multispeciality surgery teams.

Combination surgeries:

- Breast augmentation with vaginoplasty.
- Breast augmentation with Rhinoplasty.
- Breast reduction with Hysterectomy/ Oophorectomy.
- Genital surgery (Phalloplasty/vaginectomy) with Hysterectomy/Oophorectomy.

Other surgeries like laryngoplasty (to alter the voice pitch) may also be undertaken.

Risks and Complications

1. Adverse reaction to the local anaesthetic drug.
2. DVT (Deep vein thrombosis) and PE (Pulmonary Embolism).
3. PONV (Post-operative Nausea and Vomiting).
4. Sedation and respiratory distress.

30.2.2 Various Measures Are Taken to Reduce DVT/PE-Associated Risk

These include:

- Judicious perioperative anticoagulation.
- Intraoperative continuous cardiorespiratory monitoring.
- Adequate intravenous fluids, throughout the surgery.
- Perioperative application of pneumatic compression devices and compression stockings.
- Early mobilization in post-operative period.

30.2.3 Some Other Important Intraoperative Concerns Are

- Minimize traffic of staff in the operating room, as there is exposure of the genitalia or chest area. This is more significant when the patient is awake (during MAC/Regional anaesthesia).
- There can be difficult urinary catheterization due to previous surgeries like vaginoplasty/phalloplasty.
- Be ready for difficult airway due to previous surgery like laryngoplasty (to alter the voice pitch).
- Universal precautions to be taken (as many patients are HIV positive).
- Blood transfusion and intraarterial blood pressure monitoring may be planned in combined/major surgeries.

B. Special anaesthetic considerations in abdominoplasty, Butt reshaping and Liposuction

Abdominoplasty is a major surgery involving resection of a large area of abdominal tissue with making of a neo-umbilicus at a new site (Fig. 30.2). General Anaesthesia is induced with a balanced mixture of intravenous anaesthetic, analgesics and short or intermediate-acting muscle relaxants. Newer inhalational anaesthetics like desflurane or sevoflurane are used for the

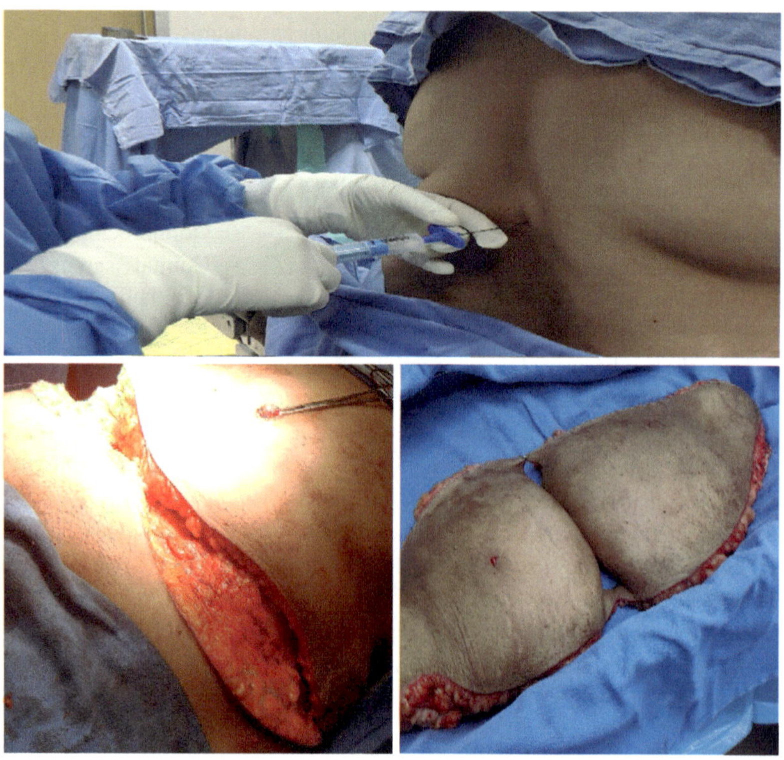

Fig. 30.2 Abdominoplasty done under GA with Lumbar Epidural Block

maintenance of anaesthesia, so as to ensure a faster recovery. Lumbar epidural anaesthesia is also combined with GA, to take care of perioperative analgesia.

Liposuction and Butt reshaping are relatively less invasive procedures in most of the patients requiring fat removal from certain areas with subsequent contouring /shaping of that area.

With the advances in tumescent and superwet liposuction techniques, removal of larger volumes of fat with less blood loss has become safer. Here, large volumes of dilute lignocaine with adrenaline are used as wetting solution to be injected subcutaneously near fat pockets.

Large volume liposuction is a more complex procedure involving, either 4 litres of fat or 5 litres of total volume (including the wetting solution) removed during the procedure.

Major complications regarding fluid and electrolyte balance, like pulmonary oedema and life-threatening arrhythmias can be associated with large volume liposuction.

The anaesthesiologist has to record the hourly balance of intake (IV fluids, wetting solution infused) and output (fat and saline aspirated, blood loss and urine output). Careful monitoring of vitals along with this fluid balance enables him to prevent fluid overload or hypovolemia. There is initial oliguria in the first few hours due to adrenaline in wetting solution used. 5–10 mg of intravenous furosemide is sometimes used to stimulate diuresis.

These patients are also more prone to hypothermia due to injection of large volumes of wetting solution, long procedure time and exposure to large body surface area during surgery.

Overnight observation in ICU is important in these patients, as it may take 18–24 hours for fluid balance and haemodynamics to return to baseline values.

C. Special anaesthesia concerns of Breast lift / reduction surgeries

For breast surgeries, there are various regional anaesthesia techniques like thoracic epidural, paravertebral or PEC1 and PEC2 blocks, which are administered with or without GA, to decrease the requirement of systemic analgesics and to enhance the patient's recovery from anaesthesia.

30.2.4 Anaesthesia for Minimally Invasive Cosmetic Gynecological Procedures

1. Labiaplasty
2. Cliteroplasty
3. Hymenoplasty
4. Vaginoplasty and perineoplasty

These surgeries can be undertaken in a day-care centre or a fully equipped hospital. Most of these surgeries last few minutes to 1–2 hours only.

Minimum investigations required are complete blood count, blood sugar and coagulation profile. The patients are told to take a light breakfast before minor procedures and avoid heavy meals. If GA is planned, 6–8 hours of NPO is maintained.

Minimum monitoring for minor procedures (like labiaplasty and cliteroplasty) is pulse oximetry/ECG, for which a dedicated nurse/healthcare worker is appointed. All resuscitative drugs should be available. NIBP and End-tidal CO_2 (if GA) are also used for relatively longer and invasive surgeries (vaginoplasty and perineoplasty).

In most of the least invasive procedures, EMLA cream 5% application to the area is only required. EMLA cream is Eutectic mixture of Local Anaesthetics namely, 2.5% prilocaine and 2.5% lignocaine. This should be applied to the normal skin of the genital area, at least 45 minutes before incision and covered with a dressing. This application can give 1–2 hours of numbness to that area. A 5% lignocaine cream also may be used.

Local anaesthesia with 2% lignocaine alone (or mixture with 0.5% bupivacaine) (Fig. 30.3) can also provide 1–2 hours of anaesthesia. Usually, 5–10 ml is only required for this purpose.

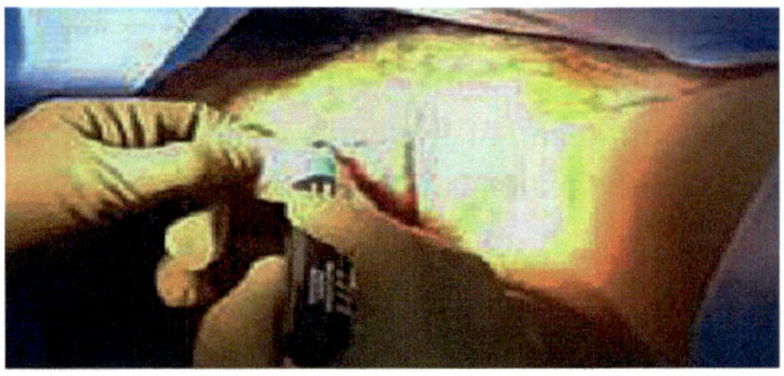

Fig. 30.3 Vaginoplasty under local anaesthesia with conscious sedation

Nowadays, there is an emerging practice of using larger volumes of dilute local anaesthetics as intumescent anaesthesia, where the surgical area is injected with 15–50 ml of 0.05–0.1% lignocaine (or mixture with 0.125–0.25% bupivacaine in longer surgeries). Care is taken, not to exceed toxic doses for lignocaine and bupivacaine.

30.2.5 Few Considerations for Successful and Less Painful LA Injections

- Apply EMLA or Lignocaine cream 45 min earlier.
- Explaining the whole procedure of anaesthesia/surgery prior to it.
- Reassurance and holding patient's hand during surgery by a nurse.
- Short and swift injection of local anaesthetics.
- Wait for 5–7 minutes for the numbing effect to come to its peak and then only allow the surgery to commence.
- Use thin needles (hypodermic/ spinal) of 26/27 Gauge.
- To reduce the initial stinging sensation of LA, combine it with sodium bicarbonate solution 8.4% (0.2 ml /10 ml of LA).

Surgeries like hymenoplasty, vaginoplasty and perineoplasty can also be undertaken under spinal anaesthesia or a short GA. Saddle block is a form of low spinal anaesthesia, which targets the required sacral segments (S 2, 3, 4) only, leading to minimal haemodynamic derangement and allows a dense block with faster recovery. If a longer duration of surgery (>2 hrs) is expected, either additive (fentanyl 10–25 mcg/clonidine 75–150 mcg) are added to LA for spinal anaesthesia, or combined spinal epidural anaesthesia (CSEA) is planned.

30.2.6 Contraindications for Neuraxial Anaesthesia (Spinal/Epidural/CSEA) Are

Absolute contraindications

- Patient refusal
- Severe coagulopathy
- Infection at injection site

Relative contraindications

- Difficult spine anatomy
- Patient on antiplatelet/anticoagulant therapy
- Thrombocytopenia (Platelet count <80,000)
- Aortic stenosis, HOCM (Hypertrophic Obstructive Cardiomyopathy)
- Postoperative plan to travel by Airplane (Pressure changes may increase CSF leakage through dural puncture hole)
- Pre-existing CNS diseases

30.2.7 Anaesthesia for Non-invasive Cosmetic Gynecological Procedures

1. Lasers
2. Radio frequency
3. HIFU and HIFEM
4. PRP therapy, fillers, chemical peels

These therapies are done for cosmetic improvement of female genitalia, treatment of menopausal abnormalities, urinary incontinence or for sexual rejuvenation. These are vaginal hair removal, scar/pigment removal, vulval whitening or vaginal tightening.

Pre-anaesthetic check-up (PAC) is done in these cases also, to rule out any comorbidities which if undetected, may lead to undesirable events. Minimum investigations required are complete blood count, blood sugar and coagulation profile.

This procedure can be undertaken in an office/daycare setting. The patients are told to take a light breakfast before the procedure and avoid heavy meals. Minimum monitoring during the procedure is pulse oximetry, for which a dedicated nurse/healthcare worker is appointed. All resuscitative drugs should be available.

The anaesthesia is usually not required in most of these procedures but proper monitoring of vitals should be done under Monitored Anaesthesia Care. Some of the patients may require Conscious Sedation, as they may be anxious because of procedure or due to exposure of their private parts. Midazolam (2–4 mg) as intermittent bolus doses can be given. Dexmedetomidine (Alfa-2 agonist) infusion can also be utilized for sedation.

30.2.7.1 Daycare Guidelines

There are some criteria for home discharge for short stay/daycare surgeries, which have to be observed for patient safety and to avoid readmissions because of complications:

- Patient should be awake, alert and well oriented to time, place and person.
- Patient should have haemodynamic stability (pulse/BP near baseline values).
- Patient should have no or tolerable pain (VAS score < 3) with/without oral analgesics.
- Patient should not have any bleeding at the operative site.
- Patient should be able to walk, with or without support.
- Patient should tolerate intake of liquids or solids, without nausea or vomiting and should be able to pass urine spontaneously.
- Patient should have a reliable escort to drive her home.

30.3 Conclusion

As the cosmetic awareness of patients is increasing with advance in fashion and technology, many aesthetic and regenerative gynecological procedures are being developed. The ultimate goal of cosmetic gynecological surgery is to make patients look, feel and function better and boost their self-confidence.

Anaesthesiologists are required to take up the challenge of providing anaesthesia to these patients for these ambulatory and short-stay surgeries, which are mostly done in office-based or day-care centres.

Some aesthetic gynecological surgeries such as GRS are more invasive and also may be done in multiple stages. These are undertaken in fully equipped hospitals, under general or regional anaesthesia.

To favour an adequate outcome, anaesthesia has to be tailored to meet individual surgery plans and patient's cardiorespiratory and metabolic status. This involves rational use of short/intermediate action drugs and adequate intraoperative monitoring. Utmost care is taken to prevent pain, nausea, vomiting and deep vein thrombosis, with the goal of reducing morbidity and achieving a shorter recovery time after anaesthesia. Minimal complication rates and maximum patient satisfaction can be achieved by adequate communication between surgeon and anaesthesiologist.

Acknowledgement The authors would like to express their gratitude to Ms. Tia Nagpal for her help to make the figures in this manuscript.

Stem Cells and Recent Advances in Aesthetic and Regenerative Gynecology

31

Asawari Bapat

31.1 Introduction (Incidence, Effect on Society, etc)

A state-of-the-art speciality that is emerging to address the women's health is Aesthetic *gynecology*. The term "Aesthetic Medicine" was unheard of until recently [1]. The basic ideology around a few simple skincare techniques, beauty, and treatments began in Europe. Now it has vastly advanced, with aesthetic medicine clinics in most cities. The practices are offering various services with many types of treatments, inclusive of laser technologies, fillers, injectables, spa treatments, and even minimally invasive cosmetic surgeries and procedures [2].

This speciality addresses the need of patients for the visual appearance, sensation, and cosmetic issues in relation to the health and functioning of the female reproductive system including the vagina, uterus, ovaries, and breasts. Aesthetic gynecology accompanies reconstructive surgeries to address some of the issues with women's health [3]. The aesthetic and cosmetic representation of the female reproductive system is different globally, and an exact description of a normal view of external genitalia is extremely difficult. The social and cultural differences in each country play a highly significant role in this matter.

A. Bapat (✉)
Nashville, TN, USA

A constant barrage of social media, glamorous lifestyles, and managing one's health as an important priority has led to a woman, who is extremely aware of her surroundings. All these reasons; and many other societal factors, has led to escalating expectations of a woman toward her appearance. To look good and feel confident, it has become an inevitable decision for a woman in addressing her aesthetic issues.

Patients are now seeking physicians and surgeons who are trained in Regenerative Medicine (RM) for their demand for aesthetic solutions. Even though Aesthetic gynecology customarily falls under the scope of obstetrics and gynecology, surgeons, plastic surgeons, and cosmetic surgeons have developed a great interest in this field. Presently, there are few if any obstetrics and gynecology residency or fellowship programs that teach the subject. A multitude of physicians and surgeons are establishing their practices through various training programs offered by educational institutions and primarily from the stem cell industry and medical devices industry.

There are inroads that have taken place in the training programs, for plastic surgery and cosmetic surgery that had the foresight to include special training in this field. At this point of time, a multitude of practitioners, gynecologists, and surgeons start by first training in various established certification and preceptorship programs based in the United States and the United

P. Jindal et al. (eds.), *Aesthetic and Regenerative Gynecology*,
https://doi.org/10.1007/978-981-16-1743-0_31

Kingdom. In recent times, there have been new programs that were launched globally to offer certified training to interested physicians in both surgical and nonsurgical treatments. The Middle East, Turkey, Spain, and South America have presented a huge interest with a steady flow of certificate programs that continues to advance. Soon we would be seeing a second generation of experts emerge.

This chapter is a review of surgical and nonsurgical techniques of what is presently called "aesthetic gynecology" and explains the various approaches of prominent gynecology societies regarding this relatively new subspecialty.

Vaginal rejuvenation (VR) is a controversial procedure that is questionable and much debated. However, this is a procedure that is most popular, and is asked often in conjunction with other procedures. It helps in treating vaginal laxity syndrome (VLS). Multiple pregnancies, infections, scars due to cancers, or other procedures, lead to laxity of the vagina. The prolapse affects the sexual function, and after the procedure has been performed, it is markedly improved. Vaginal rejuvenation surgery (vaginal tightening for sexual function) is the procedure to repair VLS that may not contain symptomatic prolapse. Previously there were fewer incidences or scientific studies to prove the rationale, but in recent times we have seen a lot of emerging scientific evidence to support these types of surgery. However, in recent years scientific studies supporting vaginal rejuvenation are surfacing and are being presented at scientific meetings throughout the world. In female urology textbooks (such as the Cardoza and Staskin), we are seeing book chapters and scientific articles that are being published as the procedure is becoming more prevalent. It is critical to have more high-level scientific studies to be completed to validate these procedures in women.

Female genital cosmetic surgery also includes aesthetic procedures to improve the cosmetic appearance of the external vulvar/vaginal region. Procedures include labiaplasty or reduction of labia minora with or without excess prepuce reduction, labia majora reduction or augmentation, vaginal introital repairs for cosmetic issues as well as reduction of lipodystrophy in the mons pubis region. Labiaplasty procedures have been reported to be the largest growing trend of plastic or cosmetic surgery procedures for women in the United States and throughout the world. This may be secondary to the increased public awareness created by the media or popular TV shows, or it may be secondary to the fact that in the past women's feelings about the appearance of their genitalia have been ignored. It has been scientifically shown that the appearance of a woman's genitalia affects her self-confidence and sexuality. Women now have been empowered with the choice of options to change the external appearance of their vulvovaginal region if they are unhappy with the cosmetic appearance. Studies have also recently shown that this is a trend driven by women themselves, and not their sexual partners. However, with this trend there have been many different procedures described with very little scientific validation and therefore there is a danger that surgeons with very little experience with vulvar or vaginal surgery (inside or outside the field of gynecology) are doing these procedures incorrectly and causing injury or damage to women.

Recently, new technology has been introduced in the field offering nonsurgical/office-based procedures to treat female sexual dysfunction (VLS), vaginal health, and vulvo/vaginal cosmetic issues for women. This technology includes non-fractional lasers as well as radio frequency treatments. Scientific studies are currently evaluating this technology for these uses as well as treatment for mild urinary incontinence, urgency/frequency issues as well as vaginal dryness.

Though uterine regeneration has the potential to be a treatment for AUFI, there have been only a few studies on uterine regeneration involving the myometrium in vivo. In the present chapter, those relevant articles are reviewed. A literature search was conducted in PubMed with a combination of keywords, and 10 articles were found, including nine in rat models and one in a mouse model. Of these studies, eight used scaffolds and two were performed without scaffolds.

Organ regeneration was established in the 1970s, based on the culture of differentiated cells, such as epidermal cells and chondrocytes [4]. Langer and Vacanti showed that the use of scaffolds allowed cells [5] to settle in transplanted sites more effectively than with injecting suspensions locally. This result led to the concept of tissue engineering. Nowadays, multipotent cells, such as embryonic stem cells (ESCs) and induced pluripotent stem cells [6], are used for studies in this field. Scaffolds can accurately develop the complexity of vascular structures and those produced by the decellularization of organs are used in tissue engineering [7] research, skin, cartilage, blood vessel, liver, heart, and kidney regeneration has been studied; skin and cartilage regeneration has been used in clinical practice. However, whole solid organs still cannot be produced and various devices for this purpose are currently under development.

Types of Stem Cells: Stem cells are the foundation for every organ and tissue in your body. There are many different types of stem cells that come from different places in the body or are formed at different times in our lives.

On the basis of origin, stem cells are divided into different categories these include:

Embryonic stem cells (ESCs): It forms all the tissue types of the developing body.

Fetal stem cells (FSCs): It forms almost all types of tissue except embryonic tissue.

Adult stem cells (ASCs): It forms only a limited number of tissue types.

1. Embryonic Stem Cells (ESCs): Embryonic stem cells are pluripotent, meaning they can give rise to every cell type in the fully formed body, but not the placenta and umbilical cord. These cells are incredibly valuable because they provide a renewable resource for studying normal development and disease, and for testing drugs and other therapies. Human embryonic stem cells have been derived primarily from blastocysts created by in vitro fertilization (IVF) for assisted reproduction that were no longer needed which they donated for research purposes with informed consent of the donors and ESCs are not derived from eggs fertilized in a woman's body. In view of the ethical and social dilemmas involved, collecting, culturing, and experimenting on embryonic and fetal Stem Cells is legally restricted in many countries.
2. Adult Stem Cells (Somatic Stem Cells) (ASCs): Our body is formed from a large number of cells, like liver cells, muscles cells, bone cells, skin cells, intestine cells, and so on. Most of these cells have to be replaced every once in a while. The cells that replace the old cells and repair the damaged tissue are the adult stem cells, also called somatic stem cells. Adult stem cells are undifferentiated (they have not specialized function to do but can be differentiated to many types of tissue but not all types of tissue. The term "Adult Stem Cell" is a little misleading, since these cells are actually found in infants and children as well as in adults and it is also called Somatic Stem Cells, which possess the same basic characteristics as all of the stem cells.

Unlike embryonic stem cells, the use of adult stem cells in research and therapy is not controversial because the production of adult stem cells does not require the destruction of an embryo. Embryonic Stem Cells, which can become any cell in the body (called pluripotent), Adult Stem Cells, can form into only a limited number of tissue types (called multipotent). ASCs exist in the tissue for decades. It may remain inactive (non-dividing) for a long period of time until they are activated to replace dying tissue to maintain the tissue as the continuous growth for new skin, intestine, and the bone marrow or to regenerate damaged or lost tissue either by diseases or injury. That means the ASCs main function is to maintain and repair the specific tissue where they reside.

Researchers on adult stem cells have found it in many more tissues than they once thought possible. Scientists now have evidence that stem cells exist in the brain and the heart, two locations where adult stem cells were never expected to reside. If the differentiation of adult stem cells can be controlled in the laboratory, these cells may become the basis of transplantation-based therapies.

The use of adult stem cells and tissues derived from the patient's own adult stem cells would mean that the cells are less likely to be rejected by the immune system. This represents a significant advantage, as immune rejection can be circumvented only by continuous administration of immunosuppressive drugs, and the drugs themselves may cause deleterious side effects.

Adult stem cells have been identified in many organs and tissues, including brain, bone marrow, cord blood, cord tissue, blood vessels, skeletal muscles, skin, teeth, heart, gut, liver, ovarian epithelium, and testis. They are thought to reside in a specific area of each tissue (called a "Stem Cell Niche"). In many tissues, current evidence suggests that some types of stem cells are pericytes, cells that compose the outermost layer of small blood vessels. Typically, there is a very small number of stem cells in each tissue and, once removed from the body, their capacity to divide is limited. Generation of large quantities of stem cells was difficult until they discovered that the adipose tissue (fat) is a rich source of stem cells. For that, adult stem cells become the simplest type to obtain, and have the greatest importance for medical purposes.

Advantages of ASCs: The following advantages are associated with the use of adult stem cells in regenerative medicine:

1. Cells are derived entirely from mature tissues and therefore it is non-controversial, and no ethical codes are breached.
2. Even though adult Mesenchymal Stem Cells are somewhat differentiated, they are still multipotential and can form a number of new tissues.
3. Adult Mesenchymal Stem Cells are less teratogenic (in other words less prone to tumor formation).
4. These cells are immunosuppressant in nature meaning they would not elicit a strong immune response after treatment and therefore rejection is unlikely.
5. Once a pure mesenchymal stem cell population has been obtained both inter-individual and inter-species transplants are possible.
6. There are many different sources from which these cells can be obtained.

Types of Adult Stem Cells (ASCs):

- Hematopoietic Stem Cells (HSCs): They are found in the bone marrow and umbilical cord blood and give rise to all the blood cell types plus all the tissues types of the developing body.
- Mesenchymal Stem Cells (MSCs): They are formed in a limited number of tissue types and have been isolated from placenta, adipose tissue, lung, bone marrow, and blood, Wharton's jelly from the umbilical cord and teeth.
- Mammary Stem Cells (MSCs): Provide the source of cells for growth of the mammary gland during puberty and gestation. It has been isolated from human and mouse tissues as well as from cell lines derived from the mammary glands.
- Intestinal Stem Cells: Can divide continuously throughout life and use a complex genetic program to produce the cells lining the surface of the small and large intestine.
- Endothelial Stem Cells: are found in the bone marrow and are rare.
- Neural Stem Cells: The presence of stem cells in the mature primate brain was first reported in 1967. Neural stem cells are commonly cultured in vitro as so-called neurosphere, it shares many properties with hematopoietic stem cells (HSCs). Remarkably, when injected into the blood, neurosphere-derived cells differentiate into various cell types of the immune system.
- Olfactory Stem Cells: Olfactory adult stem cells have been successfully harvested from the human olfactory mucosa cells, which are found in the lining of the nose and are involved in the sense of smell. Olfactory stem cells hold the potential for therapeutic applications and, in contrast to neural stem cells, can be harvested with ease without harm to the patient. This means they can be easily obtained from all individuals, including older patients who might be most in need of stem cell therapies.
- Testicular Stem Cells: Multipotent adult stem cells have been derived from the germ cells found in human testicles.

Adult Mesenchymal Stem Cells (MSCs): Mesenchymal stem cells (MSCs) are adult stem cells that can be isolated from human and animal sources. Human MSCs (hMSCs) are the non-hematopoietic, multipotent stem cells with the capacity to differentiate into mesodermal lineages such as osteocytes, adipocytes, and chondrocytes as well ectodermal (neurocytes) and endodermal lineages (hepatocytes). Since the first description of hMSCs derived from bone marrow, they have been isolated from almost all tissues including the perivascular area. Human MSCs are a favorite choice of stem cells due to its self-renewal ability, multi-potency, easily accessible, and culturally expandable in vitro with exceptional genomic stability and few ethical issues, marking its importance in cell therapy, regenerative medicine, and tissue replacement.

Sources: The first time reported in the bone marrow and till now have been isolated from various tissues, including; Adipose tissue, amniotic fluid and membrane, endometrium, menstrual blood, peripheral blood, dental tissues, salivary glands, in skin and foreskin, synovial fluid, umbilical cord, and Wharton's jelly which harbors potential MSCs.

Characteristics of Mesenchymal Stem Cells (MSCs): Today, Mesenchymal Stem Cells (MSCs) are one of the most famous stem cell types used in medicine. They have been shown to be effective for a wide range of diseases with a minimal risk profile. At the same time, several hundred clinical studies are ongoing to investigate their regenerative abilities in greater detail. From a biological point of view, they are progenitor cells of connective tissues. This means that they are important for building and maintaining the healthy status of connective tissues throughout the whole body and are known as Stromal Stem Cells, they possess the ability to differentiate into a spectrum of other cell types, including chondrocytes (cartilage cells), osteoblasts (bone cells), and adipocytes (fat cells).

This property classifies them as "multipotent stem cells" (being able to transform into many cell types).

There are several reasons why they are the leading stem cell type in medical applications and clinical trials on humans. They are relatively easy to harvest and can be isolated from:

- Fat tissue, known as ADMSCs (Adipose-Derived MSCs)
- Bone Marrow, known as BmMSCs (Bone marrow MSCs)
- Umbilical Cord, known as UcMSC (Umbilical cord MSCs)

1. MSCs are relatively stable to culture and expand in the laboratory.
2. They can go through relatively many division cycles without losing their expression profile, health, and division potential.
3. They possess little to no potential to grow into cancer cells.

Bone marrow–Mesenchymal Stem Cells (BM-MSCs): Bone Marrow Concentrate (BMC) is one of the most commonly applied sources of stem cells. Despite the fact that the actual number of stem cells in BMC is biologically limited, several other (regenerative) factors in BMC have been shown to deliver promising results in the treatment of numerous diseases.

Adipose-Derived Mesenchymal Stem Cells (ADSCs): Fat (adipose)-derived Mesenchymal Stem Cells (ADSCs I ADMSCs) are a commonly used source of stem cells. ADSCs share many of the characteristics of bone marrow-derived mesenchymal stem cells (BMSC), they can be obtained more easily with a 100–1000 times greater cellular yield, and in practice, many patients are eager for the harvest of unwanted fat. A number of preclinical and clinical trials have established the safety and efficacy of ADSCs and range from breast reconstruction and correction of defects to neural regeneration in spinal cord injuries. ADSCs have the ability to produce a range of tissue types and are an ideal cell source for the clinical application of tissue engineering.

Induced pluripotent stem cells (iPSCs): Scientists have recently discovered how to turn adult stem cells into pluripotent stem cells. These new types of cells are called induced pluripotent stem cells (iPSCs). They can differentiate into all types of specialized cells in the body. This means they can potentially produce new cells for any organ or tissue.

To create iPSCs, scientists genetically reprogram the adult stem cells so they behave like embryonic stem cells. The breakthrough has created a way to "de-differentiate" the stem cells. This may make them more useful in understanding how diseases develop. Scientists are hoping that the cells can be made from someone's own skin to treat a disease. This will help prevent the immune system from rejecting an organ transplant. Research is underway to find ways to produce iPSCs safely.

Adipose Tissue (Fatty Tissue): Adipose tissue or fat is an anatomical term for loose connective tissue composed of adipocytes (fat cells). Best known for its role in fat storage, in mammals, three types of adipose tissue exist White Adipose Tissue (WAT), Brown Adipose Tissue (BAT), and Beige or brite adipose. Brown adipose tissue is present in small mammals (e.g., mice) and in newborn humans. Anatomically located around the major organs in a neonate and known to dissipate with age. Its distinct color is due to its high vascularization and contains large numbers of intracellular mitochondria. It releases heat via oxidation of fatty acids. Classical BAT is specialized to dissipate chemical energy serving a protection function against hypothermia. It can suppress weight gain and metabolic disease through its action as specialized, heat-producing adipocytes, most of it decreases with human aging.

Beige or brite adipose tissue; has morphological and molecular characteristics of classical thermogenic brown adipocytes. *Adipocytes of white adipose tissue WAT*, is the most common and is the fat that so many of us complain of acquiring. Mainly consist of one vacuole for storage of triglycerides. Adipose tissue is primarily located beneath the skin, but is also found around internal organs. In the integumentary system, which includes the skin, it accumulates in the deepest level, the subcutaneous layer which provides insulation from heat and cold also, around organs as it provides protective padding. It also functions to reserve nutrients. WAT, developed in multiple anatomical sites, stores chemical energy and demonstrates known structural, functional, metabolic, and endocrine differences between different WAT deposits. Adipose tissue can further be classified according to macroscopic tissues type, anatomical location and structural / functional characteristics. Although both WAT and BAT are of mesodermal origin, they are believed to originate from different mesenchymal stem cells lineages.

Recent studies have described adipose tissue as a metabolic and endocrine organ producing various substances including adipocyte-derived hormones such as liptin and adipsin; bioactive peptides are known as adipokines such as adiponectin, visfatin, omentin, and resistin to name a few; as well as cortisol and various sex and steroid hormones. These substances act both locally (paracrine/autocrine) and systemically (endocrine), exerting various physiological effects (Gimble et al. 2007; Harwood 2012; Kershaw & Flier 2004). The adipose tissue contains multipotent stem cells, which can differentiate into fat, bone, cartilage, and types of tissues.

Characteristics of Tissue According to Age and Gender: Marked differences have been observed between genders in both the metabolism and the endocrine function of adipose tissue. Women are known to have a higher percentage of body fat and mainly store adipose tissue in the gluteal-femoral region. Adiposity in this region is associated with larger fat cell size with increased stimulated lipolysis and triglyceride synthesis. Adipose tissue storage in men is primarily in the visceral and abdominal regions. obesity in men is associated with increased lipoprotein lipase activity and with decrease stimulated lipolysis and triglycerides synthesis, Blaak (2001); Edens et al. (1993); Fried et al. (1993). Several studies have shown that the differences in visceral adipocytes metabolism between genders disappear with menopause. It was further suggested that female sex hormones may play a role in this gender-specific adipose deposition; this includes, for example, weight gain in the abdominal region of postmenopausal women as well as associated metabolic changes, Rebuffe-Scrive et al. (1989); Trujillo and Scherer (2006).

31.2 Historical Aspects

A french doctor, Dr. JJ Legrand, an endocrinologist from the 1970s, initiated the practice of aesthetics and established a new field for aesthetics in mainstream healthcare. The ideology grew in Europe; Spain, Italy, Belgium, then the United States and later expanding globally. It took root in the practice and science of aesthetic medicine as we see it today.

Initially, the focus of AM was on providing simple skin care and treatment for maladies. Eventually, the patient's requirements and the search for innovative techniques to satisfy the burgeoning demands, led to rapid progress in the field of aesthetic medicine. Aesthetics practitioners borrowed extensively from mainstream medicine.

Fat grafts are used as structural supports in the field of orthopedics. Aesthetic practitioners took that knowledge of fat grafts and are now routinely utilizing it to restructure, reshape, rejuvenate the female genito–urinary system. The techniques found new life in compatible long-term fillers to flush out vaginas, cervix, and G-spots in aging patients.

Rejuvenation therapy using Platelet-Rich Plasma (PRP) is in routine use to vitalize the genital area and the surrounding areas. The mechanisms of action (MoA) are growth factors, paracrine effects of the exosomes present in them.

Today, effective aesthetic medicine, as a minimally invasive practice, is based on doctors having a safe and skilled pair of hands leveraging on reliable leading-edge medical technology in new lasers, chemical peels, fillers, and injectables of natural or bio-ingredients. It spans surface treatments by chemical peels and lasers to minimally invasive procedures such as thread-lifts, botulinum type A injections, derma fillers, fat grafts, and hair transplants.

So aesthetic medicine doctors (AM doctors) are in fact treating healthy people, subjecting them to some degree of medical risks in order that these people can look better. The undeniable dilemma for aesthetic doctors, is that medical aesthetic procedures, no matter how minimally invasive, would still carry some risks. As a result, it is ethically more challenging when practicing aesthetic medicine as compared to conventional medicine.

Mental Confidence and Dysmorphia: Aesthetic medicine bridges the gap between beauty and health. It is important because beauty is not just skin deep. Beauty includes the need to feel good under one's own skin, in having a psycho-physical balance. We cannot ignore the importance of aesthetics from a psycho-sociological point of view, especially when today's society highly rates attractive appearances. So patients do seek out aesthetic doctors to improve their appearances, perhaps to improve self-confidence or simply to get a job. However, it's a thin mental line for doctors to effectively distinguish between the many consumers who are healthy both physically and mentally to undergo aesthetic treatment from the few patients who perceive their own bodies negatively as they have a form of psychosomatic disorder. The latter can become "beauty-obsessed" and seek excessive treatments. The old adage of "*too much of a good thing may become bad*" rings true for such patients.

Aesthetic Considerations Concern Conventional Medicine: Human beings wanting to look good is a universal truth. Even when a sick patient is facing a minor ailment or a life-threatening condition, he or she still seeks to come out looking as normal as possible. Aesthetic consideration is of relevance even in conventional medicine.

When doctors pause and take a look, they find daily examples of aesthetic consideration in many conventional medical disciplines. Examples such as: when overweight patients consult family doctors for slimming tips; dentists extracting a tooth are asked to whiten teeth yellowed by age or tobacco; young parents-to-be on fertility treatment ask for dermatological treatment for over-active acne; sun lovers removing suspicious moles, ask about skin treatment and wrinkle removals. Even cancer patients, ravaged by the effects of treatment, ask if something can be done to make them look better. The list goes on. Then, of course, there are medical conditions where conventional medicine itself aims a cure that delivers an improved aesthetic appearance such as—in treatment for congenital malformations, such as port-wine stains and strabismus.

Then there are examples when different conventional medical disciplines collaborate as a team for better aesthetic outcomes, such as a surgeon treating cancer of the face with surgery while a dermatologist treats skin texture with medication and creams; a dermatologist curing heavy acne infestation by working with a team of dietician, an endocrinologist, a psychologist, a cosmetologist, a beautician or a gynecologist (if the patient is female); or an accident victim with grievous wounds healing with more normal appearances with a team of vascular specialist, angiologist, phlebologist, plastic surgeon, dermatologist, dietician, endocrinologist, physiotherapist, orthopedist, and physical education instructor.

31.2.1 Spontaneous Growth of Aesthetic Medicine

The root difference that conventional medicine cures and heals the sick and the ill whereas aesthetic medicine is about helping the healthy to look better has major implications. For centuries, public welfare agencies, governments, and medical academia are prioritized, as they should be, toward conventional medicine for curing and healing the sick. Given its noble objective, conventional medicine globally is slower to play a role in standards and helping evidence-based research of aesthetic medicine, which is seen as a less noble form of "lifestyle" consumerism medicine. Conventional academia and professional medical standards bodies have for many years past developed formal board recognition in many conventional medical disciplines. They do these to nurture higher standards and promote research and professional sharing of know-how. For instance, conventional plastic surgery is a long recognized specialty with its conventional practice training on healing or curing via reconstructive surgery for war, accident, or cancer-ravaged disfigurements. The horrible war wounds of the First World War in fact led to many cutting edge plastic surgery procedures and raised the respect and standing of plastic surgeons.

Aesthetic medicine or AM came into being on its own spontaneously over the past decades and was not through the conventional medical academia. AM, being minimally invasive, and delivered in a clinic premises, was, is and can be learned by peer-to-peer, via courses sponsored by suppliers, or industry training organizations. Aesthetic medicine is practiced by doctors ranging from family doctors to many disciplines in conventional medicine. It is largely self-regulated, with doctors deciding for themselves as to their own level of competence and ability. It remains the aim of AM doctors in the aesthetic medicine community to practice responsibly and ethically to the highest of standards so that the collective performance and urging of this community may over time receive some form of formal recognition by conventional medicine.

31.2.2 For AM Doctors, It Is an Ongoing Personal Process

Until aesthetic medicine becomes a conferred medical specialty by conventional medicine, by conventional academia and institutions, or until it becomes a core curriculum in conventional medical schools, aspiring practitioners or AM doctors seeking higher competence of skills have to acquire or continuously work on their competence, on-the-job, or apprenticed or mentored by dermatologists and plastic surgeons, or by experienced aesthetic medicine peers and seniors or by courses with reputable organizations. For the foreseeable future, our community generates its own supply of aesthetic doctors from within the medical community.

However, just by taking courses or passing board assessment tests or by peer learning, or having met Board standards cannot and do not in any way imply that is constantly monitoring or taking a recurring assessment of the practice competence or standards of its past graduates. Just as no medical schools continue to undertake any ongoing assessments of their graduates after they left school. Doctors must not stop maintaining, learning, and honing their skills and they are themselves to be the judge in good faith of their own competence levels. Aesthetic medicine is in part a medical science and an art. It takes competent skills, honed repetitively, experiential feedback and continuous learning, and repetitive practice to build on one's core training in basic medical school training. That is why all participants at courses must have a basic medical school degree and a valid medical licence from their own jurisdiction.

31.2.3 Doctors Duty to Respond Responsibly and Ethically to Demand

Consumer-driven aesthetic medicine, where patients are healthy, electing for aesthetic treatment, paying with their hard-earned discretionary income, is different from conventional "heal-the-patients" obligations.

Misinformation and media hype abounds which can lead consumer-patients to make poor decisions, form unrealistic expectations and underestimate the medical risks. It is over this situation that has always felt that the aesthetic doctors community owes a duty to react responsibly and ethically when facing growing numbers of consumer-patients. Any botched cases, or even properly performed cases, but falling short of unreal consumer-patients expectations, will have adverse repercussions for the aesthetic practice community.

The aesthetic medicine discipline grew spontaneously over decades and is relatively young compared to conventional medicine. It is by and largely self-regulated. So far AM doctors have been carefully ethical and competent and regulators in many countries have generally left these doctors to self-regulate their own entry into and the conduct of their practice of aesthetic medicine. But with a growing number of aesthetic procedures and consumer-patients, some healthcare agencies are considering or have implemented some basic regulations. The aesthetic medical community welcomes these measures as being the overall good for the AM community and the general public. The AM community has to be ready for some form of basic regulations. When regulators start to see a need to set some regulations, it means at the same time that they start to see it as having a standing in medicine and to regard it as a form of medical discipline—which can only be good news for the community. One way of being ready is to develop an internationally acceptable set of ethics and protocol. Another is to raise funds for the conduct of more evidence-based trials, which is relatively lacking in this sector.

Aesthetic Medicine comprises all medical procedures that are aimed at improving the physical appearance and satisfaction of the patient, using non-invasive to minimally invasive cosmetic procedures. The Aesthetic Medicine specialty is not confined to dermatologists and plastic surgeons as doctors of all specialties seek to offer services to address their patient's aesthetic needs and desires. Some Aesthetic Medicine procedures are performed under local anesthesia while some procedures do not require anesthetics at all.

The exciting field of Aesthetic Medicine is a new trend in modern medicine. Patients not only want to be in good health, but they also want to enjoy life to the fullest, be fit, and minimize the effects of normal aging. Indeed, patients are now requesting quick, noninvasive procedures with minor downtime and very little risk. As a general rule, the needle is increasingly replacing the scalpel.

This recent trend explains the current success of Aesthetic Medicine around the globe. These aesthetic procedures consist of:

- Injections of neurotoxins and dermal fillers
- Chemical peels
- Cosmetic dermatology treatments
- Microdermabrasion
- Body contouring and treatment of cellulite
- Nutrition
- Hair transplant
- Hair reduction

- Fat grafting/platelet-rich plasma
- Laser and IPL
- Scar management
- Venous treatment
- Cosmetic gynecology

The real benefit of practicing Aesthetic Medicine is the type of care that practitioners are offering to their patients. These procedures are elective and are performed on patients who do not suffer from illness. They are usually happy and in excellent health. They simply want a minimally invasive preventative procedure to help manage the normal effects of aging.

This, along with the very lucrative business it represents, are the benefits any doctor should expect by expanding his/her own practice to an Aesthetic practice.

1. Anatomy and physiology with relevance to (topic if applicable): Anatomy of Female Reproductive System: The Female Reproductive System consists of 5 different parts.
2. From External to internal it is: Vulva: consisting of Mons Pubis, Labium Majora, Labium Minora, Clitoris, and hymen (present in virgins).

Vagina: is the muscular canal leading to the entrance of the female reproductive system. The outer walls of the anterior and posterior vagina are formed into longitudinal columns, or ridges, and the superior portion of the vagina—called the fornix—meets the protruding uterine cervix. The walls of the vagina are lined with an outer, fibrous adventitia; a middle layer of smooth muscle; and an inner mucous membrane with transverse folds called rugae. Together, the middle and inner layers allow the expansion of the vagina to accommodate intercourse and childbirth. The thin, perforated hymen can partially surround the opening to the vaginal orifice. The hymen can be ruptured with strenuous physical exercise, penile–vaginal intercourse, and childbirth. The Bartholin's glands and the lesser vestibular glands (located near the clitoris) secrete mucus, which keeps the vestibular area moist. The vagina is home to a normal population of microorganisms that help to protect against infection by pathogenic bacteria, yeast, or other organisms that can enter the vagina. In a healthy woman, the most predominant type of vaginal bacteria is from the genus *Lactobacillus*. This family of beneficial bacterial flora secretes lactic acid, and thus protects the vagina by maintaining an acidic pH (below 4.5). Potential pathogens are less likely to survive in these acidic conditions. Lactic acid, in combination with other vaginal secretions, makes the vagina a self-cleansing organ. However, douching—or washing out the vagina with fluid—can disrupt the normal balance of healthy microorganisms, and actually increase a woman's risk for infections and irritation. Indeed, the American College of Obstetricians and Gynecologists recommend that women do not douche, and that they allow the vagina to maintain its normal healthy population of protective microbial flora.

Ovaries: The ovaries are the female gonads. Paired ovals, are each about 2–3 cm in length, about the size of an almond. The ovaries are located within the pelvic cavity, and are supported by the mesovarium, an extension of the peritoneum that connects the ovaries to the broad ligament. Extending from the mesovarium itself is the suspensory ligament that contains the ovarian blood and lymph vessels. Finally, the ovary itself is attached to the uterus via the ovarian ligament.

The ovary comprises of an outer covering of cuboidal epithelium called the ovarian surface epithelium that is superficial to a dense connective tissue covering called the tunica albuginea. Beneath the tunica albuginea is the cortex, or outer portion, of the organ. The cortex is composed of a tissue framework called the ovarian stroma that forms the bulk of the adult ovary. Oocytes develop within the outer layer of this stroma, each surrounded by supporting cells. This grouping of an oocyte and its supporting cells is called a follicle. The growth and development of ovarian follicles will be described shortly. Beneath the cortex lies the inner ovarian medulla, the site of blood vessels, lymph vessels, and the nerves of the ovary. You will learn more about the overall anatomy of the female reproductive system at the end of this section.

31.2.4 The Ovarian Cycle

The ovarian cycle is a set of predictable changes in a female's oocytes and ovarian follicles. During a woman's reproductive years, it is a roughly 28-day cycle that can be correlated with, but is not the same as, the menstrual cycle (discussed shortly). The cycle includes two interrelated processes: oogenesis (the production of female gametes) and folliculogenesis (the growth and development of ovarian follicles).

Oogenesis: Gametogenesis in females is called oogenesis. The process begins with the ovarian stem cells, or oogonia. Oogonia are formed during fetal development, and divide via mitosis, much like spermatogonia in the testis. Unlike spermatogonia, however, oogonia form primary oocytes in the fetal ovary prior to birth. These primary oocytes are then arrested in this stage of meiosis I, only to resume it years later, beginning at puberty and continuing until the woman is near menopause (the cessation of a woman's reproductive functions). The number of primary oocytes present in the ovaries declines from one to two million in an infant, to approximately 400,000 at puberty, to zero by the end of menopause.

The initiation of ovulation—the release of an oocyte from the ovary—marks the transition from puberty into reproductive maturity for women. From then on, throughout a woman's reproductive years, ovulation occurs approximately once every 28 days. Just prior to ovulation, a surge of luteinizing hormone triggers the resumption of meiosis in a primary oocyte. This initiates the transition from primary to secondary oocyte. Cell division does not result in two identical cells. Instead, the cytoplasm is divided unequally, and one daughter cell is much larger than the other. This larger cell, the secondary oocyte, eventually leaves the ovary during ovulation. The smaller cell, called the first polar body, may or may not complete meiosis and produce second polar bodies; in either case, it eventually disintegrates. Therefore, even though oogenesis produces up to four cells, only one survives.

Folliculogenesis: Again, ovarian follicles are oocytes and their supporting cells. They grow and develop in a process called folliculogenesis, which typically leads to ovulation of one follicle approximately every 28 days, along with death to multiple other follicles. The death of ovarian follicles is called atresia, and can occur at any point during follicular development. Recall that, a female infant at birth will have one to two million oocytes within her ovarian follicles, and that this number declines throughout life until menopause, when no follicles remain. As you'll see next, follicles progress from primordial, to primary, to secondary and tertiary stages prior to ovulation—with the oocyte inside the follicle remaining as a primary oocyte until right before ovulation.

Folliculogenesis begins with follicles in a resting state. These small primordial follicles are present in newborn females and are the prevailing follicle type in the adult ovary. Primordial follicles have only a single flat layer of support cells, called granulosa cells, that surround the oocyte, and they can stay in this resting state for years—some until right before menopause.

After puberty, a few primordial follicles will respond to a recruitment signal each day, and will join a pool of immature growing follicles called primary follicles. Primary follicles start with a single layer of granulosa cells, but the granulosa cells then become active and transition from a flat or squamous shape to a rounded, cuboidal shape as they increase in size and proliferate. As the granulosa cells divide, the follicles now called secondary follicles—increase in diameter, adding a new outer layer of connective tissue, blood vessels, and theca cells—cells that work with the granulosa cells to produce estrogens.

Within the growing secondary follicle, the primary oocyte now secretes a thin acellular membrane called the zona pellucida that will play a critical role in fertilization. A thick fluid, called follicular fluid, that has formed between the granulosa cells also begins to collect into one large pool, or antrum. Follicles in which the antrum has become large and fully formed are considered tertiary follicles (or antral follicles). Several follicles reach the tertiary stage at the same time, and most of these will undergo atresia. The one that does not die will continue to grow and develop until ovulation, when it will expel its secondary oocyte surrounded by several layers of granulosa cells from the ovary. Keep in mind that most follicles don't make it to this point. In fact, roughly 99 percent of the follicles in the ovary will undergo atresia, which can occur at any stage of folliculogenesis.

Hormonal Regulation of Ovulation: The hypothalamus and pituitary gland regulate the ovarian cycle and ovulation. GnRH activates the anterior pituitary to produce LH and FSH, which stimulate the production of estrogen and progesterone by the ovaries.

When only the one dominant follicle remains in the ovary, it again begins to secrete estrogen. It produces more estrogen than all of the developing follicles did together before the negative feedback occurred. It produces so much estrogen that the normal negative feedback does not occur. Instead, these extremely high concentrations of systemic plasma estrogen trigger a regulatory switch in the anterior pituitary that responds by secreting large amounts of LH and FSH into the bloodstream (see Fig. 5). The positive feedback loop by which more estrogen triggers the release of more LH and FSH only occurs at this point in the cycle.

It is this large burst of LH (called the LH surge) that leads to ovulation of the dominant follicle. The LH surge induces many changes in the dominant follicle, including stimulating the resumption of meiosis of the primary oocyte to a secondary oocyte. As noted earlier, the polar body that results from unequal cell division simply degrades. The LH surge also triggers proteases (enzymes that cleave proteins) to break down structural proteins in the ovary wall on the surface of the bulging dominant follicle. This degradation of the wall, combined with pressure from the large, fluid-filled antrum, results in the expulsion of the oocyte surrounded by granulosa cells into the peritoneal cavity. This release is ovulation.

The Uterine Tubes: The uterine tubes (also called fallopian tubes or oviducts) serve as the conduit of the oocyte from the ovary to the uterus. Each of the two uterine tubes is close to, but not directly connected to, the ovary and divided into sections. The isthmus is the narrow medial end of each uterine tube that is connected to the uterus. The wide distal infundibulum flares out with slender, finger-like projections called fimbriae. The middle region of the tube, called the ampulla, is where fertilization often occurs. The uterine tubes also have three layers: an outer serosa, a middle smooth muscle layer, and an inner mucosal layer. In addition to its mucus-secreting cells, the inner mucosa contains ciliated cells that beat in the direction of the uterus, producing a current that will be critical to moving the oocyte.

Following ovulation, the secondary oocyte surrounded by a few granulosa cells is released into the peritoneal cavity. The nearby uterine tube, either left or right, receives the oocyte. Unlike sperm, oocytes lack flagella, and therefore cannot move on their own. High concentrations of estrogen that occur around the time of ovulation induce contractions of the smooth muscle along the length of the uterine tube. These contractions occur every 4–8 seconds, and the result is a coordinated movement that sweeps the surface of the ovary and the pelvic cavity. Current flowing toward the uterus is generated by coordinated beating of the cilia that line the outside and lumen of the length of the uterine tube. These cilia beat more strongly in response to the high estrogen concentrations that occur around the time of ovulation. As a result of these mechanisms, the oocyte–granulosa cell complex is pulled into the interior of the tube. Once inside, the muscular contractions and beating cilia move the oocyte slowly toward the uterus. When fertilization does occur, sperm typically meet the egg while it is still moving through the ampulla.

The Uterus and Cervix: The uterus is the muscular organ that nourishes and supports the growing embryo. Its average size is approximately 5 cm wide by 7 cm long (approximately 2 in 3) when a female is not pregnant. It has three sections. The portion of the uterus superior to the opening of the uterine tubes is called the fundus.

The middle section of the uterus is called the body of the uterus (or corpus). The cervix is the narrow inferior portion of the uterus that projects into the vagina. The cervix produces mucus secretions that become thin and stringy under the influence of high systemic plasma estrogen concentrations, and these secretions can facilitate sperm movement through the reproductive tract.

Several ligaments maintain the position of the uterus within the abdominopelvic cavity. The broad ligament is a fold of peritoneum that serves as a primary support for the uterus, extending laterally from both sides of the uterus and attaching it to the pelvic wall. The round ligament attaches to the uterus near the uterine tubes, and extends to the labia majora. Finally, the uterosacral ligament stabilizes the uterus posteriorly by its connection from the cervix to the pelvic wall.

The wall of the uterus is made up of three layers. The most superficial layer is the serous membrane, or perimetrium, which consists of epithelial tissue that covers the exterior portion of the uterus. The middle layer, or myometrium, is a thick layer of smooth muscle responsible for uterine contractions. Most of the uterus is myometrial tissue, and the muscle fibers run horizontally, vertically, and diagonally, allowing the powerful contractions that occur during labor and the less powerful contractions (or cramps) that help to expel menstrual blood during a woman's period. Anteriorly directed myometrial contractions also occur near the time of ovulation, and are thought to possibly facilitate the transport of sperm through the female reproductive tract.

The innermost layer of the uterus is called the endometrium. The endometrium contains a connective tissue lining, the lamina propria, which is covered by epithelial tissue that lines the lumen. Structurally, the endometrium consists of two layers: the stratum basalis and the stratum functionalis (the basal and functional layers). The stratum basalis layer is part of the lamina propria and is adjacent to the myometrium; this layer does not shed during menses. In contrast, the thicker stratum functionalis layer contains the glandular portion of the lamina propria and the endothelial tissue that lines the uterine lumen. It is the stratum functionalis that grows and thickens in response to increased levels of estrogen and progesterone. In the luteal phase of the menstrual cycle, special branches off of the uterine artery called spiral arteries supply the thickened stratum functionalis. This inner functional layer provides the proper site of implantation for the fertilized egg, and—should fertilization not occur—it is only the stratum functionalis layer of the endometrium that sheds during menstruation.

Recall that during the follicular phase of the ovarian cycle, the tertiary follicles are growing and secreting estrogen. At the same time, the stratum functionalis of the endometrium is thickening to prepare for a potential implantation. The post-ovulatory increase in progesterone, which characterizes the luteal phase, is key for maintaining a thick stratum functionalis. As long as a functional corpus luteum is present in the ovary, the endometrial lining is prepared for implantation. Indeed, if an embryo implants, signals are sent to the corpus luteum to continue secreting progesterone to maintain the endometrium, and thus maintain the pregnancy. If an embryo does not implant, no signal is sent to the corpus luteum and it degrades, ceasing progesterone production and ending the luteal phase. Without progesterone, the endometrium thins and, under the influence of prostaglandins, the spiral arteries of the endometrium constrict and rupture, preventing oxygenated blood from reaching the endometrial tissue. As a result, endometrial tissue dies and blood, pieces of the endometrial tissue, and white blood cells are shed through the vagina during menstruation, or the menses. The first menses after puberty, called menarche, can occur either before or after the first ovulation.

The Menstrual Cycle: Now that we have discussed the maturation of the cohort of tertiary follicles in the ovary, the build-up and then shedding of the endometrial lining in the uterus, and the function of the uterine tubes and vagina, we can put everything together to talk about the three phases of the menstrual cycle—the series of changes in which the uterine lining is shed, rebuilds, and prepares for implantation.

The timing of the menstrual cycle starts with the first day of menses, referred to as day one of a woman's period. Cycle length is determined by counting the days between the onset of bleeding in two subsequent cycles. Because the average length of a woman's menstrual cycle is 28 days, this is the time period used to identify the timing of events in the cycle. However, the length of the menstrual cycle varies among women, and even in the same woman from one cycle to the next, typically from 21 to 32 days.

Just as the hormones produced by the granulosa and theca cells of the ovary "drive" the follicular and luteal phases of the ovarian cycle, they also control the three distinct phases of the menstrual cycle. These are the menses phase, the proliferative phase, and the secretory phase.

The Breasts: Whereas the breasts are located far from the other female reproductive organs, they are considered accessory organs of the female reproductive system. The function of the breasts is to supply milk to an infant in a process called lactation. The external features of the breast include a nipple surrounded by a pigmented areola, whose coloration may deepen during pregnancy. The areola is typically circular and can vary in size from 25 to 100 mm in diameter as shown in Fig. 31.1. The areolar region is characterized by small, raised areolar glands that secrete lubricating fluid during lactation to protect the nipple from chafing. When a baby nurses, or draws milk from the breast, the entire areolar region is taken into the mouth.

Breast milk is produced by the mammary glands, which are modified sweat glands. The milk itself exits the breast through the nipple via 15 to 20 lactiferous ducts that open on the surface of the nipple. These lactiferous ducts each extend to a lactiferous sinus that connects to a glandular lobe within the breast itself that contains groups of milk-secreting cells in clusters called alveoli. The clusters can change in size depending on the amount of milk in the alveolar lumen. Once milk is made in the alveoli, stimulated myoepithelial cells that surround the alveoli contract to push the milk to the lactiferous sinuses. From here, the baby can draw milk through the lactiferous ducts by suckling. The lobes themselves are surrounded by fat tissue, which determines the size of the breast; breast size differs between individuals and does not affect the amount of milk produced. Supporting the breasts are multiple bands of connective tissue called suspensory ligaments that connect the breast tissue to the dermis of the overlying skin.

During lactation, milk moves from the alveoli through the lactiferous ducts to the nipple. During the normal hormonal fluctuations in the men-

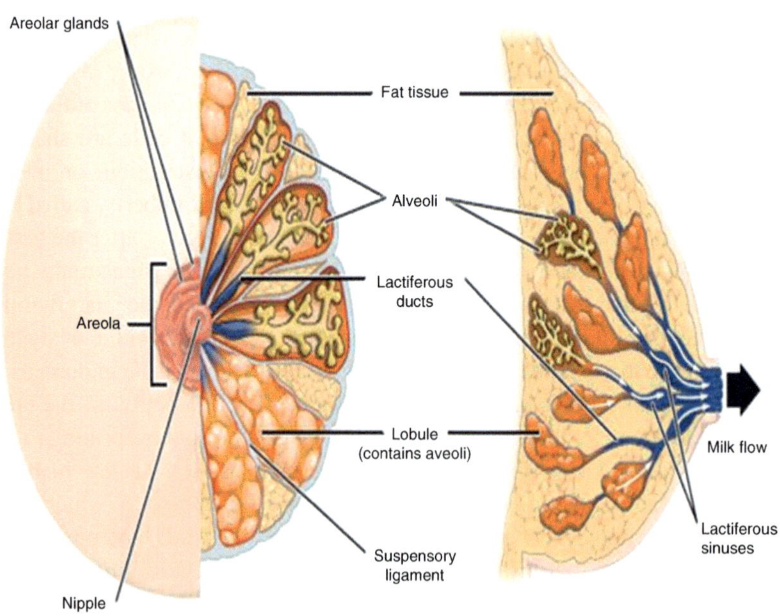

Fig. 31.1 Anatomy of the breast

strual cycle, breast tissue responds to changing levels of estrogen and progesterone, which can lead to swelling and breast tenderness in some individuals, especially during the secretory phase. If pregnancy occurs, the increase in hormones leads to further development of the mammary tissue and enlargement of the breasts.

31.2.5 Procedure Details (Investigations, Techniques, and Anesthesia)

Products used in AM: It is important to choose ethical sources of biological products for treating your patients. There are many products that are commercially available for use as fillers or as support for repair of the tissues. Mainly they can be categorized as autologous, allogeneic, and tissue scaffolding.

PRP (Platelet-Rich Plasma)—It is used in nonsurgical and surgical treatments. PRP contains high levels of growth factors such as platelet-derived growth factor, transforming growth factor-beta, and epidermal growth factor. It is nonantigenic because it is autologous, and there have been no adverse effects detected. It is used in conditions such as Lichen Sclerosus. O-Shot® is a PRP procedure that was trademarked by Charles Runels. Hymenoplasty involves many ethical issues. It is known as revirgination in Western countries and is a sociocultural issue, especially in Muslim countries.

31.2.6 Role of Fat Transfer and Adipose-Derived Stem Cells (ADSCs) in Rejuvenation and Treatment of Female Genital Area

Fat Transfer for Volume Restoration: Restore the lost volume by aging or losing weight, in order to recover the youthful and vitalize appearance to these areas. Some young women with normal genitalia but still ask for enlargement of labia majora size when it looks and feels small in comparison to surrounding to give the area a better shape and appearance to the labia majora (outer lip) and called Augmentation Labiaplasty.

31.2.6.1 Fat Injections to Improve the Changes by Weight Lost or Aging

Fat Transfer for Rejuvenation of vulvar skin and vaginal wall: Rejuvenation by injection of stem cells from your own fatty tissue (ADSCs) to rejuvenate and treat menopausal symptoms or vulvar dystrophy of outer side of the genital area to get rid of wrinkling and laxity of skin and in some case can do whitening the area if the pigmentation is secondary and not the area natural color. The stem cells also can be used for vaginoplasty due to the changes in the area. To rejuvenate and treat menopausal symptoms or vulvar dystrophy and using the injected stem cells either as SVF (Stromal Vascular Fraction) or Cultured-ADSCs, PRP (Platelet-Rich Plasma) or Nanograft may be used as a medium for SVF.

Vaginal Recalibration by Lipograft Technique: This procedure consists in reducing the vaginal caliber by thickening the vaginal walls with adipose tissue transplant. It is indicated in patients who are concerned by a sensation of a wide vagina. The causes are often post gravid in multiparous women and sometimes constitutional. Clinical and gynecological examinations are frequently normal. However, may reveal a prolapse which is a contraindication for this intervention.

31.2.6.2 Labiaplasty of Labia Majora (Augmentation and Rejuvenation)

Labiaplasty of labia majora is the plastic surgery on the outer or the larger lips of the vagina, which can be done alone or in combination with vaginoplasty. Fat injections to the labia majora which is called (Augmentation Labiaplasty), to plump out these structures to make it bigger in cases where it is considered to be too small or to correct an asymmetry between them.

Other cases are those who lack the volume secondary to weight loss or aging to give it a more youthful appearance. However, it is not always about appearance and cosmetic concerns. The candidate for labia augmentation and rejuvenation have normal vulvar tissue, normal gynecological examination but, the patient has a cosmetic concern. Normal size and good tone of tissue but, the woman seeking labial enlargement as she has a concern about the size of her labia majora and/or shape.

As with many aspects of human anatomy, there are a wide variety of shapes, sizes, and appearances of the female genitalia, all of which are within the limits of normal. Before undergoing any surgery, it is important to determine whether there is really a problem with the genitalia or whether another solution would be more rewarding.

Normal size and good tone of tissue but, the woman seeking labial symmetry, as she has a significant asymmetry in the size and/or shape. Naturally, there is asymmetry in size between the two labias as with breast in female and testes in male and other body parts but, when the asymmetry is significant and influences the woman's self-esteem, in such cases labiaplasty which is a minor surgery can be helpful. The intervention is either by doing augmentation labiaplasty for the smaller side (the most common approach) or by reducing the larger one. Before any intervention, a proper clinical evaluation should be done to exclude a pathological lesion underlying the asymmetry like hemangioma, lump, or cyst formation.

Loss of volume and tissue tone, due to involutional changes in the area with aging (labia, mons pubis, and/or vagina) or after significant fatty tissue loss from the whole body in a short period. There is an increasing number of cases in need of rejuvenation of mons pubis and external female genitals after bariatric surgeries.

Vulvar Dystrophy: Physiological and Pathological: Vulvar Dystrophy is a condition that may occur in women both in physiological and in pathological conditions. Associated symptoms may include itching, burning, dyspareunia, vaginal dryness, and bleeding.

Physiological vulvar dystrophy (Postmenopausal Female Genital Atrophy): Vaginal dystrophy commonly affects postmenopausal women, with prevalence ranging from 10% to 50% and it is estimated that up to 45% of all women are symptomatic. However, recognition of their etiology is poorly understood by women and their partners. The diagnosis of vaginal atrophy is typically based on the history of certain symptoms and specific physical findings as vaginal tissue and urethral mucosa are altered as a result of the diminished amount of circulating estrogen.

Recognized symptoms include:

Vaginal changes: Vaginal dryness due to atrophy of the vaginal mucosa and decreased vaginal lubrication, burning, pruritus, abnormal vaginal discharge, and dyspareunia Vaginal inflammation develops due to estrogen-deficient atrophy of the soft tissues. The most common etiology of estrogen deficiency is the natural process of menopause.

External genitalia signs include: atrophy of the labia majora and minora, loss of subcutaneous fat, dry labia, vulvar dermatoses, vulvar lesions, and sparse pubic hair and these external signs also occur due to decreased estrogen levels. All these annoying symptoms greatly affect women's sexuality and quality of life.

The etiology of vulvovaginal atrophy among postmenopausal women is most commonly explained by the decrease in circulating estrogen associated with the menopausal transition, which has an adverse effect on skin collagen and elasticity. Even while taking systemic estrogen, 10% to 20% of women may still have residual symptoms.

Current Concept of Management of (Postmenopausal Female Genital Atrophy): The primary goal of treating genitourinary syndrome of menopause is to relieve symptoms. For women with vulvovaginal symptoms unrelated to sexual activity, first-line therapies include long-acting vaginal moisturizers and a short course of low-dose vaginal estrogen and hormonal replacement therapies. The hormonal replacement therapy is widely available. The route of delivery can be systemic via the mouth, through the skin, via a nasal spray, and by injection. Local treatment includes vaginal ring, creams, tablets, and vaginal pessaries. Although estrogen-based treatments can be effective, according to the Women's Health Initiative and other clinical trials, However, many patients are reluctant to be treated with such formulations due to health concerns. A number of menopausal women have discontinued taking hormones, and have turned to herbs, phytoestrogens, and dietary supplements instead because they worry about their reactions to hormones.

Since many patients are reluctant to use estrogen-based treatments due to health concerns, physicians are eager to find acceptable alternatives. In addition to diminished production of estrogen, vaginal changes are affected also by the aging process like elsewhere in the body. Due to loss of tone and volume of the tissues by aging, decent and sagging of the skin may follow. These kinds of changes can cause an unattractive aged look of external genitalia. But there is no established treatment for contour correction of vagina.

31.2.6.3 Role of Stem Cell Therapy in the Management of (Postmenopausal Female Genital Atrophy)

Lipofilling is an effective and minimally invasive modality to restore tissue volume in both aesthetic and reconstructive surgery. Autologous lipofilling has become a popular procedure for soft tissue rejuvenation and body contouring in aesthetic and reconstructive surgery. Autologous fat can be harvested by a relatively simple procedure, and it is completely biocompatible and available in large quantities. But, the disadvantages of lipofilling are its unpredictability and variable reabsorption rate. In addition to the permanent volumetric effect of adipose tissue, certain dynamic phenomena of tissue regeneration occur at the recipient site after adipose tissue transfer. It has been noted by many researchers the improvement of the quality of the overlying skin. Since vaginal changes occur following the aging process, we assumed that the regenerative features of lipograft and ADSCs might play a significant role. Recently, PRP has emerged as a new matrix that can enhance fat graft survival. Using either PRP enriched lipofilling or ADSCs enriched lipofilling improves the long-term outcome. The addition of PRP to fat grafts improves graft survival in a simple and safe procedure and less cost. Stem cell-enriched fat transfer will be costlier but more efficient and longer outcome. Besides the volume correction, the patient experienced a substantial relief of symptoms.

Lipofilling is now widely accepted in expectation of both the filling and rejuvenating effect. While surgical manipulation and harvest of fatty tissue are generally simple procedures, the tissue itself is complex. Fatty tissue has been found to contain adipocytes and subpopulations of cells including adipose-derived stem cells (ADSCs), which have the potential to aid in tissue regeneration.

31.2.7 Pathological Vulvar Dystrophy

Lichen sclerosus (LS): It is a chronic immune-mediated inflammatory skin disorder of poorly understood etiology, which may be localized anywhere on the body but has a predilection for the anogenital area. The exact etiology of lichen sclerosus is unknown and probably multifactorial. Although disease onset has been reported at all ages, it occurs most commonly in women in their fifth or sixth decade of life. Lichen sclerosus is characterized by the presence of well-defined white papules and plaques. The skin affected by lichen sclerosus becomes atrophic or thickened. Genital lichen sclerosus causes both dryness and severe, persistent pruritus, and it often leads to functional impairment. Additionally, subcutaneous bleeding with ecchymosis or hemostasis and fissures with superficial ulceration and erosion may occur. Late progressive symptoms include thinning of the mucosa, edema, and fibrosis of the dermis, shrinkage of the labia, and agglutination of the labia minora that can lead to stenosis of the introitus. Moreover, while the disorder is considered benign, some women with vulvar LS may later develop squamous cell carcinoma (SCC) of the vulva and some women have concomitant SCC when initially diagnosed with LS. Vulvar dystrophy may negatively affect the entire sexual response cycle, inducing significant changes in desire, arousal, orgasm, and satisfaction at menopause and beyond. Patients may be embarrassed by the disfiguring changes that may occur and avoid sexual intimacy. Furthermore, these patients may have an increased risk to develop bacterial vaginosis due to vaginal pH changes and urinary tract infection or stress urinary incontinence. A vulvar biopsy is recommended when malignancy cannot be excluded or in those who have failed to respond to first-line treatment. But the diagnosis of lichen sclerosus is usually clinical.

Non-aging Oestrogen Deficiency: Symptoms related to estrogen reduction may occur during other stages of women's lives than menopausal period, in response to events associated with sustained decreases in estrogen such as lactation or also related to chemotherapy that induces premature ovarian failure in 14–100% of cases. These patients are at high risk of transient or permanent amenorrhea, and, for those women who continue to menstruate or who recover their cycles, there is an additional long-term risk of premature ovarian failure. In particular, breast cancer treatment increases the prevalence of vulvar dystrophy because the surgical, endocrine, and chemotherapeutic agents used in its treatment can cause or exacerbate this condition.

Iatrogenic cause of vulvar dystrophy: It is related to allogeneic hematopoietic stem cell transplant. This procedure has found a place in the treatment of a variety of malignant and non-malignant diseases of the bone marrow and immune system. However, it may be complicated by chronic graft-versus-host disease (GVHD), due to activation of donor immunological cells against host tissues.

Current Concept of Management of (pathological Vulvar Dystrophy): Vulvar dystrophy negatively impacts women's lives, but women lack knowledge of the subject and are hesitant to consult about it. Main treatment for it currently includes drug therapies and lubricants but, in the majority of patients they have limited outcomes. The goal of treatment is to reduce symptoms such as pruritus and dryness, improve the patient's quality of life, and detect any malignant transformation. Main treatments to solve urogenital and sexual dysfunction due to vulvovaginal tissue alteration currently include drug therapy and lubricants, the treatment of choice in adults and children and in both sexes is an ultrapotent topical steroid but, in the majority of patients these therapies do not imply a complete remission of symptoms.

Role of Stem Cell Therapy in the Management of (Pathological Vulvar Dystrophy Lichen Sclerosus): In this case, lichen sclerosus was steroid-resistant so that the resolution could be achieved after transfer of autologous fat mixed ADSCs and/or with PRP. As mentioned earlier, ADSCs enriched lipograft has the regenerative properties and capabilities to produce anti-inflammatory and immunomodulatory effects. Besides the presence of ADSCs in the lipograft, PRP seemed to be an essential factor influencing the healing and regeneration process in this case. PRP has been shown to enhance wound healing and has rejuvenation effects through the release of significant amounts of growth factors.

Stem Cell Therapy in Vulvar Dystrophy: Recent studies emphasized that adipose tissue is a rich source of adult stem cells, which is called Adipose-Derived Stem Cells (ADSCs). They have tested in a study published by Giuseppina et al. (2016), the hypothesis that ADSCs can enhance the trophism and tone of dystrophic tissue by applying it to dystrophic valvular tissue as the current treatments often do not imply a complete remission of symptoms. They have concluded that stem cell therapy using ADSCs injection represents a valid alternative therapy to treat vulvar dystrophies, with significant symptoms improvement compare to other modalities of treatment, since it can increase the vascularization due to their angiogenic properties and proven eutrophic effect and ability to improve tone and trophism of the involuted and dystrophic tissue.

The process of Adipose-derived Stem Cells Therapy: Stem cell therapy is a term used to define the process of isolating adult stem cells from the host tissue, and then reintroducing them to the area of the body that needs healing or regeneration and in this case is the vulvar tissue and vagina. The process gets started by locating the ideal place to harvest the adipose stem cells (stem cells that are stored in the fat deposits just below the skin). The extraction is only for a limited amount of the fat stored in the stem cell reserves and needs a local anesthetic during the procedure. Once this is harvested, the fat is transferred into a sterile container or tube and is used either to prepare the Ready-to-use ADSCs Pellet (from the SVF) or is sent to a specialized stem cells laboratory where it is expanded to get a higher number of stem cells (Cultured-ADSCs). The reinjection should be performed under local anesthesia with sedation. Recovery period is short with extra care about personal hygiene of the area.

Vaginal Tightening and Rejuvenation (Vaginoplasty): Vaginal laxity is a condition where the connective tissue that supports the vagina has become loose. It is often experienced by women going through menopause or those who have recently been through childbirth, but it can occur to young nulliparous women presenting with wide vagina without perineum lesions. Concerns about vaginal size are quite common because it can affect sexual gratification, lesser sensitivity, and urinary incontinence. Term "Vaginoplasty" is used for the procedure that aims to "tighten up" vaginal muscles to correct defects and deformities of vaginal canal, congenital deformities, acquired cause-physical trauma, or cancer. Today, vaginoplasty is most commonly used for surgical and nonsurgical approaches to manage post-childbirth and postmenopausal changes. The aim today is not just to "tighten up" the vagina but to rejuvenate it again and revitalize the tissue that has become slack or loose from vaginal childbirth or natural estrogen deficiency with aging. Postmenopausal vaginal changes include; dryness due to atrophy of the vaginal mucosa and decreased vaginal lubrication, burning, pruritus, abnormal vaginal discharge, and dyspareunia Vaginal inflammation develops due to estrogen-deficient atrophy of the soft tissues. For that, it is more important to address these symptoms too by natural way of treatment if possible and not just by having hormonal therapy and "tightening up" the vagina.

One of the most advanced technologies today is by using patients' body fat tissue to get stem cell ADSCs and inject it back to the vagina with or without fat graft. This advanced modality of treatment is showing a good promise in clinical trials in treating menopausal symptoms. Stem cell-enriched lipofilling can also help in cases where clinically not a candidate for surgical tightening as no underlying muscular defect. They are more in need to recover the histological changes that happen to their vagina lining layer. This can help to treat all uncomfortable menopausal symptoms plus helping women who feel less wide vagina. Vagina can be recalibrated to a smaller diameter but to limit by stem cell-enriched lipofilling. A combination of treatments can give a higher satisfactory outcome by adding laser therapy or Radio frequency therapy for more vaginal tightening and help with mild stress incontinence symptoms. It can also combine the surgical tightening of the vaginal wall. It starts with tightening of posterior wall muscles followed by adding the fat graft to the sidewalls of the vagina. The complications are rare, corresponding to any other surgical procedure in this very area, and do not jeopardize the final result.

Vaginal Recalibration by Lipograft Technique: This procedure consists in reducing the vaginal caliber by thickening the vaginal walls with adipose tissue transplant. It is indicated in those patients who feel concerned by a sensation of a wide vagina. The causes are often present in multiparous women and sometimes it may be constitutional. Clinical and gynecological examinations are frequently normal. However, the examination may reveal a prolapse which is contraindicated for this intervention.

31.2.8 The Scientific Evidence Behind Using Stem Cells for Female Genital Rejuvenation

For symptomatic vulvar dystrophia's: Regardless of the etiology of vulvar dystrophy, a remarkable pain and dryness reduction could be observed in all patients and for that all patients obtained an improvement of sexual function after ADSCs treatment has been documented it has been documented symptomatic improvement within one month after the first ADSCs treatment, and dramatic pain reduction after one year. Sexual function improvement after second treatment. The clinical trials documented maintained improvement at 2 years follow-up. After treatment with ADSCs, a regression of the morphological alterations of epithelial cells and full attenuation of inflammatory signs in the connective tissue were observed. Biopsy specimens of patients affected with LS displayed after treatment with ADSCs, the dermis sclerosis was significantly reduced, capillaries that resulted were less dilated, and inflammatory infiltrate was dramatically reduced.

For Menopausal Dystrophy: A loss of vaginal rugae, vaginal pallor, and petechiae are evident in patients in menopause. Histological examination after ADSCs treatment for menopausal patients has been shown the improvement of the vaginal elastic network.

31.2.9 Stem Cells and Cell Therapies

Lipotransfer (Fat Grafting) (Fat Transfer): Fat grafting, fat transfer, also referred to as fat injections, is the surgical process by which the movement of fat cells (adipocytes) from one part of the body (the donor area) to another (the recipient area). The surgical goal is to improve or augment the area where the fat is injected.

The technique of grafting a tissue involves removing it by liposuction from its blood supply entirely, then processing the fat, and then re-injecting purified fat into the area needing improvement relying on the in-growth of a new capillary network to support the transferred cells once they are at the recipient site as shown in Fig. 31.2. It is the establishment of this new blood supply that is crucial to the survival of the transferred tissue. If this fails, it will result in graft loss; in the case of adipocytes, necrosis of the cells, with resorption of their lipid content (and loss of volume at the recipient site). As the capillaries in-growth takes a few days following graft placement, the fat cells to survive must rely on diffusion of nourish to the block of tissue from surrounding recipient tissue.

If the fat particles are very small and have a large surface-to-volume ratio (micro-lipofilling technique) then, the whole particle will survive. Large particle size, meaning the inner surface will not achieve re-establishment of blood flow in time to prevent cellular death. Therefore, small particle grafting (micro-lipofilling) is essential and will have a better rate of graft take. Since the 1990s Plastic Surgeons have reliably used fat grafting as a way to improve and enhance the

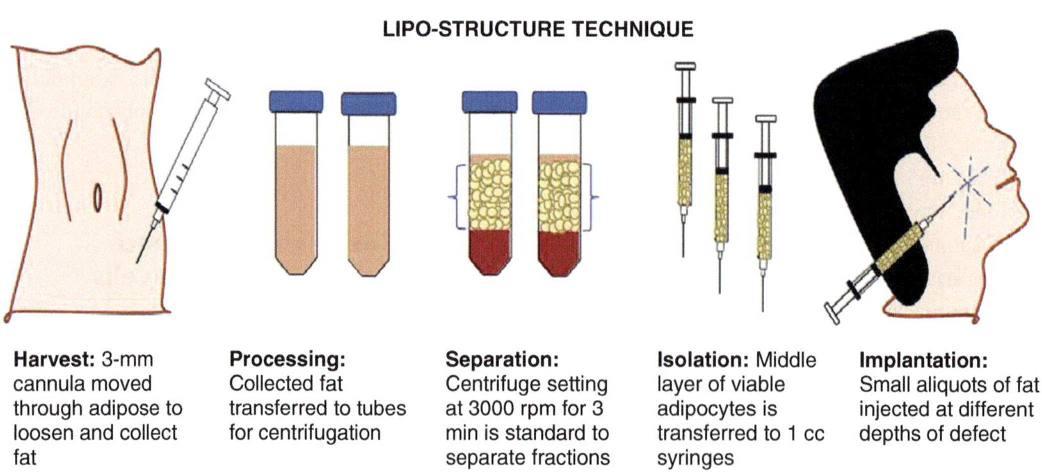

Fig. 31.2 Lipo-structure technique

cosmetic appearance of the face, breast, hands, feet, hips, and buttocks. However, more recently, clinicians have documented the therapeutic benefits of fat grafting in the healing of wounds and scars, as well as fat's ability to repair damage to breast tissue following radiation treatment.

History of Fat Transfer and ADSCs: The first "fat grafting" procedure dates back to the late 19th century, 1893, when a German Plastic Surgeon, Gustav Neuber (1850–1932) transferred fat from the arm to the orbital (eye) region to correct scars formed from osteomyelitis (bone infection). Only two years later, 1895, Dr. Viktor Czerny (1842–1916) transferred a lipoma to the breast to establish symmetry following a unilateral partial mastectomy. The first reports of fat injection to the face and breast in the human body to re-established contour deformities were carried out by Eugene Hollander in 1909.

With the advent of liposuction, fat injection was rediscovered, but the reabsorption rate was still high. In the 1990s, Sydney Colman systematized the technique for harvesting, purification, and placement of fat, so as to reduce the resorption rate. In 2001, Zuk and colleagues demonstrated that adipose tissue is the greatest source of adult mesenchymal stem cells, adipose-derived stem cells, capable of differentiating into other types of tissues. Stromal vascular fraction (SVF), a source of ADSCs, endothelial (progenitor) cells, T cells, B cells, mast cells, and adipose tissue macrophages were identified. In 2007, Gino Rigotti (Italy) applied the regenerative properties of ADSCs in a human patient. He successfully managed radiation tissue damages, with complete restitution of the affected tissues. This was one of the first examples of regenerative therapies.

Development of Lipofilling Procedures: Transfer of adipose tissue or lipofilling, is recognized as a promising and novel technique for correction of volume deficiency, skin rejuvenation, and as treatment for scars. This is strongly supported by evidence-based clinical trials as well as fundamental studies in animals and humans. The first case of fat transfer in literature dates from 1893. As soon as liposuction was further developed in the mid-1980s, interest developed in reusing the lipoaspirate subcutaneous adipose tissue. Liposuction pioneers such as Illouz and co-workers (Illouz, 1983) developed the first clinical applications and methods for lipofilling to restore or gain volume. The real breakthrough in lipofilling came with fat harvesting, subsequent processing, and subcutaneous administration as described by Coleman (2002), which allowed better survival of the fat transfer.

In the 1990s, Dr. Sydney Coleman, a New York City Plastic Surgeon, began publishing papers describing standardized techniques for fat extraction, processing, and injection. Since that point in time, the procedure's popularity has only increased along with a wider variety of clinical applications. He opened the door for lipofilling of the face and hands for both reconstructive and aesthetic purposes, especially in these applications with rather superficial lipofilling, effects described as 'more than volume alone' were an improved appearance and quality of the skin, and has subsequently been described in many case reports. Yet a mechanistic underpinning was still lacking. These clinical observations initiated a wide range of clinical applications for lipofilling (fat transfer), other than just volume adjustment (Rigotti et al., 2007). This novel idea to use lipofilling for treatment of fibrosis result from tissue damage has led to the use of lipofilling to treat burn scars (Klinger et al., 2008) and even to alleviate scar-associated pain as occurring, for example, after mastectomy Caviggioli et al., (2011). Zuk et al., (2001) demonstrated that adipose tissue had a source of endogenous mesenchymal stem cells (MSC), which were named adipose-derived stem or stromal cells (ADSC). This discovery significantly advanced the use of lipofilling as a regenerative therapy, as it had been shown that at least one of the components of adipose tissue had therapeutic potential. Since then, many of the beneficial effects observed after lipofilling have been attributed to ADSC.

Battle Against Fat Reabsorption: The grafted fatty tissue is placed under ischemia (low oxygen with low nutrition) and is temporarily nourished only by plasmatic diffusion from the surrounding host tissue for a few days until a direct capillary supply is formed. Adipocytes, known to be very sensitive to hypoxia, can die within 24 hours. ADSCs are likely to be as resistant to ischemia up to 72 hours under severe ischemia. Early cell death likely results in volume loss and histologic fibrosis and cyst formation. Models of fat graft behavior following implantation have defined three zones, with a peripheral, regenerating zone, and intermediate, inflammatory zone, and a central, necrotic zone. Diffusion of nutrients to the central zone is insufficient, resulting in necrosis and ultimate graft resorption.

In response to injury and cell death, many growth factors and proteinases are released from damaged tissue and activated platelets. Inflammatory cells are infiltrated and inflammatory cytokines such as interleukins are secreted.

Adipose grafting induces injury in the recipient tissue; bleeding from the host tissue activates platelets, which release various soluble factors activating sleeping stem cells. At the same time transplanted adipose tissue is temporarily placed under severe ischemia; factors released from the injured donor site and one released from dying cells stimulate adipose-derived stem cells (ADSCs) to release hepatocyte growth factor HGF, which promote angiogenesis and inhibits fibrogenesis. Most of the differentiated cells in graft die, but graft-resident stem cells are activated. The dead cells are partly replaced with the next generation.

Best Technique to have a Successful Fat Grafting Outcome: Although fat grafting is generally safe with good patients' satisfaction, many still experience unpredictable outcomes with varying fat cell survival as the adipocytes are less robust than stem cells and require a nontraumatic method of harvesting to prevent cellular injury and subsequent apoptosis or phagocytosis. Depending on the technique and time that is required for harvesting and lipofilling, 40–90% of the injected fat graft volume will remain, while the rest is reabsorbed within months after grafting. Oily cysts may remain in the grafted area as a consequence of this fat necrosis.

31.2.10 Best Methodology of Fat Transfer

Highest fat retention, which indicates a high percentage of fat survival which can end good surgical results and least unsatisfactory outcome.

Highest fat retention depends on the survival of highest number of fat cells plus the number of highest survival of stem cells in the fat graft to play its major role of fat survival.

High-quality fat graft depending on the fat aspiration, processing and injecting techniques that are used and need to be considered carefully.

Every step-in fat transfer that consists of harvesting, processing, and administration may have "Deleterious effects on cell viability" as illustrated in Fig. 31.3.

31.2.11 Process of Fat Transfer Lipo-Aspiration (Harvesting the Fat) Technique

Simple harvesting requires two pieces of equipment:

- A suction source
- A collection receptacle

The lipoaspiration cannula can either be connected to a syringe, so that aspiration can be carried out manually or using a liposuction machine to aspirate the fat and this is mainly for large amount fat transfer procedures. Whether the cannula is connected to a syringe as the suction forces created by retracting the plunger of a syringe or by an external suction machine, the key factor is not only by what we are suctioning but also, how much the negative pressure that we are creating during suctioning. About 10–15% of cells were lost when the suction was on or near-maximum negative pressure (-28 in. Hg (700 mmHg) or one atmosphere) (Shiffman et al. 2001).

Small amount lipo-transfer: The surgical tool to aspirate fat is highly important from design and size of cannula to the negative pressure inside

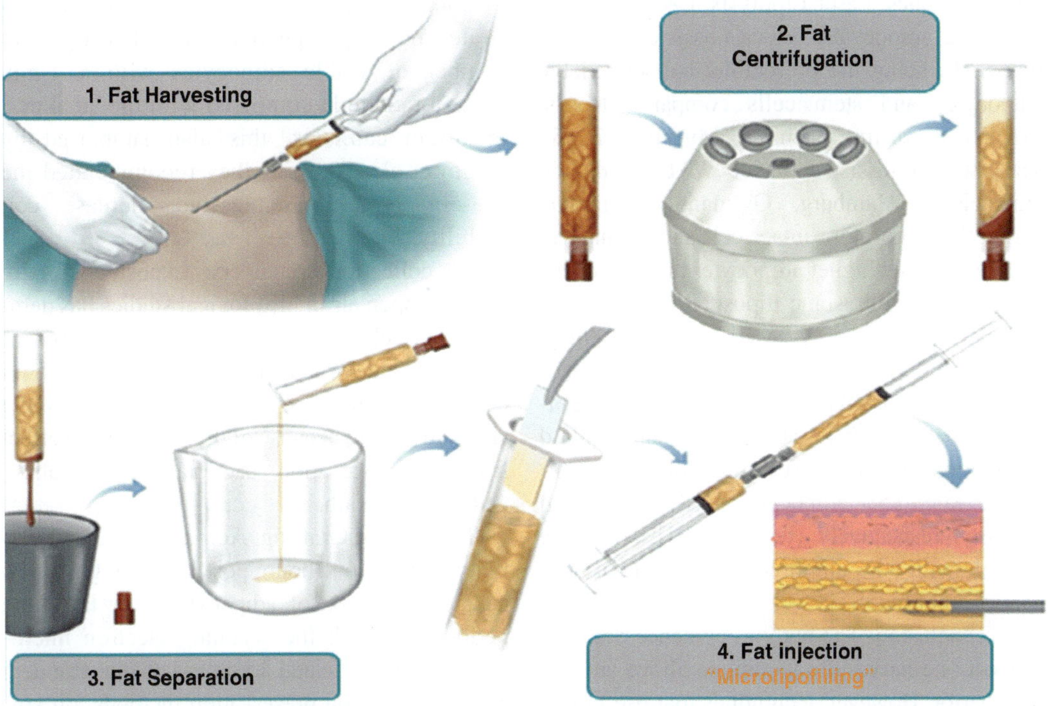

Fig. 31.3 Steps in fat transfer

the syringe. Pulling of the plunger should be carried out with caution because it will determine the negative pressure inflicted on the tissue. The negative pressure caused by syringes is a critical factor in graft survival, few who know that by syringe can cause negative pressure higher even than conventional liposuction machine, it could reach up to 700 mmHg. While any negative pressure above 500, the fat should not be used for lipotransfer. With this in mind, special syringes with a controlled plunger can be used. A prospective, randomized, comparative study demonstrated that the main negative pressure of 10, 20, 60 ml syringes were 275, 394, and 549 mmHg, respectively, and NO significant differences in adipocytes and stem cells integrity and viability were observed (Charles-de-Sa et al.2015). Pu et al. (2008) looked at the cellular function of adipose aspirates using the syringe technique and conventional liposuction on the same patients, they found a significantly higher quantity of viable adipocytes in the Coleman technique (Syringe Technique).

A lipofilling study by Witort et al. (2007) evaluated the effects of different harvesting techniques on adipocytes, the results indicated that the gentle syringe technique (Coleman technique) was less traumatic than the mechanical aspirator (680 mmHg vacuum), which uses power-assisted aspiration (Witort et al. 2007). These results were supported by Herold et al. (2011) who compared the fat graft viability of adipocytes using the Coleman technique to the Shipper technique. The Coleman technique involves manual aspiration using as syringe and centrifugation, while the shipper technique uses automatic liposuction and no centrifugation of adipose tissue and found Coleman technique was superior with significantly increased fat graft viability (Herold et al. 2011).

Large amount lipotransfer: In large amount fat transfer (Mega Lipo-transfer) like fat grafting for buttock, breast, or multiple areas; a liposuction machine has to be used to decrease the time of surgery. Most doctors worldwide currently practice, with experience in different types of liposuction techniques. Laser Lipolysis, LAL, produces nonviable adipocytes. Suction-assisted liposuction SAL, being the preferable, but less viable adipocytes and stem cells compared to the Coleman harvesting technique (Myer et al. 2015). The water jet-assisted device WAL (Body-Jet, HumanMed, Hamburg, Germany) technique, WAL, there are more intact cell and normal fat architecture and higher adipocytes and stem cells survival than ultrasonic liposuction machines. Evaluation of energy-based systems, third-generation ultrasonic-assisted liposuction (UAL) has been shown to benefit 53% better skin tightening and up to 30% less blood loss than standard liposuction. Histological examination of lipoaspirate revealed 85% adipocyte and 87% stem cell viability, respectively (the energy levels were V-mode with 60% amplitude or C-mode at 50% energy amplitude).

In conclusion: The body of evidence does not support one harvesting technique above another as superior. However, techniques that use lower-pressure suction by means of large-bore cannulas appear to increase adipocyte viability. If a higher pressure is used during harvesting, then it is best to employ a fat-processing methodology to include filters, washing, and/or centrifugation to remove dead cells and oil. Some fat collection containers contain filters, the pore size of a filter 500–800 μm (800 μm nominally better than the 500 μm filter size), this filter is very effective for removal of oil and extracellular fluid and retained viable adipocytes and stem cells.

31.2.12 Factors Affecting the Outcome (How to Get the Best Stem Cell Quality from Adipose Tissue)

Donor Site: As far as fat viability and graft outcome are concerned, several studies have been unable to demonstrate a significant difference between harvesting sites. Recently, it was confirmed that the most common source for the harvesting of fat grafts in the lower abdomen (abdominal wall is significantly higher than flanks areas).

Cannula Size: Ozsoy et al. (2006) demonstrated in his prospective study that a greater number of viable adipocytes with a 4-mm diameter cannula compared with the smaller 2 or 3-mm cannulas, this also supported by Erdim et al. (2009) as they recommended the use of larger cannulas to increase cell viability.

Fat collection containers: Exposure of fatty tissue to open air in histological studies has demonstrated cell membrane lysis of up to 50% of cells exposed for a 15-minutes duration. Closed collection containers and syringes are now the standard of care because of the air exposure effect from cell desiccation, bacterial contamination, and cellular death.

Time to Fat Grafting: Adipose tissue cell viability from 1 to 4 hours post harvesting showed an increased loss of adipocytes over this time period. Although the scanning electron microscopic assay indicated no remarkable anatomical changes based on preservation methods, oil volume significantly increased in fat preserved at room temperature for 4 hours (Matsumoto et al.2007), they concluded that adipocytes should be transplanted as soon as possible if stored at room temperature. The processing should perform as soon as there is enough fat collected to process and then the fat, with or without concentrated stem cells, is immediately grafted into the planned recipient site.

Purification Step (Processing of Harvested Fat): Aims to remove the interstitial liquid while at the same time ensuring the viability of the graft cells. When choosing a processing method for fat grafting, the volume retention, viability, and vascularity of the resulting graft are important aspects to consider. The three main widely used purification protocols are, namely:

Decantation: The decantation process is easy to perform and does not necessarily require any additional devices. It can be performed directly in syringes or in a special device designed to trap the fat tissue and to isolate the liquid fraction, it is probably the cheapest and easiest to use for surgeons. The main limitation of this process is, if the there is a significant amount of liquid fraction left in a syringe or a special device. The liquid will be the first to be reabsorbed after reinjection. The liquid fraction can be responsible for a higher concentration of proinflammatory cytokines secreted by the adipose tissue during aspiration. This may trigger inflammation at the recipient site, with the recruitment of immune cells that can eventually lead to increased graft resorption.

Centrifugation: Due to this limitation in the decantation process, since the 1990s the lipoaspirate can also be centrifuged to remove the liquid fraction and improve graft uptake. Centrifugation of adipose tissue separates fat cells from lipid, blood cells, water, and water-soluble ingredients such as proteases and lipases, but does not shift ADSCs between the adipose and fluid portions, possibly due to the strong adherence to adipose tissue or since they are resident with the adipose tissue.

The short-term survival rate of aspirated adipose grafts per volumetric unit after centrifugation increased with centrifugal forces up to 3000 g-force.

Older fat cells are relatively larger in size than younger fat cells. Squeezing the fat cells by centrifugation enables selective destruction of larger and older cells that tend to be more fragile.

Excessive centrifugation can destroy adipocytes and adipose-derived stem cells, but appropriate centrifugation concentrates them.

Centrifugation does result in increased cellular destruction; however, it is from the older, larger, or more injured fat cells.

They found that 1200 g-force (3000 rpm) is the optimal centrifugal force for obtaining good short- and long-term results in adipose transplantation. Yoshimura et al. (2006).

It is shown also that the increased centrifugal forces compacted the adipose portion more and therefore concentrated the red cell within the adipose portion rather than shifting the red blood cells into the fluid portion.

In contrast, Khater et al. (2009) found an improved clinical outcome with the washed at one year compared to centrifuged fat. Botti et al. (2011) compared one side with centrifuged fat (3000 rpm for 3 min.) with another half filtered and washed. At 12 months, there were NO subjective or objective differences in outcome.

Ferraro et al. (2011) study suggest that lowering the force of the specimens improved cellular viability and also showed less resorption of liquid fraction during the centrifugation compared to simple decanting alone.

Kurita et al. (2008) found that 1200 g-force (3000 rpm) is the optimal centrifugal force, as it leads to condensation of cell numbers per volume of adipocytes with an improved ADSCs to adipocyte ratio.

Likewise, Kim et al. (2009) evaluated the centrifugation technique on fat cell viability in autologous fat transplantation. They concluded that excessive centrifugation with 5000 rpm for more than 5 minutes increased adipocyte destruction and recommended that centrifuged force be limited to 3000 rpm (1200 g-force). Furthermore, when centrifugation is used, several articles suggest that forces greater than 3000 rpm (about 1200 g-force) cause more cellular damage.

Removal of the lipoaspirate liquid component seems to be critical to obtaining the best graft maintenance, at the same time centrifugation can be harmful to the tissue when the duration and speed are excessive.

In conclusion: The body of evidence does not support one processing technique above another as superior.

Filtration Method (Washing): One of the simplest and yet most efficacious methods for fat processing is preparing the lipoaspirates by washing with normal saline. Washing steps are crucial to reduce undesirable components such as blood, debris, free lipid, and ruptured adipocytes to purify the fat before reinjection while retaining viable adipose tissue for further use. The most common release criteria value for cell viability is above 70%.

31.2.13 According to Conde´-Green et al. (2010)

- It is the best processing method.
- Maintains adipocyte membrane integrity.

- Maintains adipocyte numbers.
- Clear most blood contaminants.
- Increases the number of ADSCs and endothelial cells.

Moreover, Conde´-Green (2010), found a greater concentration of ADSCs in the centrifuged fat than with the decantation method alone. Caution must be taken with this centrifugation step to preserve the integrity of the adipose tissue.

High-speed centrifugation resulted in consistent volume retention but lower viability. Each of these approaches is ideal under different circumstances and contributes to the versatility and reliability of fat grafting.

Some protocols also integrate multiple washing steps (e.g., Macrofill by Adip'sculpt; PureGraft by Cytori) that aim to remove the remaining infiltration solution, to get rid of most of the local anesthetic, vasoconstrictors, inflammatory molecules, and death factors (released during liposuction) that may damage the tissue and limit graft success.

It is accomplished by passing the lipoaspirate through a filter (e.g., PureGraft by Cytori) to remove particles below the chosen size. In 2013, Zhu found less oil and greater viability with this system compared to high-speed centrifugation. Coleman, (2004) states that washing, although advocated by a number of authors, is not recommended, because in this experience the mechanical action of washing subjects the reticular fibers and connective tissue septa to unnecessary trauma and can disrupt the fragile fatty tissue architecture.

Fat Administration (Reinjection of Harvested Fat): The final step of fat transfer is the reinjection of the adipose tissue. Inevitably, some degree of hypoxia occurs around the grafting of the lipoaspirate. In the recipient, the integration of the graft requires extensive (re)vascularization. Adipocytes are sensitive to hypoxia and therefore prone to death and necrosis. This step usually requires the use of different-sized syringes (from 1 cc to 50 cc) depending on the quantity of adipose tissue to graft. Too large "lumps" of fat transfer obviously develop necrotic cores due to diffusion insufficiency, as a result of which the graft "take" may be reduced due to hypoxia and fat cells death. Best methodology to avoid massive fat necrosis is required "Multi-levels Fat Grafting" in the recipient area by which, a small amount of fat tissue is injected by a retrograde movement that deposes fat cells in multiple levels by micro-tunnelling and in multiple directions as shown in Figs. 31.4 and 31.5.

Therefore, the injection of small aliquots of fat is now commonly accepted, care must be taken not to over-graft in the donor area as the amount to inject judge by the desired outcome plus tolerance of the donor area. Faced with resorption, many surgeons try to overcome this limit by "over-correction" of the recipient site. However, exceeding the volume might lead to an inverse effect with increased graft pressure resulting in tissue damage, necrosis of fat, and more chances of oil cyst formation.

31.3 More Complex Protocols

Due to unprotectable fat resorption rate, methods to increase graft viability are needed. In the last 5 years, a new approach has been developed to improve fat grafting through the addition of stem cells or growth factors

31.3.1 Platelet-Rich Plasma (PRP) Enriched Lipotransfer

In surgical fields, great interest has been demonstrated in platelet concentrates, notably in PRP, which is easy to use and only requires a blood sample. This blood-derived formula contains a concentrated number of platelets that release several growth factors. Gentile et al. (2012) reported a 1-year graft survival of 69% PRP-enriched compared to 39% for the Coleman technique alone. PRP clearly has a concentration-dependent effect, so, more studies with a higher number of patients will be required to reach a consensus on the use of PRP at a defined concentration.

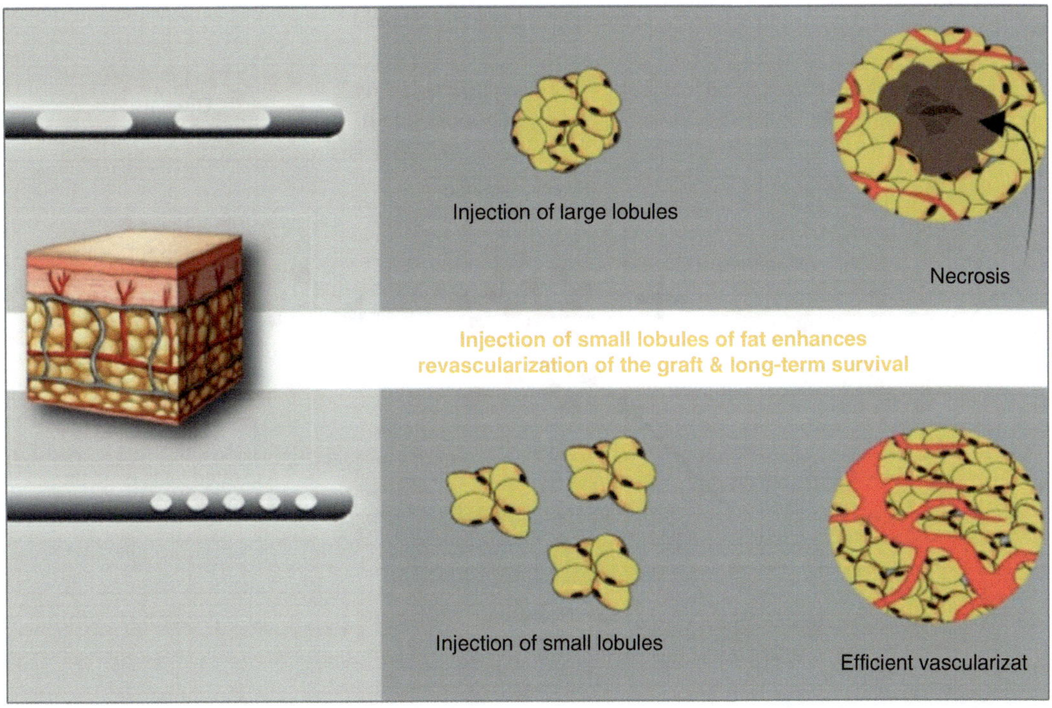

Fig. 31.4 Micro-lipofilling

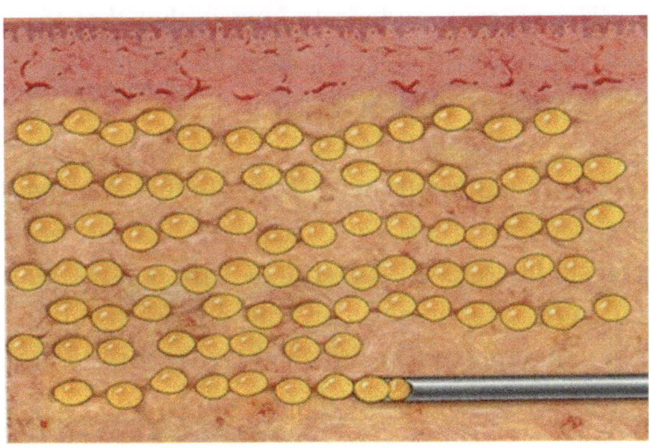

Fig. 31.5 Multi-level fat grafting

31.3.2 Stem Cell-Enriched Fat Transfer

31.3.2.1 Cell-Assisted Lipotransfer (CAL)

Over the past decade, studies have shown a great deal of interest in the SVF mainly because of its ASCs content and its ability to secrete pro-angiogenic factors. Thus, new procedures have emerged, consisting of the injection of a SVF-enriched fat graft. The common name for this protocol is Cell-Assisted Lipotransfer (CAL). The goal of this approach is to inject more SVF cells due to:

- The multipotent capacity of stem cells in SVF allows them to differentiate into endothelial cells which promote vascularization and increase blood supply to the fat and survival of the graft plus secretion of growth factors by the SVF, which can increase blood to fat.
- The multipotent capacity also enables the stem cell to differentiate into adipocytes to overcome the death of adipocytes that occur during and after fat transfer.
- Finally, the immunoregulatory potential of ASCs may decrease the inflammation that is responsible for a sub-optimal outcome, causing adipocyte death.

In the CAL protocol, the adipose tissue is harvested by liposuction, half of the harvested fat is used to prepare SVF as the lipoaspirate is subsequently enzymatically digested by collagenase

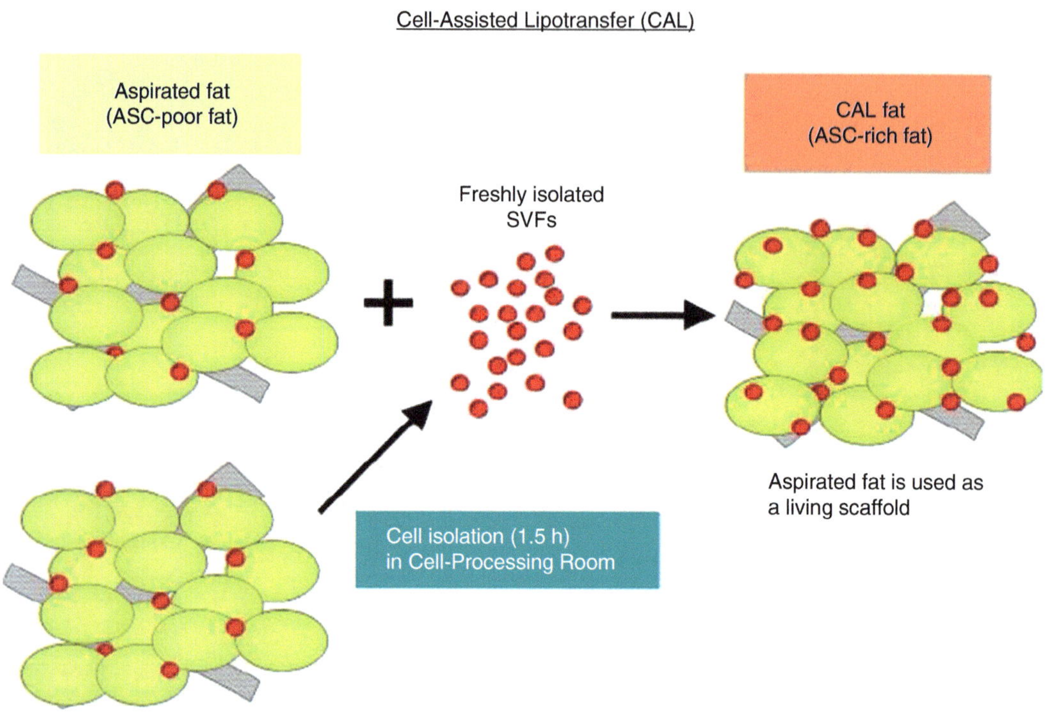

Fig. 31.6 Cell-assisted lipotransfer

and fractionated into SVF as shown in Fig. 31.6. The SVF then combined with the other half to create a stromal cell-enriched fat graft. Several devices are now used in clinical practice to isolate the adipose SVF. These devices often use enzymatic digestion to digest the tissue and allow the isolation of the SVF cells as a pellet following centrifugation of the digested tissue.

In the past 10 years, many research groups have shown interest in Cell-enriched Lipotransfer (CAL) protocol (Fat with SVF), SVF seems to improve scar healing and might be suitable for low vascularity recipient sites but still not clear if this protocol is better than conventional fat grafting. It might be promising for hostile and poorly vascularized tissue sites (an irradiated breast, for example) but not sure if it is:

- Justifying the higher cost.
- The necessity to harvest more tissue (half of it is digested).
- Supplementary operation time that is required (approximately 2 h more)

The present evidence suggests that there is a big potential for CAL in reconstructive surgery as seventeen studies examined weight/volume retention of which 15 studies favored CAL over conventional lipotransfer. One clinical study did not find any efficacy of CAL, however, the present studies are so far still of low quality with inherent weaknesses. Several aspects regarding CAL remain unknown such as the optimal degree of cell enrichment and also its safety.

31.4 ADSCs-Enriched Fat Transfer (Cultured Stem Cells)

Since the study published by Trojahn Kølle et al. which carried out with volunteers in which fat grafts were compared with fat grafts enriched with cultured ASC. The result of the study showing the ADSC-enriched fat grafts had significantly higher residual volumes (80.34%) compared to non-enriched fat grafts (16.3%) after measuring it by MRI. The difference between the group was scientifically very significant and no serious adverse events were noted. This study has raised the interest in cultured adipose-derived stem cell supplemented fat graft (ADSC-enriched fat grafting) as it had an excellent feasibility and safety.

These promising results add significantly to the prospect of stem cell use in clinical settings and indicate that ADSCs-enriched fat transfer could render lipofilling a reliable alternative to major tissue augmentation, such as breast surgery as augmentation, reconstruction especially in radiated breasts or secondary breast surgeries to correct complication from previous breast surgeries as shown in Fig. 31.7.

However, this protocol is not a one-step procedure but it includes two steps, the first one is harvesting or collecting a small amount of the patients own body fat under local anesthesia. The harvested body fat is collected in a sterile container with appropriate media and is properly labeled with patient details. The sample is then sent in a dedicated dispatch kit with optimum temperature to a licensed biological laboratory, who is approved for manufacturing or culturing the cells and has full accreditation to culture and bank the mesenchymal stem cell from the patients own adipose tissue. On the day of surgery the cultured ADSC's are again mixed with aspirated fat before re-injecting the enriched fat in the recipient area.

31.5 Indications and Contraindications

1. Indications for treatment are:
   - 1.1. Urinary dysfunction/incontinence
   - 1.2. Stress incontinence
   - 1.3. Overactive bladder
   - 1.4. Fecal incontinence
   - 1.5. Sexual dysfunction
   - 1.6. Cervical laxity due to childbirth
   - 1.7. Pelvic organ prolapse/cervical, rectal prolapse
   - 1.8. Pelvic pain

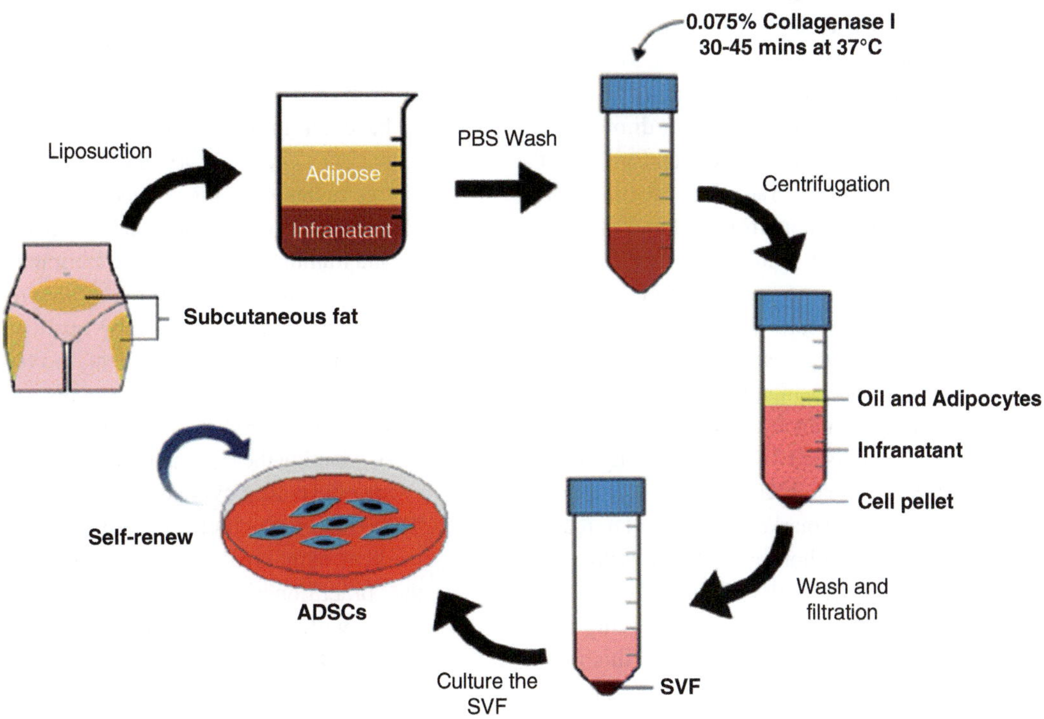

Fig. 31.7 ADSCs-enriched fat transfer

   1.9. Vaginal pain, itching
   1.10. Abnormal vaginal discharge
   1.11. Vulval pain, itching, or diseases like Lichen, Warts, and others
   1.12. Abnormal presentation of the female labia, vulva
   1.13. Pregnancy Striae, marks
   1.14. Breast sagging or abnormal shape

2. Practical tips from the author:
   2.1. Choose the patient wisely. Counsel the patient for the procedure and the options.
   2.2. Obtain informed consent.
   2.3. Success of therapy or the procedure is in good counseling of the patient to address the stress, anxieties, and overcome over-expectations. The patient should be in the right frame of realistic expectations from the procedure.
   2.4. The cell therapy may give fast results, but sometimes may take time to provide the optimum rejuvenation.
   2.5. The influence of the growth factors and the stem cells leading to the effects may last for a shorter duration or may home in and give a longer respite from the condition. Stem cells are sensitive to the surrounding environment, many of the effects are either enhanced or diminished depending on the local effect. Success of the procedure/therapy thus lies in managing the local environment and in minimizing the risk factors.
   2.6. According to a new market research report "Global Medical Aesthetics Market Size, Share, Development, Growth and Demand Forecast to 2023 – Industry Insight by Product and by End User" published in July 2017 by P&S Market Research USA, the global medical aesthetics market is projected to reach $19,419.7 million by 2023, growing at a CAGR of 12.2% during 2017–2023.

      There is a combination of driving factors that are contributing to this phenomenal growth. The first observation is that attitudes and mentalities have changed, more and more individuals are conscious of the importance of taking care of their skin. Millions of middle-class individuals with disposable income who are now reaching their early thirties, have developed a strong interest and commitment to skin care and protecting their "Capital of Youth." One of their concerns is the desire to optimize their natural skin potential and to slow the signs of skin aging, by seeking out minimally invasive and noninvasive antiaging aesthetic procedures and dermo-cosmetic solutions.

      Another driving factor is the market availability and accessibility; there is a growing trend among patients themselves to directly proactively research and purchase the latest skin treatment products. Unfortunately, it is becoming more and more difficult for consumers to cut through the clutter and decipher truth from false claims and promises. Aesthetic medicine doctors represent an expert and reassuring contact for patients in the event of skin problems. Medical practitioners often have an in-depth knowledge of the patient's skin history and skin care treatments over time. Thanks to their medical acumen, practical expertise, and training provided by the laboratory experts such as NUNII Laboratoire, aesthetic doctors are ideally qualified to understand the biological mechanisms of action of the treatments and its active ingredients, yet are capable of adapting the treatments to meet their specific patient skin needs.
   2.7. This trust factor between patients and medical skin professionals, along with the practitioner's reliability and efficiency with the patients, make for a winning combination.

      The third significant element of the international growth of aesthetic medicine is the recent fast-paced development of emerging economies, which has seen the middle-class segments of their population boom over the past decade. This is particularly true for countries such as China, South Africa, Saudi Arabia, and Brazil, to name just a few.

31.6 Conclusion

Regenerative Medicine is being used for women to live a better life. The newer techniques and the innovative regenerative therapies are bringing hope and joy to women of all ages, races, and conditions. Field has seen a diaspora of various conditions being treated for the patient's satisfaction.

Cosmetic gynecology has become the fastest growing area with expanded access. Many conditions that women suffered in silence have found a novel therapy to not only rejuvenate but many a times regenerate.

References

1. Aesthetic dermatology. ResearchGate. https://www.researchgate.net/publication/6534190_Aesthetic_dermatology. Accessed 28 Dec 2020.
2. Spatial distribution of cosmetic-procedure businesses in two.... 6 Dec 2013. https://www.ncbi.nlm.nih.gov/pmc/articles/PMC3881144/. Accessed 28 Dec 2020.
3. A mini-review of aesthetic gynecology and leading.... 21 Jun 2018. https://www.ncbi.nlm.nih.gov/pmc/articles/PMC6022427/. Accessed 28 Dec 2020.
4. The role of tissue engineering in articular cartilage repair.... 29 Jul 2011. https://www.ncbi.nlm.nih.gov/pmc/articles/PMC3146065/. Accessed 29 Dec 2020.
5. Biomaterials & scaffolds for tissue engineering. ScienceDirect. https://www.sciencedirect.com/science/article/pii/S136970211170058X. Accessed 29 Dec 2020.
6. Stem cells: past, present, and future. NCBI – NIH. 26 Feb. 2019. https://www.ncbi.nlm.nih.gov/pmc/articles/PMC6390367/. Accessed 29 Dec 2020.
7. Decellularization strategies for regenerative medicine: from.... https://www.hindawi.com/journals/bmri/2017/9831534/. Accessed 29 Dec 2020.

PRP and Exosomes in Regenerative Gynecology

32

Carolyn DeLucia

The most exciting area of medicine today is regenerative procedures. Regenerative procedures are those that involve biologic tissue to help heal injury and disease. Classifications of regenerative procedures are either autologous which is the source derived from the recipients own body such as platelet-rich plasma, adipose tissue or bone marrow. The second is allograft. An allograft is transplanted from one human to another originating mostly from prenatal tissue.

Platelet-rich plasma is derived from the recipients own blood. it contains a concentration of platelets and their growth factors that activate local stem cells to grow new tissue in an area of injury or inflammation. They are widely used in aesthetic and orthopedic medicine. The benefits of platelet-rich plasma have been clinically shown in the area of sexual medicine. Platelet-rich plasma can be created through centrifugation of a patient's own blood in either single spin or double spin systems. The platelet-rich plasma can be concentrated between two–eight times the original concentration. The history of platelet-rich plasma goes back to 1970s with it being first used in a fibrin glue used for surgery. In 1987 Dr. Ferrari used platelet-rich plasma in open heart surgery to decrease bleeding. In 1997 Dr. Marxs used platelet-rich plasma for maxillofacial surgery concentrating platelets and activating the platelets with calcium chloride 10% and seeing a much faster graft maturation in Trabecular bone. The principle of platelet-rich plasma is that it is an autologous biotherapy used for its healing properties. The platelets release a cocktail of growth factors and cytokines from the alpha granules in the cytoplasm. There have been no long-term complications from autologous use of platelet-rich plasma to date. There are contra indications to the use which include platelet dysfunction syndromes, critical thrombocytopenia, hemodynamic instability, septicemia, or local infection at the site of the procedure. It is very easy for office use requiring phlebotomy specialized kits and centrifuge. The benefits seen in female sexual health include greater sexual response, improvement in texture and quality of skin, improvement in support of the bladder neck, increased vasculature and innervation and indirectly increased sexual desire.

The credit for use of PRP for sexual purposes goes to Dr. Charles Runels in Alabama, USA. He theorized that since PRP had success in the regeneration of soft tissue in orthopedics, it may be helpful for erectile dysfunction in men and decreased sensation in women. He invented and trademarked the procedure known as the O-Shot that was the impetus to use of PRP in female genitalia. His pilot study showed a significant improvement in orgasmic function and sexual satisfaction in women. The trademarked procedure OShot is performed by numbing the clitoris

C. DeLucia (✉)
Hillsborough, NJ, USA

P. Jindal et al. (eds.), *Aesthetic and Regenerative Gynecology*,
https://doi.org/10.1007/978-981-16-1743-0_32

and vaginal introitus with topical compounded mixture of benzocaine, lidocaine, and tetracaine then using a peri-clitoral block of lidocaine and ice. The PRP is created and a small amount is injected into the clitoris and the rest in the anterior vaginal wall in the area of the g-spot.Due to the ambiguity of the g-spot, Dr Runels has identified the area of his injection as the O-spot which is the area approximately 1.5 to 2 centimeters inside the vagina in the midline of the anterior vaginal wall. It is theorized that since the PRP will extravasate wherever it is injected that if injected in this general area it is sure to hit the tissues in the peri-urethral area. The PRP in this procedure is recommended to be activated which requires the addition of 10% calcium chloride. 0.1cc is added to 4 cc of prp for the vaginal injection and 0.01cc is added to the 1cc of prp for the clitoral injection. The thought here is that activation will assure that the clotting cascade is initiated and more likely to have the desired effect. Activation of prp is controversial in the literature because it is questioned whether it is necessary as the simple act of injecting or passing through the needle is probably enough to begin the reaction. However Dr Runels feels that if activation causes no harm then why not take every measure to assure that the best product is being injected into the genitalia. In his piolet study of the OShot, two standard questionnaires were used, the female sexual distress scale (FSDS), and The female sexual function index (FSFI). After 12 and 16 weeks following one procedure 70% of women progressed from "distressed "to "not distressed." Therefore this simple fifteen minute in office procedure has been shown to improve sensation and sexual satisfaction. The Cellular Medicine Association has grown in memberships and it is estimate that five thousand procedures have been performed worldwide. The association has certified training physicians who teach the procedure internationally. Anecdotally it has been seen that the oshot alone can help mild to moderate stress incontinence, interstitial cystitis and the texture of vulvar tissues.The controversy comes in whether a medical technique can be trademarked. Historically the answer is no but the brilliance in Dr. Runels' method is that he trademarked a name that became a recognized commercial name and developed an association which supports the technique he originated and furthers the research in the field of prp. Members pay a fee in order to advertise that they do the procedure he developed. The fee goes towards research projects such as the paper published by Dr. Goldstein on Lichen Sclerosus.

Subsequently, it has been shown to improve lichen sclerosus. Lichen Sclerosus is a debilitating condition thought to be autoimmune in nature. It affects the skin of the genitalia causing leukoplakia and scarring. The tissue is thin, itchy, and burns. The anatomical land marks tend to fuse with obliteration of the labia minora and agglutination of the clitoral hood. This can cause significant pain with intercourse and loss of sensation from clitoral scarring.The protocols being used for treatment of lichen are to inject all the affected tissue with double spin PRP every six weeks for at least three sessions and may be repeated as needed to see the decreased inflammation and normalization of the skin. There are a couple of studies published demonstrating the clear benefit. The patients see significant decrease in symptoms and require less topical steroids which have been the standard of care for this condition. It is well established that long-term use of high potency topical steroids leads to dermatological changes that cause iatrogenic issues [1].

The primary efficacy variable, as measured by two masked dermatopathologists, was the change in inflammation between the pre- and post-treatment biopsies. Secondary endpoints included changes from baseline in pruritus and vulvar burning using visual analog scales (VAS) and change in Investigator's Global Assessment (IGA) of the severity of the disease (0 to 3 scale). Twelve of the 15 participants completed the protocol, 2 participants were lost to follow-up before the second PRP treatment, and another refused post-treatment biopsy. Of the 12 patients who completed the protocol, the dermatopathologists determined that 7 had decreased inflammation on their post-treatment biopsies (Fig. 32.1), 3 had no change, and 2 had a "minimal" increase in inflammation. A repeated measures ANOVA showed

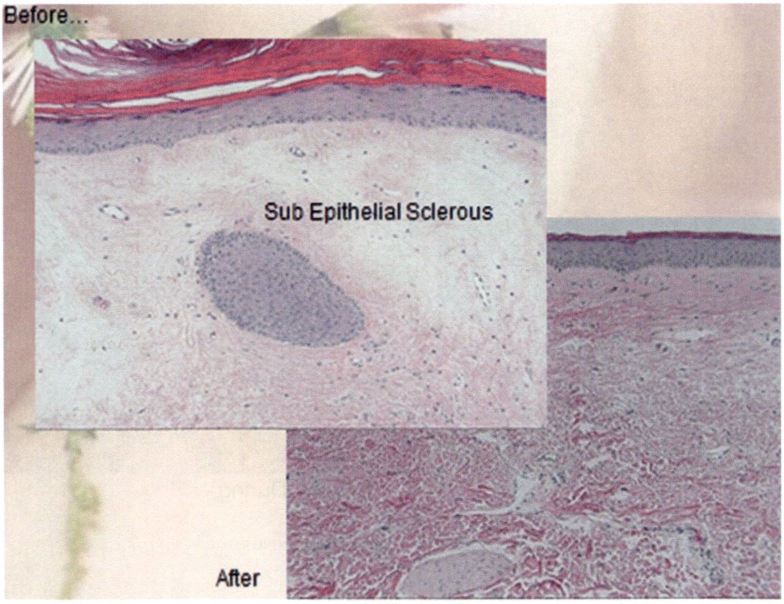

Fig. 32.1 Shows decrease inflammation following prp treatment in post-treatment biopsies

that these results were statistically significant: F (1, 11) = 6.81, $P = .024$. A paired-sample t-test comparing the pre- and post-treatment IGA scores showed a statistically significant difference: pre-treatment (M = 2.67, SD = 0.49) and post-treatment (M = 1.83, SD = 0.83); t (11) = 3.4578, $P = .0054$. Changes in subjective VAS scores for pruritus and burning were not statistically significant. No adverse reactions were reported except transient discomfort and bruising at the biopsy and injection sites.

Kathleen Posey from Louisiana published a paper as well on the use of PRP for Lichen Sclerosus [2].

She uses gentle tissue separation once the tissue has been injected with PRP and then continues the six week protocol of treatment (Fig. 32.2).

A known effect of PRP is that it can stimulate the adipocyte and build fat tissue. It is rare to desire more fat but many women lose tissue and support in the breast after breastfeeding and with natural aging. It has been well established over the past several years that autologous fat transfer with combined PRP is safe in breast tissue. No evidence of neoplasia or increased risk of recurrence has been demonstrated in patients with breast cancer. What if we leave out the fat? Dr. Runels trademarked the procedure known as The Vampire Breast Lift. The results seen from this cosmetic procedure are very satisfying to the patients.The PRP ideally is prepared from 120 cc of the patient's blood and double spun. This will result in 50cc of prp at about 4 times concentration. Then 25 cc are injected in a butterfly pattern on each breast. As is typical for prp it takes 3–6 months to see the full results of the procedure. The tissue created appears to be fat tissue and the observations are improved tone and texture of the skin as well as fullness and modest lift the breast. See in the before and after pictures below (Fig. 32.3a, b). In the first case the results were after eight months.In the second case the results are at three months. The duration of the results will vary. As with all PRP the results will last as long as it takes for the new developed tissue to age. Since the tissue developed in breast is fat, the results tend to last much longer and do not seem to be affected by weight loss.

Case #1(Fig. 32.4) Top photo is before picture with bottom or right side being after 8 months of PRP alone injected into the cleavage and lateral breast. Note the texture and tone of the skin as well as the fullness of the breast tissue.

Case #2 (Fig. 32.5) is result after 3 months of PRP alone injected into the cleavage area and lateral breast. Note the fullness and modest lift of

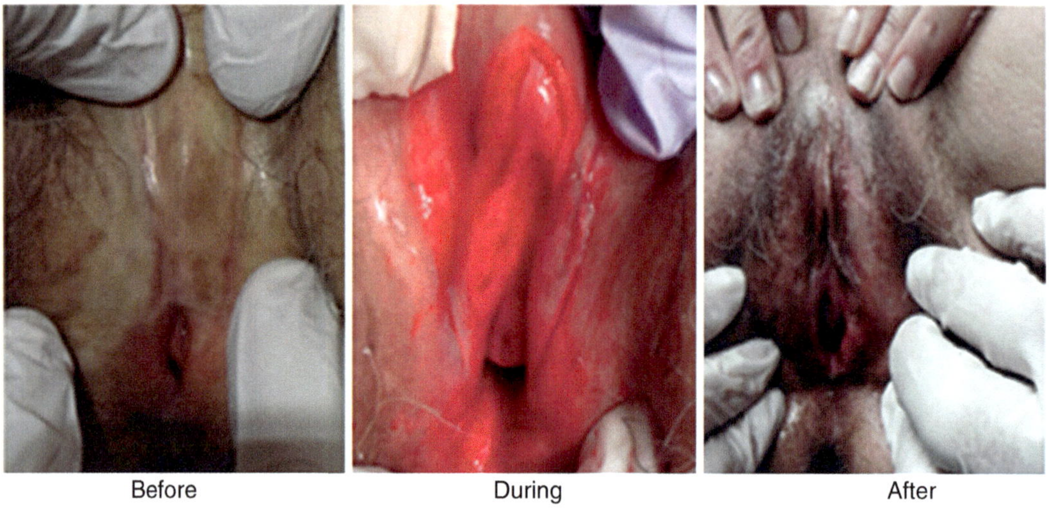

Fig. 32.2 Before and after prp treatment in Lichen Scleosus

the breast.This case is also with a ten pound weight loss [3].

This recent study in 2018 showed that autologous activated platelet-rich plasma injection into adult human ovarian tissue might increase reproductive response. The benefits of platelet-rich plasma will continue to be seen as they are used in more areas in the medicine. Premature Ovarian Insufficiency Receiving Ovarian Administration of Platelet-Rich Plasma (PRP) in Combination With Gonadotropin: A Case Report Chao-Chin Hsu[1, 2*], Leonard Hsu[3], Isabel Hsu[2], Yi-Jen Chiu[1] and Sonam Dorjee[1]

This article reports the first live birth with this novel treatment.

Stem cells are embryonic, or bone marrow derived as the longest existing source, or adipose derived first discovered in 2001. Prenatally derived stem cells are also a common source however have been surrounded by ethical issues worldwide. The function of stem cells is for self-renewal, therefore, creating more stem cells, and secondarily differentiation creating new cell lineages under appropriate conditions. Stem cells can signal molecules, cause intercellular contact, interact with neighboring extracellular matrix and seem to activate the recipients own stem cells and extracellular matrix to take action or are activated by the recipients extracellular matrix to differentiate. Stem cells are used for bone regeneration, cancer treatment, spinal cord injuries, cardiac muscle repair, Cartlidge damage and urinary tissue. The fastest-growing area is the use of adipose stem cells. They are derived by mini liposuction, they are then processed and isolated from the stromal vascular fraction and have multiple uses. There is some controversy about the use of adipose stem cells. The disadvantages are that the procedure is invasive and inconvenient due to having to have liposuction. There is pain, scarring, risk of infection, and it is time consuming. Quality control is much more difficult because they are processed at the time of the procedure in the doctor's office. Also, as a patient age or has to disease the quantity and quality of stem cells declines. The final product quality is inconsistent and is operator and equipment dependent. Stem cells are less potent with shorter telomeres with less versatility and secrete less anti-inflammatory growth factors. They age more rapidly and are less neuroprotective. There are, of course, FDA issues and many clinics are being shut down by local governments. The use of stem cells is still highly regulated worldwide.

Exosomes are the next frontier. The field of Nanotechnology, specifically, Exosomes and other micro-vesicles, are the next frontier of regenerative medicine. They fall into the allograft classification. They have transplanted from one

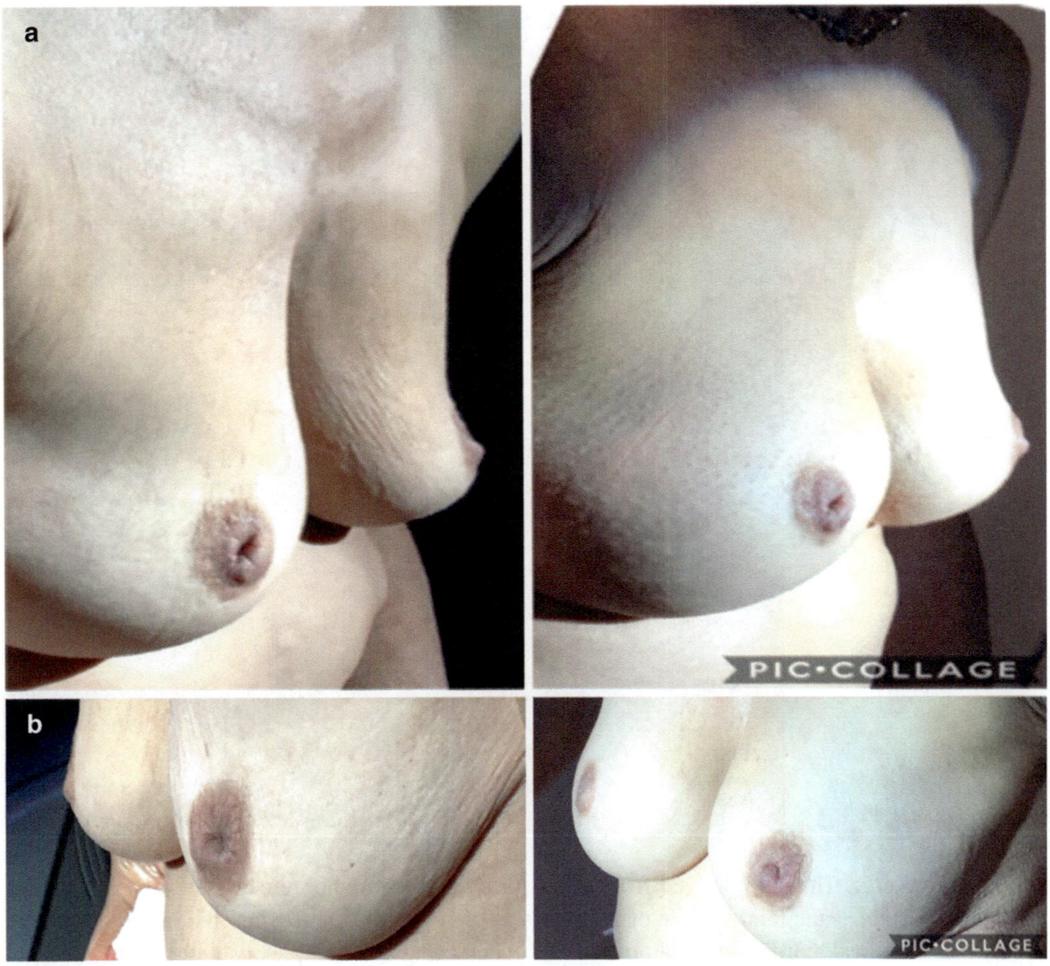

Fig. 32.3 Before (**a**) and after (**b**) autologous fat transfer combined with PRP in breast tissue

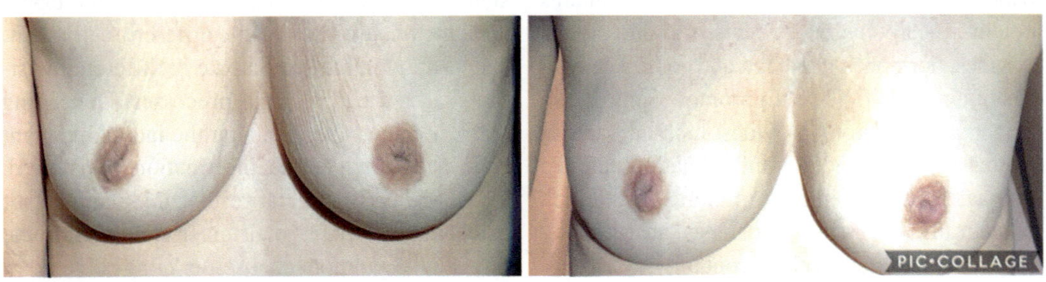

Fig. 32.4 PRP treatment in breast reconstruction

human to another the immediate advantages are that there is no matching necessary considered immune privileged, and mesenchymal stem cells actually are immunosuppressive. Cells are primitive and express little surface markers. Young cells are more potent, higher anti-inflammatory effect and secrete more growth factors and more neuroprotective effects. The cells with longer telomeres and less accumulated genetic changes. They are easy to obtain, abundant, and have no

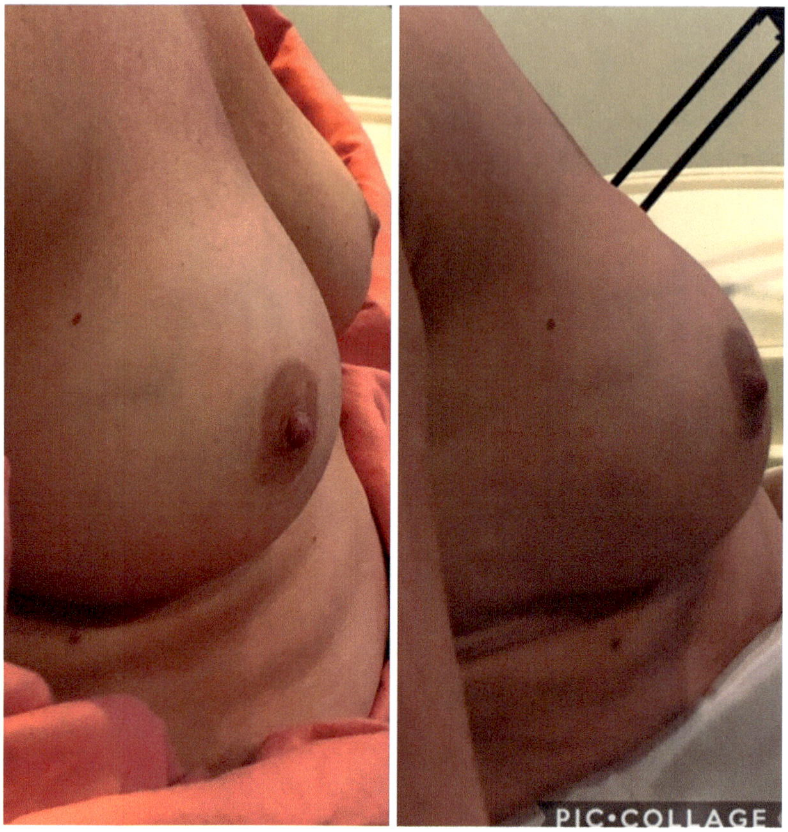

Fig. 32.5 Injection of PRP alone into cleavage and breast tissue

ethical controversy. They are derived from perinatal sources such as the fetal surface of the placenta and umbilical cord, the amnion or the amniotic fluid. What is amniotic fluid? Amniotic fluid is a protective, nourishing fluid that surrounds the embryo during pregnancy. It contains proteins, carbohydrates, fats, amino acid, enzymes, hormones, pigment, and cells. The benefits of amniotic fluid are to promote hemostasis, protect and cushion, provides lubrication, has optimal viscosity, reduces inflammation and is a synovial fluid supplement. Amniotic fluid is rich in naturally occurring Exosomes secreted by mesenchymal stem cells, contains cytokines, chemokines, growth factors, hyaluronic acid, collagen, electrolytes, proteins and carbohydrates [4]. This article explains the potential use of amniotic fluid.

The Exosomes can be easily isolated and purified. Autologous stem cell treatments traditionally have not been a consistent treatment on a patient to patient outcome. Now we can use secretions from the stem cells hence exosomes as a main focus, increasing the availability and decreasing the complexity required to incorporate stem cells themselves into medical practice. It is an indirect way of using the patient's own stem cells. Exosomes have an endosomal origin and are released by many different cell types. They subsequently participate in different physiological and pathological processes. Depending on their origin, they can alter the fate of recipient cells according to the information transferred. There are early clinical results for treatment of osteoarthritis pain, hair restoration, chronic obstructive pulmonary disease, congestive heart failure, pulmonary artery hypertension and age-related decline. The Future directions for Exosomes, clinical applications and use in current IRB studies are expanding daily. The studies fall into many categories. There are oncological studies, regenerative studies, drug delivery studies as well as future engineering studies. The use in sexual medicine is truly in its infancy.

The first paper to review is going to be an article written in 2019. The paper is called "The new relationship between molecular content of mesenchymal stem cell derived Exosomes and they are potential: Opening the way For Exosomes based therapeutics." The highlights of this paper are that Exosomes are the future of regenerative medicine. They have a myriad of content and composition. Molecular characterization could help for large-scale and synthetic exosome development.

A paper released in May 2019 in titled "mesenchymal stem cell derived extra cellular vesicles: a new impetus of promoting angiogenesis in tissue regeneration" is worth review. Highlights include that the use of extracellular vesicles to enhance angiogenesis in the development of insufficient angiogenesis such as chronic wounds, stroke, and myocardial infarction. Another few papers to highlight are called "Extracellular vesicles in personalized medicine: the input a physically triggered production loading and theragnostic properties" and"the potential of Exosomes in the therapy of the Cartlidge and bone complications: emphasis on osteoarthritis." In the realm of cardiac exorcisms, it is being shown to be protective in acute ischemia, chronic ischemia, decrease infarct size, stimulate angiogenesis, reduce fibrosis, reduce scarring and remodeling and improve cardiac contractility. It was shown that micro RNA via the exosome could go into the cardiac tissue in a rat model and repair the tissue. This article was written in 2016 entitled "More than tiny sacs stem cell Exosomes as cell free modality for cardiac repair." The article published in 2017 in the European heart journal showed how the Exosomes were able to repair the damaged cardiac tissue and regenerate the tissue. An excellent review article entitled "Circulating extracellular vesicles in human disease" from the New England Journal of Medicine showed that Exosomes could target ischemic myocardium and mesenchymal stem cell-derived mRNA in Exosomes have therapeutic effects on acute myocardial infarction. There are equally as many articles written on the effects of exosomes on pulmonary diseases and more coming in the area of renal and hepatic, and orthopedic diseases.

The use of exosomes in female sexual health has begun with the initial studies being done by Dr. Red Alinsod in California. He is using amniotic fluid as his source. He has been using amniotic fluid to treat urinary incontinence and Lichen sclerosus. His pilot studies are showing great promise. As with any new technology, the use of regenerative products is strictly monitored and regulated. It is important to emphasize that one must be aware of their local laws and expectations. At this time in the USA exosomes are only being used in IRB-approved studies.

References

1. King M, Tolson H, Runels C, Gloth M, Pfau R, Goldstein AT. Autologous platelet rich plasma (PRP) intradermal injections for the treatment of vulvar lichen sclerosus. J Low Genit Tract Dis. July 2015;19(3):S1–S25.
2. Posey K, Runels C. In office surgery and use of platelet rich plasma for treatment of vulvar lichen sclerosus to alleviate painful sexual intercourse. J Low Genit Tract Dis. 2017;21(45):S14.
3. Sfakianoudis K, Simopoulou M, Nitsos N, et al. Autologous platelet-rich plasma treatment enables pregnancy for a woman in premature menopause. J. Clin. Med. 2018;8(1):1. https://doi.org/10.3390/jcm8010001.
4. Moraghebi R, Kirkeby A, Chaves P, Rönn RE, Sitnicka E, Parmar M, Larsson M, Herbst A, Woods N-B. Term amniotic fluid: an unexploited reserve of mesenchymal stromal cells for reprogramming and potential cell therapy applications. Stem Cell Res Ther. 2017;8(1)

33 Setting Your Practice and Future of Aesthetic and Regenerative Gynecology

Preeti Jindal and Muskan Mittal

Aesthetic gynecology is the fastest growing and evolving superspecialized field of gynecology. It has seen increasing patient and physician demand. In past, it has been mainly practiced by obstetricians and gynecologist but nowadays plastic surgeons, cosmetic surgeons, dentists, and anesthetists are showing more interest in this field. Few observership, fellowship, and residency programs are teaching this advanced technology to various aspirants who want to excel in this field. More clinicians are attracted to this field as it has less complications, better patient compliance, and long-term results.

The InSARG (Indian Society of Aesthetic & Regenerative Gynecology) is in process of forming policy and guidelines for Indian Obstetricians and Gynecologists and Indian patients and thus offering a golden opportunity for doctors to expand their knowledge in this area of interest.

Various cosmetic workshops and observership courses were also offered by the editor in the year 2019–2020, in which more than 300 delegates participated. Currently, many surgeons first start training in various established certified observership programs based in the United States and other developed countries.

A steady flow of certificate programs continues to evolve in India, Turkey, the Middle East, Spain, and South America, as the second wave of experts emerge. We present a review of surgical and non-surgical techniques of what is presently called "aesthetic gynecology" and the approaches of prominent gynecologic societies regarding this relatively new subspecialty.

33.1 How to Set up Aesthetic Clinic?

Aesthetic and regenerative gynecology have become a field of interest for various clinicians. This chapter aims to guide gynecologists, dermatologists, and plastic surgeons on how to set up a professional and ethical practice.

1. **Place**: To establish an aesthetic clinic, location is one of the key elements of your successful practice. One should understand the demographics of patient in that particular location to be able to practice.
2. **Knowledge**: One should have a complete understanding of the procedure they are performing. Keep your knowledge up to date with the latest advances so as to offer your patients the best available treatment.
3. **Safety profile**: This is important to understand that cosmetic procedures should not only deal with appearance but also patient

P. Jindal (✉) · M. Mittal
The Touch, Advanced Obstetrics, IVF & Cosmetic Gynae Centre, Mohali, India

P. Jindal et al. (eds.), *Aesthetic and Regenerative Gynecology*,
https://doi.org/10.1007/978-981-16-1743-0_33

safety and convenience. Therefore, every case must be treated with professional competence and due care.

4. **Procedures**: Before performing any cosmetic procedure, make sure that patients understand the purpose, adverse effects involved, complications that can happen, outcome results, and expenses involved. A detailed informed consent should be signed by the patient mentioning all the above points.
5. **Skills**: When one is beginning aesthetic practice, It is important to develop your procedure skills and improve your confidence day by day. This can be done by keeping yourself up to date by attending the latest cosmetic conferences and workshop.
6. **Establish a rapport with patient**: Give attentive listening to patients' problems and maintain a written record of it. This will help you in gaining confidence and building a rapport with the patient.
7. **Equipment**: Proper handling of equipment is required to avoid any hazards and for smooth running of clinical practice.
8. **Good cosmetic team**: A team should consist of experienced doctors and nurses to avoid any interference with the procedure.

33.2 A Promising Future for Lasers

Energy-based devices continue to get more advance. More developments are coming from different parts of the world, especially Asia. "Better visualization of structures with optical coherence tomography allows us to improve patient outcomes," according to Dr. Pozner. "We can further customize treatments using the right settings for optimal results for each individual patient. We use combination treatments more widely today and have moved from face to off-face targets (e.g., chest, neck, hands, and legs) and often treat more than one area in a single session."

Another area of potential benefit is the ability to in vivo image tissue with lasers. "This could allow real-time diagnosis without invasively procuring tissue from the patient," Dr. Kelleher says. "This could allow for more precise and immediate confirmation of tissue margins when treating skin cancers" [1].

In today's world, many people want to look and feel their best. Cosmetic procedures can help achieve this and delay the aging process. In the future, this desire to look young and attractive may become even more common. They are painless, quick, non- invasive with very satisfactory results as people work and live longer.

Increase in trends of cosmetic gynecological procedure in the recent years indicate that the demand will continue to grow as it becomes more advanced, less invasive and more affordable. New technologies and innovative techniques will be pioneered to improve the quality of procedures even further.

The future aim of cosmetic gynecology field is associated with regenerative medicine and genetic engineering.

33.3 Newer Upcoming Technologies in Cosmetic Gynecology

1. **Stem cell therapy**: Stem cell therapy use in medicine has been there in recent years for various autoimmune disorders. Nowadays, its use has been established in various cosmetic procedures. Its applications in cosmetic gynecology include vaginal atrophy, lichen sclerosis, genitourinary syndrome of menopause, and vaginal dryness.

   The amniotic fluid-derived mesenchymal stem cell can also promote skin rejuvenation because it is a rich source of growth factors, chemokines, and cytokines, including vascular endothelial growth factor, transforming growth factor-β, interleukin (IL)-6, IL-8, and tumor necrosis factor-α, that have been shown to be upregulated during the inflammatory phase of the wound healing process.

   Rejuvenation of vagina by injection of stem cells (adipose or bone marrow) into the genital area can get rid of vaginal wrinkling, laxity, improve skin tone and texture.

The addition of PRP to fat grafts improves graft survival in a simple and safe procedure and less cost. Stem cell-enriched fat transfer will be costlier but, more efficient and longer outcome. Besides the volume correction, the patient experienced substantial relief of symptoms [2].

2. **Exosomes**: Exosomes are extracellular vesicles that exist outside of the normal cell structure in the body. These powerful membranous vesicles facilitate cell-to-cell communication and also function as modes of transportation to deliver regenerative medicine to specific areas of the body. These are one of the latest technologies in the field of regenerative medicine. Adipose tissue-derived stem cells are called ASC-exos. They act as coordinated "vital networks" of multiple growth factors, proteases, progenitors, and immune cells producing inflammatory cytokines. Exosomes have tissue regeneration capacity thus helps in plastic and cosmetic surgery, including skin anti-aging therapy, dermatitis improvement, wound healing, scar removal, flap transplantation, bone tissue repair and regeneration, obesity prevention, fat grafting, breast cancer, and breast reconstruction [3].

   Exosomes help in facial rejuvenation by delivering growth factors, cytokines, and targeted peptides to promoted rapid skin healing. Adipose-derived stem cells are an ideal stem cell source as it is easily available, provides greater yield, biosafe, biological differentiation properties, very well tolerated, and an effective carrier.
3. **Artificial intelligence (AI)**: As the consumer demands are increasing in recent times, cosmetic surgeons have started using AI in their practice to meet the better health outcomes. With the help of AI, the surgeon can decide which surgical approach is most appropriate to achieve maximal results and fewer complications. Although the development and deployment of AI in cosmetic surgery is accelerating and has tremendous potential, even advanced technologies still cannot fully replace humans. As in the healthcare industry, AI is only playing a supporting role, and as such the cost of cosmetic surgery will likely not be significantly reduced in the short term. Also, many associated AI technologies remain somewhat unstable, which restricts their application in sensitive areas like cosmetic surgery [4].
4. **Genetic engineering**: Genetic engineering is a rapidly evolving branch in today's time. Genetic engineering aims to modify the genes to enhance the capabilities of the organism beyond what is normal. CRISPR—the abbreviation stands for "clustered regularly interspaced short palindromic repeats"—was discovered as an adaptive immune mechanism in bacteria. CRISPR gene editing is revolutionizing the potential of gene therapy due to its simplicity, specificity, efficiency, low cost, and versatility," Dr. Liao and coauthors write. "Potential applications of CRISPR are numerous and will certainly impact plastic and reconstructive surgery." Gene therapy is a promising approach to enhancing wound and tissue healing. In addition to accelerated healing of skin wounds, CRISPR may lead to new approaches for repair and regeneration of bone, cartilage, nerve, and muscle [5].

References

1. Appold, K. The history of aesthetic lasers. Dermatology Times, April 2019. (Vol. 40, No. 4).
2. Cosmetic Gynecology. n.d. https://stemcells-aesthetica.com/cosmetic-gynecology/. Accessed 1 Dec 2020.
3. Xiong M, et al. Exosomes from adipose-derived stem cells: the emerging roles and applications in tissue regeneration of plastic and cosmetic surgery. Front Cell Dev Biol. 2020; https://doi.org/10.3389/fcell.2020.574223.
4. Murphy DC, Saleh DB. Artificial intelligence in plastic surgery: what is it? where are we now? What is on the horizon? Ann R Coll Surg Engl. 2020;102:577–80.
5. Wolters Kluwer Health. CRISPR gene editing will find applications in plastic and reconstructive surgery. ScienceDaily, 30 Oct 2018. Available from: www.sciencedaily.com/releases/2018/10/181030134342.htm.

Correction to: Breast Lifting and Reduction in Aesthetic and Regenerative Gynecology

Mohan Thomas and James D'silva

Correction to: Chapter 20 in P. Jindal et al. (eds.), Aesthetic and Regenerative Gynecology, https://doi.org/10.1007/978-981-16-1743-0_20

The original version of this book was inadvertently published with incorrect author name and author sequence in chapter-20 and table of contents. The author name and sequence was updated as below

Chapter 20: Breast Lifting and Reduction in Aesthetic and Regenerative Gynecology
Mohan Thomas and James D'silva

The updated online version of the chapter can be found at
https://doi.org/10.1007/978-981-16-1743-0_20

P. Jindal et al. (eds.), *Aesthetic and Regenerative Gynecology*,
https://doi.org/10.1007/978-981-16-1743-0_34

Index

P. Jindal et al. (eds.), *Aesthetic and Regenerative Gynecology*,
https://doi.org/10.1007/978-981-16-1743-0

GPSR Compliance

The European Union's (EU) General Product Safety Regulation (GPSR) is a set of rules that requires consumer products to be safe and our obligations to ensure this.

If you have any concerns about our products, you can contact us on ProductSafety@springernature.com

In case Publisher is established outside the EU, the EU authorized representative is:

Springer Nature Customer Service Center GmbH
Europaplatz 3
69115 Heidelberg, Germany

Batch number: 10732807

Printed by Printforce, the Netherlands